Epigenetic Mechanisms in Plant Stress Adaptation

Muhammad Waseem
Hainan University, China

Liu Pingwu
Hainan University, China

IGI Global
Scientific Publishing
Publishing Tomorrow's Research Today

Vice President of Editorial | Melissa Wagner
Director of Acquisitions | Mikaela Felty
Director of Book Development | Jocelynn Hessler
Production Manager | Mike Brehm
Cover Design | Jose Rosado

Published in the United States of America by
IGI Global Scientific Publishing
701 East Chocolate Avenue
Hershey, PA, 17033, USA
Tel: 717-533-8845
Fax: 717-533-7115
Website: https://www.igi-global.com E-mail: cust@igi-global.com

Library of Congress Cataloging-in-Publication Data

Names: Waseem, Muhammad, 1988- editor. | Liu, Pingwu, 1980- editor.
Title: Epigenetic mechanisms in plant stress adaptation / edited by Muhammad Waseem, Liu Pingwu.
Description: Hershey, PA : IGI Global Scientific Publishing, [2026] | Includes bibliographical references and index. | Summary: "The primary objective of book is to elucidate the role of epigenetic mechanisms in plant stress adaptation, focusing on how these regulatory pathways facilitate growth, development, and response to abiotic stress stimuli"-- Provided by publisher.
Identifiers: LCCN 2024056970 (print) | LCCN 2024056971 (ebook) | ISBN 9798369386484 (hardcover) | ISBN 9798369386491 (paperback) | ISBN 9798369386507 (ebook)
Subjects: LCSH: Plant genetics. | Plants--Effect of stress on. | Epigenetics. | Plants--Adaptation.
Classification: LCC QK981 .E65 2025 (print) | LCC QK981 (ebook) | DDC 581.3/5--dc23/eng/20250206
LC record available at https://lccn.loc.gov/2024056970
LC ebook record available at https://lccn.loc.gov/2024056971
British Cataloguing in Publication Data
A Cataloguing in Publication record for this book is available from the British Library.

Editorial Advisory Board

Table of Contents

Detailed Table of Contents

Epigenetic-based breeding offers a promising approach to developing stress-resilient crops by utilizing mechanisms like DNA methylation, histone modifications, and RNA interference (RNAi) to regulate gene expression under stress. These modifications can enhance a plant's ability to withstand drought, salinity, and extreme temperatures. The concept of epigenetic memory allows plants to "remember" past stress, aiding future resilience. Despite challenges in stability and regulation, this approach, combined with traditional breeding, holds significant potential for ensuring global food security in changing climates.

Epigenetic modifications, a biological process, can be mediated and directed by various types of RNAs (such as siRNA, sRNAs, lncRNAs, etc) in plants. These modifications comprise a tremendously interactive network of RNA-directed DNA methylations, chromatin and histone modifications and gene silencing. These epigenetic modifications are unique to plants and possess significant roles in growth, development, abiotic and biotic stresses. The latest discoveries in this field have been made possible through next-generation sequencing and omics technologies, which help design an essential blueprint for the epigenome in plants. As the researchers unrevealed multiple levels of epigenetic regulation mediated by RNAs in plants, a better understanding of the significance of epigenetic states and marks has been established, and it opens the path to constructing new plant varieties with desired traits.

Saman Zulfiqar, Government Sadiq College Women University, Pakistan
Yasir Nawaz, School of Life Sciences, Nanjing Normal University, China
Sidra Aslam, University of Sialkot, Pakistan
Fiza Rimal Butt, University of Sialkot, Pakistan
Saba Munir, University of Okara, Pakistan
Rikza Tahreem, University of Okara, Pakistan

Epigenetics mechanisms, such as histone alterations, DNA methylation, and small non-coding RNAs, play a vital role in regulating plant drought tolerance by modulating gene expression. These changes enable plants to conserve water, protect cells, and recover from stress. Epigenetic memory also helps plants respond more effectively to recurring droughts. Understanding these mechanisms offers promising approaches for improving crop flexibility by breeding and biotechnology, crucial for addressing climate change and water scarcity challenges in agriculture.

Rooma Waqar, Quaid-i-Azam University, Islamabad, Pakistan
Banzeer Ahsan Abbasi, Rawalpindi Women University, Pakistan
Abdul Samad Mumtaz, Quaid-i-Azam University, Islamabad, Pakistan
Muhammad Kaleem, Quaid-i-Azam University, Islamabad, Pakistan
Shumaila Ijaz, Shenzhen University, China
Zakir Ullah, Quaid-i-Azam University, Islamabad, Pakistan
Zahid Ullah, University of Swat, Pakistan
Ahmad Ali, University of Swat, Pakistan

Chromatin remodeling complexes are crucial for plant stress adaptation, dynamically modulating chromatin structure and regulating stress-responsive genes. Complexes like SWR1, SWI2/SNF2, and CHD family remodelers coordinate mechanisms such as histone variant deposition, histone modifications, DNA methylation, and nucleosome positioning to fine-tune gene expression. Under stress, they facilitate rapid transcriptional reprogramming of defense-related genes, with examples including SWR1-mediated H2A.Z deposition and SWI2/SNF2 remodelers like BRM and SYD modulating key defense genes. CHD remodelers, such as CHR5, enhance immune gene expression, while factors like BRHIS1 and DDM1 regulate basal defense gene activity under non-stress conditions. These complexes integrate environmental signals, supporting plant resilience and offering potential for crop improvement through genetic and epigenetic strategies.

Plants are not able to flee or communicate under unfavorable biotic or abiotic stress conditions, which can have a profound impact on their growth, development and reproduction. As a result, plants have established more sophisticated adaptation mechanisms that enable them to thrive in harsh environments. Epigenetic regulation is essential to regulate cellular processes needed for survival in adverse environments. In plants, epigenetic changes play important role in root development because they affect gene expression linked with root architecture, growth and stress response allowing plants to adapt to changing conditions. These changes include DNA methylation, histone modification and small RNA-mediated regulation. Epigenetic mechanisms change chromatin state of stress-responsive genes regulating them at transcriptional and posttranscriptional phases. These approaches emphasize significance of epigenetic alterations in connecting root systems developmental plasticity and environmental adaptability, opening door to create ways to improve crop performance under difficult circumstances.

Plant stress significantly affects plant growth, development, and productivity, creating unfavorable conditions that disrupt physiological functions and survival. Environmental pollution, climate change, and global warming intensify the impact of stressors on growth and yield. Plants adapt through a complex innate immune system. While progress has been made in identifying stress-inducible genes and using advanced sequencing tools, signaling mechanisms under combined stressors remain poorly understood. Studies reveal overlapping but distinct responses to environmental stimuli, intricate interactions between abiotic and biotic stress responses, and improper signaling crosstalk during multiple stresses. Epigenetic crosstalk integrates signals to regulate traits through interconnected pathways. We explore chemical regulators in epigenetic research and propose interdisciplinary strategies for developing new regulators.

Plant-microbe interactions are vital for ecological balance and crop growth, with epigenetics playing a key role in regulating these relationships. Epigenetics involves changes in gene expression without altering DNA, enabling plants to respond to stress and interact with microbes effectively. These mechanisms help plants adapt by activating or suppressing genes in response to microbial presence. Dynamic relationships with bacteria, fungi, and viruses significantly influence plant health and growth. Epigenetic modifications enhance plant resilience by regulating gene expression during interactions, optimizing plant function under stress. This chapter summarizes recent findings on the impact of epigenetics on plant-microbe interactions and stress adaptation, highlighting its potential to revolutionize sustainable agriculture and ecosystem management practices.

Plants as immobile organisms are perpetually subjected to environmental changes and various biotic stresses. Pathogens such as viruses, bacteria and fungi can severely hinder plant growth and development thus presenting significant challenges to global food security. The relationship between plants and pathogens has led to evolution of intrinsic system such as gene expression networks, which enable plants to protect

Today, climate change presents a significant challenge that impacts the growth and performance of plants, leading to adverse environmental conditions that disrupt their metabolic and physiological processes, ultimately threatening their survival. Plants have developed crucial genetic and epigenetic regulatory mechanisms, such as DNA methylation and histone modifications, in addition to managing stress-responsive genes to withstand thermal stress. Chromatin regulators are vital in sustaining this gene network in highly stressful conditions. While many of these modifications are often temporary, specific changes may be enduring and can be inherited by the next generation, contributing to plant stress memory. Understanding genetic mechanisms and the characteristics of chromatin regulators and histone modifications reveals the complexity of plant stress memory.

Plants are constantly exposed to various environmental stress, such as drought, salinity, extreme temperatures and pathogen attacks, which have a negative impact on their growth and development. In response to these stresses, highly complex molecular processes are activated, epigenetic mechanisms being among the most important. These mechanisms, which include DNA methylation, histone modifications and non-coding RNAs, regulate gene expression and modify the plant's response to abiotic stresses. In this chapter, we highlight progress on the epigenetic mechanisms associated with the evolution of plants to adapt to abiotic stress. Combined with knowledge gleaned from other omics approaches, these strategies will facilitate the identification of regulatory networks and their dynamics under stress, enabling the development of strategies for sustainable agriculture that will ensure yield security in a context of climate change.

The transcriptome's activity depends on both the genomic profile and epigenetic regulation. Epigenetic changes are essential for phenotypic variation and stress resistance triggered by environmental stressors. Stress signals regulate genes via histone modifications, DNA methylation, and miRNA. The epigenetics field has advanced rapidly over two decades. While modern plant breeding is efficient, it falls short of traditional crop breeding demands. Epigenetic methods, such as novel epiallele generation and RNAi techniques, can enhance crop yield and nutritional value. For instance, RNAi has improved barley, wheat, and soybean by reducing anti-nutritional factors and enhancing taste, such as silencing the Omega-3 fatty acid desaturase gene to modify linoleic acid content. HvTX1 gene expression in barley under drought stress boosts tolerance, while HAT genes in rice activate under water and heat stress. These insights aid in developing stress-tolerant crops and deepen our understanding of plant adaptation mechanisms.

The refined use of these methodologies allows for precise epigenetic manipulation, presenting a promising avenue for tailoring plants to withstand abiotic stressors, including drought, salinity, and extreme temperatures. Cutting-edge methods, such as advanced next-generation sequencing, ChIP-seq, and CRISPR-based epigenome editing, take center stage in unraveling the intricate landscape of epigenetic modifications. This chapter comprehensively examines how these tools empower researchers to decode the epigenetic language governing DNA methylation, histone modifications, and small RNA-mediated pathways. Within the context of this chapter, notable achievements in harnessing epigenetic insights to bolster abiotic stress tolerance are highlighted. These accomplishments range from the development of crops with enhanced drought resistance and salinity tolerance to those exhibiting resilience in the face of temperature extremes.

Epigenomic processes i.e. DNA methylation, histone modification and RNA-mediated regulatory processes are pivotal for the adaptive gene expression responses of plants to environmental stressors. Bisulfite sequencing, chromatin immunoprecipitation sequencing, and epiGWAS, facilitate the comprehensive mapping of dynamic modifications across the genome. These advanced technologies are elucidating the processes through which plants develop "stress memories" that bolster resilience across generations, thereby illuminating the intricate regulatory networks that enable plant survival in challenging environments. Epigenomic knowledge with breeding initiatives enable scientists to develop crop varieties exhibiting enhanced stress tolerance, thereby contributing to sustainable agricultural practices. This chapter outlines current methods, key findings, and future directions in the field, highlighting the potential of epigenomic research to revolutionize plant resilience strategies.

Epigenetics, the examination of inheritable modifications in gene function absent variations in DNA sequence, has become a crucial domain in plant biology. This chapter examines the processes of epigenetic control in model plants, including Arabidopsis thaliana, rice (Oryza sativa), and maize (Zea mays), emphasizing their distinct roles in elucidating DNA methylation, histone modifications, and RNA-directed DNA methylation (RdDM). Epigenetic systems are essential for plant development, stress responses, and adaptation to environmental changes. The investigation of epigenetic regulation is being transformed by recent developments in single-cell epigenomics, CRISPR-based epigenome editing technologies, and multi-omics integration. The chapter discusses the challenges of fully grasping transgenerational inheritance and the stability of epigenetic markers. Connecting advanced research with agricultural applications highlights the possibility of utilizing epigenetics for crop enhancement and sustainable agriculture in response to climate change.

Epigenetics is a key feature in plants responding to drought stress by regulating plant tolerance or adaptation capacity with physical water deficit. Transgenerational inheritance can occur because epigenetic changes regulate gene expression without changing the genetic code (for example, DNA methylation, histone modification and non-coding RNAs). Such adaptations, allow plants to induce stress-responsive genes, modify metabolic pathways and maintain cellular homeostasis under drought conditions. Epigenetic regulation helps in the rapid adaptation corresponding to developing drought tolerance and memory that has the potential of trans-generational expression. Knowledge of these mechanisms may be useful for breeding and biotechnological strategies to improve crop resistance in water-limited environments.

Stress by salts is one of the important environmental factors that negatively impacts plant growth, development and yields especially on cereal crops. The use of DNA methylation, histone modification and chromatin remodeling in plant responses to salinity stress cannot be overemphasized. These mechanisms regulate genes that allow plants to cope with osmotic, ionic and oxidative stresses under salt stress that can cause salt sensitivity or tolerance in crops. Additionally, cell signaling under salinity stress leads to formation of stress memory which improves current and future coping with similar stress. Finally, this chapter is focused on discussing the possibilities of epigenetics to improve salinity stress tolerance and crop adaptation as well as evolution of salt tolerant crops. Knowledge of epigenetic regulation

Preface

INTRODUCTION

In the face of an increasingly volatile climate and the growing demand for sustainable agriculture, the resilience of crop plants to environmental stresses has become a central focus of contemporary plant science. The field of epigenetics—dealing with heritable changes in gene function that do not involve changes in the DNA sequence—has emerged as a powerful tool to uncover the molecular foundations of plant stress responses and adaptation. This volume, *Epigenetic Mechanisms in Plant Stress Adaptation*, brings together cutting-edge research and expert insights into how epigenetic regulation shapes plant responses to abiotic and biotic stress conditions, and how this knowledge can be harnessed for crop improvement.

ORGANIZATION OF THE BOOK

The volume comprises 20 chapters authored by a diverse and interdisciplinary group of researchers across Pakistan, China, India, Morocco, and beyond, each contributing a unique perspective on the interplay between epigenetic processes and plant stress adaptation. Beginning with an overview of epigenetic-based breeding strategies for stress-resilient crops, the volume progresses through discussions of RNA-mediated regulation, chromatin remodeling, histone modifications, DNA methylation, and the integration of multiple stress signaling pathways. Chapters also explore epigenetic control of root development, plant-microbe interactions, and immune responses—highlighting the dynamic and interconnected nature of epigenetic regulation across various physiological processes.

Several chapters provide deep insights into specific stress conditions such as drought, salinity, thermal stress, and pathogen attacks, while others address broader concepts such as epigenetic memory, transgenerational inheritance, and the integration of epigenomic tools and high-throughput approaches in plant research. The final chapters look ahead, offering critical analyses of current achievements and future challenges in the field, with a strong emphasis on translating epigenetic understanding into practical strategies for sustainable agriculture.

Chapter 1: Epigenetic-Based Breeding Strategies for Stress-Resilient Crops

In this opening chapter, the authors present an innovative frontier in crop improvement—epigenetic-based breeding—as a means to develop varieties that can withstand the mounting pressures of environmental stress. By harnessing natural processes such as DNA methylation, histone modifications, and RNA interference to regulate gene expression, plants can acquire an adaptive advantage against drought, salinity, and temperature extremes. The discussion extends to the concept of epigenetic memory, which enables plants to "remember" prior stress events, thus equipping them for future challenges. While acknowledging current limitations in the stability and regulation of these traits, the chapter underscores the promise of integrating epigenetic tools with conventional breeding to secure global food systems in the face of climate volatility.

Chapter 2: Mechanisms of RNA-Mediated Epigenetic Regulation in Plants

This chapter offers a detailed exploration of the remarkable role RNAs play in directing epigenetic modifications in plants. Through small RNAs, long non-coding RNAs, and other RNA species, plants orchestrate a dynamic network of DNA methylation, chromatin remodeling, and gene silencing events that are integral to their development and stress responses. The authors highlight how breakthroughs in next-generation sequencing and multi-omics have unveiled the complex interplay of these pathways, providing a blueprint for targeted crop improvement. The chapter makes clear that a nuanced understanding of RNA-mediated epigenetic control could revolutionize the design of plant varieties with precisely tuned traits.

Chapter 3: Epigenetics Mechanisms in Plant Stress Adaptation: Epigenetic Mechanisms Governing Drought Tolerance in Plants

Focusing on one of agriculture's most pressing challenges, this chapter examines the epigenetic underpinnings of drought tolerance in plants. It explains how histone modifications, DNA methylation, and small non-coding RNAs modulate stress-related gene expression, enabling water conservation, cellular protection, and recovery from dehydration. The authors discuss how epigenetic memory contributes to improved resilience during recurrent drought events and how this knowledge can be strategically applied in breeding and biotechnological approaches to enhance crop adaptability in an era of water scarcity.

Chapter 4: Role of Chromatin Remodeling Complexes in Plant Stress Adaptations

Here, the emphasis shifts to the architectural rearrangements of chromatin that dictate plant responses to stress. Chromatin remodeling complexes—such as SWR1, SWI2/SNF2, and CHD families—are presented as pivotal regulators of stress-responsive genes through mechanisms including histone variant deposition, nucleosome repositioning, and epigenetic marking. Detailed case studies illustrate how these remodelers coordinate rapid transcriptional reprogramming under stress conditions, integrating environmental cues into precise gene regulation. The chapter highlights the untapped potential of targeting chromatin remodeling pathways to engineer stress-resilient crops.

Chapter 5: Epigenetic Modifications in Root Development and Stress Responses

This chapter draws attention to the central role of root systems as the plant's interface with its environment and as a critical determinant of stress resilience. The authors describe how DNA methylation, histone modifications, and small RNAs shape root architecture and function under challenging conditions, linking developmental plasticity to environmental adaptability. By explaining the transcriptional and posttranscriptional regulation of stress-responsive genes in roots, the chapter underscores the opportunity to enhance crop performance through targeted manipulation of root epigenetics.

Chapter 6: Epigenetic Crosstalk: Integrating Multiple Stress Signals in Plants

Addressing the complex reality that plants often encounter multiple stressors simultaneously, this chapter delves into the phenomenon of epigenetic crosstalk. The authors examine how abiotic and biotic stress pathways interact—sometimes synergistically, sometimes antagonistically—through interconnected signaling and regulatory networks. They present emerging insights into how epigenetic mechanisms integrate these signals to coordinate adaptive responses and discuss the potential of chemical and molecular regulators to fine-tune this process. This integrative perspective is particularly relevant in the context of climate change, where plants face increasingly multifaceted environmental pressures.

Chapter 7: Epigenetic Influences on Plant-Microbe Interactions and Stress Adaptation

This chapter illuminates the subtle yet profound ways in which epigenetic regulation shapes plant relationships with microbial communities. The authors describe how gene expression changes, triggered without alterations to DNA sequence, allow plants to manage interactions with beneficial and pathogenic microbes alike. Through case studies involving bacteria, fungi, and viruses, the chapter demonstrates how epigenetic adjustments optimize plant performance under stress and maintain ecological balance. The findings offer pathways for leveraging epigenetics in sustainable agriculture and integrated pest management.

Chapter 8: Epigenetic Regulation of Plant Immune Responses to Pathogens

The intricate interplay between plants and their pathogens is the focus of this chapter, which details how epigenetic regulation fortifies plant immunity. The authors explain how DNA methylation, histone modifications, chromatin remodeling, and non-coding RNAs regulate defense gene activation, enabling plants to mount robust responses to microbial attack. Notably, the discussion highlights the role of epigenetic memory in transmitting defense preparedness to subsequent generations. This perspective reinforces the importance of epigenetic mechanisms as both adaptive and heritable components of plant immune strategies.

Chapter 9: Epigenetic Response Underlying Drought and Salt Stress Tolerance in Plants

The final chapter expands the discussion to encompass both drought and salinity—two of the most destructive abiotic stresses to global agriculture. By synthesizing recent research, the authors show how genetic and epigenetic networks converge to detect environmental stimuli and adjust defensive mechanisms accordingly. Mechanisms including chromatin remodeling, DNA methylation, histone modification, and small RNA activity are presented as critical modulators of phenotypic plasticity. The chapter closes by positioning epigenetic insights as indispensable tools for breeders seeking to develop crops with robust multi-stress tolerance.

Chapter 10: Thermal Stress and Epigenetic Regulation in Plants

Addressing the growing impact of climate change, this chapter focuses on the genetic and epigenetic strategies plants employ to withstand high-temperature stress. DNA methylation, histone modifications, and the action of chromatin regulators are explored in the context of maintaining stress-responsive gene networks. The authors also highlight the role of stress memory—temporary or heritable—in enabling plants to survive and adapt under prolonged or repeated heat stress. This perspective deepens our understanding of how plants integrate molecular changes to safeguard metabolic and physiological stability under thermal extremes.

Chapter 11: Epigenetic Regulation and Plant Adaptation to Environmental Stress: Towards New Strategies for Sustainable Agriculture

This chapter presents epigenetic regulation as a cornerstone of plant adaptation to multiple environmental stresses, including drought, salinity, extreme temperatures, and pathogens. DNA methylation, histone modifications, and non-coding RNAs are discussed as central mechanisms controlling stress-responsive genes. Integrating these insights with data from other omics technologies, the authors outline a framework for identifying dynamic regulatory networks and applying them to crop improvement strategies. The goal is to achieve sustainable agricultural productivity and yield stability under the growing pressures of climate change.

Chapter 12: Harnessing Epigenetics for the Breeding of Stress-Tolerant Crops

Highlighting the synergy between genomic and epigenetic regulation, this chapter details how histone modifications, DNA methylation, and miRNAs influence gene expression and phenotypic traits under stress. The authors showcase specific examples—such as RNAi applications in barley, wheat, and soybean—to improve taste, nutrition, and stress tolerance. They also examine the role of genes like HvTX1 in drought tolerance and HAT in rice under water and heat stress. The discussion underscores how targeted epigenetic approaches can complement modern breeding to accelerate the development of stress-resilient, high-performing crops.

Chapter 13: Epigenetics for Developing Abiotic Stress-Tolerant Plants: Updated Methods and Current Achievements

This chapter surveys cutting-edge methodologies for precise epigenetic manipulation, including next-generation sequencing, ChIP-seq, and CRISPR-based epigenome editing. The authors explain how these tools are advancing our understanding of DNA methylation, histone modifications, and small RNA pathways in stress tolerance. By highlighting notable achievements—such as crops with improved drought, salinity, and temperature resilience—the chapter illustrates how emerging techniques are shaping a new era of targeted, epigenetically informed crop development.

Chapter 14: Epigenomics: A High Throughput Approach to Study Plant Stress Adaptations

In this chapter, the focus shifts to epigenomics as a powerful high-throughput approach for mapping and interpreting stress-induced modifications at the genome-wide level. Technologies such as bisulfite sequencing, chromatin immunoprecipitation sequencing, and epiGWAS are described as essential for understanding how plants form and retain stress memories. By integrating epigenomic insights into breeding programs, scientists can develop crop varieties with enhanced resilience to environmental challenges, paving the way for more sustainable agricultural systems.

Chapter 15: Case Studies on Epigenetic Regulation in Model Plant Species

Drawing on research in Arabidopsis thaliana, rice, and maize, this chapter uses model species to dissect the molecular details of DNA methylation, histone modifications, and RNA-directed DNA methylation. The authors emphasize how advances such as single-cell epigenomics and CRISPR-based editing are expanding our capacity to manipulate and study epigenetic states. They also address the complexity of transgenerational inheritance and the stability of epigenetic markers, highlighting opportunities to translate these insights into practical crop improvement strategies.

Chapter 16: Plant Epigenetics for Crop Improvement

This chapter looks ahead, identifying emerging priorities in plant epigenetics research. Key areas include integrating epigenetic data with genomic and phenotypic datasets, refining high-throughput and editing technologies, and understanding the context-specific nature of epigenetic changes. The authors also address the challenge of converting laboratory discoveries into real-world agricultural applications, stressing that solving these issues will be essential for boosting crop resilience and productivity in the face of environmental change.

Chapter 17: Future Directions and Challenges in Plant Epigenetics Research

Concluding the volume, this chapter offers a broader perspective on the future of plant epigenetics. It emphasizes the importance of building reference epigenomes for diverse species, navigating ethical considerations in agricultural biotechnology, and fostering interdisciplinary collaboration. By combining technological innovation with global research efforts, the authors envision a path toward using epigenetic knowledge to secure sustainable crop production in a changing climate.

Chapter 18: Epigenetic Mechanisms Underlying Drought Tolerance in Plants

Chapter 18 examines how epigenetic regulation equips plants to withstand drought stress. Mechanisms such as DNA methylation, histone modifications, and non-coding RNAs are shown to activate stress-responsive genes, adjust metabolic pathways, and maintain cellular balance under water-limited conditions. Importantly, the chapter emphasizes transgenerational inheritance, where plants pass on adaptive traits to future generations through epigenetic memory. By linking these insights to breeding and biotechnological applications, the chapter highlights promising strategies to improve crop performance and resilience in drought-prone environments.

Chapter 19: Epigenetic Mechanisms Underlying Salinity Tolerance in Plants

This chapter focuses on the epigenetic regulation of plant responses to salinity stress, a major constraint on crop yields worldwide. DNA methylation, histone modifications, and chromatin remodeling are discussed as central regulators that help plants cope with osmotic, ionic, and oxidative stress. The concept of stress memory is highlighted as a critical factor in enabling plants to mount more effective responses to repeated salt stress. By examining the potential of epigenetic strategies to enhance salinity tolerance, the chapter underscores their relevance in improving crop productivity in salt-affected soils and promoting food security.

Chapter 20: Epigenetic Memory and Transgenerational Stress Adaptation in Plants

This chapter highlight a distinctive appearance of "intelligence" within the kingdom plantae. Regardless of the lack of a nervous system, plants have developed a mature molecular machinery that assist to remember, to compose decisions, and take responses according to the stress stimulus. Being sessile organisms, plants can utilize these multifaceted systems to boost up their resource strategies and develop their strength for survival among hard environmental conditions. Along with these capabilities, these sessile plants also gain potential to transmit their acquired experience to their next generation via epigenetic modifications, thereby extending typical genetic inheritance. This chapter explores the sensory system by which plants observe, recognize, remember, and respond to their surroundings. As a result, these acquired adaptations can be transferred to their next generations to equip them more efficiently against any climatic challenges.

CONCLUSION

The breadth and depth of the topics covered in this volume are a testament to the collaborative efforts of researchers across multiple institutions and disciplines. We are particularly grateful to all contributing authors whose rigorous scientific work forms the foundation of this book. Their contributions not only enhance our understanding of plant epigenetics but also provide a roadmap for future innovations in crop stress tolerance and productivity.

This book is intended for a diverse audience, including graduate students, researchers, plant breeders, molecular biologists, and agricultural scientists. Whether the reader is seeking a foundational understanding or a detailed analysis of specific epigenetic mechanisms, this volume offers comprehensive and up-to-date coverage of a rapidly evolving field.

We hope that *Epigenetic Mechanisms in Plant Stress Adaptation* serves as both an informative reference and a source of inspiration for continued exploration into the epigenetic foundations of plant resilience.

Muhammad Waseem
Hainan University, China

Pingwu Liu
Hainan University, China

Chapter 1
Epigenetic-Based Breeding Strategies for Stress-Resilient Crops

Asad Azeem
https://orcid.org/0000-0001-6550-3571
Department of Plant Breeding and Genetics, University of Layyah, Pakistan

Muhammad Abu Bakar Zia
Department of Plant Breeding and Genetics, University of Layyah, Pakistan

Sana Khan
Department of Plant Breeding and Genetics, University of Agriculture, Faisalabad, Pakistan

Tehmina Yasmin
Department of Entomology, University of Agriculture, Faisalabad, Pakistan

Nasir Abbas
Department of Plant Breeding and Genetics, University of Agriculture, Faisalabad, Pakistan

Sonia Azeem
Department of Zoology, Government College University Faisalabad, Layyah, Pakistan

ABSTRACT

Epigenetic-based breeding offers a promising approach to developing stress-resilient crops by utilizing mechanisms like DNA methylation, histone modifications, and RNA interference (RNAi) to regulate gene expression under stress. These modifications can enhance a plant's ability to withstand drought, salinity, and extreme temperatures. The concept of epigenetic memory allows plants to "remember" past stress, aiding future resilience. Despite challenges in stability and regulation, this approach, combined with traditional breeding, holds significant potential for ensuring global food security in changing climates.

DOI: 10.4018/979-8-3693-8648-4.ch001

INTRODUCTION

The increasing global environmental pressures necessitate developing crop cultivars that display stress resilience, which determines food security. While traditional breeding methods work well, the onset of stresses such as drought, salinity, and extreme temperatures becomes a problem to keep up with changes in genomic regulation to meet the rapidly changing demands (Omony, Jimmy, et al 2020).. Epigenetics could therefore hold considerable promise for the enhancement of crop resilience. However, epigenetic mechanisms include methylation of DNA, histone modifications, and RNA interference (RNAi), which regulate stress responses in plants. It activates or silences a set of genes to increase the capability of the plant to survive adverse conditions. Methylation of DNA adds a methyl group to the DNA molecule, usually at cytosine bases (Mercé et al., 2020). It can also suppress genes, especially those that would otherwise be active under stress conditions. For example, suppressing some stress-responsive genes will allow a plant to avoid those deleterious effects which otherwise could compromise its growth and productivity. Histone modifications, on the other hand, refer to chemical changes in the proteins where DNA wraps itself. These would stabilize or destabilize DNA structure and thus can influence gene access to transcription (Samantara et al., 2021).. The expression of drought tolerant genes, including those which embrace genes for water retention and ion transport within the saline environment, can be modulated by changing histone modifications in plants. This mechanism further adds to the regulation of RNA interference, or RNAi, in which small RNAs have been shown to target specific messenger RNAs (mRNAs) for degradation or translational repression and thus prevent the formation of stress-related proteins that would be lethal in some environmental conditions. Breeding strategies that are based on epigenetic knowledge are tagged as identification of epigenetic markers with desirable traits such as drought resistance or disease resistance. Markers identified can be used in selections to accelerate the development of varieties resilient to stress. Advanced tools, such as CRISPR/dCas9, enable targeted epigenetic modifications and allow for precise editing of genes that are involved in stress responses without changing the DNA sequence. This precision breeding approach promises substantial opportunities to develop crops that not only tolerate current environmental challenges but will also likely tolerate future conditions. However, there are several key challenges to be overcome before the full potential of epigenetic-based breeding can be exploited fully: technical challenges, including questions regarding stability and heritability of epigenetic modifications (Munshi et al., 2015). Although some epigenetic modifications can be heritable and stable over generations, others may be unstable or may respond to environmental signals, making it difficult to achieve consistent expression of the trait. Ethical and regulatory considerations will also be of paramount importance, where public perception and acceptance of crops produced by means of epigenetic interventions is required. Regulation of epigenetically modified organisms remains a space whose rules and guidelines are still being developed to guide these organisms into development and use. While these are important considerations in evaluating epigenetic-based breeding strategies, their contribution to global food security is significant. Integrating epigenetics with traditional breeding and advanced genomic tools presents a meaningful opportunity for developing crops more resilient to all environmental stresses (Lämke & Bäurle, 2017). This approach covers immediate agricultural challenges but puts the world's food system more in a position to meet future uncertainties through climate change and population growth. In crux, epigenetics plays a core role in the next generation of crop improvement as research continues and technologies advance into facing growing global challenges that ensure agriculture remains productive and sustainable.

EPIGENETIC MECHANISM IN PLANTS

In eukaryotes, chromatin status and gene regulation are intimately related. Most of the time, chromatin state is preserved throughout cell division and is not influenced by nucleotide sequence. Consequently, chromatin dynamics are thought to be regulated by epigenetics (Maeji & Nishimura, 2018). The histone modification, small RNA, and DNA methylation are the three main categories into which epigenetic mechanisms can be divided. These systems influence the expression of the relevant gene and associated features via controlling epigenetic modifications (Samantara et al., 2021).

DNA Methylation

The DNA methylation referred as the DNA methyltransferases covalently add a CH_3 group to the 5th place of a cytosine ring (5 mC) without changing the sequence of DNA (Mercé et al., 2020; Zhang et al., 2018). The recognized epigenetic marker to be examined was DNA methylation. The CH_3 group attachment to cytosine's cyclic carbon-5 causes methylation, a heritable enzymatic change. The alterations and phytohormones like flowering and seed germination regulate DNA methylation in plants, which varies by age, species, organelle, and tissue (Munshi et al., 2015). Epigenetic chemical targeting of plants with histone deacetylases, H3K9 methylation (BIX-01294) or inhibitors of DNA methylation (5-AzaC, Zebularine) has been revealed to enhance in vitro regeneration and reprogramming of plant cells, to generate DHs, and to proliferate specific clonal plants in a variety of forest and crop species, including wheat, cork wood, and rapeseed (Castillo et al., 2020; Kakoulidou et al., 2021).

In rice, genotypic-specific patterns of drought-induced DNA methylation sites were found by (Wang et al., 2011). These patterns included 70% of sites where epigenetic changes were reversed to prestress status after recovery, and 29% of sites where drought-induced DNA demethylation/methylation changes continued even after recovery. Additionally, they observed that DNA methylation changes in rice plants exhibited a substantial amount of tissue and developmental specificity. They proposed that drought-induced differences in DNA methylation within the rice genome may be a crucial method of regulation for adaptation (Miryeganeh, 2021; Wang et al., 2011). The salt-responsive gene expression is regulated partially by DNA methylation. When comparing the worldwide DNA methylation levels of salt-sensitive and salt-tolerant rice genotypes in reaction to salinity stress, it was shown that gene body methylation and stress promoter are crucial in controlling gene expression in a way that is individual to each genotype and organ (Miryeganeh, 2021). The DRM2 is the gene for DNA demethylase was found to have differential expression in salt-sensitive but not salt-resistant genotypes, and hypomethylation was observed in reaction to high salinity. According to the findings, plants' ability to withstand salt stress may be impacted by variations in DNA methylation patterns (Miryeganeh, 2021). DNA methylation was found throughout the genome of maize seedlings under cold stress, with the exception of a 1.8 kb section (ZmMI1). Under cold stress, the typically methylated ZmMI1 segment which is indicative of a stress reaction gene demethylates. The hypomethylation caused by cold stress was persisted 7 days after the cold therapy was stopped. According to studies, epigenetic modifications have also been linked to low-temperature-induced dormancy in crops (Hu et al., 2012). Heat stress-induced DNA methylation mechanisms, particularly methyltransferase activity, have been extensively researched. In the *Arabidopsis thaliana* native population, (Shen et al., 2014) found a variety of alleles associated with a plastic response to climatic fluctuation with varied genome-wide CHH methylation connected to seasonality. They discovered a link among temperature seasonality and CMT2, and they demonstrated that CMT2 mutants are fewer susceptible to high temperature stress. This suggests that the genetic control of epigenetic changes may be the fundamental cause of natural acclimatization to fluctuating temperatures, possibly via distinction allelic plasticity to high temperature stress (Shen et al., 2014).

Histone Modification

In the early 1960s, the well-researched epigenetic marker known as histone modification was initially identified by Vincet Allfrey. Numerous histone-modifying enzymes and histone variations are found in plants, and these play a crucial function in the epigenetic control of plants in reaction to various environmental stimuli. The chromatin packing condition affects gene expression. Nucleosomes are the fundamental repeating unit of chromatin, comprising around 146 base pairs of DNA that enclose an octameric histone core particle (Samantara et al., 2021). Histone molecules can be changed in a variety of pathways to produce closed or open chromatin states within the nucleosome. The existence of other modifications and the geographical dissemination of a particular alteration over a gene region determine whether histone modifications have an effect on gene expression (Samantara et al., 2021).

The plant reactions to salinity stress are also significantly influenced by the dynamics of histone modifications, which result in changes in chromatin structure. Plant adaptation to shifting conditions depends on the cooperation of HATs and HDACs (Miryeganeh, 2021). Salt stress typically results in the accumulation of repressive histone marks like H3K27me3 and H3K9me2 on salt- resistance genes, while also causing the buildup of active histone marks like H3K4me3 and H3K9K-14Ac (Yolcu et al., 2016). For example, high degrees of H3K9Ac and H3K4me3 on the promoter, as well as highly enriched gene body region marks such as H3K23 and H3K27acetylation play crucial roles in induction of drought-stress genes in *A. thaliana* (Kim et al., 2012). In eukaryotes, histone core proteins H4, H3, H2B, and H2A can undergo various alterations including ubiquitination and methylation acetylation at the N-terminal tail. The pattern of modification on the histone tail is known as histone code is very closely related to gene expression and chromatin state. The GW studies of ChIP-seq exhibited that the influences of histone alteration on gene expression are often preserved among plants and mammals (Sequeira-Mendes et al., 2014). Some types of histone modifications contribute to a memory of developmental and environmental stimuli, (Lämke & Bäurle, 2017) which further suggests that systems need to exist that retain information about patterns of histone modification across rounds of DNA replication as well as over generations in the context of DNA methylation. The analyses in Arabidopsis mutants have exposed the respective enzymes controlling histone modifications in plants.

The Arabidopsis MEDEA protein which can cause a constitutively chromatin-silent state is an example of an H3K27-specific methyltransferase that is a crucial component of the PcG complex in plants ((Munshi et al., 2015). The regulation of alteration in histone H3K9 and H3K27 provides important insights into the identity setting mechanisms to produce, inherit, and remove epigenetic memory related to plant cell division and differentiation (Luo et al., 2012). Histone alterations which play a prominent part in mediator responses of plant to abiotic stressors are also appreciably affected by histone changes. Hypoxia-induced rice seedlings acetylate histone H3 and convert the hypoxia-responsive genes ADH1 and PDC1 from di-methyl H3K4 to trimethyl H3K4 (Munshi et al., 2015).

RNA Interference (RNAi)

The recent revelation that dsRNAs can proficiently suppress genes in plants and the nematode Caenorhabditis has resulted in RNA interference (RNAi), a fresh framework for understanding gene expression regulation (Munshi et al., 2015). The small RNAs (21–24 nucleotides long) function as an RNA interference (RNAi) strategy for gene downregulation through base-pairing to direct sequence-specific effector proteins to a target nucleotide sequence. The expression of the RNA or DNA target downregulates by effector protein complex. The small RNA-directed gene regulatory systems are now known to be present in human cells, fungi, flies, worms, and plants, (Lindbo, 2012). For instance, its impact on tomato growth vigour and production has shown the importance of epigenetic control in crop breeding(Yang et al., 2015). The RNA interference (RNAi) silences the MSH1 gene in tomatoes, increasing productivity and plant development even when the transgene is not present. The quantity and fruit weight are both augmented in the field (Kakoulidou et al., 2021).

In plants, miRNAs and siRNAs are examples of sRNAs, which are molecules capable of regulating gene expression at the post-transcriptional or transcriptional level. The TEs and TGS of the genes are silenced by small RNA through RdDM, which is a constituent of DNA methylation (Samantara et al., 2021). Several investigations have been undertaken to treatment of the epigenetic change in plants during biotic stress. The small RNA showed a typical pattern of reduced susceptibility to the root-knot nematode Meloidogyne incognita, beetroot cyst nematode Heterodera schachtii and Meloidogyne javanic in Arabidopsis mutants. This suggests that the widespread disturbance of small RNA biogenesis will hinder effective nematode parasitism (Medina et al., 2017; Ruiz-Ferrer et al., 2018). (Díaz-Manzano et al., 2018) revealed that the total number of galls and magnitude of giant cells/ plant on the same plant can be reduced by silencing certain miRNA genes. For instance, plant growth in adult plants may be impacted by the epigenetic regulation of HSP expression, which happens in germ cells. Heat stress controls the synthesis of many short RNAs in pollen and modifies chromatin conformation through necessary chromatin modifications (McCue et al., 2015). The ability of short RNAs to rewire the male germline and prevent TEs controls the pollen's resistance to temperature stress. The mutations in the DDM1 enzyme cause some of these TEs to become reactivated, which increases the production of siRNA. The siRNAs reduce TE mobility, protecting the epigenetic evidence in male gametes (Kakoulidou et al., 2021).

Table 1. Some epigenetics instances for improvement of crop

Species	Topic	Epigenetic Alteration	Key Findings	Reference
Rice & Arabidopsis	During plant gametogenesis, RdDM led to the discovery of DN Amethylation signatures exclusive to sexual lineages.	DNA methylation	It will be crucial to identify the genes and oligonucleotides that regulate RNA-directed DNA methylation and map them to sequencing genomes in order to create novel molecular markers linked to self-incompatibility, male sterility, and fertility.	(Chen et al., 2016)
Arabidopsis, tomato	The comparative study of rootstock epigenetic variation in tomato and Arabidopsis.	Small RNA	They illustrated the potential of epigenetic disruption techniques in crop production by reproducing the improved plant vigor phenotypic traits of the MSH1 system in studies on tomato field	(Kundariya et al., 2020)
White clover and Arabidopsis	DNA demethylating chemicals reduce the distinctions between wild-type and memorized plants.	DNA methylation	Research aimed to describe the function of DNA methylation in the phenomenon of stress memory.	(Yang et al., 2020)
Spinach, Basket willow and Arabidopsis	After treatment increased flowering with a substance that demethylates DNA.	DNA methylation	Stimulated flowering	(Cheng et al., 2019)
Canola	The researchers demonstrated that in an isogenic canola population, energy consumption efficiency may be artificially chosen through an epigenetic trait to boost output in hybrids.	Histone alterations and DNA methylation	The modification of the epigenome has the ability to artificially boost agricultural output.	(Kakoulidou et al., 2021)
Cotton	Allotetraploid cottons' lack of photoperiod sensitivity during domestication is attributed to epialleles, according to epigenomic and functional investigations.	DNA methylation	Influence photoperiodic flowering time and seed dormancy	(Song et al., 2017)
Grapevine	Stress-induced epigenetic memory.	DNA methylation	The conservation of DNA methylation in regenerated plants is affected by medium-to-high temperatures.	(Baránek et al., 2015)

continued on following page

Table 1. Continued

Species	Topic	Epigenetic Alteration	Key Findings	Reference
Maize	The maize methylome exhibits broad paramutation-like switches mediated by short RNA and affects mRNA splice sites.	DNA methylation	the methylation map will be vital instrument for epigenetic research and the prediction of important traits through methylation patterns.	(Regulski et al., 2013)
Maize	The hybrids' grain yield can be accurately predicted by parental sRNA divergence.	Small RNA	Measurements of epigenetics could be employed as supplementary biomarkers in crops.	(Seifert et al., 2018)
Rapeseed, barley	H3K9 de-methylation is produced by small compounds that encourage somatic embryogenesis and cell reprogramming in crop species.	Histone alterations	Different tiny compounds that reduce histone H3K9 methylation degrees indorse cell reprogramming in microspore embryogenesis and double haploid output, hence accelerating crop breeding efforts.	(Berenguer et al., 2017)
Soybean	patterns of DNA methylation in soybean root hairs	DNA methylation	DMRs exhibit different methylation patterns under high temperature stress and across stripped roots and root hairs in each methylation setting.	(Hossain et al., 2017)
Wheat	Three TaEXPA1 homoeologs' divergent expression in hexaploid heat and the role of epigenetic changes.	DNA methylation	Three TaEXPA1 homoeologs exhibit divergent expression during wheat growth, which is partly due to epigenetic changes.	(Hu et al., 2013)
Tomato	Connections among ripening tomato fruit's metabolites, non-coding RNA levels, mRNAs, and genome methylation.	DNA methylation	The ripening processes and the ethylene pathway were connected to several alterations in gene methylation.	(Zuo et al., 2020)
Tobacco, potato	Silenced transgenes can be reactivated using compounds that demethylate DNA.	DNA methylation	More effective plant genetic modification	(Yamagishi & Kikuta, 2021)

EPIGENETIC VARIABILITY AND INHERITANCE

Induced Epigenetic Variation

The phenotypic variances in wild plant populations may result from induced epigenetic modifications, which is inherited to offspring for several generations (Miryeganeh & Saze, 2020). The phenotypic variances are naturally caused by genetic and environmental differences, as well as their interactions. The altered phenotypic qualities may result from a similar genetic structure, with homozygous alleles activating in distinct manner in reaction to abiotic or biotic stress (Fortes & Gallusci, 2017; Zhi & Chang, 2021). The adaptation may be impacted by epigenetic changes, particularly when there is little genetic variety in the population such as in clonal or totally inbred plant populations (Latzel et al., 2016). Although epigenetic Arabidopsis mutants have been helpful in testing the importance of epigenetic roles in stress reactions, we must exercise caution when inferring a precise role for epigenetic functions, particularly when employing epigenetic mutants that exhibit a variety of phenotypes as a consequence of secondary effects (Popova et al., 2013). For example, MET1 gene mutation results in the formation of RNA induced methylation marks and histone H3K9 methylation in new genomic areas while also suppressing the production of DNA demethylases. Many of the phenotypes and stochastic epi-mutations that result from this are likely to indicate randomly formed unique epigenetic marks rather than direct MET1 targets (Meyer, 2015).

In *Nicotiana benthamiana*, the Geminivirus Rep protein lower the expression levels of the NbMET1 and NbCMT3 methyltransferase genes, while in Arabidopsis, biotic stress or salicylic acid causes the transcript levels of MET1 and DDM1 to decrease, while other stressors cause the transcript levels of the histone deacetylases HDA6 and HDA19 to increase (Rodríguez-Negrete et al., 2013). If temporary exposure to stressors affects the effectiveness and fidelity of epigenetic maintenance, it may increase epigenetic diversity (Di Ruscio et al., 2013).

Transgenerational Epigenetic Inheritance

The epigenetic pathways can cause phenotypic plasticity and are crucial for a variety of biological processes as well as interactions between phenotype and environment in response to different stressors. They include a number of processes that can be passed down through the germline to succeeding generations because they are mitotically and occasionally meiotically heritable (Fallet et al., 2023). TEI refers to the transmission of environmentally induced phenotypic parameters such as altered fecundity, metabolic function, or behavior elicited by parental care imprinting, pathogen exposure, pollution, or climate change to descendants via epigenetic modulators. Additionally, Research on TEI has also contributed to our insight of the function that epigenetic mechanisms may play in adaptability, adversity and evolution (Fallet et al., 2023). One of the most prominent instances of TI of epigenetic states is the stress-induced methylation of certain Arabidopsis genes that occurs through a heritable mechanism mediated by changes in cytosine methylation (Tricker et al., 2013). In this research reduced humidity induced DNA methylation in two loci controlling stomatal development and the transcription of many short RNAs corresponding to this methylation. However, when the stressor was not reapplied and hence acquired methylation was reversible, it appeared in many generations of progeny (Brukhin & Albertini, 2021).

Transgenerational epigenetic inheritance can theoretically contribute to adaptability in several ways analogous to phenotypic plasticity, the alteration of (some) phenotypes in response to environmental factors (Lind & Spagopoulou, 2018). First, phenotypic plasticity can move a population toward an adaptive peak because they survive in new conditions while their genomes are still adapting (Kronholm & Collins, 2016). The offspring's environment is more important for adaptation since the environmental predictability across generations should determine the adaptive importance of TEI of a phenotype produced by environmental factors (Lind & Spagopoulou, 2018). According to theoretical models, when environmental changes are gradual and last longer than generation times, and when parents and children can live in similar surroundings, TEI of the parental phenotype can be adaptive (Uller et al., 2015).

The use of ddm1, a chromatin-binding protein necessary for the preservation of CG methylation, as an epimutator in Arabidopsis provided the most insightful information regarding the stability of epialleles. The subset of epiRILs underwent whole-genome bisulfite sequencing following one generation and eight generations of selfing (Sarkies, 2020). This showed that after eight generations, some parts of the genome stayed variously methylated in relation to the wild type parent line, even though many other parts of the genome returned to their wild type epigenetic state. The small RNA-targeted regions tended to revert more than non-targeted regions, indicating a possible mechanism governing the stability of distinct epialleles. This clearly shows that certain parts of the genome can work together to maintain an alternate methylation state (Cortijo et al., 2014). It was revealed that the epiRILs showed phenotypic variations, which may possibly be linked to distinct areas with variations in DNA methylation (Cortijo et al., 2014).

Stability of Epigenetic Marks

The key prerequisite for a plant population to acquire durable, selective advantages is the environmental inheritance of epigenetic markers and their conservation in subsequent generations. In Arabidopsis demonstrates that spontaneous nucleotide mutations are incredibly uncommon, with only twenty SNPs per genome after thirty plant generations deriving from a single plant, and that millions of cytosines have novel methylation (Brukhin & Albertini, 2021). However, it was also uncommon to find consistent methylation changes in sizable genomic areas within the same generational series. The homeostasis of methylation maintenance remained relatively steady because methylases, methyltransferases, and demethylases worked in unison. It is assumed that the formation of new alleles is linked to difference methylation of transposons, while transposons are crucial to the genesis of novel epialleles and their inheritance (Brukhin & Albertini, 2021).

The type of change and the location of the epigenetic marks on the genes, for instance, whether they are on the gene body or the promoter region determine their ability to impact gene expressions. For example, if H3K9me2 and DNA methylation are found in the promoter area, they may inhibit transcription and thereby suppress gene expression. For example, these heterochromatin epigenetic marks exist over the gene body and then they promote transcription of the whole gene (Lei et al., 2014). Th epigenetic marks such as histone alterations and DNA methylation are crucial regulators of gene expression, changing in a dynamic fashion to respond to different stimuli (Miryeganeh & Saze, 2020). The constant turnover of epigenetic marks on stress-activated genes governs chromatin accessibility and transcriptional or translational activation of select genes. Many of the regulatory mechanisms that control abiotic stress responses are likely to be epigenetically modified, including chromatin remodeling, histone modifications and histone variants, DNA methylation and long noncoding RNAs. The epigenetic alterations are a unique acclimatization strategy for plants owing to their significant function in gene expression control and ability to transmit towards next generation (Miryeganeh & Saze, 2020).

BREEDING STRATEGIES USING EPIGENETICS

For investigating the relationship between epigenetic modifications, such as DNA methylation or histone modifications and various traits or disease Epigenetic wide Association Studies are powerful tool (Khouly, Ismael, et al. 2020). EWAS target the genome to uncover patterns of epigenetic changes that might be linked to environmental exposures, lifestyle factors, or disease states unlike Genome Wide Association Studies which focus on genetic variants (Campagna, Maria Pia, et al2021).

The most studies epigenetic modification in EWAS is DNA methylation (Ebrahimi, Vida, et al. 2020). Typically, at CpG sites, it involves the addition of methyl group to the cytosine base in DNA.Including cancer, diabetes, and cardiovascular diseases, DNA methylation patterns have been associated with various diseases (Lobos, P2021).

In EWAS histone modifications also less studied compared to DNA methylation, histone modification such as acetylation, methylation, and phosphorylation play role in chromatin structure and gene expression regulation RNA molecules do not code for proteins but regulate gene expression at transcriptional and post transcriptional levels including microRNAs and long non-coding RNAs.In EWAS research their roe in epigenetic regulation is gaining interest (Zhang, Yanjun, et al2021). On the Epigenome the feature of EWAS is the focus on the impact of environmental factors. Different environmental factors such as smoking, diet, pollution, and stress can lead to epigenetic changes that may influence disease susceptibility and other traits (Ogunjobi, Taiwo Temitope, ET al2024).

Methodology of EWAS

EWAS typically involves the collection of sample from tissue from individuals with or without the trait of interest (Hillary, Robert F., et al. 2023). In EWAS the most common method is the measurement of DNA methylation levels across the genome using technologies like Illumina's infinium Methylation EPIC Bead Chip or whole genome bisulfite sequencing (WGBS)(Kaur, Diljeet, et al2023). Statistical analysis is performed to identify differently methylated reigions(DMRs)or CpG sites that are associated with the trait or exposure after obtaining epigenetic data. In independent cohorts or through functional studies to confirm their biological relevance identified epigenetic associations are typically validated (Bernabeu, Elena, et al. 2023).To gain more comprehensive understanding of biological mechanism underlying the trait, EWAS results are often integrated with other omics data, such as transcriptomics(gene expression) or genomics(genetic variants),to gain more comprehensive understanding of the biological mechanisms underlying the trait(-Mai, Jialin, et al2023).

Epigenetic Editing Techniques

Without altering the underlying DNA sequence epigenetic editing techniques involve targeted modifications of the Epigenome (Ueda, Jun, 2023). These techniques are used to control cellular states, study the role of epigenetic marks in development, disease, used to modify gene expression patterns, and therapeutic interventions. Main techniques used in epigenetic editing are following Castro-Muñoz, Leonardo Josué, et al. 2023).

CRISPER-based Epigenome Editing

CRISPER-dCas9 System: Without cutting the DNA the deactivated Cas9 protein from the CRISPER/Cas9 system is modified to target specific DNA sequence(Arthi et al).Researchers can add or remove epigenetic marks such as DNA methylation or histone modifications at specific genomic loci by fusing dCas9 to epigenetic modifiers(Cai, Ruijie, et al2023). CRISPER-dCas9-TET: Reversing the DNA methylation this fusion is used to demethylate specific CpG sites in the genome (Tan, Jiajun, et al. 2024). CRISPER-dCas9-DNMT3A: To increase methylation and suppress gene expression dCas9can be fused with DNA methyltransferase like DNMT3A (Sapozhnikov, 2024).

TALEN-based Epigenetic Editing

To target specific DNA sequences TALEN (Transcription Activator-Like Effector Nuclease) are proteins (Mehra, Sweety, et al. 2024). To target specific genes and modify their epigenetic landscape TALENs can be fused with epigenetic modifiers such as histone acetylytransferase or DNA methyltransferase (Parker, Matthew P. 2021).

Zinc Finger Protein-based Editing

To bind specific DNA sequences Zinc finger proteins can be engineered to bind to specific DNA sequences (Dhanasekaran, Muthu, Shigeru Negi, and Yukio Sugiura. 2006). They are used to induce or remove epigenetic marks to target loci, affecting gene expression through histone modifications, methylation, or chromatin remodeling (Clark, Jeliyah, and Julia E. Rager2020).

RNA-Guided Epigenetic Editing

To specific genomic loci short RNA molecules such as small inferring RNAs or short hairpin RNAS can guide the recruitment of epigenetic enzyme to specific genomic loci(Hajieghrari, Behzad, and Naser Farrokhi. 2022). To modify gene expression epigenetically these RNA molecules can interact with chromatin-modifying complexes like polycomb group proteins (Mangiavacchi, Arianna, Gabriele Morelli, and Valerio Orlando2023).

Synthetic DNA-Binding Proteins

To change chromatin states and influence gene expression researchers design synthetic DNA-binding proteins that specifically target genes or regulatory elements often coupled with epigenetic modifiers or histone acetylytransferase to change chromatin states and influence gene expression(Parker, Matthew P. 2021).

Epigenetic Selection in Breeding Programs

Based on genetic modifications epigenetic selection in breeding programmes refers to the process of selecting and breeding plants, gene expression changes that do not involve changes in underlying DNA sequence. Key Concepts in Epigenetic Selection:

Epigenetic Mechanisms:

DNA Methylation: Typically silence gene expression, the addition of methyl groups to DNA.

- **Histone Modification**: Affecting gene accessibility, chemical changes to histone proteins which influence how tightly DNA is wound around them.
- **Non-coding RNAs**: At the post-transcriptional level small RNA molecules that regulate gene expression.

Difference from Genetic Selection:

- *Genetic Selection*
- **Epigenetic selection targets heritable changes in gene expression without altering the DNA sequence focus on DNA sequence variations.**

2. **Benefits in Breeding**: providing a faster response to environmental changes than genetic mutatios, epigenetic changes can be induced by environmental factors like temperature, water stress or nutrition
3. **Rapid Adaptation**: Epigenetic changes can be induced by environmental factors like temperature, water stress, or nutrition, providing a faster response to environmental changes than genetic mutations.

- **Transgenerational Inheritance**: offering stable way to enhance traits overtime, some epigenetic modifications can be passed down to future generations

- **Plasticity**: Allowing organisms to adapt to changing environments without altering their genetic code epigenetic modifications offer flexibility in gene expression.

Applications in Breeding Programs

- **Stress Tolerance**: Epigenetic modifications can improve tolerance to environmental stresses like drought, salinity, or heat.
- **Disease Resistance**: Epigenetic modifications can develop resistance to specific pathogens also help organisms to develop immunity.
- **Yield and Quality Enhancement**: By epigenetic modifications traits like flowering time, fruit size and nutritional quality can be influenced leading to improved crop yield and quality.

CHALLENGES AND FUTURE DIRECTIONS

Epigenetics breeding is an emerging field with great potential, but it faces several technical challenges:

Epigenetic Markers Identification:

In identifying reliable genetic makers, that consistently correlate with desired traits such as methylation or histone modifications (Li, Yuanyuan. 2021). The complexity of epigenetic modifications makes it difficult to distinguish meaningful markers from background noise, high throughput technologies like bisulfite sequencing are often used (Halabian, Raheleh, et al. 2021).

Heritability of Epigenetic Modifications:

Epigenetic changes can be reversible ad may not always be stably inherited across generations unlike genetic mutations (Bellver-Sanchis, Aina, Mercè Pallàs2021). Aimimg for long term stability, this poses a challenge in breeding programmes, breeders must ensure that the desired epigenetic modifications are transmitted consistently (Singer, Stacy D., et al. 2021)

Environmental Influence

Epigenetic changes which can complicate the breeding process are often affected by environmental conditions (Donkin, Ida, and Romain Barrès. 2018). Making it had to predict or control outcomes in different settings traits driven by epigenetic mechanisms might fluctuate depending on the growing environment (Ashe et al., 2021).

Complex Trait Regulation

Many agronomically important traits and yield or stress resistance involve complex interactions between genetic and epigenetic factors and are controlled by multiple genes (Sun, Chao, ET al2021).

High-Cost and Specialized Technologies

Different techniques such as bisulfite sequencing, ChIP-seq (Chromatin Immunopreciptation sequencing, ATAC-seq (Assay for transposase-Accessible chromatin) are expensive and required special expertise (Ma, Shaoqian, and Yongyou Zhang. 2020). In breeding programs this limits their widespread use, for smaller research groups especially in breeding programs.

Limited Understanding of Mechanisms

Mechanistic understanding and the influence phenotype of epigenetic modifications is still incomplete. Making targeted breeding difficult (Dalakouras, Athanasios, and Dimitrios Vlachostergios2021). Making targeted breeding difficult modifications such as non-coding RNAs or histone modifications are less understood (Summanwar, Aarohi, et al 2020).

Off-Target Effects and Unintended Consequences

Interventions aimed at modifying epigenetic marks may inadvertently affect non-target genes leading to unintended phenotypic sequences (Bertheau, Yves 2021). When using technologies like CRISPER/dCas9-based epigenetic editing this challenge is particularly relevant.

Ethical and Regulatory Considerations in Epigenetic Breeding

Without altering the underlying DNA sequence epigenetic breeding which involves modifying heritable gene expression brings unique ethical and regulatory considerations due to its innovative nature.

Ethical Considerations

Uncertainty of Long-Term Effects

For ecosystems, biodiversity and human health Epigenetic changes might have unforeseen consequences. Epigenetic modifications may impact future generations and can be, especially in wild species and ecosystems that's concern about unintended consequences (Olivares-Castro, Gabriela, et al. 2021).

Naturalness and Public Perception

With genetically modified organisms public concerns about manipulating epigenetic states may parallel issues seen (Shah, Esha, David Ludwig, 2021). Raising ethical concerns about consent and transparency in the food supply people may question whether this form of breeding is natural (Mankad, Aditi, et al. 2022).

Equity and Access

Epigenetic technologies may be accessible to well-funded organizations or countries and costly. Potentially widening existing agriculture inequalities, this raises ethical concerns about equitable access for small scale farmers or developing nations (Resnik, David B. 2021).

Consent and Labeling

For products derived from epigenetic breeding consumers may demand transparency and labelling. Whether consumers are adequately informed about the nature of such products ethical concerns arise around informed consent (Cohen, I. Glenn, et al., eds. 2021).

Impact on Biodiversity

Epigenetic breeding could impact biodiversity if use on large scale leading to monocultures or offering non target.to avoid environmental harm the ethical implications of such changes must be considered (Larsen, Jørgen Bo, et al. 2022).

Regulatory Considerations

Classification as GMOs or Non-GMOs

Epigenetic modifications do not involve changes to the DNA sequence although many countries have strict regulations on GMOs(Turnbull, Crystal, Morten Lillemo, and Trine AK Hvoslef-Eide. 2021).

Risk Assessment

Including unintended environmental or health impacts regulatory bodies must evaluate potential risks of epigenetic breeding (Eckerstorfer, Michael F., et al. 2021). From traditional breeding changes this could involve new risk assessment protocols since epigenetic modifications may behave differently (Singer, Stacy D., et al. 2021).

Intellectual Property

As companies may seek IP protection for specific epigenetic traits, patents and intellectual property rights around epigenetic modifications could become contentious (Kock, Michael Andreas. 2023). Regulatory frameworks need to balance innovation with fair access to technologies, especially for smaller entities (Reimsbach-Kounatze, Christian. 2021).

Cross-Border Regulations

Leading to regulatory fragmentation, different countries may develop different stances on epigenetic breeding (Singer, Stacy D., ET al2021).Requiring global co-operation to avoid trade disruptions or regulatory loopholes harmonizing regulations especially for international trade may be challenging (Eruaga 2024).

Monitoring and Enforcement

To ensure that epigenetically modified crops meet safety and ethical standards, regulatory agencies may need to develop new monitoring techniques (Turnbull, Crystal, Morten Lillemo,et al 2021). Across generations this could involve tracking both intended and unintended epigenetic changes.

The field of **epigenetics breeding** is rapidly gaining attention as it offers new opportunities for enhancing crop traits beyond the scope of traditional genetic modification. Here are some future prospects and research directions in epigenetics breeding:

Epigenetic Markers for Trait Selection

To select for desirable traits like yield, stress tolerance and disease resistance DNA methylation and histone modifications can serve as markers (Samantara, Kajal, et al. 2021). **DNA methylation** and **histone modifications** can serve as markers to select for desirable traits like yield, stress tolerance, and disease resistance (Gupta, Chainika, and Romesh K. Salgotra. 2022). Future research may focus on developing **epigenetic maps** for key crop species (Kakoulidou, Ioanna, et al. 2021).

- Advancing **epigenome-wide association studies (EWAS)** to identify epigenetic variants linked to important agricultural traits (Hoang, Thanh T., et al2020).

Stress-Induced Epigenetic Memory

- Epigenetics could allow crops to "remember" past stress events, such as drought or pathogen attacks, through heritable epigenetic changes. Understanding how these stress memories are retained and exploited for breeding **climate-resilient crops** is a promising research direction.

Epigenetic Editing Tools

- With the development of precision tools like **CRISPR-dCas9** fused with enzymes that can modify the epigenome, it will be possible to **directly edit epigenetic marks** without altering the underlying DNA sequence. Research could explore methods to apply this technology to induce desired epigenetic modifications in crops.

Epialleles and Their Stability

- Identifying and stabilizing beneficial **epialleles**—alleles that arise from epigenetic changes—will be key (Kakoulidou, Ioanna, and et al. 2021). Current challenges involve ensuring the **heritable stability** of these epialleles across generations, so future research will likely focus on methods to maintain these advantageous epigenetic states (Husby, Arild2022).

Integration with Conventional Breeding Programs

- Epigenetics can complement conventional breeding by introducing an extra layer of control over gene expression (Kakoulidou, Ioanna, et al. 2021).

Future research may develop **hybrid breeding models** that integrate both genetic and epigenetic data for more precise and effective selection of crop traits (Crisp, Peter A., ET al2022).

Non-Coding RNAs in Crop Breeding

- Non-coding RNAs, including **microRNAs** and **siRNAs**, play significant roles in regulating epigenetic changes (Pathania, Anup S., et al., 2022). Future research may focus on how manipulating these RNAs can lead to epigenetic modifications that improve traits like disease resistance or drought tolerance (Samantara, Kajal, ET al2021).

Epigenetics in Heterosis and Hybrid Vigor

- Research might focus on how **epigenetic mechanisms** contribute to **hybrid vigor (heterosis)**, which is a key driver of higher yield in hybrid crops (Liu, Wenwen, et al. 2022). Understanding how epigenetic regulation influences Heterosis could lead to novel breeding strategies (Das, Abhijit Kumar, et al. 2021).

Epigenetic Response to Environmental Changes

- Understanding how epigenetic mechanisms enable plants to adapt to **changing environmental conditions** like temperature fluctuations and water availability will be crucial for breeding crops resilient to climate change (Kakoulidou, Ioanna, et al. 2021).

Epigenetic Modifications in Clonal Propagation

- Epigenetic modifications could improve the propagation of crops through **tissue culture** and **clonal reproduction** (Ghosh, Amrita, 2021). Research may aim to stabilize beneficial epigenetic states during clonal propagation processes (Rajpal, Vijay Rani, et al2022).

Bioinformatics and Computational Approaches

- As more data on crop epigenome become available, **bioinformatics tools** will be essential for analyzing complex epigenetic datasets (Omony, Jimmy, et al 2020). Future research might focus on developing **epigenomics data-**

bases and computational models to predict epigenetic outcomes in breeding programs (Arora & Tollefsbol. 2021).

Transgenerational Epigenetic Inheritance

- Investigating the mechanisms of **transgenerational epigenetic inheritance**, where epigenetic changes induced by environmental factors are passed down to future generations, will help exploit epigenetics for long-term crop improvement (Kakoulidou, Ioanna, et al2021).

Epigenetic Regulation of Secondary Metabolites

- Epigenetic control of **secondary metabolite production** can be harnessed to enhance the nutritional and pharmaceutical properties of crops (Singh, Harshara 2023). This area of research holds promise for developing plants with higher concentrations of beneficial compounds, (Haruna, Abdurrashid, 2021).

CONCLUSION

In developing stress-resilient crops by harnessing the heritable changes in gene expression that do not involve alterations in DNA sequence epigenetic based breeding strategies offer a promising frontier. To modulate plant responses to environmental stresses like drought, salinity, and extreme temperatures, these approaches leverage mechanisms such as DNA methylation, histone modifications and non-coding RNAs. Addressing both the need for sustainability and public acceptance, breeders can introduce or enhance traits related to stress tolerance without restoring to traditional transgenic methods by targeting these epigenetic regulators. Opens new avenues for precise breeding the application of epigenetic markers in selection combined with advanced genomic tools like epigenome-wide association studies (EWAS) and these strategies can increase the development of crop varieties under adverse conditions also retain high yields that are not only stress-resilient. However, there remain challenges in fully understanding the long-term stability of epigenetic modifications and their heritability across generations. Exploring the integration of epigenetic variation into conventional breeding programs, and addressing regulatory concerns future research should focus on improving the predictability of epigenetic changes. to enhance global food security in the face of climate change and environmental stressors epigenetic based breeding holds immense potential.

REFERENCES

Arora, I., & Tollefsbol, T. O. (2021). Computational methods and next-generation sequencing approaches to analyze epigenetics data: Profiling of methods and applications. *Methods (San Diego, Calif.)*, *187*, 92–103. PMID: 32941995

Arthi, B., Senthilkumar, N., Kalaiselvi, S., Krishnapriya, S., & Anand, A. V. (2024). Gene editing using CRISPR/Cas9 system: methods and applications. In *CRISPR and Plant Functional Genomics* (pp. 258–270). CRC Press.

Ashe, A., Colot, V., & Oldroyd, B. P. (2021). How does epigenetics influence the course of evolution?. Philosophical transactions of the Royal Society B, 376(1826), 20200111.

Baránek, M., Čechová, J., Raddová, J., Holleinová, V., Ondrušíková, E., & Pidra, M. (2015). Dynamics and reversibility of the DNA methylation landscape of grapevine plants (Vitis vinifera) stressed by in vitro cultivation and thermotherapy. *PLoS One*, *10*(5), e0126638. DOI: 10.1371/journal.pone.0126638 PMID: 25973746

Berenguer, E., Bárány, I., Solís, M.-T., Pérez-Pérez, Y., Risueño, M. C., & Testillano, P. S. (2017). Inhibition of histone H3K9 methylation by BIX-01294 promotes stress-induced microspore totipotency and enhances embryogenesis initiation. *Frontiers in Plant Science*, *8*, 1161. DOI: 10.3389/fpls.2017.01161/full PMID: 28706533

Brukhin, V., & Albertini, E. (2021). Epigenetic modifications in plant development and reproduction. *Epigenomes*, *5*(4), 25. DOI: 10.3390/epigenomes5040025 PMID: 34968249

Castillo, A. M., Valero-Rubira, I., Burrell, M. Á., Allué, S., Costar, M. A., & Vallés, M. P. (2020). Trichostatin A affects developmental reprogramming of bread wheat microspores towards an embryogenic route. *Plants*, *9*(11), 1442. DOI: 10.3390/plants9111442 PMID: 33114625

Chen, Y., Müller, F., Rieu, I., & Winter, P. (2016). Epigenetic events in plant male germ cell heat stress responses. *Plant Reproduction*, *29*, 21–29. DOI: 10.1007/s00497-015-0271-5 PMID: 26639000

Cheng, Y.-H., Peng, X.-Y., Yu, Y.-C., Sun, Z.-Y., & Han, L. (2019). The effects of DNA methylation inhibition on flower development in the dioecious plant Salix viminalis. *Forests*, *10*(2), 173. DOI: 10.3390/f10020173

Cortijo, S., Wardenaar, R., Colomé-Tatché, M., Gilly, A., Etcheverry, M., Labadie, K., Caillieux, E., Hospital, F., Aury, J.-M., & Wincker, P. (2014). Mapping the epigenetic basis of complex traits. *Science*, *343*(6175), 1145–1148. PMID: 24505129

Di Ruscio, A., Ebralidze, A. K., Benoukraf, T., Amabile, G., Goff, L. A., Terragni, J., Figueroa, M. E., De Figueiredo Pontes, L. L., Alberich-Jorda, M., & Zhang, P. (2013). DNMT1-interacting RNAs block gene-specific DNA methylation. *Nature*, *503*(7476), 371–376. DOI: 10.1038/nature12598 PMID: 24107992

Díaz-Manzano, F. E., Cabrera, J., Ripoll, J. J., Del Olmo, I., Andrés, M. F., Silva, A. C., Barcala, M., Sánchez, M., Ruíz-Ferrer, V., & de Almeida-Engler, J. (2018). A role for the gene regulatory module microRNA172/TARGET OF EARLY ACTIVATION TAGGED 1/FLOWERING LOCUS T (mi RNA 172/TOE 1/FT) in the feeding sites induced by Meloidogyne javanica in Arabidopsis thaliana. *The New Phytologist*, *217*(2), 813–827. DOI: 10.1111/nph.14839 PMID: 29105090

Fallet, M., Blanc, M., Di Criscio, M., Antczak, P., Engwall, M., Bosagna, C. G., Rüegg, J., & Keiter, S. H. (2023). Present and future challenges for the investigation of transgenerational epigenetic inheritance. *Environment International*, *172*, 107776. DOI: 10.1016/j.envint.2023.107776 PMID: 36731188

Fortes, A. M., & Gallusci, P. (2017). Plant stress responses and phenotypic plasticity in the epigenomics era: Perspectives on the grapevine scenario, a model for perennial crop plants. *Frontiers in Plant Science*, *8*, 82. DOI: 10.3389/fpls.2017.00082 PMID: 28220131

Hossain, M. S., Kawakatsu, T., Kim, K. D., Zhang, N., Nguyen, C. T., Khan, S. M., Batek, J. M., Joshi, T., Schmutz, J., & Grimwood, J. (2017). Divergent cytosine DNA methylation patterns in single-cell, soybean root hairs. *The New Phytologist*, *214*(2), 808–819. DOI: 10.1111/nph.14421 PMID: 28106918

Hu, Y., Zhang, L., He, S., Huang, M., Tan, J., Zhao, L., Yan, S., Li, H., Zhou, K., & Liang, Y. (2012). Cold stress selectively unsilences tandem repeats in heterochromatin associated with accumulation of H3K9ac. *Plant, Cell & Environment*, *35*(12), 2130–2142. DOI: 10.1111/j.1365-3040.2012.02541.x PMID: 22616903

Hu, Z., Han, Z., Song, N., Chai, L., Yao, Y., Peng, H., Ni, Z., & Sun, Q. (2013). Epigenetic modification contributes to the expression divergence of three T a EXPA 1 homoeologs in hexaploid wheat (T riticum aestivum). *The New Phytologist*, *197*(4), 1344–1352. DOI: 10.1111/nph.12131 PMID: 23360546

Kakoulidou, I., Avramidou, E. V., Baránek, M., Brunel-Muguet, S., Farrona, S., Johannes, F., Kaiserli, E., Lieberman-Lazarovich, M., Martinelli, F., & Mladenov, V. (2021). Epigenetics for crop improvement in times of global change. *Biology (Basel)*, *10*(8), 766. DOI: 10.3390/biology10080766 PMID: 34439998

Kim, J.-M., To, T. K., Ishida, J., Matsui, A., Kimura, H., & Seki, M. (2012). Transition of chromatin status during the process of recovery from drought stress in Arabidopsis thaliana. *Plant & Cell Physiology*, *53*(5), 847–856. DOI: 10.1093/pcp/pcs053 PMID: 22505693

Kronholm, I., & Collins, S. (2016). Epigenetic mutations can both help and hinder adaptive evolution. *Molecular Ecology*, *25*(8), 1856–1868. DOI: 10.1111/mec.13296 PMID: 26139359

Kundariya, H., Yang, X., Morton, K., Sanchez, R., Axtell, M. J., Hutton, S. F., Fromm, M., & Mackenzie, S. A. (2020). MSH1-induced heritable enhanced growth vigor through grafting is associated with the RdDM pathway in plants. *Nature Communications*, *11*(1), 5343. DOI: 10.1038/s41467-020-19140-x PMID: 33093443

Lämke, J., & Bäurle, I. (2017). Epigenetic and chromatin-based mechanisms in environmental stress adaptation and stress memory in plants. *Genome Biology*, *18*, 1–11. DOI: 10.1186/s13059-017-1263-6 PMID: 28655328

Latzel, V., Rendina González, A. P., & Rosenthal, J. (2016). Epigenetic memory as a basis for intelligent behavior in clonal plants. *Frontiers in Plant Science*, *7*, 1354. DOI: 10.3389/fpls.2016.01354 PMID: 27630664

Lei, M., La, H., Lu, K., Wang, P., Miki, D., Ren, Z., Duan, C.-G., Wang, X., Tang, K., & Zeng, L. (2014). Arabidopsis EDM2 promotes IBM1 distal polyadenylation and regulates genome DNA methylation patterns. *Proceedings of the National Academy of Sciences of the United States of America*, *111*(1), 527–532. DOI: 10.1073/pnas.1320106110 PMID: 24248388

Lind, M. I., & Spagopoulou, F. (2018). *Evolutionary consequences of epigenetic inheritance* (Vol. 121). Nature Publishing Group.

Lindbo, J. A. (2012). A historical overview of RNAi in plants. Antiviral Resistance in Plants: Methods and Protocols, 1-16. https://doi.org/ DOI: 10.1007/978-1-61779-882-5_1

Luo, M., Liu, X., Singh, P., Cui, Y., Zimmerli, L., & Wu, K. (2012). Chromatin modifications and remodeling in plant abiotic stress responses. Biochimica et Biophysica Acta (BBA)-. *Gene Regulatory Mechanisms*, *1819*(2), 129–136. DOI: 10.1016/j.bbagrm.2011.06.008 PMID: 21708299

Maeji, H., & Nishimura, T. (2018). Epigenetic mechanisms in plants. In *Advances in botanical research* (Vol. 88, pp. 21–47). Elsevier., DOI: 10.1016/bs.abr.2018.09.014

McCue, A. D., Panda, K., Nuthikattu, S., Choudury, S. G., Thomas, E. N., & Slotkin, R. K. (2015). ARGONAUTE 6 bridges transposable element m RNA-derived si RNA s to the establishment of DNA methylation. *The EMBO Journal*, *34*(1), 20–35. DOI: 10.15252/embj.201489499 PMID: 25388951

Medina, C., Da Rocha, M., Magliano, M., Ratpopoulo, A., Revel, B., Marteu, N., Magnone, V., Lebrigand, K., Cabrera, J., & Barcala, M. (2017). Characterization of microRNAs from Arabidopsis galls highlights a role for miR159 in the plant response to the root-knot nematode Meloidogyne incognita. *The New Phytologist*, *216*(3), 882–896. DOI: 10.1111/nph.14717 PMID: 28906559

Mercé, C., Bayer, P. E., Tay Fernandez, C., Batley, J., & Edwards, D. (2020). Induced methylation in plants as a crop improvement tool: Progress and perspectives. *Agronomy (Basel)*, *10*(10), 1484. DOI: 10.3390/agronomy10101484

Meyer, P. (2015). Epigenetic variation and environmental change. *Journal of Experimental Botany*, *66*(12), 3541–3548. DOI: 10.1093/jxb/eru502 PMID: 25694547

Miryeganeh, M. (2021). Plants' epigenetic mechanisms and abiotic stress. *Genes*, *12*(8), 1106. DOI: 10.3390/genes12081106 PMID: 34440280

Miryeganeh, M., & Saze, H. (2020). Epigenetic inheritance and plant evolution. *Population Ecology*, *62*(1), 17–27. DOI: 10.1002/1438-390X.12018

Munshi, A., Ahuja, Y., & Bahadur, B. (2015). Epigenetic mechanisms in plants: an overview. Plant biology and biotechnology: volume II: plant genomics and biotechnology, 265-278. https://doi.org/ DOI: 10.1007/978-81-322-2283-5_12

Popova, O. V., Dinh, H. Q., Aufsatz, W., & Jonak, C. (2013). The RdDM pathway is required for basal heat tolerance in Arabidopsis. *Molecular Plant*, *6*(2), 396–410. PMID: 23376771

Regulski, M., Lu, Z., Kendall, J., Donoghue, M. T., Reinders, J., Llaca, V., Deschamps, S., Smith, A., Levy, D., & McCombie, W. R. (2013). The maize methylome influences mRNA splice sites and reveals widespread paramutation-like switches guided by small RNA. *Genome Research*, *23*(10), 1651–1662. DOI: 10.1101/gr.153510.112 PMID: 23739895

Rodríguez-Negrete, E., Lozano-Durán, R., Piedra-Aguilera, A., Cruzado, L., Bejarano, E. R., & Castillo, A. G. (2013). Geminivirus R ep protein interferes with the plant DNA methylation machinery and suppresses transcriptional gene silencing. *The New Phytologist*, *199*(2), 464–475. DOI: 10.1111/nph.12286 PMID: 23614786

Ruiz-Ferrer, V., Cabrera, J., Martinez-Argudo, I., Artaza, H., Fenoll, C., & Escobar, C. (2018). Silenced retrotransposons are major rasiRNAs targets in Arabidopsis galls induced by Meloidogyne javanica. *Molecular Plant Pathology, 19*(11), 2431–2445. DOI: 10.1111/mpp.12720 PMID: 30011119

Samantara, K., Shiv, A., de Sousa, L. L., Sandhu, K. S., Priyadarshini, P., & Mohapatra, S. R. (2021). A comprehensive review on epigenetic mechanisms and application of epigenetic modifications for crop improvement. *Environmental and Experimental Botany, 188*, 104479. DOI: 10.1016/j.envexpbot.2021.104479

Sarkies, P. (2020). Molecular mechanisms of epigenetic inheritance: Possible evolutionary implications. *Seminars in Cell & Developmental Biology*. PMID: 31228598

Seifert, F., Thiemann, A., Schrag, T. A., Rybka, D., Melchinger, A. E., Frisch, M., & Scholten, S. (2018). Small RNA-based prediction of hybrid performance in maize. *BMC Genomics, 19*, 1–14. DOI: 10.1186/s12864-018-4708-8 PMID: 29783940

Sequeira-Mendes, J., Aragüez, I., Peiró, R., Mendez-Giraldez, R., Zhang, X., Jacobsen, S. E., Bastolla, U., & Gutierrez, C. (2014). The functional topography of the Arabidopsis genome is organized in a reduced number of linear motifs of chromatin states. *The Plant Cell, 26*(6), 2351–2366. DOI: 10.1105/tpc.114.124578 PMID: 24934173

Shen, X., De Jonge, J., Forsberg, S. K., Pettersson, M. E., Sheng, Z., Hennig, L., & Carlborg, Ö. (2014). Natural CMT2 variation is associated with genome-wide methylation changes and temperature seasonality. *PLOS Genetics, 10*(12), e1004842. DOI: 10.1371/journal.pgen.1004842 PMID: 25503602

Song, Q., Zhang, T., Stelly, D. M., & Chen, Z. J. (2017). Epigenomic and functional analyses reveal roles of epialleles in the loss of photoperiod sensitivity during domestication of allotetraploid cottons. *Genome Biology, 18*, 1–14. DOI: 10.1186/s13059-017-1229-8 PMID: 28558752

Tricker, P. J., Rodríguez López, C. M., Gibbings, G., Hadley, P., & Wilkinson, M. J. (2013). Transgenerational, dynamic methylation of stomata genes in response to low relative humidity. *International Journal of Molecular Sciences, 14*(4), 6674–6689. DOI: 10.3390/ijms14046674 PMID: 23531533

Uller, T., English, S., & Pen, I. (2015). When is incomplete epigenetic resetting in germ cells favoured by natural selection? Proceedings of the Royal Society B: Biological Sciences, 282(1811), 20150682. https://doi.org/DOI: 10.1098/rspb.2015.0682

Wang, W.-S., Pan, Y.-J., Zhao, X.-Q., Dwivedi, D., Zhu, L.-H., Ali, J., Fu, B.-Y., & Li, Z.-K. (2011). Drought-induced site-specific DNA methylation and its association with drought tolerance in rice (Oryza sativa L.). *Journal of Experimental Botany, 62*(6), 1951–1960. DOI: 10.1093/jxb/erq391 PMID: 21193578

Yamagishi, K., & Kikuta, Y. (2021). Nucleoside derivatives of 5-methylcytosine suppress 5-azacytidine-induced reactivation of a silent transgene in suspension-cultured tobacco cells. *Plant Biotechnology (Sheffield, England), 38*(1), 173–178. DOI: 10.5511/plantbiotechnology.20.1126b PMID: 34177339

Yang, X., Kundariya, H., Xu, Y.-Z., Sandhu, A., Yu, J., Hutton, S. F., Zhang, M., & Mackenzie, S. A. (2015). MutS HOMOLOG1-derived epigenetic breeding potential in tomato. *Plant Physiology, 168*(1), 222–232. DOI: 10.1104/pp.15.00075 PMID: 25736208

Yang, X., Sanchez, R., Kundariya, H., Maher, T., Dopp, I., Schwegel, R., Virdi, K., Axtell, M. J., & Mackenzie, S. A. (2020). Segregation of an MSH1 RNAi transgene produces heritable non-genetic memory in association with methylome reprogramming. *Nature Communications, 11*(1), 2214. DOI: 10.1038/s41467-020-16036-8 PMID: 32371941

Yolcu, S., Ozdemir, F., Güler, A., & Bor, M. (2016). Histone acetylation influences the transcriptional activation of POX in Beta vulgaris L. and Beta maritima L. under salt stress. *Plant Physiology and Biochemistry, 100*, 37–46. DOI: 10.1016/j.plaphy.2015.12.019 PMID: 26773543

Zhang, H., Lang, Z., & Zhu, J.-K. (2018). Dynamics and function of DNA methylation in plants. *Nature Reviews. Molecular Cell Biology, 19*(8), 489–506. DOI: 10.1038/s41580-018-0016-z PMID: 29784956

Zhi, P., & Chang, C. (2021). Exploiting epigenetic variations for crop disease resistance improvement. *Frontiers in Plant Science, 12*, 692328. DOI: 10.3389/fpls.2021.692328 PMID: 34149790

Zuo, J., Grierson, D., Courtney, L. T., Wang, Y., Gao, L., Zhao, X., Zhu, B., Luo, Y., Wang, Q., & Giovannoni, J. J. (2020). Relationships between genome methylation, levels of non-coding RNAs, mRNAs and metabolites in ripening tomato fruit. *The Plant Journal, 103*(3), 980–994. DOI: 10.1111/tpj.14778 PMID: 32314448

Chapter 2
Mechanisms of RNA-Mediated Epigenetic Regulation in Plants

Asia Maqbool
University of the Punjab, Pakistan

Naila Mir Baz
https://orcid.org/0009-0005-8719-7655
Hebei Agriculture University, China

Ammara Saleem
University of the Punjab, Pakistan

Javed Iqbal
https://orcid.org/0000-0002-9032-2622
Bacha Khan University, Charsadda, Pakistan

ABSTRACT

Epigenetic modifications, a biological process, can be mediated and directed by various types of RNAs (such as siRNA, sRNAs, lncRNAs, etc) in plants. These modifications comprise a tremendously interactive network of RNA-directed DNA methylations, chromatin and histone modifications and gene silencing. These epigenetic modifications are unique to plants and possess significant roles in growth, development, abiotic and biotic stresses. The latest discoveries in this field have been made possible through next-generation sequencing and omics technologies, which help design an essential blueprint for the epigenome in plants. As the researchers unrevealed multiple levels of epigenetic regulation mediated by RNAs in plants, a

DOI: 10.4018/979-8-3693-8648-4.ch002

better understanding of the significance of epigenetic states and marks has been established, and it opens the path to constructing new plant varieties with desired traits.

INTRODUCTION

Plants, being fascinating organisms, can fine-tune their developmental processes and adjust them according to their surroundings via epigenetic modifications. These modifications add a layer of epigenetic regulations beyond the domain of genetically imprinted components (Hemenway and Gehring 2023; X. Lu and Hyun 2021). These modifications refer to the variation at the gene level (gene expression variations) without any alteration in DNA sequence. These dynamic and reversible modifications assist the plant in the context of growth and face environmental stresses. DNA methylation (addition of methyl group at cytosine that can overexpress or silence the gene) (Suelves et al. 2016), histone modifications (alteration in chromatin structure that can play a role in the control of gene accessibility) (Prasad ET AL., 2021) and some RNAs (such as mRNAs) that guide chromatin modifying enzymes and silence the gene are well-known types of epigenetic modifications (Li et al. 2019). All these pathways are essential for the maintenance and stability of the genome and for ensuring the transfer of these functional states towards upcoming generations.

Various mechanisms and factors lead to the formation of these epigenetic modifications. RNA-directed epigenetic modifications are well-known mechanisms in plant science. Multiple types of RNAs, such as small interfering RNAs (siRNAs) and microRNAs (miRNAs), take part in DNA methylation (RNA-directed DNA methylation-RdDM), modify the chromatin and histone structure, silence the genes and sometimes interact with some transcriptional factors (TFs) to increase the plant tolerance against pathogens. RdDM refers to the methyl group addition at cytosine, and sometimes, this methylation occurs due to some external stimuli such as stress cues and holds great significance in contexts to the regulation of transposable elements in embryos and gametes and modulates the transgenerational genetic imprinting (Chen et al. 2024). Histone modifications are another epigenetic modification mediated by RNAs, and two histone modifications (acetylation and methylation) occur in plants that take part in gene expression regulation (Portela and Esteller 2010)(Sun et al. 2016). Sometimes, under stress cues (biotic or abiotic stress), RNA interacts with TFs, modulates gene expression, and induces tolerance against these stresses (Samad et al. 2017, Ahmad et al., 2022a).

Epigenetic modifications mediated by RNA have vast implications in fundamental biology and hold crucial promise for agriculture. Knowledge about these regulatory processes can assist scientists in creating stress-tolerant, high-yielding plant varieties and address the current challenges of climate change and the world's growing population. This chapter focuses on the mechanism of RNA-mediated epigenetic modifications in plants. We also elaborate on the biological functions of these modifications and their applications in plant biotechnology.

MECHANISMS OF RNA-MEDIATED EPIGENETIC REGULATION

Molecular Pathway of RNA-Directed DNA Methylation

The only process that adds a methyl group to cytosine in DNA (regardless of sequence context) is known as RNA-directed DNA methylation (RdDM). DNA methylation is divided into CHH, CGH and CG based on the sequence context of methylated cytosine. These context-specialized pathways are involved in DNA methylation maintenance. For example, methyltransferase 1 (MET1) is highly conserved and maintains methylation in CG context. Chromomethylase 2 (CMT2) and CMT3 maintain CHH and CHG methylation (Wendte and Pikaard 2017). It consists of two key processes: the generation of siRNA and the downstream chromatin phase. During this process, double-stranded RNAs (dsRNA) convert into small interfering RNAs (siRNAs) via dicer-like enzymes. Then, these siRNAs attach to AGO4 proteins, leading to RNA-induced silencing complex (RISC) and forming DNA methylation. In the second phase of chromatin regulation, siRNA and AGO4 form a complex to guide RNA polymerase V towards specific regions to produce scaffold RNAs.

These types of scaffold RNAs make a platform for creating de novo DNA methyltransferases such as DRM1 and DRM2, which further methylate cytosine deposits in target DNA. As a result of this methylation, transcriptional silencing occurs, which is an essential part of epigenetic regulation. Methylation produces supplementary regulatory proteins that help maintain these marks through DNA replication. Maintenance methyltransferases (MET1) confirm that the methylation patterns are authentically replicated into newly made DNA, conserving the epigenetic information crosswise generations (Ferrari et al. 2023). AGO4 is crucial for helping epigenetic modifications at new loci by initiating the RdDM process at un-methylated sites. RNA Polymerases II and IV are critical suppliers: Pol II-derived transcripts produce DNA recognition signals, whereas IV generates siRNAs (Arce et al. 2023). Furthermore, chromatin remodelling and hetero-chromatin creation through histone modifications carry a repressive chromatin state that is vital for gene silencing.

The RdDM pathway includes a complex network of RNA molecules, proteins, and enzymatic processes which combine to regulate gene expression and DNA methylation. The RdDM pathway, as shown in **Figure 1**, is crucial for genomic constancy, adaptive responses to environmental stimuli, and long-term gene regulation in plants by directing definite genomic sites for methylation and preserving these modifications through feedback loops (Arce et al. 2023).

Figure 1. The RdDM pathway: The siRNA generation, driven by Pol IV, RDR2, and DCL3. DNA methylation process, guided by Pol V, DRM1/DRM2, and maintained by MET1

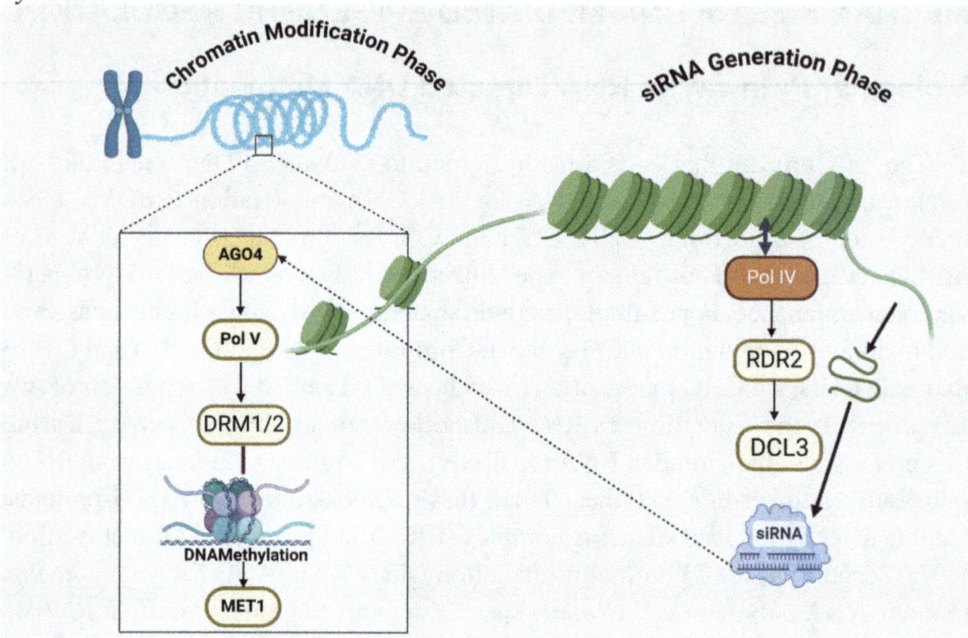

Role of siRNAs in DNA Methylation

Small interfering RNAs (siRNAs) facilitate DNA methylation through the RdDM pathway. They function as molecular guides, leading Argonaute (AGO) proteins to specific genomic sites. This interaction helps engage the DNA methylation machinery (Ali and Tang 2024).

The siRNAs (essential for RNA-mediated DNA-methylation) are derived from long double-stranded RNAs, which frequently initiate from repetitive sequences, transposable elements, or viral RNAs, processed by dicer-like enzymes into 21-24 nucleotide fragments. The siRNAs then are loaded on Argonaute proteins, mainly Argonaute 4 (AGO4), establishing RNA-induced silencing multiplexes (RISC). The employment of transcriptional machinery, with RNA polymerase V (Pol V) and de-novo DNA methyltransferases like DRM1 and DRM2, is made likely by the twenty-one nucleotide siRNAs that direct AGO4 to complementary DNA sequences

(Arce et al. 2023). This targeted employment leads to cytosine methylation in CpG, CHG, and CHH contexts, silencing transcription and creating stable epigenetic marks. siRNAs also play a crucial role in preserving DNA methylation among the generations. Once established, the siRNAs will support the silenced state by leading other rounds of methylation during DNA replication, safeguarding the persistent inheritance of epigenetic modifications. This process is vital for suppressing the repetitive sequences and transposable elements, protecting the genome from volatility and harmful genetic activity.

Gene Silencing (GS)

Various types of RNAs, such as siRNA and lncRNAs that cause DNA methylation, can take part in gene silencing. These RdDMs play a significant role in gene silencing. The RdDM with additional RNA polymerase V complex directs the *DRM1/2*-dependent methylation (Matzke, Kanno, and Matzke 2015). Chromatin dynamics and post-translational modifications of histones add a layer to DNA methylation. HISTONE DEACETYLASE 6 protein (HDA6) interacts with the MET1 gene to maintain gene silencing. RdDM and RdRM also take part in gene silencing (S. Liu et al. 2021). It is a regulatory process that helps to sustain the genome stability without any alterations in the DNA sequence.

Post-Transcriptional gene Silencing (PTGS)

At the post-transcriptional level, a phenomenon of homology-dependent gene silencing, which depends on dsRNA during gene silencing, is RNA interference. The whole factory of RNAi gene silencing contained some important components: RDR, DCL, AGO and the dsRBP (dsRNA-binding domain) (Wassenegger 2002). Further the biosynthesis of sRNA (Zheng et al. 2018) is of vital importance that also includes the miRNAs and some types of siRNA such as nat-siRNA (natural-antisense siRNA), ta-siRNA, hc-siRNA (heterochromatic siRNA) or ra-siRNAs (repeated-associated siRNAs). The action of DCL1, dsRBP and HYL1 (hyponastic leaves 1) in the nucleus is mainly involved in the cleavage process of pri-miRNAs (primary miRNAs) into the pre-miRNAs (precursor miRNAs). So, for the final conversion of pre-miRNA into miRNA duplex, DCL1 and HYL1 (Vazquez et al. 2004) are involved in the second cleave process of pre-miRNA (C. Lu and Fedoroff 2000). Subsequently, the sRNA-specific methyltransferase HEN1 methylated the duplex miRNA and then exported it into the cytoplasm with the help of HST (exportin-5 ortholog HASTY). Inside the cytoplasm, for activation of the RISC, the single-stranded mature miRNA binds with the AGO1, which causes cleavage of all homologous transcripts. dsRNAs cleaved into siRNAs via DCL4 and methylated by HEN1 (Hua Enhancer 1). This methylated siRNA further attaches to AGO1 and makes RISC (Vazquez et al. 2004). RISC plays a role in mRNA cleavage and transgene silencing (Li et al. 2005, Ahmad et al., 2022b). It helps to regulate the genome stability and development of plants under stress.

RNA-MEDIATED REGULATION OF HISTONE MODIFICATIONS: ACETYLATION, METHYLATION, AND DEACETYLATION

Epigenetics refers to the regulatory evidence beyond the genetic DNA sequence across generations. Studies reveal that histone modifications, DNA methylation, and RNA-based processes significantly contribute to epigenetic regulation. Histone alterations are epigenetic modifications which regulate the gene expression (S. Liu et al. 2021). There are two types: histone modifications happen mainly at lysine residues in histone H3, recognized as methylation and acetylation (Wassenegger 2002). Acetylation is a specific phenomenon with a stable state facilitated by the contrasting activities of histone acetyltransferases (HATs) and deacetylases, as shown in **Figure 2**. These actions include large regulatory complexes responsible for responding to particular DNA sequences and comprise transcription factors (TFs), signal transduction and cell cycle proteins (Ying Wang et al. 2011). Acetylation plays an important role in nucleosome assembly. Likely, the particular acetylation forms of emerging histones are essential for their assembly *in vitro* (Zheng et al. 2018). The acetylation pattern in the mature nucleosomes varies from that of newborn fused histones (C. Lu and Fedoroff 2000). Improved acetylation is usually linked with transcriptionally active or controlled genes. Histone acetylation within nuclei is affected by A-type histone acetyltransferases (HAT A) that are likely to participate in gene activation (Vazquez et al. 2004).

On the other hand, Histone methylation is crucial for epigenetic alteration, which controls many nuclear mechanisms, including transcription, replication, and repair of the DNA (J. Li et al. 2005). Particular histone methyl marks may either activate or inhibit the transcription liable on their position and methylation state (Shvedunova and Akhtar 2022). All histone methylation regions establish functional sequences in the genome, throughout a gene and inside non-coding and regulatory elements, such as enhancers and promoters. Histone methylation is actively regulated by two proteins, histone methyltransferase and histone demethylase that deposit and eliminate methylation marks (J. Li et al. 2005). They primarily occur on histone lysine methylation sites such as H3K and two arginine methylation sites (H3R17 and H4R3), altered by mono-methylation, di-methylation or tri-methylation (Bottomley 2004). Significantly, the genomic allocation of histone methylation alters intensely throughout the growth development and in response to exogenic signals, to bring about widespread transcriptional reprogramming (Albini, Zakharova, and Ait-Si-Ali 2019). Despite its chief regulatory role, there is little knowledge about how histone methylation as a structure receives data from upstream signalling pathways.

Figure 2. Histone modifications: methylation and acetylation. HATs enzymes regulate histone acetylation. HATs change the chromatin structure in the nucleus by comforting the chromatin and permitting transcriptional activation. RNAPII is involved in the opening of chromatin

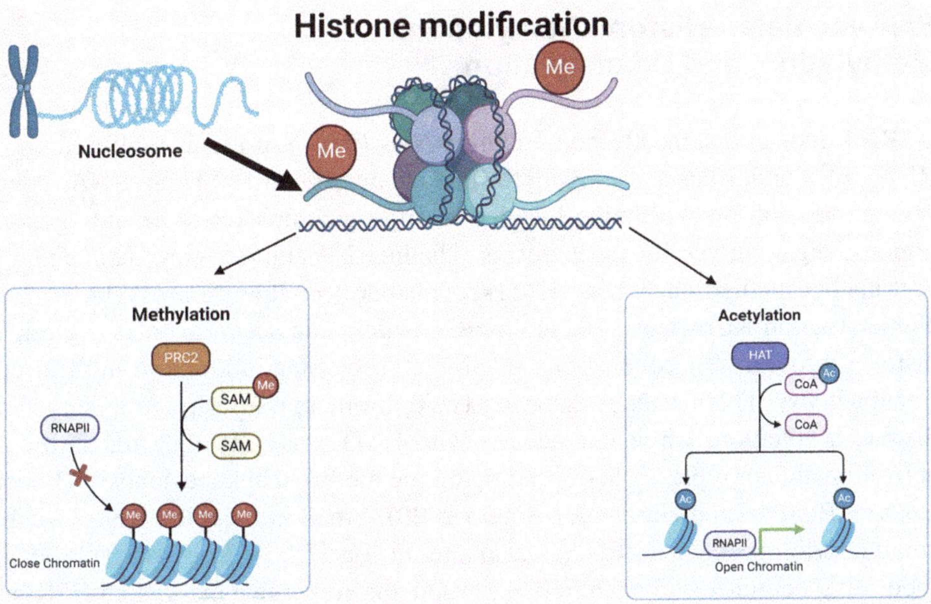

Histone Modifications

Histone modifications are critical regulators of gene expression, and chromatin structure is essential in regulating DNA accessibility for transcriptional machinery. The chemical alterations include acetylation, methylation, phosphorylation, and ubiquitination, promote or inhibit gene expression, depending on their type and location on histone tails (Shvedunova and Akhtar 2022). Histone acetylation is related to active transcription; adding acetyl groups neutralizes the positive charge of histones. It decreases their affinity for the negatively charged DNA, resulting in a more relaxed chromatin structure that helps transcription (Bottomley 2004). Moreover, histone methylation has various effects; Modifications such as H3K4me3 are linked with transcriptional activation, while others, H3K27me3, are related to gene repression.

Histone modifications also act as molecular signals, engaging effector proteins and facilitating downstream chromatin structure and function variations. H3K4me3 marks fascinate proteins which promote transcriptional activation, emphasizing gene expression at dynamically transcribed loci (Albini, Zakharova, and Ait-Si-Ali 2019). Histone modifications' flexible and dynamic nature allows cells to respond quickly to environmental and developmental stimuli, highlighting their role in cellular flexibility and adaptability.

RNA-Mediated Histone Acetylation, Methylation, and Deacetylation

RNA molecules, mostly non-coding RNAs (lncRNAs) and small non-coding RNAs, play vital roles in arranging the histone modifications. These RNAs control gene expression by facilitating histone-modifying complexes to definite genomic regions, engaging various mechanisms. The lncRNAs bind to chromatin and unite with histone methyltransferases (HMTs) or histone acetyltransferases (HATs) (Bure, Nemtsova, and Kuznetsova 2022). LncRNAs help add acetyl or methyl groups to histones through these connections, manipulating chromatin structure and transcriptional activity. The lncRNAs directing HATs promote transcription by enhancing histone acetylation, while interactions with HMTs may facilitate transcriptional activation and, or repression, depending on the modified histone deposits (Loaeza-Loaeza, Beltran, and Hernández-Sotelo 2020). Small non-coding RNAs, such as small interfering RNAs (siRNAs), also alter histone. These RNAs lead chromatin-modifying complexes to specific loci, producing repressive marks like H3K9me3 deposition (L.-C. Li 2014). Such alterations promote hetero-chromatin establishment, helping long-term gene silencing and preserving genomic stability. The active relationship between RNA species and histone-modifying enzymes underscores the complexity of epigenetic regulation. This multifaceted network allows the precise gene expression switch, responding to developmental and environmental indicators while sustaining genomic integrity.

Role of lncRNAs in Chromatin Remodeling

Long non-coding RNAs (lncRNAs) are crucial regulators of chromatin remodelling, exerting their impact by networking with transcription factors (TFs) and chromatin-remodeling. Additionally, lncRNAs help chromatin architecture by facilitating long-range interactions among distant genomic regions (X. Li and Fu 2019). COLD ASSISTED INTRONIC NON-CODING RNA (COLDAIR) is a renowned example of IncRNA in *Arabidopsis thaliana*. COLDAIR boosts the epigenetic regulation of the *FLOWERING LOCUS C (FLC)*, a crucial gene that controls flowering time. During the vernalization period, COLDAIR facilitates the Polycomb Repressive Complex 2 (PRC2) to the *FLC* locus, leading to tri-methylation of histone H3 at lysine 27 (H3K27me3) and resulting in transcriptional silencing of *FLC* which is crucial for the plant's response to temperature changes, allowing the plants to flower in the suitable conditions (Zhao et al. 2022). HIDDEN TREASURE 1 (HID1) is IncRNAs responsible for regulating photo-morphogenesis in *Arabidopsis thaliana*. It acts as a trans-acting lncRNA, which binds to the promoter region of the *phytochrome-interacting factor 3 (PIF3)* gene, silencing the transcriptional activity. HID1 interacts with chromatin remodelers and helps modify the chromatin structure at the PIF3 loci, limiting its activity and accessibility (Sánchez-Retuerta, Suaréz-López, and Henriques 2018). The transcriptional factories are formed by these interactions, which are nuclear hubs that collect co-regulated genes and increase transcription. Furthermore, lncRNAs stabilize the binding of chromatin-remodeling complexes to their targets, allowing transcriptional activation or knockout depending on the chromatin context.

BIOLOGICAL SIGNIFICANCE OF RNA-MEDIATED EPIGENETIC MODIFICATIONS

The Biological Significance of RdDM

Epigenetic modifications mediated by various types of RNAs have many significant functions in plants. In flowering plants, RdDM plays an important role in reproduction and normal development. In Arabidopsis, RdDM represses a FLOWERING WAGENINGEN (FWA) gene. *FWA* is accountable for flowering at the appropriate time, and the promoter of this gene contains tandem repeats, which are methylated via RdDM which leads to transcriptional repression (Soppe et al. 2000) (Kinoshita et al. 2007). Phenotype with late-flowering occurs when this methylation loss reactivates the expression level *FWA* (Soppe et al. 2000) (Kinoshita et al. 2007). Mutations in the pathway of RdDM can significantly affect seed viability and gamete formation(Han et al. 2019), especially in *Brassica rapa* and maize, due to higher TE content (Dew-Budd et al. 2024). During gamete formation, RdDM plays an important role in reinforcing TE silencing in germ cells (Erdmann and Picard 2020). A cell referred to as a support cell undergoes epigenetic reprogramming in ovules and pollen, shedding epigenetic markers (including DNA methylation) and TEs at various loci (Slotkin et al. 2009). It results in TE reactivations and boost-up the RdDM-derived sRNAs synthesis to combat these TEs. To boost the TE silencing in the following generations, sRNAs are believed to leave the support cells and migrate towards germ cells (Erdmann et al. 2017)(Martínez et al. 2016). RdDM also plays a significant role in fruit ripening, seed dormancy, plant development and flowering (Cheng JingFei et al. 2018; Guo et al. 2017; Iwasaki et al. 2019). Parental genome balance is crucial for seed development as it helps to avoid seed abortion and developmental disorders (Lafon Placette 2020). In *Capsella grandiflora*, maternal RdDM is an effective approach for balancing the parental genome (Dew-Budd et al. 2024). At some genes, The RdDM is also responsive to the regulations of some imprinted expressions (Vu et al. 2013).

RdDM plays a significant role during heat, salt, drought stress and phosphate starvation (Erdmann et al. 2017). During these stresses, the upregulation in TEs occurs (Fortes and Gallusci 2017; Ito et al. 2011, Ahmad et al., 2024a), and RdDM helps in this activation. For example, ONSEN (a retrotransposon) remains suppressed in normal conditions due to RdDM-associated sRNAs. Still, it upregulates during heat stress and can transpose in plants that suffer heat stress and lack sufficient RdDM (Cavrak et al. 2014; Ito et al. 2011, Ahmad et al., 2024b). Upregulation in various aspects and mutations in a few components of the RdDM pathway eliminates heat tolerance. These findings suggested that RdDM acts as a vital phenomenon during this stress. To initiate proper stress responses, RdDM regulates the genes along with its upregulation. RdDM functions in the downregulation of genes related to stomatal development during low humidity and produces fewer stomata (Tricker et al. 2012). Similar results were observed during salt stress in which the expression level of transcription factor related to the resistance of salt stress triggers (Xu et al. 2015, Ali et al., 2022). Recently demonstrated that in maize (Wang et al. 2025), *ZmKTF1 (*KOW DOMAIN-CONTAINING TRANSCRIPTION FACTOR1) is involved in the RdDM pathway related to salt tolerance and *ZmKTF1* can be used as an epigenetic resource for the enhancement of salt tolerance.

Furthermore, various RNAs control the biotic stress caused by viroids and bacteria. The machinery of RNAi and RdDM recognize the foreign viral RNAs and then process them into sRNAs. Both pathways further use the sRNAs to degrade silent viral DNA and viral RNA (RNAi) (Huang et al. 2012; Jackel et al. 2016, Ijaz et al., 2023). However, little knowledge regarding the RNAi machinery and RdDM is known in biotic stress. Defective mutations in RdDM and some other methylations are hypersensitive against viral infection (Gouveia et al. 2017). RdDM loss exhibits an opposing impact on resistance for various pathogens, such as the increased susceptibility against *Agrobacterium tumefaciens* was observed in some RdDM mutants. Still, RdDM mutants exhibited lower susceptibility against *Pseudomonas syringae* (Dowen et al. 2012; Gohlke et al. 2013, (Huang et al. 2012; Jackel et al. 2016, Ijaz et al., 2024a). These findings express the pathogen susceptibility towards RdDM.

Biological Significance of RNA-Associated Chromatin and Histone Modifications

RNA associated adenosine and chromatin modifications are reported to be involved in plant growth and stress response. RNA-associated chromatin and adenosine modifications have been reported to modulate plant growth and some environmental adaptations. In *Arabidopsis, HIDDEN TREASURE 1* (*HID1*) (chromatin-associated non-coding RNA) regulates various light-mediated biological processes (Ying Wang et al. 2018; Yuqiu Wang et al. 2014). ARABIDOPSIS TRITHORAX-RELATED 7 (ATXR7), a methyltransferase for H3K4me3 repress the abscisic acid (ABA) biosynthesis gene transcription and promotes the germination of seeds because *HID1* associates with and repels chromatin occupancy on ATXR7 (L.-X. Wang et al. 2023; Zhang et al. 2023). Another study revealed similar results in which *LHPI1* is decoyed by lncRNA *MARNERAL SILENCING* (*MARS*) just to limits the level of H3K27me3 as well as promotes an ABA related enhancer such as formation of chromatin loop around genes (e.g. *MARNERAL SYNTHASE 1* (*MRN1*), *CYP71A16* and *CYP705A12*). It stimulates the transcription in response to ABA (Roulé et al. 2022, (Huang et al. 2012; Jackel et al. 2016, Ijaz et al., 2024b).

N6-methyladenosine (m6A) mRNA methylase poses a significant role in embryonic development in Arabidopsis. m^6A is positioned on the 3′ end of transcripts in a region 100–150 bp (before the poly(A) tail). Reduction in the levels of m^6A leads to lessened apical dominance and altered growth during different growth stages. It also affects the flowering with defects in identity, size and number of floral organs. Thus, adenosine methylations on mRNAs with reduced levels lead to developmental defects in *Arabidopsis* (Bodi et al. 2012). RNA (IncRNAs) associated chromatin and histone modifications in the genome also take part in the silencing of the *FLOWERING LOCUS C* (*FLC*) gene (a gene involved in floral induction) under cold stress. These IncRNAs are present at the *FCL* locus. *COLD OF WINTER-INDUCED NON-CODING RNA FROM THE PROMOTER* (*COLDWRAP*) and *COLD ASSISTED INTRONIC NON-CODING RNA* (*COLDAIR*) are transcribed from the 1st intron and proximal promoter of *FLC*. COLDAIR recruits Polycomb Repressive Complex 2 (PRC2) at the *1st* intron of *FLC,* and it results in *FLC* repression via interacting with CURLY LEAF (CLF)- a key part of PRC2 for H3K27me3 methylation (Heo and Sung 2011). Cold stress induces *COLDWRAP that* facilitates the chromatin loop formation between the FLC promotor and the 1st intron, which mediates stable repression of *FLC* mediated by PRC2 (Kim and Sung 2017). During vernalization in *Arabidopsis*, *COLDWRAP* synergistically functions with *COLDAIR* in the PRC2-mediated silencing of *FLC* (Swiezewski et al. 2009, Ullah et al., 2023a).

Biological Significance of RND-Mediated Gene Regulation

Gene regulation mediated by RNA is a crucial mechanism that plays a significant role in overcoming environmental stress and modulating plant growth. Previous studies show that some genes only express themselves under specific conditions; under the central dogma, genes that code proteins host genetic data. Previously, RNA was limited as a messenger between genetic materials and translation machinery. Recently, researchers have focused on elucidating the RNA role in cellular processes (as they regulate the expression of many genes in plants). Many non-protein-coding RNAs are involved in regulating many genes via RNA silencing pathways. Gene regulation mediated by RNA has been a significant part of many cellular and biological pathways, such as development, growth, morphogenesis, vegetative phase variations and nutrient homeostasis (Prabhakaran 2024, Ullah et al., 2023b). The diversity in biological purposes is due to the changes in abundance and expression level of small RNAs. However, at some time, these functionalities occur at tissue and organismal levels.

Moreover, RNA-mediated gene regulation can also turn over many regions of the transcriptome during physiological transition in plants. For example, long non-coding RNAs (lncRNAs) related to alkalinity stress were identified in bread wheat. *lncRNA11310, lncRNA13472* and *lncRNA2846* showed upregulation under salinity stress and helped the plant to encounter this stress. Another study revealed that *FLC* (a regulatory locus) serves as a repressor in *Arabidopsis* during flowering under cold stress and, epigenetically, controls the flowering (Bastow et al. 2004; Whittaker and Dean 2017, Haider et al., 2021a). In comparison, vernalization inhibits the *FLC* gene via lncRNA through polycomb-mediated epigenetic regulation during cold stress (Sachs 1887; Q. Sun et al. 2013; T.-T. Sun et al. 2016; Swiezewski et al. 2009). A lincRNA in *Arabidopsis* (trans-acting siRNA3 (*TAS3*), regulates the transport of nitrogen in nitrogen-deficient stress via targeting the nitrate transporter2 gene (Fukuda et al. 2019).

On the other hand, many TFs and miRNAs regulate gene expression via direct and indirect regulatory relationships. For example, miR399 is regulated by MYB TF under phosphate in *Arabidopsis* (Baek et al. 2013). MYB2 overexpression results in miR399f higher expression and higher phosphate contents. All these variations are due to the higher expression of phosphate starvation-induced genes (Baek et al. 2013; Baldoni, Genga, and Cominelli 2015, Haider et al., 2021b). The regulatory relationship between TF and miRNAs also plays a crucial role in heat stress. Rice plants with arsenic treatment exhibit downregulation in miR396, resulting in WRKY TF upregulation (Q. Liu and Zhang 2012). RNA (miRNAs) mediated gene regulation significantly enhances plant resistance against pathogens (Bazin et al. 2013). Various types of RNA are shown in the following Table:

Table 1. RNA-mediated epigenetic modifications and their role in plants

RNA types	Function	Epigenetic modifications	Reference
ncRNAs	Modulates vernalization and flowering inhibitor proteins	Recruit histone methylation on *FLC* locus	(Heo and Sung 2011)
lncRNA LAIR	Enhance grain yield	*LAIR* attaches to *WDR5* and *MOF* (chromatin modifying proteins) at the genomic locus of *LRK1* and activates the transcription of *LRK1*	(Ying Wang et al. 2018)
siRNA	Promotes growth of plants and enhances stress tolerance	Recruit DNA methylation	(Chan 2008; Pikaard et al. 2008)
lncRNA	Enhance stress tolerance	Recruit DNA methylation	
siRNA	At several loci, it regulates genomic imprinting in *Arabidopsis*'s parental plants.	Recruit DNA methylation	(Vu et al. 2013)
siRNA	Seed development	Recruit DNA methylation	(Grover et al. 2018)
sRNAs	During gametogenesis, it regulates genome reprogramming	Recruit DNA methylation	(Martinez and Köhler 2017)
siRNA	Enable the plants to withstand salt stress	Recruit DNA methylation and then transcriptionally regulate *MYB74*	(Xu et al. 2015)
sRNA	Regulate epigenetic silencing	AGO3 recruits sRNAs	(Zhang et al. 2016)

APPLICATIONS IN PLANT BREEDING AND BIOTECHNOLOGY

Epigenetic modifications caused by plant RNAs provide novel methods for improving crop traits and producing resilient varieties. Without altering the DNA sequences, these modifications help control the genes and enable the plants to adapt to agricultural needs and environmental challenges. Below are some applications of epigenetic modifications mediated by RNA in biotechnology and plant breeding.

RdDM induced modifications can be balanced and inherited in various generations without outside manipulation or intervention. These findings suggested it can be a significant tool for targeted epigenome editing. Recently, using CRISPR or zinc finger nucleases (Gallego-Bartolomé et al. 2019, Haider et al., 2022a), RdDM has been bypassed altogether by artificially tethering DRM2 (or some other RdDM pathway's components) directly to specific target loci. Tethering the machinery of RdDM to a locus results in DNA methylation at the target position, sometimes heritable for multiple generations (Papikian et al. 2019, Haider et al.,

2022b). However, more research is required to minimize the off-target effects and enhance the efficiency of DNA methylation. RdDM can be utilized for Genetically Modified Organisms (GMOs) production. No doubt, GMOs played a significant part in agriculture research, but they also faced many regulatory barriers regarding implementation. GMOs can be produced by foreign genetic material inclusion into the genome. The plant treatment with engineered RNAs led to the triggering of the RdDM, but it did not cause any changes in the DNA sequence of that plant. In short, it only alters the epigenetic state of portions of the DNA sequence already present, and consequently, these plants are regarded as GMOs. Thus, these efforts led to using RdDM for agriculturally beneficial traits such as herbicide and pathogen susceptibility, speeding up plant breeding by inducing beneficial characteristics (Bonfim et al. 2007; Gohlke et al. 2013). Although this field is of interest, currently, there are few widely used applications in plant science.

Gene silencing induced by RNA can also be used to produce transgenic plants. The bean golden mosaic virus (BGMV) caused the loss of many vegetables and bean crops. A construct of intron-spliced hpRNA interference is used for silencing products of the *ACI* gene (it is enabled for regulation of its own expression to inhibit the expression of viral genes in resistant transgenic plants of beans) [84]. Approximately 93% of one transgenic line progeny showed no symptoms against this virus. Moreover, gene silencing based on siRNA is used for the improvement of crops via designing the antisense RNA transcripts by inverting the coding sequence of the target gene in a transgenic cassette to hybridization with its target mRNA, subjecting to siRNA process machinery and leading to rapid RNA degradation (Frizzi and Huang 2010; Zhou and Luo 2013).

CONCLUSION

In plant life, RNA-mediated epigenetic modifications are actively regulatory pathways that carry out significant roles in diverse aspects. Various genetic factors take part in the mechanism of RNA-mediated epigenetic modifications. RNA molecules assemble these complex epigenetic alterations with DNA, chromatin, and TFs to maintain a specific control over genomic stability and gene expression. This understanding of epigenetic modifications paved new avenues in plant breeding and biotechnology. Advanced techniques such as CRISPR-Cas, RNAi and other gene editing methods help researchers to engineer plants with suitable qualities such as disease-free, stress-resistant and improved yield.

REFERENCES

Ahmad, Y., Haakim, Z., Iqbal, J., Abbasi, B. A., Mahmood, T., & Kazi, M. (2024b). Technological Innovations for Abiotic Stress Resistance in Horticultural Crops. OMICs-based Techniques for Global Food Security, 233-244.

Ahmad, Y., Haider, S., Iqbal, J., Abbasi, B. A., Yaseen, T., & Mahmood, T. (2022a). The mechanisms of genome editing technologies in crop plants. In *Principles and Practices of OMICS and Genome Editing for Crop Improvement* (pp. 295–313). Springer International Publishing.

Ahmad, Y., Haider, S., Iqbal, J., Abbasi, B. A., Yaseen, T., & Mahmood, T. (2022b). The mechanisms of genome editing technologies in crop plants. In *Principles and Practices of OMICS and Genome Editing for Crop Improvement* (pp. 295–313). Springer International Publishing.

Ahmad, Y., Haider, S., Iqbal, J., Naseer, S., Attia, K. A., Mohammed, A. A., & Mahmood, T. (2024a). In-silico analysis and transformation of OsMYB48 transcription factor driven by CaMV35S promoter in model plant–Nicotiana tabacum L. conferring abiotic stress tolerance. *GM Crops and Food: Biotechnology in Agriculture and the Food Chain, 15*(1), 130–149. PMID: 38551174

Albini, S., Zakharova, V., & Ait-Si-Ali, S. (2019). Histone Modifications. In *Epigenetics and Regeneration* (pp. 47–72). Elsevier.

Ali, I., Salah, K. B. H., Sher, H., Ali, H., Ullah, Z., Ali, A., & Mahmood, T. (2022). Drought stress enhances the efficiency of floral dip method of Agrobacterium-mediated transformation in Arabidopsis thaliana. *Brazilian Journal of Biology, 84*, e259326. PMID: 35703626

Ali, S., & Tang, Y. (2024). Noncoding RNA-Mediated Regulation of DNA Methylation: Insights into Plant Epigenetic Mechanisms. *Journal of Plant Growth Regulation*, •••, 1–16.

Arce, A. L., Mencia, R., Cambiagno, D. A., Lang, P. L., Liu, C., Burbano, H. A., & Manavella, P. A. (2023). Polymorphic inverted repeats near coding genes impact chromatin topology and phenotypic traits in Arabidopsis thaliana. *Cell Reports, 42*(1).

Baek, D., Kim, M. C., Chun, H. J., Kang, S., Park, H. C., Shin, G., & Park, J.. (2013). Regulation of MiR399f Transcription by AtMYB2 Affects Phosphate Starvation Responses in Arabidopsis. *Plant Physiology, 161*(1), 362–373. PMID: 23154535

Baldoni, E., Genga, A., & Cominelli, E. (2015). Plant MYB Transcription Factors: Their Role in Drought Response Mechanisms. *International Journal of Molecular Sciences*, *16*(7), 15811–15851. PMID: 26184177

Bastow, R., Mylne, J. S., Lister, C., Lippman, Z., Martienssen, R. A., & Dean, C. (2004). Vernalization requires epigenetic silencing of FLC by histone methylation. *Nature*, *427*(6970), 164–167.

Bazin, J., Khan, G. A., Combier, J.-P., Bustos-Sanmamed, P., Debernardi, J. M., Rodriguez, R., & Sorin, C.. (2013). MiR396 Affects Mycorrhization and Root Meristem Activity in the Legume M Edicago Truncatula. *The Plant Journal*, *74*(6), 920–934. PMID: 23566016

Bodi, Z., Zhong, S., Mehra, S., Song, J., Graham, N., Li, H., May, S., & Fray, R. G. (2012). Adenosine Methylation in Arabidopsis MRNA Is Associated with the 3′ End and Reduced Levels Cause Developmental Defects. *Frontiers in Plant Science*, *3*, 48. PMID: 22639649

Bonfim, K., Faria, J. C., Nogueira, E. O., Mendes, É. A., & Aragão, F. J. (2007). RNAi-mediated resistance to Bean golden mosaic virus in genetically engineered common bean (Phaseolus vulgaris). *Molecular Plant-Microbe Interactions*, *20*(6), 717–726.

Bottomley, M. J. (2004). Structures of Protein Domains That Create or Recognize Histone Modifications. *EMBO Reports*, *5*(5), 464–469. PMID: 15184976

Bure, Irina V, Marina V Nemtsova, and Ekaterina B Kuznetsova. (2022). Histone Modifications and Non-Coding RNAs: Mutual Epigenetic Regulation and Role in Pathogenesis. *International Journal of Molecular Sciences*, *23*(10), 5801. PMID: 35628612

Cavrak, V. V., Lettner, N., Jamge, S., Kosarewicz, A., Bayer, L. M., & Scheid, O. M. (2014). How a Retrotransposon Exploits the Plant's Heat Stress Response for Its Activation. *PLOS Genetics*, *10*(1), e1004115. PMID: 24497839

Chan, S. W. (2008). Inputs and outputs for chromatin-targeted RNAi. *Trends in Plant Science*, *13*(7), 383–389.

Chen, B., Wang, M., Guo, Y., Zhang, Z., Zhou, W., Cao, L., & Zhang, T.. (2024). Climate-related Naturally Occurring Epimutation and Their Roles in Plant Adaptation in A. Thaliana. *Molecular Ecology*, •••, 17356. PMID: 38634782

Cheng, J., Niu, Q., Zhang, B., Chen, K., Yang, R., Zhu, J. K., & Lang, Z. (2018). Downregulation of RdDM during strawberry fruit ripening. *Genome Biology*, *19*(1), 212.

Dew-Budd, K. J., Chow, H. T., Kendall, T., David, B. C., Rozelle, J. A., Mosher, R. A., & Beilstein, M. A. (2024). Mating system is associated with seed phenotypes upon loss of RNA-directed DNA methylation in Brassicaceae. *Plant Physiology*, *194*(4), 2136–2148.

Dowen, R. H., Pelizzola, M., Schmitz, R. J., Lister, R., Dowen, J. M., Nery, J. R., & Ecker, J. R. (2012). Widespread dynamic DNA methylation in response to biotic stress. *Proceedings of the National Academy of Sciences of the United States of America*, *109*(32), E2183–E2191.

Erdmann, R. M., & Picard, C. L. (2020). RNA-directed DNA methylation. *PLOS Genetics*, *16*(10), e1009034.

Erdmann, R. M., Hoffmann, A., Walter, H.-K., Wagenknecht, H.-A., Groß-Hardt, R., & Gehring, M. (2017). Molecular Movement in the Arabidopsis Thaliana Female Gametophyte. *Plant Reproduction*, *30*, 141–146. PMID: 28695277

Ferrari, M., Muto, A., Bruno, L., & Cozza, R. (2023). DNA Methylation in Algae and Its Impact on Abiotic Stress Responses. *Plants*, *12*(2), 241. PMID: 36678953

Fortes, A. M., & Gallusci, P. (2017). Plant Stress Responses and Phenotypic Plasticity in the Epigenomics Era: Perspectives on the Grapevine Scenario, a Model for Perennial Crop Plants. *Frontiers in Plant Science*, *8*, 82. PMID: 28220131

Frizzi, A., & Huang, S. (2010). Tapping RNA Silencing Pathways for Plant Biotechnology. *Plant Biotechnology Journal*, *8*(6), 655–677. PMID: 20331529

Fukuda, M., Nishida, S., Kakei, Y., Shimada, Y., & Fujiwara, T. (2019). Genome-Wide Analysis of Long Intergenic Noncoding RNAs Responding to Low-Nutrient Conditions in Arabidopsis Thaliana: Possible Involvement of Trans-Acting SiRNA3 in Response to Low Nitrogen. *Plant & Cell Physiology*, *60*(9), 1961–1973. PMID: 30892644

Gallego-Bartolomé, J., Liu, W., Kuo, P. H., Feng, S., Ghoshal, B., Gardiner, J., & Zhao, J. M.-C.. (2019). Co-Targeting RNA Polymerases IV and V Promotes Efficient de Novo DNA Methylation in Arabidopsis. *Cell*, *176*(5), 1068–1082. PMID: 30739798

Gohlke, J, C J Scholz, S Kneitz, D Weber, and J Fuchs. 2013. "DNA Methylation Mediated Control of Gene Expression Is Critical for Development Of."

Gouveia, B. C., Calil, I. P., Machado, J. P. B., Santos, A. A., & Fontes, E. P. (2017). Immune receptors and co-receptors in antiviral innate immunity in plants. *Frontiers in Microbiology*, *7*, 2139.

Grover, Jeffrey W, Kendall, Timmy, Baten, Abdul, Burgess, Diane, & Freeling, Michael, Graham J King, and Rebecca A Mosher. (2018). Maternal Components of RNA-directed DNA Methylation Are Required for Seed Development in Brassica Rapa. *The Plant Journal*, *94*(4), 575–582. PMID: 29569777

Guo, X., Ma, Z., Zhang, Z., Cheng, L., Zhang, X., & Li, T. (2017). Small RNA-Sequencing Links Physiological Changes and RdDM Process to Vegetative-to-Floral Transition in Apple. *Frontiers in Plant Science*, *8*, 873. PMID: 28611800

Haider, S., Iqbal, J., Naseer, S., Shaukat, M., Abbasi, B. A., Yaseen, T., & Mahmood, T. (2022b). Unfolding molecular switches in plant heat stress resistance: A comprehensive review. *Plant Cell Reports*, *41*(3), 775–798. PMID: 34401950

Haider, S., Iqbal, J., Naseer, S., Yaseen, T., Shaukat, M., Bibi, H., & Mahmood, T. (2021a). Molecular mechanisms of plant tolerance to heat stress: Current landscape and future perspectives. *Plant Cell Reports*, •••, 1–25. PMID: 33890138

Haider, S., Iqbal, J., Shaukat, M., Naseer, S., & Mahmood, T. (2021b). The epigenetic chromatin-based regulation of somatic heat stress memory in plants. *Plant Gene*, *27*, 100318.

Haider, S., Raza, A., Iqbal, J., Shaukat, M., & Mahmood, T. (2022a). Analyzing the regulatory role of heat shock transcription factors in plant heat stress tolerance: A brief appraisal. *Molecular Biology Reports*, *49*(6), 5771–5785. PMID: 35182323

Han, Q., Bartels, A., Cheng, X., Meyer, A., An, Y.-Q. C., Hsieh, T.-F., & Xiao, W. (2019). Epigenetics Regulates Reproductive Development in Plants. *Plants*, *8*(12), 564. PMID: 31810261

Hemenway, E. A., & Gehring, M. (2023). Epigenetic Regulation during Plant Development and the Capacity for Epigenetic Memory. *Annual Review of Plant Biology*, *74*(1), 87–109. PMID: 36854474

Heo, J. B., & Sung, S. (2011). Vernalization-Mediated Epigenetic Silencing by a Long Intronic Noncoding RNA. *Science*, *331*(6013), 76–79. PMID: 21127216

Huang, H., Sha, S. H., Wang, Y., Xin, Z. Z., & Zhen, H. H. (2012). Variations in Leaf Morphology and DNA Methylation Following in Vitro Culture of Malus Xiaojinensis. [PCTOC]. *Plant Cell, Tissue and Organ Culture*, *111*, 153–161.

Ijaz, S., Iqbal, J., Abbasi, B. A., Tufail, A., Ullah, Z., Sharifi-Rad, J., & Uddin, S. (2023). Biofortification: Lessons from the Past and Strategies for Future Food Security. In *Legumes Biofortification* (pp. 521–545). Springer International Publishing.

Ijaz, S., Iqbal, J., Abbasi, B. A., Yaseen, T., Rehman, S., Kazi, M., & Mahmood, T. (2024a). Role of OMICS-Based Technologies in Plant Sciences. OMICs-based Techniques for Global Food Security, 45-66.

Ijaz, S., Iqbal, J., Abbasi, B. A., Yaseen, T., Rehman, S., Kazi, M. & Mahmood, T. (2024b). Role of OMICS-Based Technologies in Plant Sciences. OMICs-based Techniques for Global Food Security, 45-66.

Ito, H., Gaubert, H., Bucher, E., Mirouze, M., Vaillant, I., & Paszkowski, J. (2011). An SiRNA Pathway Prevents Transgenerational Retrotransposition in Plants Subjected to Stress. *Nature*, *472*(7341), 115–119. PMID: 21399627

Iwasaki, M., Hyvärinen, L., Piskurewicz, U., & Lopez-Molina, L. (2019). Non-Canonical RNA-Directed DNA Methylation Participates in Maternal and Environmental Control of Seed Dormancy. *eLife*, *8*, e37434. PMID: 30910007

Jackel, J. N., Storer, J. M., Coursey, T., & Bisaro, D. M. (2016). Arabidopsis RNA polymerases IV and V are required to establish H3K9 methylation, but not cytosine methylation, on geminivirus chromatin. *Journal of Virology*, *90*(16), 7529–7540.

Kim, D.-H., & Sung, S. (2017). Vernalization-Triggered Intragenic Chromatin Loop Formation by Long Noncoding RNAs. *Developmental Cell*, *40*(3), 302–312. PMID: 28132848

Kinoshita, Y., Saze, H., Kinoshita, T., Miura, A., Soppe, W. J., Koornneef, M., & Kakutani, T. (2007). Control of FWA gene silencing in Arabidopsis thaliana by SINE-related direct repeats. *The Plant Journal*, *49*(1), 38–45.

Li, J., Yang, Z., Yu, B., Liu, J., & Chen, X. (2005). Methylation Protects MiRNAs and SiRNAs from a 3′-End Uridylation Activity in Arabidopsis. *Current Biology*, *15*(16), 1501–1507. PMID: 16111943

Li, L.-C. (2014). Chromatin Remodeling by the Small RNA Machinery in Mammalian Cells. *Epigenetics*, *9*(1), 45–52. PMID: 24149777

Li, S., Li, M., Li, Z., Zhu, Y., Ding, H., Fan, X., Li, F., & Wang, Z. (2019). Effects of the Silencing of CmMET1 by RNA Interference in Chrysanthemum (Chrysanthemum Morifolium). *Plant Biotechnology Reports*, *13*, 63–72.

Li, X., & Fu, X.-D. (2019). Chromatin-Associated RNAs as Facilitators of Functional Genomic Interactions. *Nature Reviews. Genetics*, *20*(9), 503–519. PMID: 31160792

Liu, Q., & Zhang, H. (2012). Molecular Identification and Analysis of Arsenite Stress-Responsive MiRNAs in Rice. *Journal of Agricultural and Food Chemistry*, *60*(26), 6524–6536. PMID: 22712679

Liu, S., de Jonge, J., Trejo-Arellano, M. S., Santos-González, J., Köhler, C., & Hennig, L. (2021). Role of H1 and DNA methylation in selective regulation of transposable elements during heat stress. *The New Phytologist, 229*(4), 2238–2250.

Loaeza-Loaeza, J., Beltran, A. S., & Hernández-Sotelo, D. (2020). DNMTs and impact of CpG content, transcription factors, consensus motifs, lncRNAs, and histone marks on DNA methylation. *Genes, 11*(11), 1336.

Lu, C., & Fedoroff, N. (2000). A Mutation in the Arabidopsis HYL1 Gene Encoding a DsRNA Binding Protein Affects Responses to Abscisic Acid, Auxin, and Cytokinin. *The Plant Cell, 12*(12), 2351–2365. PMID: 11148283

Lu, X., & Hyun, T. K. (2021). The Role of Epigenetic Modifications in Plant Responses to Stress. *Botanica Serbica, 45*(1), 3–12.

Martinez, G., & Köhler, C. (2017). Role of Small RNAs in Epigenetic Reprogramming during Plant Sexual Reproduction. *Current Opinion in Plant Biology, 36*, 22–28. PMID: 28088028

Martínez, G., Panda, K., Köhler, C., & Keith Slotkin, R. (2016). Silencing in Sperm Cells Is Directed by RNA Movement from the Surrounding Nurse Cell. *Nature Plants, 2*(4), 1–8. PMID: 27249563

Matzke, M. A., Kanno, T., & Matzke, A. J. (2015). RNA-directed DNA methylation: The evolution of a complex epigenetic pathway in flowering plants. *Annual Review of Plant Biology, 66*, 243–267.

Papikian, Ashot, Liu, Wanlu, & Gallego-Bartolomé, Javier, and Steven E Jacobsen. (2019). Site-Specific Manipulation of Arabidopsis Loci Using CRISPR-Cas9 SunTag Systems. *Nature Communications, 10*(1), 729. PMID: 30760722

Pikaard, C. S., Haag, J. R., Ream, T., & Wierzbicki, A. T. (2008). Roles of RNA polymerase IV in gene silencing. *Trends in Plant Science, 13*(7), 390–397.

Placette, Lafon, Clément. (2020). Endosperm Genome Dosage, Hybrid Seed Failure, and Parental Imprinting: Sexual Selection as an Alternative to Parental Conflict. *American Journal of Botany, 107*(1), 17–19. PMID: 31797356

Portela, A., & Esteller, M. (2010). Epigenetic Modifications and Human Disease. *Nature Biotechnology, 28*(10), 1057–1068. PMID: 20944598

Prabhakaran, Sharat. 2024. "RNA in Plant Nutrient Availability." *Noncoding RNA: The Dark Matter of the Genome*: 165.

Roulé, T., Christ, A., Hussain, N., Huang, Y., Hartmann, C., Benhamed, M., & Gutierrez-Marcos, J.. (2022). The LncRNA MARS Modulates the Epigenetic Reprogramming of the Marneral Cluster in Response to ABA. *Molecular Plant, 15*(5), 840–856. PMID: 35150931

Sachs, J. (1887). *Lectures on the Physiology of Plants.* Clarendon Press.

Samad, A. F., Sajad, M., Nazaruddin, N., Fauzi, I. A., Murad, A. M., Zainal, Z., & Ismail, I. (2017). MicroRNA and transcription factor: Key players in plant regulatory network. *Frontiers in Plant Science, 8*, 565.

Sánchez-Retuerta, C., Suaréz-López, P., & Henriques, R. (2018). Under a New Light: Regulation of Light-Dependent Pathways by Non-Coding RNAs. *Frontiers in Plant Science, 9*, 962. PMID: 30140270

Shvedunova, M., & Akhtar, A. (2022). Modulation of Cellular Processes by Histone and Non-Histone Protein Acetylation. *Nature Reviews. Molecular Cell Biology, 23*(5), 329–349. PMID: 35042977

Slotkin, R. K., Vaughn, M., Borges, F., Tanurdžić, M., Becker, J. D., Feijó, J. A., & Martienssen, R. A. (2009). Epigenetic reprogramming and small RNA silencing of transposable elements in pollen. *Cell, 136*(3), 461–472.

Soppe, W. J., Jacobsen, S. E., Alonso-Blanco, C., Jackson, J. P., Kakutani, T., Koornneef, M., & Peeters, A. J. (2000). The late flowering phenotype of fwa mutants is caused by gain-of-function epigenetic alleles of a homeodomain gene. *Molecular Cell, 6*(4), 791–802.

Suelves, M., Carrió, E., & Núñez-Álvarez, Y. & Peinado, M.A. (2016). DNA Methylation Dynamics in Cellular Commitment and Differentiation. *Briefings in Functional Genomics, 15*(6), 443–453. PMID: 27416614

Sun, Qianwen, Csorba, Tibor, & Skourti-Stathaki, Konstantina, Nicholas J Proudfoot, and Caroline Dean. (2013). R-Loop Stabilization Represses Antisense Transcription at the Arabidopsis FLC Locus. *Science, 340*(6132), 619–621. PMID: 23641115

Sun, T.-T., He, J., Liang, Q., Ren, L.-L., Yan, T.-T., Yu, T.-C., & Tang, J.-Y.. (2016). LncRNA GClnc1 Promotes Gastric Carcinogenesis and May Act as a Modular Scaffold of WDR5 and KAT2A Complexes to Specify the Histone Modification Pattern. *Cancer Discovery, 6*(7), 784–801. PMID: 27147598

Swiezewski, S., Liu, F., Magusin, A., & Dean, C. (2009). Cold-Induced Silencing by Long Antisense Transcripts of an Arabidopsis Polycomb Target. *Nature, 462*(7274), 799–802. PMID: 20010688

Tricker, P. J., Gibbings, J. G., Rodríguez López, C. M., Hadley, P., & Wilkinson, M. J. (2012). Low relative humidity triggers RNA-directed de novo DNA methylation and suppression of genes controlling stomatal development. *Journal of Experimental Botany, 63*(10), 3799–3813.

Ullah, Z., Iqbal, J., Abbasi, B. A., Akhtar, W., Kanwal, S., Ali, I., & Mahmood, T. (2023b). Assessment of Gus Expression Induced by Anti-Sense Os PPO Gene Promoter and Antioxidant Enzymatic Assays in Response to Drought and Heavy Metal Stress in Transgenic Arabidopsis thaliana. *Sustainability, 15*(17), 12783.

Ullah, Z., Iqbal, J., Abbasi, B. A., Ijaz, S., Sharifi-Rad, J., Yaseen, T., & Mahmood, T. (2023a). How Genome Editing Can Be Helpful in the Biofortification of Legumes. In *Legumes Biofortification* (pp. 207–232). Springer International Publishing.

Vazquez, Franck, Vaucheret, Hervé, Rajagopalan, Ramya, Lepers, Christelle, & Gasciolli, Virginie, Allison C Mallory, Jean-Louis Hilbert, David P Bartel, and Patrice Crété. (2004). Endogenous Trans-Acting SiRNAs Regulate the Accumulation of Arabidopsis MRNAs. *Molecular Cell, 16*(1), 69–79. PMID: 15469823

Vu, T. M., Nakamura, M., Calarco, J. P., Susaki, D., Lim, P. Q., Kinoshita, T., & Berger, F. (2013). RNA-directed DNA methylation regulates parental genomic imprinting at several loci in Arabidopsis. *Development, 140*(14), 2953–2960.

Wang, J., Zheng, L., Peng, Y., Lu, Z., Zheng, M., Wang, Z., & Luo, J. (2025). ZmKTF1 promotes salt tolerance by mediating RNA-directed DNA methylation in maize. *The New Phytologist, 245*(1), 200–214.

Wang, Y., Fan, X., Lin, F., He, G., Terzaghi, W., Zhu, D., & Xing, W. D. (2014). Arabidopsis Noncoding RNA Mediates Control of Photomorphogenesis by Red Light. *Proceedings of the National Academy of Sciences of the United States of America, 111*(28), 10359–10364. PMID: 24982146

Wang, Y., Fan, Y., De Fan, X. Z., Jiao, Y., Xing, W. D., & Zhu, D. (2023). The Noncoding RNA HIDDEN TREASURE 1 Promotes Phytochrome B-Dependent Seed Germination by Repressing Abscisic Acid Biosynthesis. *The Plant Cell, 35*(2), 700–716. PMID: 36423345

Wang, Y., Itaya, A., Zhong, X., Wu, Y., Zhang, J., van der Knaap, E., Olmstead, R., Qi, Y., & Ding, B. (2011). Function and Evolution of a MicroRNA That Regulates a Ca2+-ATPase and Triggers the Formation of Phased Small Interfering RNAs in Tomato Reproductive Growth. *The Plant Cell, 23*(9), 3185–3203. PMID: 21917547

Wang, Y., Luo, X., Sun, F., Hu, J., Zha, X., Su, W., & Yang, J. (2018). Overexpressing LncRNA LAIR Increases Grain Yield and Regulates Neighbouring Gene Cluster Expression in Rice. *Nature Communications*, *9*(1), 3516. PMID: 30158538

Wassenegger, M. (2002). Gene Silencing-Based Disease Resistance. *Transgenic Research*, *11*, 639–653. PMID: 12509140

Wendte, J. M., & Pikaard, C. S. (2017). The RNAs of RNA-directed DNA methylation. Biochimica et Biophysica Acta (BBA)-. *Gene Regulatory Mechanisms*, *1860*(1), 140–148.

Whittaker, C., & Dean, C. (2017). The FLC Locus: A Platform for Discoveries in Epigenetics and Adaptation. *Annual Review of Cell and Developmental Biology*, *33*(1), 555–575. PMID: 28693387

Xu, R., Wang, Y., Zheng, H., Lu, W., Wu, C., Huang, J., Yan, K., Yang, G., & Zheng, C. (2015). Salt-Induced Transcription Factor MYB74 Is Regulated by the RNA-Directed DNA Methylation Pathway in Arabidopsis. *Journal of Experimental Botany*, *66*(19), 5997–6008. PMID: 26139822

Zhang, Z., Liu, X., Guo, X., Wang, X.-J., & Zhang, X. (2016). Arabidopsis AGO3 Predominantly Recruits 24-Nt Small RNAs to Regulate Epigenetic Silencing. *Nature Plants*, *2*(5), 1–7. PMID: 27243648

Zhang, M. L., Ye, Y. U. A. N., Sadeghnezhad, E., & Li, D. A. I. (2023). A cyclic effect of cAMP and calcium signaling contributes to jujube growth and development. *Journal of Integrative Agriculture*, *22*(7), 2094–2110.

Zhao, Z., Zang, S., Zou, W., Pan, Y.-B., Yao, W., You, C., & Que, Y. (2022). Long Non-Coding RNAs: New Players in Plants. *International Journal of Molecular Sciences*, *23*(16), 9301. PMID: 36012566

Zheng, X., Yang, L., Li, Q., Ji, L., Tang, A., Zang, L., Deng, K., Zhou, J., & Zhang, Y. (2018). MIGS as a Simple and Efficient Method for Gene Silencing in Rice. *Frontiers in Plant Science*, *9*, 662. PMID: 29868104

Zhou, M., & Luo, H. (2013). MicroRNA-mediated gene regulation: Potential applications for plant genetic engineering. *Plant Molecular Biology*, *83*(1), 59–75.

Chapter 3
Epigenetics Mechanisms in Plant Stress Adaptation:
Epigenetic Mechanisms Governing Drought Tolerance in Plants

Saman Zulfiqar
https://orcid.org/0000-0003-3677-2531
Government Sadiq College Women University, Pakistan

Yasir Nawaz
https://orcid.org/0000-0002-4969-4758
School of Life Sciences, Nanjing Normal University, China

Sidra Aslam
University of Sialkot, Pakistan

Fiza Rimal Butt
University of Sialkot, Pakistan

Saba Munir
University of Okara, Pakistan

Rikza Tahreem
University of Okara, Pakistan

ABSTRACT

Epigenetics mechanisms, such as histone alterations, DNA methylation, and small non-coding RNAs, play a vital role in regulating plant drought tolerance by modulating gene expression. These changes enable plants to conserve water, protect cells, and recover from stress. Epigenetic memory also helps plants respond more effectively to recurring droughts. Understanding these mechanisms offers promising approaches for improving crop flexibility by breeding and biotechnology, crucial for addressing climate change and water scarcity challenges in agriculture.

DOI: 10.4018/979-8-3693-8648-4.ch003

INTRODUCTION

Waddington created the term "epigenetics" to connect the two disciplines. Apart from to being a hot area for basic research, plant epigenetics has emerged as a potential new source of useful characteristics for plant breeding. The wide definition of epigenetics would be the culmination of each procedure required for the creation of the genetic program. A few decades later, specific techniques for superimposing information on DNA sequences were developed. During past ten years, attention in the topic of epigenetics has surged significantly, and the term is now more widely used in biomedical research, scientific disciplines other than the genetic sciences, like ecology and physiology, and even popular culture. Specifically, it was proposed that the methylation pattern was heritable and that 5-methyl cytosine served a factor in regulating gene expression.

Since methylation is closely linked to gene silence in a number of biological circumstances, an extensive amount of evidence now supports these predictions. Nowadays, there are numerous instances of germ-line epigenetic inheritance as well.

Other significant epigenetic mechanisms include the growing field of regulatory RNAs and chromatin and histone changes. In order to ascertain whether the regulator of gene expression occurs at level of chromatin, DNA, or both, the human epigenome project will analyze the forms of DNA methylation across multiple tissues (Holliday, 2006). The genetic and epigenetic approaches to development are not the same. At first, the term "epigenetics" signified the study of how genes and their byproducts shape phenotype. These days, its main focus is on the processes by which cells commit to a specific function and then pass that structural or functional state on to subsequent cell lineages (Jablonka, 2002) .The idea that epigenetics controls development has gained more and more attention in recent years. Heritable modifications in gene activity that take place apart from fluctuations in the main DNA sequence are the subject of the research of epigenetics. The epigenetic modifications that have been investigated the most includes DNA methylation, histone changes that alters chromatin structure, and histone exchange. A new study is the alteration of gene expression caused by long-distance chromosomal transactions. Epigenetic variations are important modulators of important developmental processes, including brain development, Hox gene patterning, genomic imprinting, and X-inactivation (Kiefer, 2007).

MECHANISM

The economic sector most vulnerable to damage from this swift climate change is agriculture (Sheoran et al., 2022).Drought is a hypothetical instance of a natural disaster that impairs efficiency of crops (Adel, 2023). Adequate plant responses to stress also depend on epigenetic processes. Plants can perceive stress during vegetative growth and might preserve a memory of these challenges, presumably through epigenetic mechanisms. This is because to the late differentiation of their germ due to their sessile lifestyle. Like other higher species, plants use three systems—DNA methylation, histone changes, and RNA interference—to start and control epigenetic gene regulation. In order to demonstrate how these epigenetic elements interact and stabilize one another, new ideas are being developed. Apart from under scrutiny are the function of epigenetic processes in hybrid vitality and epigenetic transgene silencing (Munshi et al., 2015).Chromatin status, which is frequently established by epigenetic markers like DNA methylation, histone modification, and histone variations, is linked to gene expression. Plenty of factors involved in the control of epigenetic marks have been found, and the functions of these marks have been elucidated by genetic and genomic research conducted primarily in Arabidopsis thaliana. New molecular breeding methods will surely be created and used for commercially significant plant species depends on the results of these probes (Maeji, 2018).

DROUGHT STRESS TOLERANCE PLANTS

One of most hazardous abiotic stress is drought, which has an effect on crop output and human health in around a third of the global populace. FAO future projections that between 2005 and 2015, drought adversely cost developing nations' agriculture USD 29 billion. (Sheoran et al., 2022) .Crop plants frequently face biotic and abiotic stressors, and with the expression of particular genes, they develop complex coping mechanisms. Since agriculture is a vital economic sector that provides jobs and food, its sustainability is imperative.

Nevertheless, local meteorological conditions and especially water availability affect the amount and quality of crops grown. The main function of chromatin, histone, and DNA changes is to help plants overcome obstacles and survive in adverse circumstances. Plants can recall a previous stress experience to cope with comparable challenges in the future because to epigenetic modifications brought on by environmental stress. This heritable memory, known as "plant stress memory," helps plants respond to stressors more effectively and efficiently for both present and future generations of the plant (Sun et al., 2021).

Wheat

With a global production of over 700 million tons, wheat is very noteworthy cereal crop. It supplies 4.5 billion people with 20% of their daily protein and calorie needs. A few numbers of salt-tolerant bread wheat genotypes found, which has limited the success of developing wheat on salt-affected soils. The potential mechanisms behind the wheat genotypes' varying responses to salt stress were revealed by genetic and epigenetic inquiries. Kharchia-65 was determined to be the best salt-tolerant cultivar under 200 mM NaCl stress due to its superior antioxidant potential, membrane stability, a bigger growth of osmolytes/phytophenolics, and higher K+/Na+ ratio. However, because of the elevated MDA level, reduced antioxidant capacity, and decreased levels of accessible sugar, proline, total chlorophyll, and total phenolics, HD-2329 was found to be stress-sensitive. Genetic and bioinformatics studies demonstrated that HKT1;4 of opposing genotypes (Kharchia-65 and HD-2329) displayed deletions, transitions, and transversions that led to altered organization, loss of conserved motifs (Ser-Gly-Gly-Gly and Gly-Arg), and functions in the salt-sensitive (HD-2329) genotype. HKT expressions analysis explained the reaction. Variance expression of HKT2;1 and HKT2;3 by tissue and genotype has been justified by epigenetic alterations in cytosine methylation (Kumar et al., 2017).

Wheat responds to drought stress through epigenetic methods that includes DNA methylation, histone modifications, chromatin makeover, and ncRNAs. These modifications in chromatin, histone, and DNA primarily help plants remember previous stressful events and cope with climate and flourish in tense conditions. Different short- and long-term responses are used by plants depending on whether the stress is temporary or permanent. While transgenerational variations involving the emergence of heritable gene expressions variations are long-term strategies, short-term methods involve adaptations to plant homeostasis. This process involves boosting the expressions of some genes while suppressing the expression of others, as well as producing new epigenetic marks while deleting old ones. Phenotypic variety and tolerance to environmental stresses are facilitated by epigenetic reconfiguration in response to these difficulties. There are three phases to the mechanism of an epigenetic process: (i) Epigenators are triggers, like food, poisons, radiation, or hormones, that change the environment of cells to create an epigenetic phenotype. The epigenetic process can be started by these brief but effective triggering events. The epigenetic initiator converts the epigenator signal into a chromatin epigenetic modification (ii). After being primed by the epigenator, an epigenetic initiator chooses which chromosome site needs to be marked. The chromatin environment in present and future is maintained by the epigenetic maintainer (iii). It might be necessary for the initiator and maintainer to work together for the chromatin configuration to persist.

Through the application of selective breeding, research should help produce cultivars that can withstand drought. This study highlights the progress made in the selective breeding of drought-tolerant wheat cultivars through the use of QTL visualization, synchrony, genome analysis, epigenetics, and transgenics. The msh1 mutation and synthesized the apomixis pave the door for the induction and stabilization of the epigenome in crops, potentially speeding up selective breeding for drought tolerance in semi-arid and arid regions (Adel, 2023) .

Rice

One prevalent abiotic stressor that negatively impacts plant growth, development, and output is drought. Animals are well recognized for having both short- and long-term memory, but it is still unclear if plants have this kind of retentivity. Just before flowering, drought stress was applied to various rice genotypes, and the plants were re-watered to help them recover. For three consecutive seasons, the indica pyramiding line DK151 and its recurrent parent IR64 underwent evaluation in drought-stressed and non-stressed environments.

The most significant was its genotypic specificity, which was made clear by the significant variations in the DNA methylation/demethylation sites found between IR64 and DK151. These variations were caused by introgressed genomic segments in DK151. Second, a great deal of drought-induced methylation/demethylation sites fell into one of two main categories based on their reversibility: 29% of sites where drought-induced DNA demethylation/methylation changes persisted even after recovery, and 70% of sites where drought-induced epigenetic changes were restored to their initial state.Third, the DNA methylation modification caused by drought showed a significant level of tissue and developmental specificity. The responsiveness and adaptability of rice to drought stress are expected to have been significantly influenced by each of these traits. Thus, induced epigenetic changes in the rice genome could be viewed as an essential regulatory mechanism that aids in the adaptation of rice plants to drought and possibly other environmental stresses (Wang et al., 2011) .

The function of genes in crop plants is mostly governed by cytosine methylation polymorphism (Gayacharan, 2013) .Seeds from stress-treated plant was used to grow plants for two generations in a similar experimental setup. The physio-biochemical and epigenetic [5-methylcytosine (5-mC)] traits of plants grown in stress and those grown after recovery were compared in their leaves. The amount of chlorophyll decreased significantly (>9%) under the stress, but the amount of total phenolic (>19%) and chlorophyll (>25%) as well as the antioxidant activity (>7%) and genome-wide 5-mC level (>56%) increased. It's interesting to see that some of the increased 5-mC levels, proline content, total phenolics, and antioxidant activity remained after the stress was eliminated. These improved characteristics were also discovered to be inherited by the next generation. These discoveries could enhance food production and worldwide food security by advancing the improvement of stress-resistant yields and improving agricultural production in face of a changing global climate (Kumar et al., 2023).

Maize

Drought poses a major challenge to maize production, with nearly 80% of maize in the wet season grown under rainfed conditions, and dry-season maize facing significant water shortages. Throughout their lifespan, plants are exposed to various stressors. The plant's response and adaptation to these stresses are regulated by dynamic changes in chromatin and associated transcriptional shifts. Epigenetic memory mechanisms help integrate environmental cues with the plant's developmental programs, allowing for long-term adaptation. The transcript library and genome-wide histone changes of maize plants subjected to mild, protracted drought stress shortly before the blooming transition were explored in this work (Sheoran et al., 2022). Seventeen inbred lines of maize were exposed to drought while they were seedlings. According to Lu et al. (2018), Yu 882 had the strongest drought tolerance among the 17 maize inbred lines. We put Yu 882 under stress for 60 and 96 hours in order to examine the phenotype and physiological variations to drought stress and rewatering (Cao et al., 2021).Using12% PEG 6000 for 0–7–24 hours of drought therapy, the physical in nature, biochemical, and transcriptome alterations in maize variety "M8186" during the seedling stage were investigated (Jiao et al., 2022).

Maize is very used grain in world (Hasibuzzaman et al., 2021). SNPs from agronomic attributes connected to genes linked to drought tolerance in subtropical maize lines were found using genome-wide association analysis employing two models and various locations. These genes revealed biochemical pathways and physiological reactions that relate to drought tolerance. Numerous functional features, such as stomatal closure, decreased water potential, root development, signaling pathways, and blooming, were shown to be governed by genes.In order to help the plant withstand drought, these genes interact extensively. Clarifying the mechanism of drought resistance in subtropical regions will be made easier with the use of SNPs and their functional interactions with a number of genes that respond to drought. These processes, which improve drought resistance, include stomatal closure, detoxification signaling, root development, decreased water potential, and blooming (Thirunavukkarasu et al., 2014) .

Millions of households and economies around the world have been negatively impacted by the substantial disruption in maize production caused by drought stress. Unpredictable weather occurrences will become more often in the future due to climate change, hence it will be necessary to deploy adaptive methods. High-throughput phenotyping (HTP), innovative breeding methods, and omics technology could be combined to greatly improve maize's resistance to drought stress. According to physiological principles, when plants are under stress, they consume a lot of adenosine triphosphate (ATP), which saps their energy and can cause long-term damage or even plant death By activating a pyrophosphate, which is required for the conversion of poly (ADP-ribose) into ATP molecules, plants, on the other hand, utilize "metabolic flexibility" processes to protect themselves. Drought is a multifaceted trait, and phenotyping for drought tolerance traits takes a lot of time and resources. To adequately depict the connected features, years of testing at many sites are needed. Morphologically, disturbances in the cell cycle machinery impede the growth and development of plants. Direct selection for yield under drought stress is often ineffective. In order to develop drought-tolerant crop cultivars, omics research offers more valuable and superior data.

This final memory system is explained in connection with drought response for the first time. Furthermore, applied drought stress altered floral patterning, possibly due to modifications in the expression of flowering regulatory genes and chromatin. Together, we provided a genome-wide picture of how chromatin marks and genes cooperate to help plants cope with unfavorable conditions, clarifying how this serves as the basis for stress memory creation(Forestan et al., 2020).Because conventional strategies only detected a small amount of allelic variation for drought tolerance traits, the genome-wide association studies (GWAS) methodology has been utilized to identify beneficial genes for drought tolerance in maize. Using omics methods, the drought responses of maize seeds, leaves, and roots have been intentionally examined.

(Sheoran et al., 2022). Several omics technologies must be integrated in order to control plant tolerance processes and breed maize for drought resistance (Forestan et al., 2018).

Cotton

A significant crop with several uses and significant economic worth is cotton (Gossypium spp.). In this work, we examined memory gene expression patterns and established G. hirsutum drought memory after cycles of mild drought and rewatering treatments. the molecular, physiological, and metabolic processes that underlie G. hirsutum's development of drought stress memory. In particular, the histone modification H3K4me3 is essential for controlling [+ /+] transcriptional memory. Moreover the intergenerational inheritance of drought stress memory in G. hirsutum (Forestan et al., 2020).By decreasing leaf area and restricting the formation of new leaves during the photosynthetic machinery stage, plants adopt an adaptive tolerance mechanism.

Plants can adapt to water shortages thanks to these changes. Furthermore, an exomorphic property that permits plants to withstand drought stress under dehydrated conditions is the formation of trichomes, which are hair-like structures on the leaf surface. Under challenging conditions, however, ascorbate peroxidase (APX) and glutathione reductase (GR) activities also rose concurrently, ensuring the continuous removal of ROS and sustaining the process until the plant reverted to its normal state. The experiment included the drought-tolerant CCRI-27 and CCRI-60 cotton cultivars. The CCRI-60 cultivar showed drought resilience, according to the researchers, because of several traits. Moreover, CCRI-60 demonstrated elevated proline content and antioxidant enzyme activity, particularly glutathione reductase (GR). By effectively scavenging free radicals, this cultivar demonstrated better protection against oxidative damage than CCRI-27. CCRI-60 consequently shown higher growth and more tolerance to drought.However, the researchers discovered that down-regulating the particular gene in G. barbadense, GbMYB5, resulted in a drop in the activity of antioxidant enzymes like SOD, POD, CAT, and GST. This down-regulation is controlled to increase oxidative stress when there is a water deficit. The findings indicate that a number of genes in cotton are connected to pathways involving antioxidant enzymes, which needs more investigation, particularly in drought-tolerant cultivars(Kamran et al.).The capacity of a plant cultivar to tolerate drought stress is determined by antioxidant activity. The cotton cultivar "M-503," for example, is resistant to osmotic stress and drought stress and possesses active antioxidant enzymes (APX, SOD, POX, and CAT) (Mahmood et al., 2019).

The molecular mechanisms of cotton's salt tolerance and heterosis, capillary electrophoresis, was employed for the DNA methylation levels in a cotton hybrid CCRI 29 and its 2 parents using a methylation-sensitive amplification polymorphism method. The principal result First, CCRI 29 was shown to be more salt-tolerant than either of its parents based on the findings of the test. Although there was no significant difference between the two parents under salt treatment and control, CCRI 29's global DNA methylation level rose sharply as a result. Salt-treated cotton showed a variety of DNA methylation changes, with CCRI 29 showing hypermethylation at a significantly higher rate than hypomethylation but not in its two parents. The results showed an increase in DNA methylation globally.

Second, during salt stress, some genes sustained the same amount of methylation, whereas other genes underwent both hypomethylation and hypermethylation. This implies that when cotton was exposed to salt stress, multiple alterations in gene expression occurred, emphasizing the complicated functions that certain genes performed in salt tolerance. Third, though most of the cytosine methylation sites in the hybrid CCRI 29 had same status as at least one of the parents, the number of hypomethylation sites in CCRI 29 is substantially greater than the number of hypermethylation sites under both control and salt stress, indicating that demethylation may be the mechanism underlying heterosis in cotton hybrids (Wang et al., 2016).

Potato

The potato *(Solanum tuberosum L).*, is today the world's 3rd most significant food crop and is playing a bigger role in emerging nations' local economies. The wild relatives of the cultivated potato are still present in the New World, spanning from Argentina and Chile (41°S) to the southern United States (38°N).Spanish conquistadors brought potatoes to Europe for the first time during the Columbian exchange in the sixteenth century (Hill et al., 2021). We still don't fully understand how potatoes can withstand drought, and we don't know much about how this crop stores stress memory.

In the upcoming decades, it is anticipated that the issue of water scarcity in agricultural production areas would worsen (25). We evaluated the tuber production and leaf greenness of plants from primed and non-primed seeds, as well as the responses of the plants to watering treatments in terms of antioxidant activity (AA) and tuber carbon isotope discrimination (Δ13C). Primed Sarnav plants yielded more tubers under both constrained and non-restricted water regimes. Primed Unica plants showed less reduction in tuber yield and Δ13C with water limitation. Consequently, long-term stress memory seems to be significantly(Ramírez et al., 2015). The pyramiding of numerous divergent traits will be necessary for effective crop modification for drought resistance, with different configurations being optimal for various growing situations.

It is crucial to comprehend the relationship between the roots' underground water intake and the shoot system's aboveground water loss. A fascinating new tool for precision screening is being made available by the development of high throughput precision phenotyping platforms. By combining creative screening techniques, this tool can help select and pyramid drought-related genes that are suitable for particular settings. Proteomics, metabolomics, genomics, and bioengineering results will surely enhance traditional breeding methods and offer a different path to the creation of drought-tolerant potatoes (Obidiegwu et al., 2015).

Tolerance is demonstrated by sweet potatoes. The seventh most important food crop is thought to be starchy, fresh, and vibrant tubers. An average of about 8000 novel genes and transcripts were discovered in each sample. However, no comprehensive study of the mechanisms behind the drought tolerance of a number of sweet potato cultivars has been executed as of yet.In this article we examined the mechanism behind the drought response of seven drought-tolerant sweet potato cultivars using transcriptome sequencing, physiological markers, and drought coefficients.

The seven cultivars of sweet potatoes were divided into four drought-related groups. Drought stress had no discernible effect on the first exon and last exon alternative splicing events that dominated sweet potato and were not preserved among cultivars. Meanwhile, functional annotation and analysis of genes with differential expression indicated distinct drought-tolerance mechanisms. The main way that two drought-sensitive cultivars, Xushu-22 and Shangshu-9, fought drought stress was by increasing plant signal transduction. The nitrogen/carbohydrate metabolism and isoquinoline alkaloid production were down-regulated in Jishu-26, other cultivar that is prone to drought in response to drought stress.

Only 9% of the genes that were differentially expressed were different between the drought-tolerant cultivar Chaoshu-1 and the drought-preferred breed Z15-1, yet they both shared a number of competing metabolic pathways in response to drought. While Z15-1 enhanced photosynthesis and carbon fixation ability, they primarily controlled the production and metabolism of flavonoids and carbohydrates in response to dryness.

The other able to withstand cultivar, Xushu-18, regulated the production of isoquinoline alkaloids and the metabolism of nitrogen and carbohydrates under drought stress. Drought stress had virtually no effect on the highly drought-tolerant cultivar Xuzi-8, which only adjusted its cell wall in response to the drought environment. These results offer vital information for choosing sweet potatoes for certain uses, (Liu et al., 2023).

NON-CODING RNAS IN DROUGHT STRESS REGULATION

Plants have evolved sophisticated mechanisms to endure the rigors of drought stress. A cornerstone of this adaptive repertoire is the precise regulation of stomatal aperture, which governs vital processes such as gas exchange, transpiration, and photosynthesis. Plants can mitigate the harmful effects of dehydration by controlling water loss and internal temperature. Furthermore, the ability to withstand severe water deficits is often linked to osmotic adjustment and the accumulation of specialized proteins designed to protect cellular structures under desiccation stress (Bhattacharjee, 2016).

Hormonal signaling is pivotal in managing the plant's response to drought. Subtle alterations in phytohormone levels can trigger a cascade of physiological and morphological changes, such as delayed senescence and modified root architecture. Concurrently, plants activate antioxidant defense systems to counteract the oxidative stress induced by drought. Enzymes like superoxide dismutase, peroxidase, and ascorbate peroxidase work in concert to scavenge harmful reactive oxygen species (ROS), safeguarding essential organelles such as chloroplasts and preserving the integrity of the photosynthetic apparatus (Bhargava, 2013). To fully comprehend the intricacies of drought tolerance, it is imperative to delve into the genetic underpinnings of these adaptive strategies and identifying the genes and regulatory networks involved in drought response will be instrumental in developing crops with enhanced resilience to this environmental challenge.

Non-coding RNAs (ncRNAs) are RNA molecules without protein-coding abilities that are found in various organisms. They can be linear or circular in structure. A specific type of linear ncRNA is called long non-coding RNA (lncRNA), which is longer than 200 nucleotides and is involved in many plant processes (Wang et al., 2018). Researchers use RNA sequencing (RNA-seq) to identify lncRNAs in different plants. Many lncRNAs have been discovered in rice, including those that help the plant fight diseases and cope with environmental stress i.e., drought, salt, nutrients deficiency, and heavy metals contamination (Mattick, 2009).

Mechanism of Action of LncRNAs in Drought Resistance

Through several methods, LncRNAs play a crucial regulatory role in the drought response of plants.:

Mimicking the Target

LncRNAs can trick microRNAs (miRNAs). Usually, these miRNAs attach to messenger RNAs (mRNAs) and either hinder or deteriorate their translation (Deng et al., 2018). LncRNAs affect gene expression by binding to miRNAs and preventing them from addressing their mRNA targets. Examples of this process include lncRNA_1231 in pigeon peas and lnc_253 and lnc_973 in cotton (Jiang et al., 2019).

Competing Endogenous RNAs (ceRNAs)

LncRNAs can function as ceRNAs by competing with mRNAs for common miRNA binding sites (Lu et al., 2020). LncRNAs indirectly increase the expression of target mRNAs by sequestering miRNAs. This process is demonstrated by lncRNA23468 in tomatoes, and linRNA159, and linRNA340 in cassava (Li et al., 2017).

Control of Gene Expression

LncRNAs can affect gene expression through direct interactions with proteins, RNA, or DNA. They can function as scaffolds, uniting RNA and protein molecules to create useful complexes. Examples of this mechanism include lncRNAs in winter wheat (Lu et al., 2020) and lncRNA39026 in tomatoes (Jiang et al., 2019). Overall, lncRNAs are essential for adjusting gene expression networks in response to drought stress. LncRNAs help plants survive and adapt in water-limited environments by influencing miRNA activity, vying for miRNA binding sites, and interacting with other cellular constituents (Figure 1).

Figure 1. This flowchart shows how non-coding RNAs (ncRNAs) control gene expression in plants under drought stress. The process starts when the plant is stressed by drought, which causes ncRNAs to become active. These non-coding RNAs (ncRNAs) fall into two primary categories: long non-coding RNAs (lncRNAs) and microRNAs (miRNAs). Gene regulation, which eventually aids in the plant's adaptation to drought circumstances, is influenced by both miRNAs and lncRNAs.

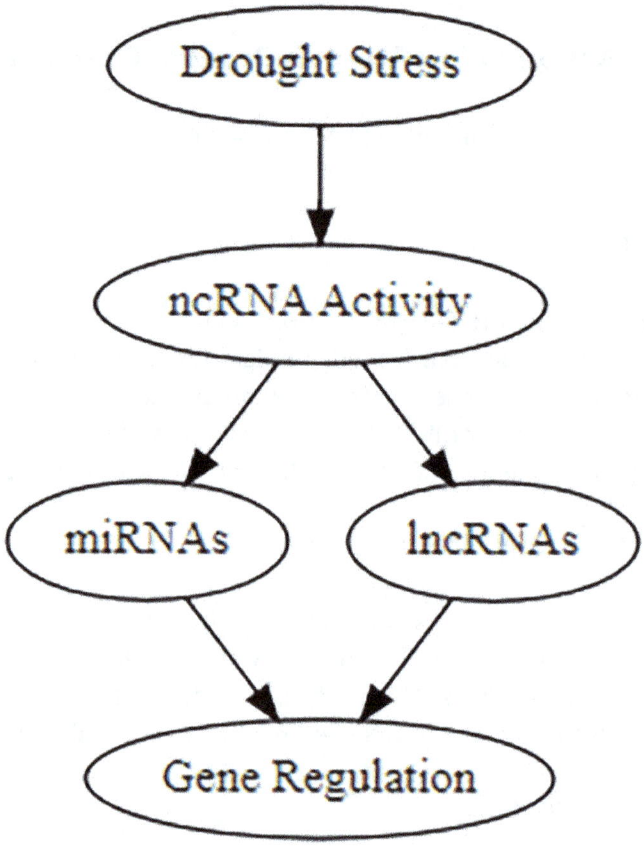

TRANSGENERATIONAL EPIGENETIC INHERITANCE OF DROUGHT TOLERANCE

The faithful transmission of features from generation to generation is a fundamental aspect of life. While DNA is unquestionably the essential blueprint for heredity, it is not the only factor determining an organism's features. A complex interplay of epigenetic information, which consists of layers of chemical alterations that control genes expression without changing the underlying DNA sequence, has a considerable impact on the manifestation of inherited characteristics (Hauser et al., 1809). John Gurdon's pioneering nuclear transfer experiments proved that its genetic code does not simply determine an organism's development. These studies revealed the important significance of epigenetic elements in developing and sustaining cellular identities. Plants are stationary and thus unable to actively avoid the effects of aggressive environmental conditions as drought, severe temperatures, insect pests, parasitic plants, and pathogenic microorganisms. To mitigate the negative experiences, plants have developed a highly adaptable phenotype. Adaptive phenotypic alterations frequently take place in environments that are changing too quickly for natural selection to favor adaptive changes. Environmental adaptation in these circumstances is most likely supported by some form of epigenetic variability (Ashapkin et al., 2020).

Plants that are stressed by drought release abscisic acid, a biochemical messenger that controls genes expression and speeds up several physiological and cellular functions, including stomatal function due to reduce stress and increase tolerance. The adjacent chromatin must be changed from a repressive to an active phase by epigenetic modification to either stimulate or inhibit gene transcription under conditions of drought stress. This is accomplished through the dynamic interaction of histone modifications, DNA methylation, and non-coding RNA generation. Plants can learn to recognize their chromatin state, which helps them cope with repeated stress under dry conditions, Additionally, plants can pass on their drought tolerance to the following generation. Stressful environments might cause memory genes' epigenetically altered chromatin structures to be transferred to freshly generated (Nguyen et al., 2022).

EPIGENETIC CHANGE IN PLANTS TO DROUGHT STRESS

One of the most extensively researched epigenetic modifications in plants, especially concerning drought tolerance, has been shown to cause changes in the sequence of DNA methylation that are securely inherited. For example, a study on rice showed that a significant number of drought-induced modifications in DNA methylation were retained in the following generations, perhaps enhancing progeny's adaptation to drought. In particular, approximately 45 percent of modified DNA methylation states were transmitted to unstressed progeny, indicating that these epimutations are critical for boosting drought adaptability in subsequent generations (Nguyen et al., 2022).

Histone alterations play a role in epigenetic transmission across generations in alongside DNA methylation. Histone marker alterations can alter DNA's accessibility to transcriptional machinery, which in turn can alter gene expression (Zhao et al., 2019). Histone H3 lysine 4 trimethylation (H3K4me3) and H3K9 acetylation (H3K9ac), for instance, have been linked to the activation of genes that respond to drought. Because these alterations are dynamic, plants can quickly modify the way their genes are expressed in response to stress. Additionally, some of these adjustments have the potential to last for several generations, which could serve as the foundation for stress memory.

Transgenerational epigenetic inheritance has also been linked to non-coding RNAs, especially long non-coding RNAs and microRNAs. These molecules can control gene appearance on several levels, which affects how plants react to drought stress. For example, it has been demonstrated that particular lncRNAs affect the expression of genes involved in the crucial drought response pathway known as abscisic acid (ABA) signaling (Yajnik et al., 2024). Such regulating ncRNAs have been shown to have a function in transgenerational epigenetic mechanisms by improving the drought tolerance of offspring when inherited as shown in Figure 2. For plants to respond to drought stress, transgenerational epigenetic inheritance is essential. Plants can pass on stress tolerance to their offspring, making them more resistant to future drought experiences, through processes like DNA methylation, histone changes, and the activity of non-coding RNAs and enhance agricultural resilience to drought as well as other environmental challenges and, eventually, support the supply of food in an environment that is changing, additional research in this field is needed.

Figure 2. The mechanism by which plants pass on their drought resistance through epigenetic inheritance is depicted in this flowchart. It starts when the parent generation is under stress from drought, which causes a variety of epigenetic alterations, such as methylation of DNA, modifications to histones, and the participation of non-coding RNAs (ncRNAs). The progeny then inherits these epigenetic markers, which contribute to their increased resistance to drought. These epigenetic changes are passed down to the progeny, improving their resistance to drought

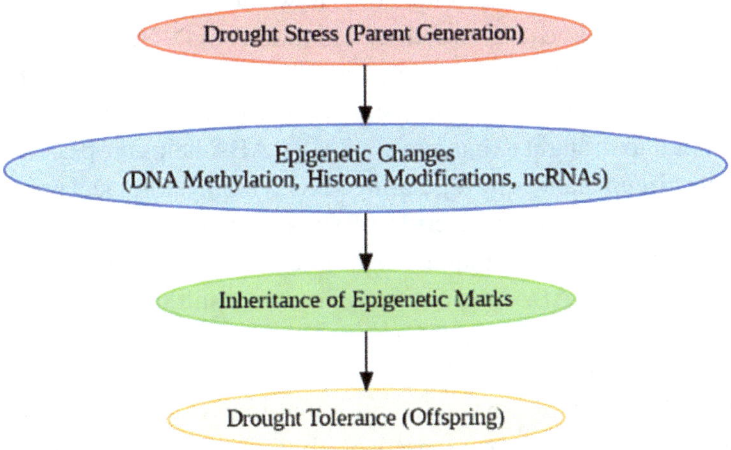

EPIGENETIC MECHANISMS IN DROUGHT TOLERANCE ACROSS DIFFERENT PLANT SPECIES

Trans-generational epigenetic inheritance is essential for plants to respond to drought stress, Plants can pass on stress tolerance to their offspring, making them more resistant to future drought experiences, through processes like DNA methylation, histone changes, and the activity of non-coding RNAs as shown in Figure 6.1. Further research is essential to strengthen agricultural resilience to drought and other environmental challenges, ultimately sustaining food supply in an ever-changing environment.

DNA Methylation and Adaptation to Drought in Different Species

DNA methylation is the process of adding methyl groups, usually at CpG sites, to the cytosine bases of DNA. According to the situation, this epigenetic change can either stimulate or inhibit genes, changing the expression of those genes. Regarding tolerance to drought, distinct plant species' patterns of methylation of DNA might differ greatly from one another, impacting how these varieties react to drought stress. For example, in rice (Oryza sativa), alterations in DNA methylation brought on by drought have been linked to the activation of genes involved in the hormone system known as abscisic acid (ABA) signaling, which is essential for the plant's ability to respond to drought (Zhang et al., 2018). ABA helps crops conserve water during drought circumstances by regulating several of physiological processes, such as stomatal closure and the production of stress-responsive genes. Research has demonstrated that based on the requirements of a plant during stress, methylation in the promoter regions of ABA-responsive genes can result in either their activation or inhibition (Chinnusamy et al., 2007).

In contrast, DNA methylation has a distinct function in other species, such as Arabidopsis thaliana. Studies have demonstrated that drought circumstances may prompt specific genomic regions—particularly those linked to genes involved in water uptake and root development—to become demethylated. These genes may get activated as a result of this demethylation, which would improve the plant's capacity to withstand drought circumstances by increasing soil water absorption. DNA methylation is also essential for drought tolerance in maize. Research has shown that drought stress can cause some genes that are responsive to drought to become hypermethylated, which ironically leads to the suppression of such genes. According to (Eichten, 2015), this suppression is thought to be a mechanism for downregulating non-essential activities during water stress to conserve energy and resources. The diversity of DNA methylation patterns among species emphasizes the intricacy of plant epigenetic regulation and shows how many species have developed unique adaptations to drought.

HISTONE VARIATIONS REGARDING ADAPTATION TO DROUGHT IN DIFFERENT SPECIES

Acetylation, methylation, phosphorylation, and ubiquitination are examples of histone modifications that are essential for controlling the structure of the chromatin and the accessibility of genes. The particular histone residues involved, these changes can either increase or decrease the expression of a gene depending on their type and Histone modifications are mechanisms that are conserved across several plant species and are essential in regulating drought tolerance. It has been demonstrated that drought stress in maize causes particular histone changes that cause drought-responsive genes to be upregulated. For instance, the promoters of genes implicated in stress reactions frequently include histone H3 lysine 9 acetylation (H3K9ac) and histone H3 lysine 4 trimethylation (H3K4me3), both of which are linked to active transcription. Due to these alterations, the chromatin structure becomes more relaxed, making it simpler for transcription factors and RNA polymerase to access DNA. This makes it possible for important drought-responsive genes to be expressed quickly (Shen et al., 2021).

Rice has been observed to undergo histone deacetylation, particularly at the promoters of genes that are less critical for immediate survival under conditions of drought (Fang et al., 2014). This deacetylation causes chromatin condensation and gene repression, which conserves resources such as energy during periods of water shortages. The species-specific differences in histone modifications are indicative of the adaptive approaches that plants have developed to regulate the distribution of resources and expression of genes under stressful circumstances. Another important alteration that has been thoroughly researched in Arabidopsis is histone methylation. Trimethylation of histone H3 at lysine 27 (H3K27me3), a mark commonly linked to gene regulation, is induced by drought stress. The plant can concentrate its energy on expressing the genes required for drought resistance by suppressing non-essential genes thanks to this change. It's interesting to note that different species may target different genes via H3K27me3 during drought stress, indicating a degree of species-specific regulation in response to environmental stressors (Saleh et al., 2008).

NON-CODING RNAS IN DROUGHT TOLERANCE ACROSS SPECIES

Across a variety of plant species, non-coding RNAs, especially microRNAs, are key moderators of the expression of genes under conditions of drought. MiR159 in Arabidopsis regulates the signaling of ABA by inhibiting MYB transcription factors, which is essential for preserving water balance in dry environments (Liu Hanhua et al., 2008). Superoxide dismutase (SOD) enzymes are the target of miR398 in wheat, which regulates the oxidative stress response to avoid cellular damage during water scarcity. The miR164 and miR167 regulate genes associated with water uptake and root development and enhance drought tolerance in maize. (Ding et al., 2013).

Furthermore, it has been discovered that miR319 in rice regulates TCP transcription factor, which affects leaf shape and improves drought tolerance. Additionally, SPL transcription factor, in several developmental processes, including modifications to root architecture during drought circumstances, are targeted by miR156 in soybeans (Arshad et al., 2017). These miRNAs play a critical role in plant stress tolerance processes, as evidenced by their conservation and specificity across many species. This is shown in Figure 3.

Figure 3. The graphic shows how numerous epigenetic mechanisms, such as histone changes, non-coding RNAs, and DNA methylation, help plants adapt to drought stress. These pathways are illustrated in several plant species, including rice, wheat, Arabidopsis, and maize. They are associated with distinct adaptive responses, including the control of ABA signaling genes, improved root development, downregulation of non-essential genes, gene upregulation, chromatin condensation, preservation of the water balance, modulation of the oxidative stress response, and increased resistance to drought

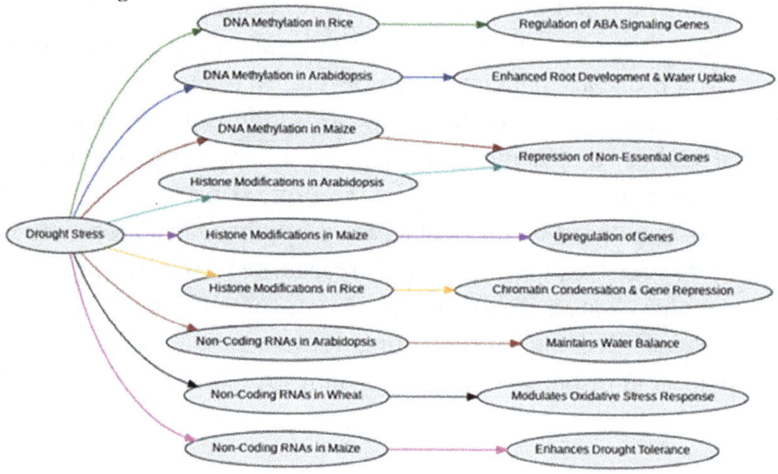

EPIGENETIC REGULATION OF GENES EXPRESSION IN DROUGHT TOLERANCE

Recently, there was an important increase in attention in eukaryotic epigenetic regulation, with scientists delving into its intricate mechanisms and impacts on gene expression. Epigenetics involves semi-permanent modifications that do not alter the DNA sequence or the structural configuration of chromosomal regions. (Gibney, 2010). In the epigenetic management of gene expression, histone and DNA regulators are essential, especially when the environment is stressed, as in the case of drought. Histone modifications and alterations in patterns of DNA methylation along with the non-coding RNAs are involved in this regulation; these factors collectively affect structures of chromatin and gene accessibility enabling the plants to adapt to the environmental changes without any change in their genetics (Desaulniers et al., 2021).

Furthermore, the epigenetic basis of activities like vernalization which triggers flowering in response to cold stress is being studied, they can be used to create chromatin regulator null mutants and have easily accessible genomes, plants are great models for studying epigenetics. Plants are sessile creatures, they must constantly adapt to environmental stresses, which calls for strong epigenetic defenses to sustain healthy growth and development(Yaish et al., 2011). The interaction of gene expression and chromatin control under abiotic stressors, like drought and salinity, emphasizes the significance of epigenetic mechanisms in plant resilience. Additionally, it will address future research directions and the potential implications of epigenetic research for breeding drought-tolerant crops and optimizing agricultural practices, ultimately contributing to enhanced food security in the face of climate change (Banerjee, 2017).

EPIGENOMIC TOOLS AND TECHNIQUES FOR STUDYING DROUGHT TOLERANCE

Plants use many epigenetic pathways for gene expression in response to drought conditions. These strategies include chromatin remodeling, short RNA-mediated gene silencing, DM, and histone alterations (Figure 4). Both controlling gene expression and preserving genomic stability are important tasks, and each strategy has a specific and vital role to perform (Sun et al., 2021).

Figure 4. Epigenetic mechanism in response to stress management in plants

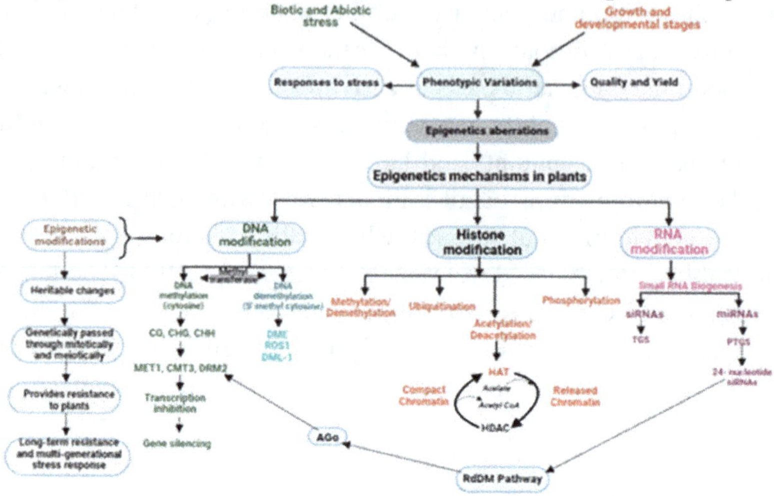

(Abdulraheem et al., 2024)

Epigenetics Methodology

DNA methylation, histone modifications, and RNA modifications are example of epigenetic alterations that are essential for controlling genes expression without changing underlying DNA sequences. Many biological procedures rely on these changes, such as cellular differentiation, development, and response to external stimuli (Li, 2021).

DNA Methylation

One of important epigenetic mechanism that profoundly affects plant development and stress responses is DNA methylation (DM). DNA methyltransferase enzymes are involved in this intricate process, which adds a methyl group to cytosine base of DNA, mainly at CpG dinucleotides. The identification of these CpG sites sets off a series of alterations that modify the structure of chromatin and decrease the transcription of genes. The regulation of fundamental plant characteristics, such as leaf shape, disease resistance, and stress tolerance, depends on this repression(Kurdyukov, 2016).

The correlation between drought resistance and DNA methylation in several rice varieties. Varieties that are susceptible to drought typically have hypermethylation, whereas genotypes that are resistant to drought display hypomethylation. DNA methylation plays a crucial role as an epigenetic marker in determining how drought-tolerant plants are, affecting not only gene expression but also reproductive characteristics like spikelet sterility and total output (Li, 2021).

Mechanism of DNA Methylation

One important epigenetic mechanism is DNA methylation, which is the adding of methyl group (CH_3) to DNA molecule at the 5-carbon position of cytosine ring, forming 5-methylcytosine (5-mC). This alteration generally takes place in the setting of CpG dinucleotides, in which a guanine comes after a cytosine. While methylation can also happen in non-CpG settings in plants, it mostly happens at specific CpG sites in mammals. DNMT1, DNMT3A, and DNMT3B are members of the DNA methyltransferases family of enzymes that control DNA methylation. During DNA replication, DNMT1 plays a crucial function in maintaining established methylation patterns, while DNMT3A and DNMT3B help create new methylation markers (Rajendran et al., 2011).

When DNMTs attach to the DNA strand, they catalyze the transfer of a methyl group from S-adenosylmethionine (SAM) to the cytosine base, which starts the process of DNA methylation. This alteration may prevent transcription factors from binding to DNA or may attract proteins that encourage chromatin compression and transcriptional silence, both of which can result in the repression of gene expression. Additionally, methylated cytosines can function as binding sites for proteins known as methyl-CpG-binding domains (MBDs), which in turn attract more repressive complexes to enhance chromatin remodeling and inhibit transcription (Gao et al., 2020).

In plant *Arabidopsis thaliana*, DM is essential for regulating flowering time because it represses certain genes linked to floral development (Becker et al., 2011). Flowering is delayed due to methylation of promoter regions of these genes, indicating the significance of DM in developmental timing. Moreover, DM affects leaf morphology because it represses genes that determine leaf size and shape, causing smaller leaves in specific situations. Lastly, DM is important for plant defense mechanisms. For example, *Pseudomonas syringae pv.* tomato can activate DM pathways that activate defense and hormone signaling, strengthening the plant's resistance to diseases (Bordiya, 2017).

With development, (DM) is also involved in stress responses. The addition of methyl group to cytosine residues produces 5-methylcytosine (5 mC), which acts as a restrictive marker for gene expression.This alteration is essential for controlling transcriptional inactivity and preserving genomic stability, especially under stressful environmental conditions. Research has demonstrated that changes in global DM concentrations can have a substantial impact on how genes express themselves in response to abiotic stressors like drought and salt. For example, studies on different types of rice have shown that DM levels change in response to salt stress; cultivars that are sensitive to salt had lower DM levels and more DNA demethylase gene expression (Umer, 2013).

Environmental stressors can also cause long-lasting alterations in DM patterns, which can be passed on to succeeding generations. These changes may continue even after the stressor is eliminated, suggesting that plants may have a way to adjust to their changing surroundings. But the effects of DM vary depending on the type of plant and the stress level. For instance, although drought stress causes global hypomethylation in Arabidopsis, comparable circumstances may cause hypermethylation in other species (Angers et al., 2010). The dynamic and vital process of DNA methylation controls the expression of genes in plants during growth and in response to stress. Because it can change chromatin structure and gene transcription, allowing plants to adapt to a range of environmental conditions, research in this area is essential to increasing crop resilience and productivity. Additionally, understanding the intricate mechanisms of drought tolerance (DM) will help develop new agricultural strategies aimed at improving plant performance in stressful environments (Dahl, 2003).

HISTONE MODIFICATION

The protein components of chromatin called histones are subjected to a variety of post-translational changes, such as acetylation and methylation, which have a profound impact on the expression of genes. For example, active transcription is linked to the trimethylation of histone H3 at lysine 4 (H3K4me3). Drought-responsive genes like Rd20 and Rd29A have been shown to have elevated levels of H3K4me3 and acetylation at H3K9 (H3K9ac) in response to drought stress. These changes improve the transcription of stress-responsive genes by facilitating the recruitment of RNA polymerase II (Li, 2021).

Arabidopsis trithorax-like1 (ATX1) protein acts as a histone methyltransferase to facilitate H3K4me3 modification, which in turn upregulates genes related to the manufacture of abscisic acid (ABA)during drought possession, reduced expression of genes responding to drought and lower levels of H3K4me3 are observed in mutants devoid of ATX1, highlighting the significance of these histone modifications in stress reactions (Ding et al., 2011). Additionally, during the post-drought recovery phase, the dynamics of histone changes are crucial. The chromatin state is reset by quick deacetylation reactions, which permits the restoration of baseline gene expression levels. This entails the elimination of H3K9ac and a more gradual loss of H3K4me3, suggesting a highly developed regulatory system that modifies gene expression in response to shifting environmental factors(Lennartsson, 2009).

Types of Histone Modifications

Another crucial component of epigenetics is histone modification, which is the covalent alteration of histone proteins, which encircles DNA to form nucleosomes. By changing the accessibility and structure of the chromatin, these alterations can affect how genes are expressed and so control the transcriptional potential of related genes. Acetylation, methylation, phosphorylation, and ubiquitination are the main forms of histone modifications. Each of these modifications can have a unique impact on the dynamics of chromatin and the expression of genes.

Acetylation: Histidine acetyltransferases (HATs) are the primary enzymes responsible for introducing acetyl groups to lysine residues on histones and acetylation loosens the chromatin structure and makes it easier for transcription factors and RNA polymerase to bind to DNA, it increases the expression of genes (Weinhold, 2006).

Methylation: The methyl group addition that histone methyltransferases (HMTs) catalyze to lysine or arginine residues on histones. Histone methylation, as opposed to acetylation, has the ability to either activate or repress gene expression based on the particular residue that is altered and the number of methyl groups that are added (mono-, di-, or trimethylation) (Rice, 2001).

Phosphorylation: The adding of phosphate group to serine, threonine, or tyrosine residues on histones; this is frequently linked to chromatin remodeling and transcriptional activation during processes like cell division and DNA repair (Rossetto et al., 2012).

Ubiquitination: The binding of ubiquitin molecules to histone lysine residues, which can change the structure of the chromatin to impact transcription or signal histone breakdown (Shilatifard, 2006).

Mechanism of Histone Modification

There are various ways in which histone changes might affect how genes are expressed. For example, acetylation neutralizes the positive charge of lysine residues, reduces the interactions in histones and negatively charged DNA. The chromatin structure becomes more open, increasing transcription machinery accessibility. On the other hand, depending on the circumstances surrounding the alteration, methylation might draw in particular binding proteins that either stimulate or impede transcription. In plant stress reactions, histone alterations are essential for controlling gene expression (Arora, 2021).

Stress-responsive genes can attract transcription factors or co-regulators through specific histone modifications. For instance, in Arabidopsis, the expression of genes associated to stress is activated when the WRKY transcription factor binds to H3K4me3 marks. Histone changes that affect nucleosome stability or the recruitment of chromatin remodeling complexes can have an impact on chromatin accessibility and structure. A more open chromatin structure is the result of acetylation, which reduces the positive charge of lysine residues and decreases the interaction between histones and DNA. Methylation, depending on the context, may attract specific binding proteins that either promote or inhibit transcription. Combined, different histone modifications can regulate gene expression, establishing what is often called a "histone code." (Çelik et al., 2019).

For example, during stress responses, the interaction between H3K4me3 and H3K27me3 markers has been linked to the regulation of gene activation and repression. In Arabidopsis, a number of abiotic conditions, including heat, salt, and low water availability, can enhance global histone modification. H3K4me3 and H3K9ac in the promoter regions of stress-responsive genes, such as RD20, RD29A, RAP2.4, and RD29B, are improved in Arabidopsis under drought stress. 4837 genes in rice seedlings stressed by drought show differently changed H3K4me3. Plants that have developed transgenerational stress memory are better able to respond swiftly to repeated stresses due in part to histone changes (Mishra et al., 2022).

RNA MODIFICATIONS

An important aspect of gene control, especially in eukaryotic species, is represented by RNA modifications. Of these, N6-methyladenosine (m6A) has become the most widely distributed and extensively researched alteration, with important functions in splicing, stability, translation, and degradation of RNA (Kazimierczyk, 2021).

Mechanism of RNA Modification

Plant stress processes rely on small RNA-mediated suppression of genes and chromatin remodeling complexes, such as miRNAs and siRNAs, which control genes expression under stress. The RNA-induced silencing complex is directed by non-coding RNA molecules to target mRNAs for translational repression or destruction. Whereas siRNAs are formed from endogenous lengthy double-stranded RNA precursors or exogenous double-stranded RNA precursors, miRNAs are derived from endogenous hairpin-shaped precursors. By base pairing with target mRNAs, both kinds of short RNAs cause translational inhibition or mRNA cleavage. Studies show that plants produce (ROS) in their peroxisomes and chloroplasts when they are under drought stress. However, plants produce antioxidative enzymes including glutathione reductase, ascorbate peroxidase, peroxidase, superoxide dismutase, and catalase to counteract the harmful effects of ROS on cells. It's interesting to note that some plant miRNAs, such miR528 and miR398, are known to control networks of oxidative stress (Liu in Pan, 2015).

Studies have demonstrated the important function of small RNAs in controlling plant stress responses, especially microRNAs (miRNAs) and long non-coding RNAs (lncRNAs). These chemicals target genes linked to stress, which affects many physiological processes and modulates stress tolerance. For example, messenger RNAs (mRNAs) can attach to stress-responsive miRNAs, which can either inhibit or degrade mRNA translation into proteins. In order for plants to respond to abiotic stimuli like heat and drought, this regulatory mechanism is essential. One such instance is miR159, which has been demonstrated to alter genes related to drought tolerance in *Arabidopsis thaliana* (Zhai et al., 2008).

Plant adaptability to environmental stressors depends on chromatin remodeling and small RNA-mediated gene silencing. For instance, miR398 targets the transcripts of the copper superoxide dismutase (CSD) gene, which increases plant sensitivity to oxidative stress by suppressing CSD production. When exposed to oxidative stress, it is active. Small RNAs are involved in chromatin remodeling complexes, which alter the structure of chromatin and regulate the expression of genes. They function at both the transcriptional and post-transcriptional levels. To be more precise, transposable elements generate small interfering RNAs (siRNAs), which direct DNA methyltransferase DRM2 to particular genomic regions, thereby silencing genes (Drongitis et al., 2016).

Small RNAs have the capacity to control genes expression in addition to their function in gene silencing through interactions with histone modifiers and chromatin remodeling complexes. MiR156 affects histone changes at flowering loci, which in turn affects flowering time. It targets transcription factors that resemble Squamosa promoter-binding proteins. The intricate nature of plant stress responses is shown by the interaction between short RNAs and chromatin dynamics. In addition to being necessary for healthy cellular operation, RNA alterations are important for many biological processes, such as stress reactions, differentiation, and development. For example, m6A alteration can attract particular binding proteins that affect RNA processing, stability, and translation. These proteins are called m6A readers. Additionally, the presence of m6A can signal RNA for degradation by the RNA exosome complex, controlling the amounts of particular transcripts during embryonic development and impacting lineage specification and cell fate decisions (Hu et al., 2024).

MODERN TOOLS AND TECHNIQUES OF EPIGENETICS

The development of diverse epigenomic methods and approaches has led to considerable breakthroughs in the study of epigenetic control in plants, especially in the context of drought tolerance. The main epigenomic methods and instruments used to investigate drought tolerance—NGS, WGBS, chIP, bisulfite sequencing, and MSAP will be covered in detail in this part, along with their significance and uses in plant research (Villota-Salazar et al., 2016).

Next Generation Sequencing (NGS)

The use of (NGS) has transformed the fields of epigenomics and genomics by offering high-throughput, economically viable techniques for genetic material analysis. Scientists can rapidly sequence entire genomes, using (NGS) to identify epigenetic changes across various plant species. NGS can be used to examine fluctuations to DNA methylation patterns and histones in result to water stress in the of drought tolerance (Yadav et al., 2018).

Whole Genome Bisulfite Sequencing (WGBS)

A potent NGS-based method called (WGBS) makes it possible to examine the methylation of DNA patterns throughout the entire genome at single-base precision. By applying sodium bisulfite to DNA, it is possible to selectively deaminate unmethylated cytosines to uracils while leaving methylated cytosines unaltered. After the DNA has been altered, it is sequenced, which enables scientists to deduce each cytosine's methylation state from the sequence information. The Whole-genome bisulfite sequencing (WGBS) provides scientists with a detailed view of the DNA methylome, enabling the identification of differentially methylated regions linked to drought resistance (Troyee et al., 2023).

Chromatin Immunoprecipitation Sequencing (ChIP-seq)

It is a vital technique for studying histone modifications and protein-DNA interactions. By combining chromatin immunoprecipitation with high-throughput sequencing, scientists can map specific histone modifications across the genome and identify genes responsive to drought. Research in various plant species, such as Arabidopsis thaliana and Zea mays (maize), has shown that increased expression of drought tolerance genes is associated with higher histone acetylation levels at certain gene promoters.

Researchers can determine the binding locations of transcription factors and the presence of particular histone markers like H3K4me3 and H3K9ac, that are frequently connected to active transcription using ChIP-seq. Through the examination of these alterations during a drought, researchers can learn more about the regulatory systems that manage gene expression in reaction to water constraint. Our knowledge of the epigenetic mechanisms underlying plant responses to drought has improved as a result of this technique's ability to connect particular histone changes to the expression of genes involved in drought tolerance (Kaufmann et al., 2010).

Bisulfite Sequencing

This is a method used to analyze DNA methylation patterns at single-base resolution. During this method, DNA is treated with bisulfite, that converts unmethylated cytosine to uracil, allowing methylated and unmethylated cytosines to be distinguished during subsequent sequencing.The dynamics of DNA methylation in response to drought stress have been clarified thanks in large part to bisulfite sequencing, which has also shed light on how these changes control gene expression in many plant species.

This technology allows scientists to examine specific genes undergoing methylation changes in response to drought stress, potentially identifying targets for breeding programs aimed at enhancing drought resistance. For example, rice research has shown that specific developmental phases and tissue types can be linked to alterations in DNA methylation patterns brought on by drought, underscoring the significance of epigenetic control in the plant's adaptive response to water scarcity (Jiménez-Arias et al., 2015).

Methylation-Sensitive Amplification Polymorphism (MSAP)

DNA methylation differences throughout the genome can be found using a technique called methylation-sensitive amplification polymorphism, or MSAP. Differentially methylated loci (DMLs) can be identified using this technique since restriction enzymes are sensitive to methylation state.Researchers have utilized MSAP to investigate the impact of drought stress on DNA methylation patterns across various plant species.

MSAP enhances our understanding of how methylation affects stress tolerance by providing a valuable tool for assessing the impact of drought stress on epigenetic variation. Researchers can learn more about the genetic foundation of drought tolerance and create plans for enhancing crop resilience by finding DMLs linked to drought (González-Benito et al., 2020).

A thorough grasp of the epigenetic landscape under drought stress can be obtained by integrating data from several epigenomic approaches, through the integration of NGS, ChIP-seq, and bisulfite sequencing data, scientists are able to build intricate gene regulatory models and pinpoint the essential epigenetic changes that support drought resistance. Untangling the intricate regulatory networks that control how plants react to environmental stress requires an integrative approach (Agius et al., 2023).

EPIGENETIC VARIATION AND NATURAL SELECTION IN DROUGHT-PRONE ENVIRONMENTS

Epigenetic Variation in Drought Condition

This phenomenon is critical in how organisms, particularly plants, adapt to environmental stresses such as drought. In these challenging conditions, epigenetic mechanisms, including DNA methylation, histone modifications, and the action of small RNAs facilitate rapid adaptations that enhance survival and reproductive success. For instance, in drought-prone environments, epigenetic changes can discuss immediate advantages, allowing plants to tolerate and flourish under stressful conditions. Natural selection then acts on these epigenetically influenced traits, shaping the evolution of plant populations. Case studies involving genetically uniform organisms, such as various plant species, illustrate how epigenetic variation can enhance drought tolerance through specific mechanisms like altered DNA methylation patterns. This intricate interplay between epigenetics and environmental factors highlights the complexity of adaptation and evolution in response to changing conditions (Meyer, 2015).

Plants frequently encounter harsh conditions in drought-prone areas, which might result in less water being available, this calls for quick and efficient solutions. Plants can respond to these stresses by modifying their phenotypic features through heritable modifications mediated by epigenetic progressions. The ability of epigenetic variation to adapt quickly to changes in the environment is one of its main characteristics. In contrast, short term adaptations are possible because epigenetic alterations may be swiftly induced and reversed, differentiating from genetic mutations, which take longer to arise in the evolutionary process. For example, studies have demonstrated that plants can respond to drought by displaying phenotypic plasticity, whereby epigenetic modifications enable the development of features that improve survival, including modified root and shoot architectures and plants cannot move away from bad environments, their adaptability is very significant. Researchers found that drought stress significantly altered DNA patterns that linked to the expression of genes that respond to drought in a study including *Arabidopsis thaliana*. These changes enabled the plants to adjust their physiological functions, such as osmotic control and stomatal closure, to cope with reduced water availabe. The significance of epigenetic diversity in plant drought tolerance is highlighted by the speed with which epigenetic mechanisms can alter gene expression. Furthermore, populations can preserve adaptive features in the absence of genetic variation thanks to the inheritance of epigenetic variation across generations (Kooke et al., 2015).

Natural Selection Relation with the Epigenetics

The idea of "epigenetic memory" demonstrates the connection between natural selection and epigenetic variation. These enduring effects of epigenetic modifications may impact phenotypic features in subsequent generations. Plants possessing epigenetic memory may have increased resilience in drought-prone areas, making them better able to withstand and recover from consecutive drought occurrences. Increased fitness and reproductive success may result from this adaptive advantage, supporting the idea that epigenetic variation plays a significant role in determining the course of evolution. Although epigenetic modifications can have immediate adaptive advantages, they also have the risk of introducing phenotypic instability if environmental circumstances change. For example, if a population of plants depends on particular epigenetic changes to withstand drought, a change in the surrounding environment may make those changes maladaptive. This demonstrates how dynamic epigenetic diversity is and how intricately it interacts with natural selection. Furthermore, the possibility that environmental influences could affect epigenetic variation raises concerns regarding the predictable nature of evolutionary events. Comprehending the significance of epigenetic variation in natural selection is imperative in forecasting plant species responses to continuous environmental modifications and formulating tactics to fortify crop resilience against drought (Ashe et al., 2021).

An essential evolutionary mechanism that forces plant's environmental adaption is natural selection. Natural selection can influence both genetic and epigenetic differences in drought-prone habitats, hence influencing the features that provide advantages in survival. When genetic variety is scarce or quick adaptation is required, the interaction between natural selection and epigenetic variation is very important. Research has demonstrated that epigenetic changes can affect the heritability of features beneficial in drought-prone environments. The studies on epimutation accumulation lines in rice have shown that drought stress can result in stable epimutations that improve osmotic stress resistance in offspring. These investigations showed that plants exposed to drought produced progeny that were more drought tolerant than their parent lines, demonstrating the contribution of epigenetic variation to adaptive evolution (Spadafora, 2023).

Conservation biology and the agriculture industry stand to gain a great benefit from our growing understanding of epigenetic variation and its function in natural selection. The knowledge of how epigenetic mechanisms may aid in adaptation in vulnerable plant populations might help guide management tactics in the field of conservation. Conservationists can prioritize protecting environments that support epigenetic variety, which strengthens plant species' ability to adapt to environmental shocks. Breeders can make selective choices for features that enhance crop production in water limited environments by identifying important epigenetic markers linked to drought tolerance. Furthermore, more focused management techniques are made possible by the incorporation of epigenetic insights into precision agriculture. Farmers can improve crop resilience through the optimization of irrigation systems and other management practices by comprehending the epigenetic responses of crops to certain environmental situations. This method helps to maintain sustainable farming practices in the face of climate change while also conserving water resources. The significance of epigenetic variation will be imperative in formulating efficacious conservation tactics and enhancing crop resilience, given the persistent obstacles that climate change presents to plant species (Lamka et al., 2022).

CONCLUSION

Epigenetic mechanisms, includes DNA methylation, histone modifications, and small RNAs, are crucial in regulating plant drought tolerance by altering gene expression. These modifications enable plants to adapt to stress and potentially "remember" previous droughts, offering a promising avenue for developing drought-resilient crops through breeding and biotechnology.

REFERENCES

Abdulraheem, M. I., Xiong, Y., Moshood, A. Y., Cadenas-Pliego, G., & Zhang, H. in Hu, J. (2024). Mechanisms of Plant Epigenetic Regulation in Response to Plant Stress: Recent Discoveries and Implications. *Plants, 13*(2), 163. https://www.mdpi.com/2223-7747/13/2/163

Adel, S. (2023). Plant tolerance to drought stress with emphasis on wheat. *Plants, 12*(11), 2170. PMID: 37299149

Agius, D. R., Kapazoglou, A., Avramidou, E., Baranek, M., Carneros, E., Caro, E., & Ebejer, J.-P. (2023). Exploring the crop epigenome: A comparison of DNA methylation profiling techniques. *Frontiers in Plant Science, 14*, 1181039. PMID: 37389288

Angers, B., & Castonguay, E. in Massicotte, R. (2010). Environmentally induced phenotypes and DNA methylation: How to deal with unpredictable conditions until the next generation and after. *Molecular Ecology, 19*(7), 1283–1295. PMID: 20298470

Arora, I. (2021). Computational methods and next-generation sequencing approaches to analyze epigenetics data: Profiling of methods and applications. *Methods (San Diego, Calif.), 187*, 92–103. PMID: 32941995

Arshad, M., Feyissa, B. A., Amyot, L., & Aung, B. in Hannoufa, A. (2017). MicroRNA156 improves drought stress tolerance in alfalfa (Medicago sativa) by silencing SPL13. *Plant Science, 258*, 122–136. PMID: 28330556

Ashapkin, V. V., Kutueva, L. I., & Aleksandrushkina, N. I. in Vanyushin, B. F. (2020). Epigenetic mechanisms of plant adaptation to biotic and abiotic stresses. *International Journal of Molecular Sciences, 21*(20), 7457. PMID: 33050358

Ashe, A., & Colot, V. (2021). How does epigenetics influence the course of evolution? V (št. 376, str. 20200111): The Royal Society.

Banerjee, A. (2017). Epigenetic regulation during salinity and drought stress in plants: Histone modifications and DNA methylation. *Plant Gene, 11*, 199–204.

Becker, C., Hagmann, J., Müller, J., Koenig, D., Stegle, O., & Borgwardt, K. in Weigel, D. (2011). Spontaneous epigenetic variation in the Arabidopsis thaliana methylome. *Nature, 480*(7376), 245–249. PMID: 22057020

Bhargava, S. (2013). Drought stress adaptation: Metabolic adjustment and regulation of gene expression. *Plant Breeding, 132*(1), 21–32.

Bhattacharjee, S. (2016). *Drought stress tolerance in plants* (prevajalec, Trans.). Springer.

Bordiya, Y. (2017). Epigenetic regulation of the defense gene induction in Arabidopsis thaliana in response to Pseudomonas syringae.

Cao, L., Lu, X., Wang, G., Zhang, P., Fu, J., Wang, Z., & Wang, T. (2021). Transcriptional regulatory networks in response to drought stress and rewatering in maize (Zea mays L.). *Molecular Genetics and Genomics*, •••, 1–17. PMID: 34601650

Çelik, Ö., Meriç, S., & Ayan, A. in Atak, Ç. (2019). Epigenetic analysis of WRKY transcription factor genes in salt stressed rice (Oryza sativa L.) plants. *Environmental and Experimental Botany*, *159*, 121–131.

Chinnusamy, V., & Zhu, J. (2007). Cold stress regulation of gene expression in plants. *Trends in Plant Science*, *12*(10), 444–451. PMID: 17855156

Dahl, C. (2003). DNA methylation analysis techniques. *Biogerontology*, *4*, 233–250. PMID: 14501188

Deng, F., Zhang, X., Wang, W., & Yuan, R. in Shen, F. (2018). Identification of Gossypium hirsutum long non-coding RNAs (lncRNAs) under salt stress. *BMC Plant Biology*, *18*, 1–14. PMID: 29370759

Desaulniers, D., Vasseur, P., Jacobs, A., Aguila, M. C., & Ertych, N. in Jacobs, M. N. (2021). Integration of Epigenetic Mechanisms into Non-Genotoxic Carcinogenicity Hazard Assessment: Focus on DNA Methylation and Histone Modifications. *International Journal of Molecular Sciences, 22*(20), 10969. https://www.mdpi.com/1422-0067/22/20/10969

Ding, Y., & Avramova, Z. (2011). The Arabidopsis trithorax-like factor ATX1 functions in dehydration stress responses via ABA-dependent and ABA-independent pathways. *The Plant Journal*, *66*(5), 735–744. PMID: 21309869

Ding, Y., & Tao, Y. (2013). Emerging roles of microRNAs in the mediation of drought stress response in plants. *Journal of Experimental Botany*, *64*(11), 3077–3086. PMID: 23814278

Drongitis, D., Rainone, S., Piscopo, M., Viggiano, E., Viggiano, A., De Luca, B., & Donizetti, A. (2016). Epigenetics and cortical spreading depression: Changes of DNA methylation level at retrotransposon sequences. *Molecular Biology Reports*, *43*, 755–760. PMID: 27169424

Eichten, S. R. (2015). Minimal evidence for consistent changes in maize DNA methylation patterns following environmental stress. *Frontiers in Plant Science*, *6*, 308. PMID: 25999972

Fang, H., Liu, X., Thorn, G., & Duan, J. in Tian, L. (2014). Expression analysis of histone acetyltransferases in rice under drought stress. *Biochemical and Biophysical Research Communications*, *443*(2), 400–405. PMID: 24309107

Forestan, C., Farinati, S., Lunardon, A. in Varotto, S. (2018). Integrating transcriptome and chromatin landscapes for deciphering the epigenetic regulation of drought response in maize. *The Maize Genome*, 97-112.

Forestan, C., Farinati, S., Zambelli, F., Pavesi, G., & Rossi, V. in Varotto, S. (2020). Epigenetic signatures of stress adaptation and flowering regulation in response to extended drought and recovery in Zea mays. *Plant, Cell & Environment*, *43*(1), 55–75. PMID: 31677283

Gao, L., Emperle, M., Guo, Y., Grimm, S. A., Ren, W., Adam, S., & Yin, J. (2020). Comprehensive structure-function characterization of DNMT3B and DNMT3A reveals distinctive de novo DNA methylation mechanisms. *Nature Communications*, *11*(1), 3355. PMID: 32620778

Gayacharan (2013). Epigenetic responses to drought stress in rice (Oryza sativa L.). *Physiology and Molecular Biology of Plants : an International Journal of Functional Plant Biology*, *19*, 379–387. PMID: 24431506

Gibney, E. R. (2010). Epigenetics and gene expression. *Heredity*, *105*(1), 4–13. DOI: 10.1038/hdy.2010.54 PMID: 20461105

González-Benito, M. E., Ibáñez, M. Á., Pirredda, M., & Mira, S. in Martín, C. (2020). Application of the MSAP technique to evaluate epigenetic changes in plant conservation. *International Journal of Molecular Sciences*, *21*(20), 7459. PMID: 33050382

Hasibuzzaman, A. S. M., Akter, F., Bagum, S. A., Hossain, N., Akter, T. in Uddin, M. S. (2021). Morpho-physiological mechanisms of maize for drought tolerance. *Plant Stress Physiology, 229.*

Hauser, M.-T., Aufsatz, W., & Jonak, C. (1809). in Luschnig, C. (2011). Transgenerational epigenetic inheritance in plants. *Biochimica et Biophysica Acta (BBA)-. Gene Regulatory Mechanisms*, (8), 459–468.

Hill, D., Nelson, D., & Hammond, J. in Bell, L. (2021). Morphophysiology of potato (Solanum tuberosum) in response to drought stress: Paving the way forward. *Frontiers in Plant Science*, *11*, 597554. PMID: 33519850

Holliday, R. (2006). Epigenetics: A historical overview. *Epigenetics, 1*(2), 76–80. PMID: 17998809

Hu, J., & Xu, T. (2024). Crosstalk between RNA m6A modification and epigenetic factors for gene regulation in plants. *Plant Communications.*

Jablonka, E. (2002). The changing concept of epigenetics. *Annals of the New York Academy of Sciences, 981*(1), 82–96. PMID: 12547675

Jiang, N., Cui, J., Shi, Y., Yang, G., Zhou, X., Hou, X., & Luan, Y. (2019). Tomato lncRNA23468 functions as a competing endogenous RNA to modulate NBS-LRR genes by decoying miR482b in the tomato-Phytophthora infestans interaction. *Horticulture Research,* •••, 6. PMID: 30729018

Jiao, P., Ma, R., Wang, C., Chen, N., Liu, S., Qu, J., & Ma, Y. (2022). Integration of mRNA and microRNA analysis reveals the molecular mechanisms underlying drought stress tolerance in maize (Zea mays L.). *Frontiers in Plant Science, 13,* 932667. PMID: 36247625

Jiménez-Arias, D., Borges, A. A., Luis, J. C., Valdés, F., & Sandalio, L. M. in Pérez, J. A. (2015). Priming effect of menadione sodium bisulphite against salinity stress in Arabidopsis involves epigenetic changes in genes controlling proline metabolism. *Environmental and Experimental Botany, 120,* 23–30.

KAMRAN, K., KHAN, B. A., RASHEED, A., HASSAN, M. U., HASHEM10, A., MECHRI11, M. in ABD ALLAH12, E. F. Battling arid adversity: unveiling the resilience of cotton in the face of drought and innovative mitigation approaches.

Kaufmann, K., Muino, J. M., Østerås, M., Farinelli, L., & Krajewski, P. in Angenent, G. C. (2010). Chromatin immunoprecipitation (ChIP) of plant transcription factors followed by sequencing (ChIP-SEQ) or hybridization to whole genome arrays (ChIP-CHIP). *Nature Protocols, 5*(3), 457–472. PMID: 20203663

Kazimierczyk, M (2021). Long non-coding RNA epigenetics. *International Journal of Molecular Sciences, 22*(11), 6166. PMID: 34200507

Kiefer, J. C. (2007). Epigenetics in development. *Developmental Dynamics, 236*(4), 1144–1156. PMID: 17304537

Kooke, R., Johannes, F., Wardenaar, R., Becker, F., Etcheverry, M., Colot, V., & Keurentjes, J. J. (2015). Epigenetic basis of morphological variation and phenotypic plasticity in Arabidopsis thaliana. *The Plant Cell, 27*(2), 337–348. PMID: 25670769

Kumar, S., Beena, A., & Awana, M. in Singh, A. (2017). Physiological, biochemical, epigenetic and molecular analyses of wheat (Triticum aestivum) genotypes with contrasting salt tolerance. *Frontiers in Plant Science*, *8*, 1151. PMID: 28713411

Kumar, S., & Seem, K. in Mohapatra, T. (2023). Biochemical and epigenetic modulations under drought: Remembering the stress tolerance mechanism in rice. *Life (Chicago, Ill.)*, *13*(5), 1156. PMID: 37240801

Kurdyukov, S. (2016). DNA methylation analysis: Choosing the right method. *Biology (Basel)*, *5*(1), 3. PMID: 26751487

Lamka, G. F., Harder, A. M., Sundaram, M., Schwartz, T. S., Christie, M. R., & DeWoody, J. A. in Willoughby, J. R. (2022). Epigenetics in ecology, evolution, and conservation. *Frontiers in Ecology and Evolution*, *10*, 871791.

Lennartsson, A. (2009). Histone modification patterns and epigenetic codes. *Biochimica et biophysica acta (BBA)-general subjects, 1790*(9), 863-868.

Li, S., Yu, X., Lei, N., Cheng, Z., Zhao, P., He, Y., & Peng, M. (2017). Genome-wide identification and functional prediction of cold and/or drought-responsive lncRNAs in cassava. *Scientific Reports*, *7*(1), 45981. PMID: 28387315

Li, Y. (2021). Modern epigenetics methods in biological research. *Methods (San Diego, Calif.)*, *187*, 104–113. PMID: 32645449

Liu, E., Xu, L., Luo, Z., Li, Z., Zhou, G., Gao, H., & Zhou, Z. (2023). Transcriptomic analysis reveals mechanisms for the different drought tolerance of sweet potatoes. *Frontiers in Plant Science*, *14*, 1136709. PMID: 37008495

Liu HanHua, L. H., Tian Xin, T. X., Li YanJie, L. Y., Wu ChangAi, W. C. in Zheng ChengChao, Z. C. (2008). Microarray-based analysis of stress-regulated microRNAs in Arabidopsis thaliana.

Liu, N. (2015). RNA epigenetics. *Translational Research; the Journal of Laboratory and Clinical Medicine*, *165*(1), 28–35. PMID: 24768686

Lu, Q., Guo, F., & Xu, Q. in Cang, J. (2020). LncRNA improves cold resistance of winter wheat by interacting with miR398. *Functional Plant Biology*, *47*(6), 544–557. PMID: 32345432

Maeji, H. (2018). Epigenetic mechanisms in plants (prevajalec, Trans.). V *Advances in botanical research* (št. 88, str. 21-47). Elsevier.

Mahmood, T., Khalid, S., Abdullah, M., Ahmed, Z., Shah, M. K. N., & Ghafoor, A. in Du, X. (2019). Insights into drought stress signaling in plants and the molecular genetic basis of cotton drought tolerance. *Cells*, *9*(1), 105. PMID: 31906215

Mattick, J. S. (2009). The genetic signatures of noncoding RNAs. *PLOS Genetics*, *5*(4), e1000459. PMID: 19390609

Meyer, P. (2015). Epigenetic variation and environmental change. *Journal of Experimental Botany*, *66*(12), 3541–3548. PMID: 25694547

Mishra, P., Singh, P., Rai, A., Abhishek, K., Shanmugam, V., Aamir, M., & Singh, S. K. (2022). *Abiotic stress-mediated transcription regulation, chromatin dynamics, and gene expression in plants: Arabidopsis as a role model (prevajalec, Trans.). V Mitigation of Plant Abiotic Stress by Microorganisms (str.* Elsevier.

Munshi, A., Ahuja, Y. (2015). Epigenetic mechanisms in plants: an overview. *Plant biology and biotechnology: volume II: plant genomics and biotechnology*, 265-278.

Nguyen, N. H., & Vu, N. T. (2022). Transcriptional stress memory and transgenerational inheritance of drought tolerance in plants. *International Journal of Molecular Sciences*, *23*(21), 12918. PMID: 36361708

Obidiegwu, J. E., Bryan, G. J., & Jones, H. G. in Prashar, A. (2015). Coping with drought: Stress and adaptive responses in potato and perspectives for improvement. *Frontiers in Plant Science*, *6*, 542. PMID: 26257752

Rajendran, G., Shanmuganandam, K., Bendre, A., Mujumdar, D., & Goel, A. in Shiras, A. (2011). Epigenetic regulation of DNA methyltransferases: DNMT1 and DNMT3B in gliomas. *Journal of Neuro-Oncology*, *104*, 483–494. PMID: 21229291

Ramírez, D., Rolando, J., Yactayo, W., Monneveux, P., & Mares, V. in Quiroz, R. (2015). Improving potato drought tolerance through the induction of long-term water stress memory. *Plant Science*, *238*, 26–32. PMID: 26259171

Rice, J. C. (2001). Histone methylation versus histone acetylation: New insights into epigenetic regulation. *Current Opinion in Cell Biology*, *13*(3), 263–273. PMID: 11343896

Rossetto, D., & Avvakumov, N. (2012). Histone phosphorylation: A chromatin modification involved in diverse nuclear events. *Epigenetics*, *7*(10), 1098–1108. PMID: 22948226

Saleh, A., & Alvarez-Venegas, R. (2008). Dynamic and stable histone H3 methylation patterns at the Arabidopsis FLC and AP1 loci. *Gene*, *423*(1), 43–47. PMID: 18638531

Shen, Q., Lin, Y., & Li, Y. in Wang, G. (2021). Dynamics of H3K27me3 modification on plant adaptation to environmental cues. *Plants*, *10*(6), 1165. PMID: 34201297

Sheoran, S., Kaur, Y., Kumar, S., Shukla, S., & Rakshit, S. in Kumar, R. (2022). Recent advances for drought stress tolerance in maize (Zea mays L.): Present status and future prospects. *Frontiers in Plant Science, 13*, 872566. PMID: 35707615

Shilatifard, A. (2006). Chromatin modifications by methylation and ubiquitination: Implications in the regulation of gene expression. *Annual Review of Biochemistry, 75*(1), 243–269. PMID: 16756492

Spadafora, C. (2023). The epigenetic basis of evolution. *Progress in Biophysics and Molecular Biology, 178*, 57–69. PMID: 36720315

Sun, C., Ali, K., Yan, K., Fiaz, S., Dormatey, R., & Bi, Z. in Bai, J. (2021). Exploration of epigenetics for improvement of drought and other stress resistance in crops: A review. *Plants, 10*(6), 1226. PMID: 34208642

Thirunavukkarasu, N., Hossain, F., Arora, K., Sharma, R., Shiriga, K., Mittal, S., & Rani, T. S. (2014). Functional mechanisms of drought tolerance in subtropical maize (Zea mays L.) identified using genome-wide association mapping. *BMC Genomics, 15*, 1–12. PMID: 25539911

Troyee, A. N., Peña-Ponton, C., Medrano, M., & Verhoeven, K. J. in Alonso, C. (2023). Herbivory induced methylation changes in the Lombardy poplar: A comparison of results obtained by epiGBS and WGBS. *PLoS One, 18*(9), e0291202. PMID: 37682835

Umer, M. (2013). Deciphering the epigenetic code: An overview of DNA methylation analysis methods. *Antioxidants & Redox Signaling, 18*(15), 1972–1986. PMID: 23121567

Villota-Salazar, N. A., & Mendoza-Mendoza, A. (2016). Epigenetics: From the past to the present. *Frontiers in Life Science, 9*(4), 347–370.

Wang, B., Zhang, M., Fu, R., Qian, X., Rong, P., Zhang, Y., & Wang, D. (2016). Epigenetic mechanisms of salt tolerance and heterosis in Upland cotton (Gossypium hirsutum L.) revealed by methylation-sensitive amplified polymorphism analysis. *Euphytica, 208*, 477–491.

Wang, W.-S., Pan, Y.-J., Zhao, X.-Q., Dwivedi, D., Zhu, L.-H., Ali, J., & Li, Z.-K. (2011). Drought-induced site-specific DNA methylation and its association with drought tolerance in rice (Oryza sativa L.). *Journal of Experimental Botany, 62*(6), 1951–1960. PMID: 21193578

Wang, Y., Luo, X., Sun, F., Hu, J., Zha, X., & Su, W. in Yang, J. (2018). Overexpressing lncRNA LAIR increases grain yield and regulates neighbouring gene cluster expression in rice. *Nature Communications, 9*(1), 3516. PMID: 30158538

Weinhold, B. (2006). *Epigenetics: the science of change. V.* National Institute of Environmental Health Sciences.

Yadav, C. B., Pandey, G., Muthamilarasan, M. in Prasad, M. (2018). Epigenetics and epigenomics of plants. *Plant Genetics and Molecular Biology*, 237-261.

Yaish, M. W., & Colasanti, J. (2011). The role of epigenetic processes in controlling flowering time in plants exposed to stress. *Journal of Experimental Botany*, *62*(11), 3727–3735. PMID: 21633082

Yajnik, K. N., & Singh, I. K. (2024). lncRNAs and epigenetics regulate plant's resilience against biotic stresses. *Plant Physiology and Biochemistry*, 108892.

Zhai, J., Liu, J., Liu, B., Li, P., Meyers, B. C., & Chen, X. in Cao, X. (2008). Small RNA-directed epigenetic natural variation in Arabidopsis thaliana. *PLOS Genetics*, *4*(4), e1000056. PMID: 18437202

Zhang, H., & Lang, Z. (2018). Dynamics and function of DNA methylation in plants. *Nature Reviews. Molecular Cell Biology*, *19*(8), 489–506. PMID: 29784956

Zhao, T., & Zhan, Z. (2019). Histone modifications and their regulatory roles in plant development and environmental memory. *Journal of Genetics and Genomics*, *46*(10), 467–476. PMID: 31813758

Chapter 4

Role of Chromatin Remodeling Complexes in Plant Stress Adaptations

Rooma Waqar
https://orcid.org/0009-0009-1654-2860
Quaid-i-Azam University, Islamabad, Pakistan

Muhammad Kaleem
https://orcid.org/0009-0003-6548-3920
Quaid-i-Azam University, Islamabad, Pakistan

Banzeer Ahsan Abbasi
https://orcid.org/0000-0003-3036-4481
Rawalpindi Women University, Pakistan

Shumaila Ijaz
Shenzhen University, China

Abdul Samad Mumtaz
https://orcid.org/0000-0002-7034-9203
Quaid-i-Azam University, Islamabad, Pakistan

Zakir Ullah
Quaid-i-Azam University, Islamabad, Pakistan

Zahid Ullah
University of Swat, Pakistan

Ahmad Ali
University of Swat, Pakistan

ABSTRACT

Chromatin remodeling complexes are crucial for plant stress adaptation, dynamically modulating chromatin structure and regulating stress-responsive genes. Complexes like SWR1, SWI2/SNF2, and CHD family remodelers coordinate mechanisms such as histone variant deposition, histone modifications, DNA methylation, and nucleosome positioning to fine-tune gene expression. Under stress, they facilitate rapid

DOI: 10.4018/979-8-3693-8648-4.ch004

transcriptional reprogramming of defense-related genes, with examples including SWR1-mediated H2A.Z deposition and SWI2/SNF2 remodelers like BRM and SYD modulating key defense genes. CHD remodelers, such as CHR5, enhance immune gene expression, while factors like BRHIS1 and DDM1 regulate basal defense gene activity under non-stress conditions. These complexes integrate environmental signals, supporting plant resilience and offering potential for crop improvement through genetic and epigenetic strategies.

INTRODUCTION

In eukaryotes, genomic DNA is intricately organized within a highly structured chromatin framework, with nucleosomes serving as its fundamental repeating units. A canonical nucleosome comprises a 147-bp segment of DNA coiled around a histone octamers consisting of two copies each of histones H3, H2B, H2A, and H4. The structure of chromatin directly influences critical cellular processes, including the maintenance of genome integrity, and the regulation of gene expression. These processes are tightly associated with epigenetic regulatory mechanisms such as, histone variants, histone modifications, DNA methylation, and chromatin remodelling (Chang et al., 2020). As sessile organisms, plants rely on precise gene expression regulation and genome integrity to adapt to their environment and ensure survival under various stress conditions. While the roles of histone variants DNA methylation, and histone modifications in plant stress responses have been extensively reviewed (Chang et al., 2020). Chromatin remodelling involves the dynamic compaction and relaxation of chromatin, enabling processes such as, DNA repair, DNA replication and selective gene expression. This remodelling is driven by ATP-dependent nucleosome repositioning, orchestrated by chromatin remodelling factors (CHRs), which are ATPases from the Snf2 family (Hopfner, Gerhold, Lakomek, & Wollmann, 2012, Ahmad et al., 2022a).

Conserved RecA-like domains are characteristics of Snf2 family ATPases, which function as the catalytic "engine" of the chromatin remodelling machinery in eukaryotes (Flaus, Martin, Barton, & Owen-Hughes, 2006). Additionally, flanking domains can be used to categorise members of this family (Clapier & Cairns, 2009). Members of the Snf2 family have been categorised into 24 subfamilies in eukaryotes based on full-length alignment of their ATPase/helicase-like regions. These subfamilies are then further divided into six functional groupings (Flaus et al., 2006). The evolutionary relationships and domain structures of the 41, 40, and 45 Snf2 family genes encoded by the genomes of *Arabidopsis thaliana* and *Oryza sativa* (rice), have been thoroughly studied (Knizewski, Ginalski, & Jerzmanowski, 2008).

The precise contributions of Chromatin remodeling factors (CHRs) to plant stress responses are still poorly understood, despite the fact that numerous recent studies have highlighted the involvement of CHRs in plant growth and development (S. Singh et al., 2020). This chapter highlights the key role that CHRs play in stress adaptation by concentrating on important developments in our knowledge of how they operate during plant responses to stress and analysing how they interact with other epigenetic mechanisms. List of the four major chromatin remodeler families is presented in Table 1. Figure 1 illustrates chromatin remodeling in plants in response to environmental stressors.

Table 1. A list of the four major chromatin remodeler families along with their structural details

Chromatin Remodelers (Family)	Subunits	Domains	References
Inositol requiring 80	Tip60/p400, INO80	HELICc	(Clapier & Cairns, 2009)
Chromodomain, Helicase, DNA binding (CHD)	HELICc, Chromo, DEx,	CHD1, CHD3, CHD2, CHD9, CHD4, NuRD subunits	(Boyer, Shao, Ebright, & Peterson, 2000)Boyer 2000
Imitation switch (ISWI)	CHRAC, NoRC, WICH, b-WICH	HELICc, DExx, SANT, HAND, SLIDE	(Boyer et al., 2000)
Switching defective/Sucrose nonfermenting (SWI/SNF)	BAF, PBAF	HELICc, DExx HSA, Bromo	(Peterson & Workman, 2000)

Figure 1. Chromatin Remodeling in response to environmental stressors in plant

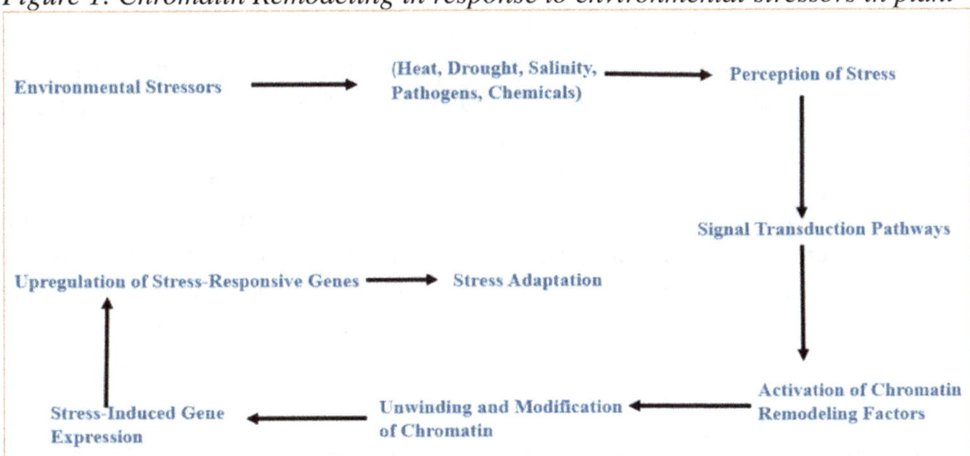

BIOTIC STRESS

Plants have evolved advanced immune mechanisms to defend against biotic stress caused by pathogenic microorganisms. These mechanisms include pathogen-associated molecular pattern (PAMP)-triggered immunity (PTI) and effector-triggered immunity (ETI). Both PTI and ETI are activated to reprogram the transcription of defense-related genes, enhancing plant survival under pathogen attack (Ramirez-Prado, Abulfaraj, Rayapuram, Benhamed, & Hirt, 2018). In normal growth conditions, however, plant immune responses and defense gene transcription are repressed. This repression involves chromatin remodelling complexes such as the SWR1 complex and the SWI2/SNF2 complex. Under normal growing conditions, *Arabidopsis* mutants like hta9, hta11 (which lack two of the three H2A.Z genes), and pie1 exhibit constitutive expression of defense-related genes in addition to increased resistance to *Pseudomonas syringae* (March-Díaz et al., 2008, Ahmad et al., 2022b). Unlike SWR1, which regulates histone variants, BRM interacts with histone modifications, indicating that some defense genes may be co-regulated by multiple chromatin remodellers through histone modifications and variant deposition. In rice, BRHIS1 (also called CHR725), an ortholog of *Arabidopsis* SNF2-RING-HELICASE LIKE2 (FRG2), suppresses innate immunity by binding to mono-ubiquitinated H2A and H2B histone variants near defense gene promoters. While mono-ubiquitination enhances the expression of these genes, BRHIS1 binding limits further mono-ubiquitination, maintaining basal expression levels in healthy rice plants (X. Li, Jiang, Ji, Liu, & Zhang, 2015). This mechanism illustrates how chromatin remodelers like BRHIS1 antagonize transcriptional activators to repress gene expression. The SWI2/SNF2 subfamily member SPLAYED (SYD) represses SNC1 transcription by counteracting its transcriptional activators. Loss-of-function syd mutants show increased resistance to *P. syringae* but heightened susceptibility to *Botrytis cinerea* (Johnson, Xia, Feng, & Li, 2015). Similarly, DDM1, which regulates DNA methylation, suppresses SNC1 expression, as demonstrated by reduced DNA methylation upstream of SNC1 in ddm1 mutants (Y. Li, Tessaro, Li, & Zhang, 2010)

In contrast to SYD and DDM1, the *Arabidopsis* CHD1 subfamily protein CHR5 positively regulates immune responses, including SNC1 activation. The chr5 mutant is more susceptible to bacterial pathogens and exhibits globally higher nucleosome occupancy near transcription start sites (TSS), which correlates with reduced transcription rates (Zou et al., 2017). However, nucleosome occupancy near SNC1 remains similar between WT and chr5 mutants (Zou et al., 2017). CHR5 also reduces H3K27me3 levels to modulate gene expression (Shen, Devic, Lepiniec, & Zhou, 2015). Thus, SNC1 transcription appears to be co-regulated by various chromatin remodellers through mechanisms such as DNA methylation, histone modifications, and the inhibition of transcriptional activators.

ROLES OF CHROMATIN REMODELING FACTORS IN GENE EXPRESSION CHANGES TRIGGERED BY ABIOTIC STRESS

Abiotic stressors like excessive heat, drought, salt, and soil contamination with chemicals, as well as biotic stressors like pathogen infections, are all examples of environmental stressors. When plants encounter stress, they rapidly reprogram gene expression to adapt to adverse conditions. Furthermore, as stress conditions are often chronic or recurrent, plants frequently acquire increased tolerance to subsequent stress exposures (Zhu, 2016, Ahmad et al., 2024a)). Figure 2 presents a list of stress priming in various plant species.

Table 2. List of stress priming in different species of plants

Treatment/Stress	Target Species	Result	References
Dehydration	*Zea mays*	Enhancement of water retention	(Virlouvet & Fromm, 2015)
Drought	*Triticum aestivum*	Improved grain filling under drought conditions	(Wang, Vignjevic, Jiang, Jacobsen, & Wollenweber, 2014)
Methyl jasmonate	*Nicotiana sylvestris*	Quick nicotine accumulation	(Baldwin & Schmelz, 1996)
ABA	*Arabidopsis thaliana*	Enhanced sensitivity of stomatal opening in response to light	(Goh, Nam, & Park, 2003)
Dehydration	*Arabidopsis thaliana*	Enhanced ability to retain water	(Ding et al., 2014)
Salt	*Solanum lycopersicum*	Improved resilience to salt stress.	(Cayuela, Pérez-Alfocea, Caro, & Bolarin, 1996)

ENVIRONMENTAL TEMPERATURE STRESS

Rising global temperatures have made heat stress (HS) a significant challenge to plant growth and development. To counter HS, plants have developed two key protective strategies: basal thermotolerance and acquired thermotolerance. Basal thermotolerance offers rapid but limited defense against immediate heat exposure. In contrast, acquired thermotolerance is triggered by a mild heat stress event, priming plants for enhanced tolerance to subsequent heat exposure. This acquired thermotolerance is maintained as heat stress memory, mediated by epigenetic mechanisms (Friedrich, Faivre, Bäurle, & Schubert, 2019, Ahmad et al., 2024b, Ali et al., 2022). Heat stress memory involves a transcriptional memory component, enabling sustained and enhanced re-induction of gene expression during recurrent heat stress episodes

(Lämke, Brzezinka, Altmann, & Bäurle, 2016). Remarkably, this memory can persist for extended periods and even be transmitted to progeny without stress exposure. Thermotolerance mechanisms rely on the activation of heat stress-responsive genes, particularly those that produce heat shock proteins (HSPs). These proteins safeguard cellular functions by maintaining proper protein folding and preventing the clumping of misfolded proteins (Friedrich et al., 2019, Ijaz et al., 2023).

Chromatin remodelling is necessary for heat-induced transcription, especially at promoter regions close to transcription start sites (TSS). After heat stress, members of the ISWI subfamily proteins, such as CHR11 and CHR17, and SWI2/SNF2 subfamily members, such as BRAHMA (BRM), help maintain the expression of genes linked to HS memory, such as (HSA32) (Brzezinka et al., 2016). Furthermore, chromatin memory is mediated by the histone H3-binding protein FORGETTER1 (FGT1), which recruits ISWI and BRM proteins to maintain low nucleosome occupancy close to the target HS memory genes' TSS. According to (Brzezinka et al., 2016) mthis recruitment facilitates the long-term activation of HS memory genes. Through its recruitment via the H3K27me3 demethylase RELATIVE OF EARLY FLOWERING6 (REF6), BRM is also essential for the transgenerational activation of HS memory genes (J. Liu et al., 2019).

The Lsh1 subfamily CHR protein DECREASE IN DNA METHYLATION 1 (DDM1) inhibits heat stress-induced epigenetic memory, in contrast to BRM's function in preserving HS memory. According to (Iwasaki & Paszkowski, 2014), the ddm1 mom1 double mutant and the ddm1 single mutant both show extended transcriptional activation of heterochromatic loci in response to HS, with the double mutant passing on this hyperactivation to offspring. Although it is unclear if BRM and DDM1 target the same loci, their opposing roles demonstrate how chromatin remodelling and other epigenetic processes work together to control transgenerational memory. To control basal thermotolerance, chromatin remodelling also interacts with histone modifications and variations.

For example, the SWR1 complex makes it easier for the variation H2A.Z to replace canonical histone H2A, guaranteeing that HS response genes like HSP70 are expressed correctly (Kumar and Wigge, 2010). Furthermore, interactions between the SWI2/SNF2 complex and HISTONE DEACETYLASE 2C (HD2C) show how chromatin remodelling and histone acetylation interact with HS reactions (Buszewicz et al., 2016), Chromatin remodelling affects thermomorphogenesis, a developmental reaction to moderately high temperatures, in addition to thermotolerance. For instance, by repressing H3K27me3 deposition and activating auxin-related genes, the Mi-2/CHD3 subfamily CHR protein PICKLE (PKL) encourages hypocotyl elongation (Zha, Jing, Xu, & Lin, 2017). Table 3 presents chromatin dynamics in plants in response to abiotic stresses.

Table 3. Chromatin dynamics in response to abiotic stresses like salinity, drought and cold in plants

Stress	Species	Genes Involved	Chromatin Modifications	References
Salinity	*Arabidopsis thaliana*	IBR5, MKP1, PTP1, PHS1, DsPTP1	H2Bub	(Zhou, Chen, Sun, & Li, 2017)
Salinity	*Glycine max*	Glyma11g02400, Glyma08g41450	H3K4me3,	(Feng et al., 2016)
Salinity	*Ricinus communis*	RSM1	H3K4/27me3	63(Han et al., 2020)
Drought	*Arabidopsis thaliana*	RD29a, AtGOLS2 RD20, ProDH	H3K9ac, H3K4me3	48(Kim et al., 2012)
Drought	*Arabidopsis thaliana*	ROP6/10/11	H3/H4ac	53(Lee & Seo, 2019)
Drought	*Populus trichocarpa*	PtrNAC006, PtrNAC007, PtrNAC120	H3K9ac	49(S. Li et al., 2019)
Cold	*Brassica rapa*	BramMDH1, Bra4CL2	5-mC in promoter	44(T. Liu, Li, Duan, Huang, & Hou, 2017)
Cold	*Solanum tuberosum*	Cold-responsive genes	H3K4/27me3	43(Zeng et al., 2019)
Cold	*Oryza sativa*	OsOST1 (Os03g0610900)	5-mC in promoter	45(Zeng et al., 2019)
Heat	*Arabidopsis thaliana*	HSFA2	H3K27me3	33(J. Liu et al., 2019)
Heat	*Arabidopsis thaliana*	Alternative splicing related genes	H3K36me3	36(Pajoro, Severing, Angenent, & Immink, 2017)
Heat	*Arabidopsis thaliana*	HSP18.2 and APX2	H3K4me2/3	(Lämke, Brzezinka, Altmann, et al., 2016; Lämke, Brzezinka, & Bäurle, 2016)

FUNCTIONS OF CHROMATIN REMODELING FACTORS IN THE DNA DAMAGE RESPONSE PATHWAY

DNA damage can result from various exogenous factors, including reactive oxygen species (ROS) produced during stress or UV radiation, as well as endogenous issues such as replication fork stalling. Environmental stresses, such as irradiation or soil contamination with excess chemicals, frequently cause DNA strand breaks or replication stress, which trigger DNA repair mechanisms. These mechanisms contribute to growth arrest, enabling plants to manage and repair damage under adverse conditions (Nisa, Huang, Benhamed, & Raynaud, 2019, Ijaz et al., 2024a).

CHROMATIN CHANGES IN SALT STRESS MEMORY

Genome-wide profiling has also uncovered chromatin changes associated with salt stress priming. For instance, a priming dose of mild salt treatment administered 10 days prior to a triggering salt stress leads to enhanced drought tolerance and reduced salt uptake, even though the priming dose does not affect plant morphology. At the chromatin level, this priming is associated with a shortening and fractionation of H3K27me3-enriched regions in the genome (Sani, Herzyk, Perrella, Colot, & Amtmann, 2013, (Nisa, Huang, Benhamed, & Raynaud, 2019, Ijaz et al., 2024b).

NUCLEOSOME OCCUPANCY AND TRANSCRIPTIONAL MEMORY

Heat stress memory is influenced by nucleosome remodelling as well as histone methylation. A putative helicase is encoded by the FORGETTER1 (FGT1) gene, which was found during a search for elements necessary for the long-term activation of memory genes. The SWI/SNF remodeler BRM is one of the chromatin remodelling complexes that this protein interacts with. While not necessary for instantaneous heat shock reactions, FGT1 and BRM are both necessary for heat stress memory (Brzezinka et al., 2016, Haider et al., 2021a). Heat-responsive FGT1 binding is necessary for decreased nucleosome occupancy at transcriptional start sites, which is necessary for the sustained activation of memory genes. Given its great species conservation, FGT1 may play roles other than heat stress memory (Mozgova et al., 2015, Haider et al., 2021b). It is yet unclear how H3K4 hypermethylation, FGT1 function, and other stress memory events are related. Chemical priming investigations have shown similar nucleosome dynamics. For instance, the fasciata 2 mutant has high H3K4me3 levels at primed genes and poor nucleosome occupancy due to a deficiency in the CHROMATIN ASSEMBLY FACTOR-1 (CAF-1) complex. Under non-sterile circumstances, this molecular phenotype causes erroneous activation of primed genes by mimicking constitutive priming (Mozgova et al., 2015).

TRANSCRIPTIONAL MEMORY IN DESICCATION STRESS

Desiccation stress in *A. thaliana* provides another example of transcriptional memory. Genes that respond differently to repeated desiccation stress compared to the initial exposure fall into distinct classes. These include genes with enhanced re-induction (*e.g., RD29B* and *RAB18*), enhanced repression or loss of repression (Ding et al., 2013). Enhanced re-induction is associated with sustained H3K4me3 hypermethylation during recovery, as well as paused transcription elongation mediated by RNA polymerase II. This paused transcription state involves hyperphosphorylation of the polymerase at serine 5 in its C-terminal domain, and the associated modifications persist for 5–7 days(Kim et al., 2012) . ABA signaling is necessary for desiccation memory, although it is insufficient on its own. Cis-regulatory elements such as adjacent ABA-response elements (ABREs) have been identified as essential for memory, and these are bound by ABRE-binding factors (ABFs). However, an *abf2 abf3 abf4* triple mutant showed compromised induction of memory genes without a complete loss of memory, suggesting that transcriptional induction and memory are mediated by distinct factors. While ABA treatment can prime stress memory, it is insufficient as a triggering cue to induce transcriptional memory (Pien et al., 2008). This indicates that additional, unidentified factors contribute to desiccation memory. In summary, somatic stress memory in plants involves complex interactions between histone modifications, nucleosome remodeling, transcriptional regulation, and signaling pathways. These mechanisms enable plants to adapt and respond more effectively to recurring environmental stresses.

DOUBLE-STRANDED DNA BREAKS (DSBS)

In order for the repair machinery to reach the damaged DNA, the processes involved in DNA damage repair require significant chromatin remodelling. Double-stranded breaks (DSBs) are the most cytotoxic of the several kinds of DNA damage and constitute a serious risk to the integrity of the genome (Nair, Shoaib, & Sørensen, 2017). Genotoxic chemicals and γ-irradiation are frequently used to study DSBs since they cause their creation. According to (Hirakawa & Matsunaga, 2019), the chromatin remodelling component RADIATION SENSITIVE 54 (RAD54) is essential for DSB repair in plants. In Arabidopsis, RAD54 co-localizes with γH2A.X, a phosphorylated histone variation linked to DSB sites and generates DNA repair foci within nuclei after γ-irradiation. RAD51, a protein necessary for homologous recombination (HR) in DNA repair, is necessary for the formation of RAD54 foci (Hirakawa & Matsunaga, 2019). Furthermore, pie1 mutants show comparable sensitivity to γ-irradiation and other DNA-damaging chemicals (Rosa, Von Harder, Aiese Cigliano, Schlögelhofer, & Mittelsten Scheid, 2013). INO80 chromatin remodelling complexes work directly with γH2A.X in yeast (Saccharomyces cerevisiae) to help repair DNA damage (Morrison et al., 2004). It is unknown, though, if γH2A.X directly recruits CHR complexes in plants.

Excess boron, frequently from overfertilization of the soil, can also cause DSBs in plants. Oxidative stress and DSB generation are the main symptoms of boron poisoning (Aquea et al., 2012). Research shows that exposed plants with severe boron stress have higher amounts of BRM protein(Sakamoto et al., 2018). In order to maintain an open chromatin state that destabilises chromatin and worsens oxidative DNA damage, BRM binds widely to acetylated histone residues (Sakamoto et al., 2018). Tolerance to DSBs brought on by boron stress is granted by a partial loss-of-function mutation in BRM (brm-3) (Sakamoto et al., 2018). Therefore, rather than encouraging repair, the presence of CHR proteins like BRM may worsen DSB formation under some circumstances, like as excessive boron exposure. Furthermore, DSB repair is regulated by the non-ATPase component SWI3B of the SWI2/SNF2 complex. When DNA is damaged or repaired, structural maintenance complexes 5 (SMC5) and 6 (SMC6), which normally bind with chromatin, move to DSB sites. In order to aid in DNA repair, SWI3B makes it easier for SMC5 to separate from its original chromatin site. Nevertheless, there is no proof that BRM or any other SWI2/SNF2 complex subunits are connected to this SWI3B function(Jiang et al., 2019).

DNA REPLICATION STRESS

Endogenous or exogenous variables that disrupt DNA synthesis or halt replication forks, such as environmental stress, rapid replication fork progression, or DNA damage, are known as DNA replication stress In order to repair DNA and resume replication, cells activate mechanisms in response to replication stress (Poole et al., 2015). DNA crosslinking is a major problem during replication. Replication fork regression is made possible in Arabidopsis by the DNA translocase activity of the Rad5/16 subfamily member RAD5A. Mitomycin C (MMC), a replication stressor that induces DNA interstrand crosslinking, causes hypersensitivity in the rad5a mutant (Klemm et al., 2017, Haider et al., 2022a). According to (Kang, Zhang, An, Shen, & Zhu, 2019), ino80 mutants also exhibit hypersensitivity to hydroxyurea (HU), which inhibits DNA replication. According to (Rosa et al., 2013), mutants in SWR1 complex subunits (such as pie1-3, arp6-3, and swc6-1) similarly exhibit hypersensitivity to MMC and HU.

HISTONE MODIFICATIONS FLEXIBILITY UNDER STRESS

Histone modifications dynamically regulate chromatin accessibility and gene expression during stress. Stress-responsive genes often exhibit transient changes in histone acetylation and methylation. For instance, drought-responsive genes in Arabidopsis are enriched with activating marks such as H3K4me3 and H3K9ac during stress (Kim et al., 2012). Cold stress induces transcriptional activation of maize chromatin knobs, associated with increased H3K9 acetylation and reduced

H3K9 methylation (X. Li et al., 2015, Haider et al., 2022b). These histone modifications often intersect with DNA methylation changes, highlighting a complex network coordinating genome function in response to environmental challenges. In summary, chromatin remodeling, DNA methylation, and histone modifications constitute an intricate regulatory system that governs plant responses to stress, facilitating adaptation and survival in dynamic environments.

SOMATIC STRESS MEMORY

Most chromatin-related responses to pathogen attacks or abiotic stress are transient, with chromatin modifications (Conrath, 2011) of cold exposure can be retained for weeks or even months after the cold period ends (Bouché, Woods, & Amasino, 2017). In *Arabidopsis thaliana*, vernalization is a process where the FLOWERING LOCUS C (FLC) gene is epigenetically silenced through H3K27me3, the trimethylation of histone H3 at lysine 27. This silencing is facilitated by a polycomb group complex activated by cold and directed to the FLC locus via cis-regulatory elements and non-coding RNAs. While the epigenetic modifications are initiated by cold exposure, they function as a cue for developmental progression (Qüesta, Song, Geraldo, An, & Dean, 2016). Figure 2 shows how chromatin modifications, histone changes, DNA methylation, and non-coding RNAs regulate transcription in response to abiotic stress. Examples of stress memory in both somatic and transgenerational contexts are presented in Table 4.

Figure 2. The interplay among chromatin modifications, histone modifications, DNA methylation and demethylation, and non-coding RNAs drives transcriptional reprogramming in response to abiotic stress

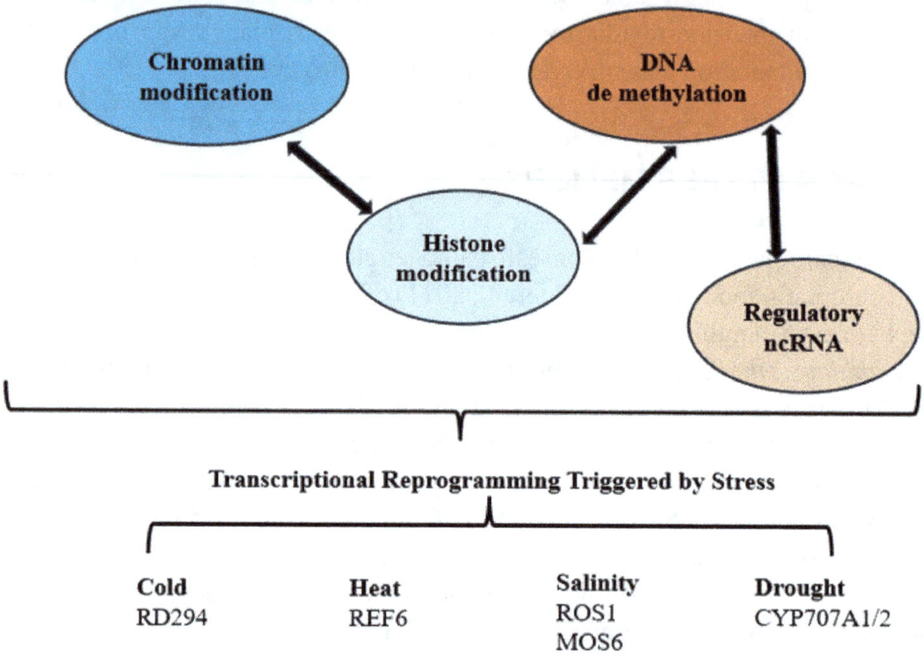

Table 4. Example of stress memory in both somatic and transgenerational contexts

Stress cue	Chromatin marks associated with priming	Maximal duration of memory (as analyzed)	Plant-level effect	References
Heat	Histone occupancy	3 d	Yes	(Brzezinka et al., 2016)
Heat	H3K4me2, H3K4me3	3d	Yes	(Lämke, Brzezinka, Altmann, et al., 2016)
Desiccation	H3K4me3, paused RNA Pol I	5–7 d	Yes	(Ding et al., 2013) Ding 2012
Hyperosmotic	H3K27me3	10d	Yes	(Sani et al., 2013)
Heat, cold, or salt	H3K14ac, H3K4me2, H3K4me3	7d	Yes	(P. Singh et al., 2014)

ROLE OF HISTONE METHYLATION

The earliest evidence connecting chromatin modifications to stress priming was found in systemic acquired resistance (SAR). This process, triggered by bacterial pathogens or salicylic acid (SA) analogs like acibenzolar-S-methyl (BTH), is linked to persistent changes in histone modifications. Specifically, histones were enriched in primed leaves, including those that were not directly exposed to the priming stress. This modification created a transcriptional memory for several loci, allowing enhanced transcriptional responses after a lag period of several days (Jaskiewicz, Conrath, & Peterhänsel, 2011). Genome-wide, H3K4me3 strongly correlates with gene expression, while H3K4me2 appears to mark chromatin in a poised state for transcription (Zhang et al., 2019). These histone modifications are also linked to other stress memory phenomena and defense-related gene priming, although mutations in the transcriptional regulator *HEAT SHOCK FACTOR B1* (HSFB1) impair the priming response (Pick, Jaskiewicz, Peterhänsel, & Conrath, 2012). Heat stress, a highly fluctuating environmental challenge, also triggers stress memory in plants. Sublethal heat exposure primes plants to tolerate subsequent lethal heat stress through mechanisms collectively termed heat shock responses. While these acute responses are conserved across eukaryotes, plants exhibit an additional phenomenon known as heat stress memory. This memory enables plants to maintain acquired thermotolerance for days and involves transcriptional memory characterized by either sustained gene activation or enhanced gene reactivation upon recurring heat stress (Bäurle, 2016). Genes classified as "memory genes" exhibit either sustained activation following an initial priming stress or stronger induction during recurring stress episodes (Lämke, Brzezinka, Altmann, et al., 2016).

Interestingly, not all genes activated by heat are classified as memory genes, suggesting that H3K4 hypermethylation is not merely a byproduct of heat-induced transcription. For instance, although the HSP70 gene is activated during heat stress, it does not exhibit H3K4 hypermethylation (Lämke, Brzezinka, Altmann, et al., 2016, Ullah et al., 2023a). Genetic studies reveal that the transcription factor *HSFA2* plays a crucial role in maintaining H3K4 hypermethylation. In *HSFA2* mutants, memory genes are initially activated but are rapidly silenced, and H3K4 hypermethylation levels are diminished. *HSFA2* operates in a "hit-and-run" mode, binding to target genes early after heat stress and then dissociating within hours. However, it is required for the sustained accumulation of H3K4me2/3, which persists for several days (Charoensawan, Martinho, & Wigge, 2015, Ullah et al., 2023b). The mechanism by which *HSFA2* promotes sustained chromatin modification remains unclear. Epigenetic factors like histone variants, remodelers, DNA modifiers, and non-coding RNAs control gene expression under abiotic stress are illustrated in Figure 3. Examples of chromatin flexibility under to abiotic stress responses presented in Table 5.

Table 5. Examples of chromatin flexibility linked to abiotic stress responses

Plant Species	Stress Type	Chromatin Alterations	References
Pinus silvestris	Ionizing radiation	DNA methylation across the entire genome	(Kovalchuk et al., 2003)
Zea mays	Cold	DNA methylation across the entire genome	(Hu et al., 2012)
Triticum aestivum	Heat	rDNA chromatin de condensation	35
Solanum lycopersicum	Water-deficit	DNA methylation	(González, Ricardi, & Iusem, 2013)
Nicotiana tabacum	Aluminium; Salt Cold; Oxidative	Methylation of the NtGPDL gene in DNA	(Choi & Sano, 2007)
Oryza sativa	Salt	rDNA chromatin de condensation	(Santos, Ferreira, Maroco, & Oliveira, 2011)

Figure 3. Chromosomes store genetic information, and chromatin structure changes regulate gene expression. Epigenetic factors like histone variants, remodelers, DNA modifiers, and non-coding RNAs control gene expression under abiotic stress

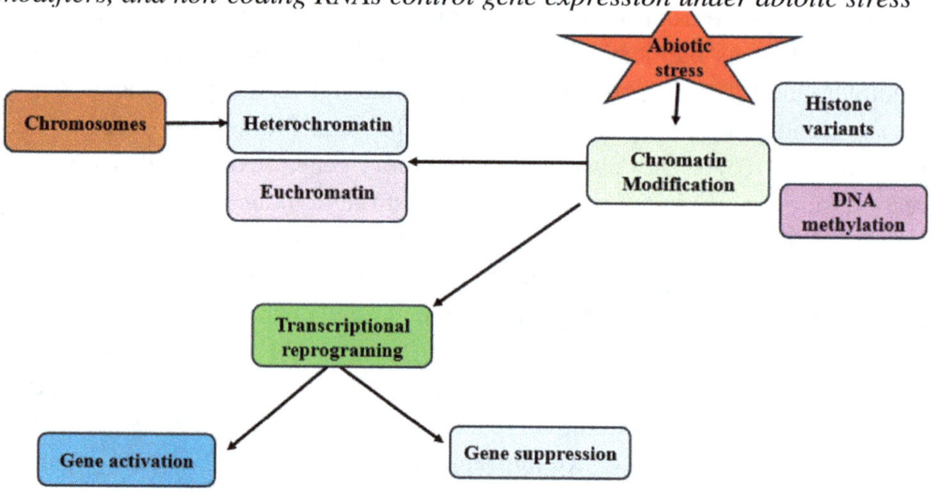

FUTURE RECOMMENDATION

Deciphering Mechanistic Interactions

Further studies should explore the intricate interactions between different chromatin remodeling complexes and other epigenetic regulators, such as DNA methylation and non-coding RNAs, to uncover their coordinated role in stress adaptation.

Integration of Biotic and Abiotic Stress Responses

Investigations should focus on how chromatin remodelers' mediate crosstalk between biotic and abiotic stress responses, providing insights into shared and unique regulatory pathways.

Temporal and Spatial Regulation

Research should examine the temporal dynamics and tissue-specific roles of chromatin remodeling complexes under stress conditions to better understand their context-dependent functions.

Functional Characterization in Crops

Extending studies to crop species beyond model organisms like *Arabidopsis* and rice will help identify key chromatin remodeling components that can be targeted to enhance stress resilience in agriculturally important plants.

Epigenome Editing Tools

The development of targeted epigenome editing technologies, such as CRISPR-based tools, should be utilized to manipulate chromatin remodeling complexes and their associated modifications for improving stress tolerance.

Network Modeling and Systems Biology Approaches

Systems biology and computational modeling should be employed to map the complex networks of chromatin remodelers and their downstream targets, providing a holistic view of stress-responsive regulatory mechanisms.

Exploring Energy Trade-offs

Future research should address how plants balance the energy costs of activating stress responses regulated by chromatin remodelers with growth and development under resource-limited conditions.

Translational Applications

Efforts should focus on translating knowledge of chromatin remodeling into crop improvement programs, including breeding strategies and genetic engineering for enhanced resilience to pathogens and environmental stresses.

These recommendations aim to deepen our understanding of chromatin remodeling in plant stress adaptation and leverage this knowledge for sustainable agriculture and food security.

CONCLUSION

Chromatin remodeling complexes are central to the dynamic regulation of plant stress responses, balancing defense gene expression under both normal and stress conditions. By modulating chromatin structure through histone variant deposition, histone modifications, DNA methylation, and nucleosome positioning, these complexes orchestrate the transcriptional reprogramming necessary for effective stress adaptation. The dual roles of chromatin remodelers, acting as both repressors and activators of defense-related genes, highlight their importance in maintaining basal immunity while enabling rapid responses to environmental challenges. Understanding the multifaceted mechanisms of these complexes provides valuable insights into plant stress biology and opens avenues for developing innovative strategies to enhance crop resilience in the face of biotic and abiotic stresses.

REFERENCES

Ahmad, Y., Haakim, Z., Iqbal, J., Abbasi, B. A., Mahmood, T., & Kazi, M. (2024b). Technological Innovations for Abiotic Stress Resistance in Horticultural Crops. OMICs-based Techniques for Global Food Security, 233-244.

Ahmad, Y., Haider, S., Iqbal, J., Abbasi, B. A., Yaseen, T., & Mahmood, T. (2022a). The mechanisms of genome editing technologies in crop plants. In *Principles and Practices of OMICS and Genome Editing for Crop Improvement* (pp. 295–313). Springer International Publishing.

Ahmad, Y., Haider, S., Iqbal, J., Abbasi, B. A., Yaseen, T., & Mahmood, T. (2022b). The mechanisms of genome editing technologies in crop plants. In *Principles and Practices of OMICS and Genome Editing for Crop Improvement* (pp. 295–313). Springer International Publishing.

Ahmad, Y., Haider, S., Iqbal, J., Naseer, S., Attia, K. A., Mohammed, A. A., & Mahmood, T. (2024a). In-silico analysis and transformation of OsMYB48 transcription factor driven by CaMV35S promoter in model plant–Nicotiana tabacum L. conferring abiotic stress tolerance. *GM Crops and Food: Biotechnology in Agriculture and the Food Chain, 15*(1), 130–149. PMID: 38551174

Ali, I., Salah, K. B. H., Sher, H., Ali, H., Ullah, Z., Ali, A., & Mahmood, T. (2022). Drought stress enhances the efficiency of floral dip method of Agrobacterium-mediated transformation in Arabidopsis thaliana. *Brazilian Journal of Biology, 84*, e259326. PMID: 35703626

Aquea, F., Federici, F., Moscoso, C., Vega, A., Jullian, P., & Haseloff, J., & ARCE-JOHNSON, P. (2012). A molecular framework for the inhibition of Arabidopsis root growth in response to boron toxicity. *Plant, Cell & Environment, 35*(4), 719–734. PMID: 21988710

Baldwin, I. T., & Schmelz, E. A. (1996). Immunological" memory" in the induced accumulation of nicotine in wild tobacco. *Ecology, 77*(1), 236–246.

Bäurle, I. (2016). Plant heat adaptation: Priming in response to heat stress. *F1000 Research*, 5. PMID: 27134736

Bouché, F., Woods, D. P., & Amasino, R. M. (2017). Winter memory throughout the plant kingdom: Different paths to flowering. *Plant Physiology, 173*(1), 27–35. PMID: 27756819

Boyer, L. A., Shao, X., Ebright, R. H., & Peterson, C. L. (2000). Roles of the Histone H2A-H2B Dimers and the (H3-H4) 2Tetramer in Nucleosome Remodeling by the SWI-SNF Complex. *The Journal of Biological Chemistry*, *275*(16), 11545–11552. PMID: 10766768

Brzezinka, K., Altmann, S., Czesnick, H., Nicolas, P., Gorka, M., Benke, E., . . . Kappel, C. (2016). Arabidopsis FORGETTER1 mediates stress-induced chromatin memory through nucleosome remodeling. *elife, 5*, e17061.

Buszewicz, D., Archacki, R., Palusiński, A., Kotliński, M., Fogtman, A., Iwanicka-Nowicka, R., & Olędzki, J. (2016). HD2C histone deacetylase and a SWI/SNF chromatin remodelling complex interact and both are involved in mediating the heat stress response in Arabidopsis. *Plant, Cell & Environment*, *39*(10), 2108–2122. PMID: 27083783

Cayuela, E., Pérez-Alfocea, F., Caro, M., & Bolarin, M. (1996). Priming of seeds with NaCl induces physiological changes in tomato plants grown under salt stress. *Physiologia Plantarum*, *96*(2), 231–236.

Chang, Y. N., Zhu, C., Jiang, J., Zhang, H., Zhu, J. K., & Duan, C. G. (2020). Epigenetic regulation in plant abiotic stress responses. *Journal of Integrative Plant Biology*, *62*(5), 563–580. PMID: 31872527

Charoensawan, V., Martinho, C., & Wigge, P. A. (2015). "Hit-and-run": Transcription factors get caught in the act. *BioEssays*, *37*(7), 748–754. PMID: 26010075

Choi, C.-S., & Sano, H. (2007). Abiotic-stress induces demethylation and transcriptional activation of a gene encoding a glycerophosphodiesterase-like protein in tobacco plants. *Molecular Genetics and Genomics*, *277*, 589–600. PMID: 17273870

Clapier, C. R., & Cairns, B. R. (2009). The biology of chromatin remodeling complexes. *Annual Review of Biochemistry*, *78*(1), 273–304. PMID: 19355820

Conrath, U. (2011). Molecular aspects of defence priming. *Trends in Plant Science*, *16*(10), 524–531. PMID: 21782492

Ding, Y., Liu, N., Virlouvet, L., Riethoven, J.-J., Fromm, M., & Avramova, Z. (2013). Four distinct types of dehydration stress memory genes in Arabidopsis thaliana. *BMC Plant Biology*, *13*, 1–11. PMID: 24377444

Ding, Y., Virlouvet, L., Liu, N., Riethoven, J.-J., Fromm, M., & Avramova, Z. (2014). Dehydration stress memory genes of Zea mays; comparison with Arabidopsis thaliana. *BMC Plant Biology*, *14*, 1–15. PMID: 24885787

Feng, X. J., Li, J. R., Qi, S. L., Lin, Q. F., Jin, J. B., & Hua, X. J. (2016). Light affects salt stress-induced transcriptional memory of P5CS1 in Arabidopsis. *Proceedings of the National Academy of Sciences of the United States of America, 113*(51), E8335–E8343. PMID: 27930298

Flaus, A., Martin, D. M., Barton, G. J., & Owen-Hughes, T. (2006). Identification of multiple distinct Snf2 subfamilies with conserved structural motifs. *Nucleic Acids Research, 34*(10), 2887–2905. PMID: 16738128

Friedrich, T., Faivre, L., Bäurle, I., & Schubert, D. (2019). Chromatin-based mechanisms of temperature memory in plants. *Plant, Cell & Environment, 42*(3), 762–770. PMID: 29920687

Goh, C. H., Nam, H. G., & Park, Y. S. (2003). Stress memory in plants: A negative regulation of stomatal response and transient induction of rd22 gene to light in abscisic acid-entrained Arabidopsis plants. *The Plant Journal, 36*(2), 240–255. PMID: 14535888

González, R. M., Ricardi, M. M., & Iusem, N. D. (2013). Epigenetic marks in an adaptive water stress-responsive gene in tomato roots under normal and drought conditions. *Epigenetics, 8*(8), 864–872. PMID: 23807313

Haider, S., Iqbal, J., Naseer, S., Shaukat, M., Abbasi, B. A., Yaseen, T., & Mahmood, T. (2022b). Unfolding molecular switches in plant heat stress resistance: A comprehensive review. *Plant Cell Reports, 41*(3), 775–798. PMID: 34401950

Haider, S., Iqbal, J., Naseer, S., Yaseen, T., Shaukat, M., Bibi, H., & Mahmood, T. (2021a). Molecular mechanisms of plant tolerance to heat stress: Current landscape and future perspectives. *Plant Cell Reports*, 1–25. PMID: 33890138

Haider, S., Iqbal, J., Shaukat, M., Naseer, S., & Mahmood, T. (2021b). The epigenetic chromatin-based regulation of somatic heat stress memory in plants. *Plant Gene, 27*, 100318.

Haider, S., Raza, A., Iqbal, J., Shaukat, M., & Mahmood, T. (2022a). Analyzing the regulatory role of heat shock transcription factors in plant heat stress tolerance: A brief appraisal. *Molecular Biology Reports, 49*(6), 5771–5785. PMID: 35182323

Han, B., Xu, W., Ahmed, N., Yu, A., Wang, Z., & Liu, A. (2020). Changes and associations of genomic transcription and histone methylation with salt stress in castor bean. *Plant & Cell Physiology, 61*(6), 1120–1133. PMID: 32186723

Hirakawa, T., & Matsunaga, S. (2019). Characterization of DNA repair foci in root cells of Arabidopsis in response to DNA damage. *Frontiers in Plant Science, 10*, 990. PMID: 31417598

Hopfner, K.-P., Gerhold, C.-B., Lakomek, K., & Wollmann, P. (2012). Swi2/Snf2 remodelers: Hybrid views on hybrid molecular machines. *Current Opinion in Structural Biology*, *22*(2), 225–233. PMID: 22445226

Hu, Y., Zhang, L., He, S., Huang, M., Tan, J., Zhao, L., & Liang, Y. (2012). Cold stress selectively unsilences tandem repeats in heterochromatin associated with accumulation of H3K9ac. *Plant, Cell & Environment*, *35*(12), 2130–2142. PMID: 22616903

Ijaz, S., Iqbal, J., Abbasi, B. A., Tufail, A., Ullah, Z., Sharifi-Rad, J., & Uddin, S. (2023). Biofortification: Lessons from the Past and Strategies for Future Food Security. In *Legumes Biofortification* (pp. 521–545). Springer International Publishing.

Ijaz, S., Iqbal, J., Abbasi, B. A., Yaseen, T., Rehman, S., Kazi, M., & Mahmood, T. (2024a). Role of OMICS-Based Technologies in Plant Sciences. OMICs-based Techniques for Global Food Security, 45-66.

Ijaz, S., Iqbal, J., Abbasi, B. A., Yaseen, T., Rehman, S., Kazi, M. & Mahmood, T. (2024b). Role of OMICS-Based Technologies in Plant Sciences. OMICs-based Techniques for Global Food Security, 45-66.

Iwasaki, M., & Paszkowski, J. (2014). Identification of genes preventing transgenerational transmission of stress-induced epigenetic states. *Proceedings of the National Academy of Sciences of the United States of America*, *111*(23), 8547–8552. PMID: 24912148

Jaskiewicz, M., Conrath, U., & Peterhänsel, C. (2011). Chromatin modification acts as a memory for systemic acquired resistance in the plant stress response. *EMBO Reports*, *12*(1), 50–55. PMID: 21132017

Jiang, J., Mao, N., Hu, H., Tang, J., Han, D., Liu, S., & Lai, J. (2019). A SWI/SNF subunit regulates chromosomal dissociation of structural maintenance complex 5 during DNA repair in plant cells. *Proceedings of the National Academy of Sciences of the United States of America*, *116*(30), 15288–15296. PMID: 31285327

Johnson, K. C., Xia, S., Feng, X., & Li, X. (2015). The chromatin remodeler SPLAYED negatively regulates SNC1-mediated immunity. *Plant & Cell Physiology*, *56*(8), 1616–1623. PMID: 26063389

Kang, H., Zhang, C., An, Z., Shen, W. H., & Zhu, Y. (2019). AtINO80 and AtARP5 physically interact and play common as well as distinct roles in regulating plant growth and development. *The New Phytologist*, *223*(1), 336–353. PMID: 30843208

Kim, J.-M., To, T. K., Ishida, J., Matsui, A., Kimura, H., & Seki, M. (2012). Transition of chromatin status during the process of recovery from drought stress in Arabidopsis thaliana. *Plant & Cell Physiology*, *53*(5), 847–856. PMID: 22505693

Klemm, T., Mannuß, A., Kobbe, D., Knoll, A., Trapp, O., Dorn, A., & Puchta, H. (2017). The DNA translocase RAD 5A acts independently of the other main DNA repair pathways, and requires both its ATP ase and RING domain for activity in Arabidopsis thaliana. *The Plant Journal*, *91*(4), 725–740. PMID: 28509359

Knizewski, L., Ginalski, K., & Jerzmanowski, A. (2008). Snf2 proteins in plants: Gene silencing and beyond. *Trends in Plant Science*, *13*(10), 557–565. PMID: 18786849

Kovalchuk, O., Burke, P., Arkhipov, A., Kuchma, N., James, S. J., Kovalchuk, I., & Pogribny, I. (2003). Genome hypermethylation in Pinus silvestris of Chernobyl—A mechanism for radiation adaptation? *Mutation Research. Fundamental and Molecular Mechanisms of Mutagenesis*, *529*(1-2), 13–20. PMID: 12943916

Lämke, J., Brzezinka, K., Altmann, S., & Bäurle, I. (2016). A hit-and-run heat shock factor governs sustained histone methylation and transcriptional stress memory. *The EMBO Journal*, *35*(2), 162–175. PMID: 26657708

Lämke, J., Brzezinka, K., & Bäurle, I. (2016). HSFA2 orchestrates transcriptional dynamics after heat stress in Arabidopsis thaliana. *Transcription*, *7*(4), 111–114. PMID: 27383578

Lee, H. G., & Seo, P. J. (2019). MYB96 recruits the HDA15 protein to suppress negative regulators of ABA signaling in Arabidopsis. *Nature Communications*, *10*(1), 1713. PMID: 30979883

Li, S., Lin, Y.-C. J., Wang, P., Zhang, B., Li, M., Chen, S., & Wang, Z. (2019). The AREB1 transcription factor influences histone acetylation to regulate drought responses and tolerance in Populus trichocarpa. *The Plant Cell*, *31*(3), 663–686. PMID: 30538157

Li, X., Jiang, Y., Ji, Z., Liu, Y., & Zhang, Q. (2015). BRHIS 1 suppresses rice innate immunity through binding to monoubiquitinated H2A and H2B variants. *EMBO Reports*, *16*(9), 1192–1202. PMID: 26202491

Li, Y., Tessaro, M. J., Li, X., & Zhang, Y. (2010). Regulation of the expression of plant resistance gene SNC1 by a protein with a conserved BAT2 domain. *Plant Physiology*, *153*(3), 1425–1434. PMID: 20439546

Liu, J., Feng, L., Gu, X., Deng, X., Qiu, Q., Li, Q., & Wang, E. (2019). An H3K27me3 demethylase-HSFA2 regulatory loop orchestrates transgenerational thermomemory in Arabidopsis. *Cell Research*, *29*(5), 379–390. PMID: 30778176

Liu, T., Li, Y., Duan, W., Huang, F., & Hou, X. (2017). Cold acclimation alters DNA methylation patterns and confers tolerance to heat and increases growth rate in Brassica rapa. *Journal of Experimental Botany*, *68*(5), 1213–1224. PMID: 28158841

March-Díaz, R., García-Domínguez, M., Lozano-Juste, J., León, J., Florencio, F. J., & Reyes, J. C. (2008). Histone H2A. Z and homologues of components of the SWR1 complex are required to control immunity in Arabidopsis. *The Plant Journal*, *53*(3), 475–487. PMID: 17988222

Morrison, A. J., Highland, J., Krogan, N. J., Arbel-Eden, A., Greenblatt, J. F., Haber, J. E., & Shen, X. (2004). INO80 and γ-H2AX interaction links ATP-dependent chromatin remodeling to DNA damage repair. *Cell*, *119*(6), 767–775. PMID: 15607974

Mozgova, I., Wildhaber, T., Liu, Q., Abou-Mansour, E., L'Haridon, F., Métraux, J.-P., & Hennig, L. (2015). Chromatin assembly factor CAF-1 represses priming of plant defence response genes. *Nature Plants*, *1*(9), 1–8. PMID: 27250680

Nair, N., Shoaib, M., & Sørensen, C. S. (2017). Chromatin dynamics in genome stability: Roles in suppressing endogenous DNA damage and facilitating DNA repair. *International Journal of Molecular Sciences*, *18*(7), 1486. PMID: 28698521

Nisa, M.-U., Huang, Y., Benhamed, M., & Raynaud, C. (2019). The plant DNA damage response: Signaling pathways leading to growth inhibition and putative role in response to stress conditions. *Frontiers in Plant Science*, *10*, 653. PMID: 31164899

Pajoro, A., Severing, E., Angenent, G., & Immink, R. (2017). Histone H3 lysine 36 methylation affects temperature-induced alternative splicing and flowering in plants. *Genome Biology*, *18*, 1–12. PMID: 28566089

Peterson, C. L., & Workman, J. L. (2000). Promoter targeting and chromatin remodeling by the SWI/SNF complex. *Current Opinion in Genetics & Development*, *10*(2), 187–192. PMID: 10753786

Pick, T., Jaskiewicz, M., Peterhänsel, C., & Conrath, U. (2012). Heat shock factor HsfB1 primes gene transcription and systemic acquired resistance in Arabidopsis. *Plant Physiology*, *159*(1), 52–55. PMID: 22427343

Pien, S., Fleury, D., Mylne, J. S., Crevillen, P., Inze, D., Avramova, Z., & Grossniklaus, U. (2008). ARABIDOPSIS TRITHORAX1 dynamically regulates FLOWERING LOCUS C activation via histone 3 lysine 4 trimethylation. *The Plant Cell*, *20*(3), 580–588. PMID: 18375656

Poole, L. A., Zhao, R., Glick, G. G., Lovejoy, C. A., Eischen, C. M., & Cortez, D. (2015). SMARCAL1 maintains telomere integrity during DNA replication. *Proceedings of the National Academy of Sciences of the United States of America*, *112*(48), 14864–14869. PMID: 26578802

Qüesta, J. I., Song, J., Geraldo, N., An, H., & Dean, C. (2016). Arabidopsis transcriptional repressor VAL1 triggers Polycomb silencing at FLC during vernalization. *Science*, *353*(6298), 485–488. PMID: 27471304

Ramirez-Prado, J. S., Abulfaraj, A. A., Rayapuram, N., Benhamed, M., & Hirt, H. (2018). Plant immunity: From signaling to epigenetic control of defense. *Trends in Plant Science*, *23*(9), 833–844. PMID: 29970339

Rosa, M., Von Harder, M., Aiese Cigliano, R., Schlögelhofer, P., & Mittelsten Scheid, O. (2013). The Arabidopsis SWR1 chromatin-remodeling complex is important for DNA repair, somatic recombination, and meiosis. *The Plant Cell*, *25*(6), 1990–2001. PMID: 23780875

Sakamoto, T., Tsujimoto-Inui, Y., Sotta, N., Hirakawa, T., Matsunaga, T. M., Fukao, Y., & Fujiwara, T. (2018). Proteasomal degradation of BRAHMA promotes Boron tolerance in Arabidopsis. *Nature Communications*, *9*(1), 5285. PMID: 30538237

Sani, E., Herzyk, P., Perrella, G., Colot, V., & Amtmann, A. (2013). Hyperosmotic priming of Arabidopsis seedlings establishes a long-term somatic memory accompanied by specific changes of the epigenome. *Genome Biology*, *14*, 1–24. PMID: 23767915

Santos, A. P., Ferreira, L., Maroco, J., & Oliveira, M. M. (2011). Abiotic stress and induced DNA hypomethylation cause interphase chromatin structural changes in rice rDNA loci. *Cytogenetic and Genome Research*, *132*(4), 297–303. PMID: 21307636

Shen, Y., Devic, M., Lepiniec, L., & Zhou, D. X. (2015). Chromodomain, helicase and DNA-binding CHD1 protein, CHR5, are involved in establishing active chromatin state of seed maturation genes. *Plant Biotechnology Journal*, *13*(6), 811–820. PMID: 25581843

Singh, P., Yekondi, S., Chen, P.-W., Tsai, C.-H., Yu, C.-W., Wu, K., & Zimmerli, L. (2014). Environmental history modulates Arabidopsis pattern-triggered immunity in a HISTONE ACETYLTRANSFERASE1–dependent manner. *The Plant Cell*, *26*(6), 2676–2688. PMID: 24963055

Singh, S., Singh, A., & Singh, A., Mahima, Yadav, S., Bajaj, I., . . . Sarkar, A. K. (2020). Role of chromatin modification and remodeling in stem cell regulation and meristem maintenance in Arabidopsis. *Journal of Experimental Botany*, *71*(3), 778–792. PMID: 31793642

Ullah, Z., Iqbal, J., Abbasi, B. A., Akhtar, W., Kanwal, S., Ali, I., & Mahmood, T. (2023b). Assessment of Gus Expression Induced by Anti-Sense Os PPO Gene Promoter and Antioxidant Enzymatic Assays in Response to Drought and Heavy Metal Stress in Transgenic Arabidopsis thaliana. *Sustainability*, *15*(17), 12783.

Ullah, Z., Iqbal, J., Abbasi, B. A., Ijaz, S., Sharifi-Rad, J., Yaseen, T., & Mahmood, T. (2023a). How Genome Editing Can Be Helpful in the Biofortification of Legumes. In *Legumes Biofortification* (pp. 207–232). Springer International Publishing.

Virlouvet, L., & Fromm, M. (2015). Physiological and transcriptional memory in guard cells during repetitive dehydration stress. *The New Phytologist*, *205*(2), 596–607. PMID: 25345749

Wang, X., Vignjevic, M., Jiang, D., Jacobsen, S., & Wollenweber, B. (2014). Improved tolerance to drought stress after anthesis due to priming before anthesis in wheat (Triticum aestivum L.) var. Vinjett. *Journal of Experimental Botany*, *65*(22), 6441–6456. PMID: 25205581

Zeng, Z., Zhang, W., Marand, A. P., Zhu, B., Buell, C. R., & Jiang, J. (2019). Cold stress induces enhanced chromatin accessibility and bivalent histone modifications H3K4me3 and H3K27me3 of active genes in potato. *Genome Biology*, *20*, 1–17. PMID: 31208436

Zha, P., Jing, Y., Xu, G., & Lin, R. (2017). PICKLE chromatin-remodeling factor controls thermosensory hypocotyl growth of Arabidopsis. *Plant, Cell & Environment*, *40*(10), 2426–2436. PMID: 28771755

Zhang, D., Gao, S., Yang, P., Yang, J., Yang, S., & Wu, K. (2019). Identification and expression analysis of Snf2 family proteins in tomato (Solanum lycopersicum). *International Journal of Genomics*, *2019*(1), 5080935. PMID: 31049349

Zhou, S., Chen, Q., Sun, Y., & Li, Y. (2017). Histone H2B monoubiquitination regulates salt stress-induced microtubule depolymerization in Arabidopsis. *Plant, Cell & Environment*, *40*(8), 1512–1530. PMID: 28337773

Zhu, J.-K. (2016). Abiotic stress signaling and responses in plants. *Cell*, *167*(2), 313–324. PMID: 27716505

Zou, B., Sun, Q., Zhang, W., Ding, Y., Yang, D.-L., Shi, Z., & Hua, J. (2017). The Arabidopsis chromatin-remodeling factor CHR5 regulates plant immune responses and nucleosome occupancy. *Plant & Cell Physiology*, *58*(12), 2202–2216. PMID: 29048607

Chapter 5
Epigenetic Modifications in Root Development and Stress Responses

Sanam Ashraf
https://orcid.org/0009-0000-7401-8236
Quaid-I-Azam University, Islamabad, Pakistan

Zubaria Tul Ain
Quaid-i-Azam University, Islamabad, Pakistan

Sania Saleem
Quaid-i-Azam University, Islamabad, Pakistan

Ishtiaq Hussain
University of Baltistan, Skardu, Pakistan

ABSTRACT

Plants are not able to flee or communicate under unfavorable biotic or abiotic stress conditions, which can have a profound impact on their growth, development and reproduction. As a result, plants have established more sophisticated adaptation mechanisms that enable them to thrive in harsh environments. Epigenetic regulation is essential to regulate cellular processes needed for survival in adverse environments. In plants, epigenetic changes play important role in root development because they affect gene expression linked with root architecture, growth and stress response allowing plants to adapt to changing conditions. These changes include DNA methylation, histone modification and small RNA-mediated regulation. Epigenetic mechanisms change chromatin state of stress-responsive genes regulating them at transcriptional and posttranscriptional phases. These approaches emphasize

DOI: 10.4018/979-8-3693-8648-4.ch005

significance of epigenetic alterations in connecting root systems developmental plasticity and environmental adaptability, opening door to create ways to improve crop performance under difficult circumstances.

INTRODUCTION

Living things are constantly exposed to environmental stresses, which can have a profound impact on their growth, development, and reproduction. However, they have evolved defense and adaptation mechanisms to cope with these stresses. Despite these efforts, stress can still have a significant and detrimental effect on organisms (Boyko and Kovalchuk, 2008, Ahmad et al., 2022a, Ahmad et al., 2022b). Plants, being immobile, are constantly vulnerable to various abiotic and biotic stressors, which can significantly impact agricultural productivity. Therefore, enhancing plant stress resilience is essential for maintaining crop yields. Recent research has shed light on the key stress signaling pathways in plants (Zhu *et al.*, 2008, Ahmad et al., 2024a). Molecular adjustments are requiring for plants in order to adapting to changing environmental conditions (Singroha and Sharma, 2019). Recent research highlights the crucial role of epigenetic controls in enhancing plant resilience to stress, thereby boosting their survival capacity. These regulators are potent and effective in mitigating stress impacts (Kim *et al.*, 2015, Ahmad et al., 2024b).

Epigenetic modifications refer to changes in gene expression that are inherited through cell division, but don't alter the DNA sequence itself. These changes can be thought of as chemical 'tags' on DNA or histone proteins that influence gene activity (Bird, 2002, Haider et al., 2022b). Epigenetic alterations, including DNA methylation, histone modifications, and minor RNA-mediated changes, collectively impact gene expression in response to stress. DNA methylation, in particular, plays a crucial role in plant adaptability and resilience by regulating stress-response genes, either activating or silencing them as needed (Slotkin and Martienssen, 2007, Ali et al., 2022). Histone modifications, like acetylation and methylation, alter chromatin structure and accessibility, thereby influencing the expression of stress-responsive genes (Zhang and Reinberg, 2001). Short RNAs, including microRNAs and siRNAs, regulate gene expression during stress by targeting specific mRNAs for degradation or translational repression (Guo *et al.*, 2016).

Epigenetic mechanisms play a crucial role in helping plants cope with short-term stress and retain long-term stress memory, enabling adaptation. Understanding these mechanisms can provide valuable insights for enhancing crop resilience and productivity (Chinnusamy and Zhu, 2009). Epigenetic mechanisms change the chromatin state of stress-responsive genes, making it possible to regulate them at the transcriptional and posttranscriptional phases. Stress treatments may result in modifications to the chromatin structure. (Kim *et al.*, 2015). Various epigenetic modifications triggered by abiotic and biotic stresses have been identified in Figure 1.

Figure 1. Multiple types of epigenetic modifications under stressful condition

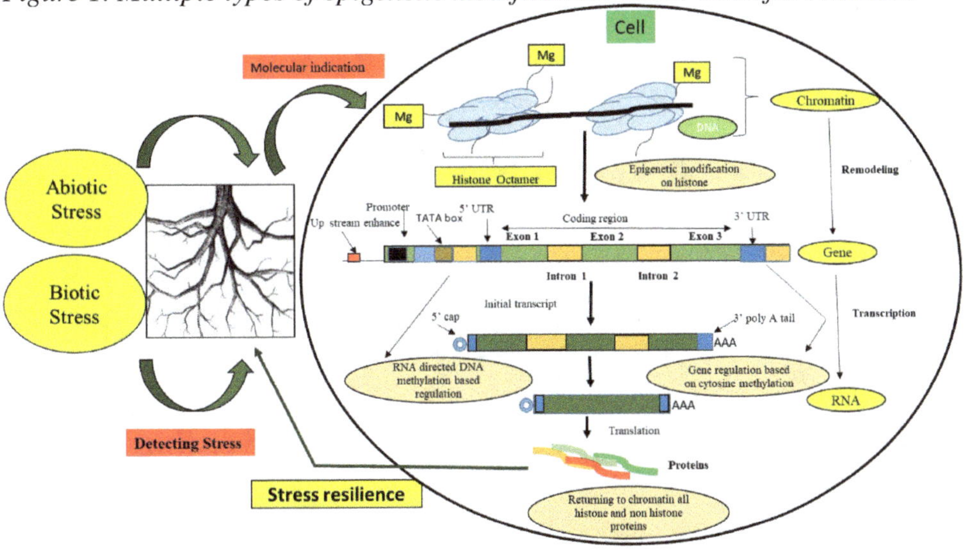

In plants, epigenetic changes play an important role in root development because they affect gene expression linked with root architecture, growth, and stress response, allowing plants to adapt to changing conditions. These changes, which regulate gene expression without altering the underlying DNA sequence, include small RNA-mediated regulation, histone modification, and DNA methylation. Additionally, it has been determined that root meristem activity and root system design are controlled by DNA methylation patterns (Shemer *et al.*, 2015). While the expression of critical developmental genes can be influenced by histone changes (Kouzarides, 2007). Furthermore, microRNAs and other short RNAs are required to regulate gene expression throughout root growth and in response to surrounding pressure (Chen *et al.*, 2021). The ability of epigenetic changes to affect cellular responses to environmental perturbations, control developmental processes, and contribute to pathogenic mechanisms makes them essential. In addition, they are critical for maintaining cellular identity and function throughout generations, which may lead to new therapeutic treatment options and improvements in personalized medicine (Jaenisch and Bird, 2003). Learning about these epigenetic modifications may enhance agricultural methods and plant resilience by revealing how plants govern root growth and respond to their surroundings.

TYPES OF EPIGENETIC MODIFICATIONS

Numerous types of epigenetic alterations are important for controlling the expression of certain genes. DNA methylation, Histone modifications, and non-coding RNAs, are the three main forms of epigenetic alterations as showed in Figure 2.

Figure 2. Types of epigenetic modifications

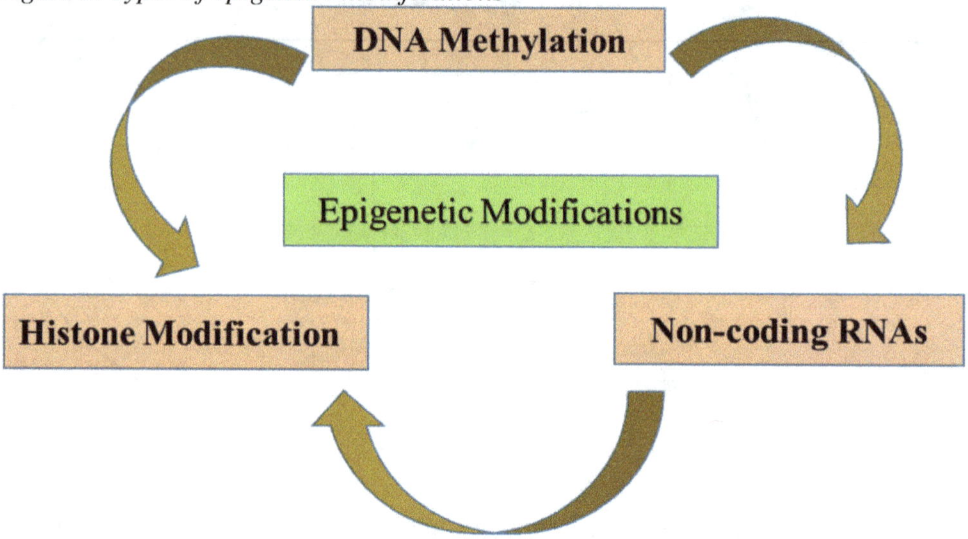

DNA Methylation

The attachment of a methyl group to the fifth carbon of the cytosine ring, usually at CpG sites, results in DNA methylation, an important epigenetic modification that forms 5-methylcytosine (Harris et al.,2018). Methylation is categorized as either symmetrical or asymmetrical based on the target sequence. Methylation that is CG and CHG symmetrical and CHH asymmetrical is described. Three different sequences CHG, CHH, and CG in plants are where DNA methylation occurs (where H stands for nucleotides C, T, or A) (Li and Tollefsbol, 2011). Transposable element sequences typically exhibit maximum methylation at the majority of their CG, but only a small percentage of genes in plants are methylated, and if they do, it's usually limited to CGs found within the gene body. whereas methylation within the gene body causes gene expression, methylation within a gene's transposable elements and the promoter region results in gene suppression (Chan et al., 2005). Therefore, the following outcomes of DNA methylation occurs: methyl cytosines in the gene body play a crucial role in controlling the expression of the gene, and methyl cytosines in repetitive sequences (transposable elements) are believed to protect normal genome function from recurrent DNA sequences (Miura et al., 2001).

The expression of genes declines when genomic DNA becomes more methylated. Plants with down-regulated gene expression are able to store energy in the event of either abiotic or biotic stress. On the other hand, the downregulation of tolerance-associated gene methylation promotes genome activation and the production of new genes that offer permanent or prolonged stress tolerance (Singroha and Shaema, 2019). According to latest research, a range of abiotic stresses make plants more vulnerable to methylation, which reduces the function of their genome. Conversely, optimal development conditions and the lack of stress lead to great expression with reduced methylation. (Li and Tollefsbol, 2011).

Histone Modification

The chemical alterations that histone proteins encounter during translation are referred to as "histone modification" and are vital for regulating gene expression. These changes, which mainly impact histone ends that extend from the nucleosome core, include acetylation, methylation, phosphorylation, and ubiquitination (Bannister and Kouzarides, 2011). The nucleosome, a compact structure made up of histone octamers made up of two copies of each of the bases H2A, H2B, H3, and H4, and 147 base pairs of DNA surrounding the histone core, is formed by histone proteins folding around DNA. Histone proteins' N-terminal tails can get methylated or acetylated, two chemical changes linked to either activation or repression of genes (Jang et al., 2011; Roudier et al., 2011). While acetylation and methylation of H3K4 and H3K36 are associated with higher gene expression and gene stimulation in plants, methylation and deacetylation of H3K9 and H3K27 result in gene repression (Lauria and Rossi, 2011). The distribution of different post-translational changes and histone variations throughout the genome has been demonstrated with immune precipitation of chromatin and deep arranging (Berger, 2007). Cells can accurately control gene expression in response to environmental and developmental signals due to these dynamic and reversible modifications (Bannister and Kouzarides, 2011). Histone changes in plants are linked to vital processes such fruit development, vernalization, nodulation, and the cell cycle (Sharma et al.,2023).

Non-Coding RNAs

RNA molecules known as non-coding RNAs (ncRNAs) are essential for controlling gene expression via transcriptional and post-transcriptional processes even though they do not encode proteins (Esteller, 2011). Nutrient uptake, stress responses, and root development are among the processes that ncRNAs in plants are essential for regulating gene expression in roots (Kar and Raichaudhuri, 2021).

MicroRNAs (miRNAs) and long non-coding RNAs (lncRNAs) are two of the most investigated non-coding RNAs. MiRNAs are 22 nucleotide length short RNAs that attach to identical sequences in target messenger RNAs' (3') untranslated regions (UTRs). This binding usually causes translation to be inhibited or mRNA to be degraded, on which affects responses to food availability, lateral root growth, and root patterning (Esteller, 2011; Sumkar, 2012). More than 200 nucleotides, or lncRNAs, control the expression of genes via a variety of methods, such as chromatin transforming, transcriptional disruption and molecular framework. Their multiple functions encompass shaping chromatin structure and regulating other RNA expressions, so helping in root growth and stress tolerance (Liu et al.,2015; Rinn and Chang 2012; Nejat and Mantri, 2018). Plant roots have a complex network of non-coding RNAs (ncRNAs) that play a crucial role in the regulatory processes that underpin root function by allowing for the fine-tuning of growth and responses to external changes (Reinhart et al., 2002).

EPIGENETIC MODIFICATIONS IN ROOT DEVELOPMENT

For an accurate regulation of root production in the initial phases of plant growth, epigenetic changes are essential. These processes control the patterns of gene expression that are necessary for the initial development and growth of roots in plants. They include DNA methylation, histone modifications, and non-coding RNAs.

DNA Methylation in Root Development

DNA methyltransferase enzymes enable the critical epigenetic change known as DNA methylation (DM), which is the addition of a methyl group to the cytosine residue of CpG dinucleotides. This change affects the expression of genes necessary for root development and stress responses by changing the structure of chromatin, which suppresses gene transcription (Bird, 2002; Ceder and Bergman, 2012). Significant effects of DNA methylation are seen in important growth-related genes connected to root formation. DNA methylation, for instance, affects the AUXIN RESPONSE FACTOR (ARF) family, which controls the development and elongation of lateral roots. According to (Ikeuchi et al., 2015), ARF gene expression is changed by DNA methylation, which has an impact on root architecture.

Another research also demonstrated the impact of methylation modifications at ARF gene promoters on the commencement and growth of lateral roots (Vangheluwe and Beeckman, 2021). likewise, roots' ability to adapt to their surroundings is modified by DNA methylation. According to Yun et al 2024, genes involved in root adaptability are impacted by changes in DNA methylation patterns that occur in response to stress and nutrition availability. For example, in Arabidopsis, modifications in root development and growth were associated with altered DNA methylation patterns in response to stress from drought (Varriale, 2017). The control of genes that alter cell walls by DNA methylation is one prominent example. According to Li et al., 2014, root elongation in Arabidopsis is influenced by DNA methylation at the promoter regions of expansion genes, which are essential for cell wall loosening. The demethylation of these areas highlights the significance of DNA methylation in regulating root architecture by increasing gene expression and encouraging root growth. Auxin signaling pathways, which are essential for root development, are similarly impacted by DNA methylation. Auxin signaling is modulated by DNA methylation in auxin-responsive sites, which affects the development of roots, according to research by Shahzad et al., 2020 likewise, it was discovered that lateral root development in rice is impacted by DNA methylation at the OsARF16 gene's promoter site (Wang et al., 2022).

Furthermore, interactions exist between DNA methylation and other epigenetic processes such small RNA pathways and histone changes. A researcher Chen et al., 2018 talked about how changes in histone and DNA methylation work together to control the sequences of gene expression that are essential for root growth. For instance, in maize, histone changes and DNA methylation cooperate to control response to stress and root formation (Ding et al., 2014). DNA methylation modifies gene expression in the course of lateral root development by changing chromatin access, which usually results in gene silencing. DNA methylation influences the expression of auxin-responsive genes that are essential for the start of lateral roots, as (Shahzad et al., 2020) showed. Studies on Arabidopsis have demonstrated that alterations in DNA methylation patterns affect the number and growth of lateral roots (Yao et al., 2013).

Histone Modifications in Root Development

During root development, histone changes like as acetylation and methylation have a major impact on the organization of chromatin and gene function. These changes affect the portability of chromatin, which affects the transcription of genes related to root morphogenesis. Certain histone methylation marks are linked to gene suppression during early root development, whereas the acetylation of histone, for instance, usually results in a more relaxed chromatin state, promoting gene activation (Kouzarides, 2007; Bannister and Kouzarides, 2011). Root architecture-controlling genes are expressed in a way determined by histone alterations. The arrangement of root tissues and the differentiation of distinct types of root cells are dependent on an open chromatin framework and active transcription, which are often connected with histone acetylation. For example, active transcription is associated with trimethylation of histone H3 at lysine 4 (H3K4me3), which is frequently seen close to the promoters of genes involved in root growth (Kouzarides, 2007; Santos et al., 2002). On the other hand, in certain developmental circumstances, trimethylation of histone H3 at lysine 27 (H3K27me3) is linked to gene silencing and aids in maintaining repression (Wiles and Selker, 2017).

A regulatory network that affects the regulation of genes necessary for cell division, elongation, and differentiation in root development is created by balancing between these and additional changes to histone (Be, 2006; Van et al., 2010). Histone changes like as acetylation, methylation, and phosphorylation are important in controlling the structure of the chromatin and the functioning of genes during root growth. For instance, (Tsilimigka et al., 2022) showed that activation of genes involved in cell division and elongation within the root meristem has been associated with histone acetylation at H3K9 and H3K27. Because of this acetylation, chromatin is relaxed, enabling transcription variables to attach to DNA and enhance the expression of genes. On the other hand, H3K27 trimethylation (H3K27me3) functions as a restrictive mark by silencing genes that prevent root expansion (Wu et al., 2019).

Maintaining gene expression during root extension requires the interaction of DNA methylation and histone alterations. To activate crucial genes involved in root development, both DNA demethylation and the removal of repressive histone markers need to occur (Zanetti et al., 2024). Lateral root development is also significantly affected by histone alterations. By encouraging a more open chromatin configuration, histone acetylation especially of H3 and H4 is linked to the activation of genes essential to the start of lateral roots. This makes DNA susceptible to transcription factors, which in response activate the genes involved in the creation of roots primordium (Kouzarides, 2007; Bannister and Kouzarides, 2011). Moreover, genes involved in cell division, differentiation, and elongation are influenced by histone methylation marks like H3K4me3 (associated with gene activation) and H3K27me3 (associated with gene repression) all of which are essential for the development of lateral roots (Berger, 2007; Van Dijk et al., 2010; Santos et al., 2002). Additionally, it has been demonstrated that histone methylation affects auxin-related genes, which are essential for the development of lateral roots (Singh et al., 2020).

NON-CODING RNAS IN ROOT DEVELOPMENT

For the development of roots, non-coding RNAs, such as long non-coding RNAs (lncRNAs) and microRNAs (miRNAs), are crucial regulators of gene expression. By targeting particular mRNAs for translational suppression or destruction, microRNAs (miRNAs) control the expression of genes. Auxin sensitivity and root elongation, for instance, are impacted by miR393, which targets auxin receptor mRNAs such TIR1 (Bian et al., 2012). By modulating, root growth is made to respond to both internal and exterior stimuli. Another important miRNA that controls the expression of ARF genes (AUXIN RESPONSE FACTOR) is miR167. These genes are essential for the development and elongation of lateral roots (Meng et al., 2010). Furthermore, miR396 targets genes involved in growth hormone signaling pathways, which affects the activity of the root meristem and the general architecture of the root (Bazin et al., 2013). According to Meng et al., 2010, also showed that miR164 targets genes involved in auxin biosynthesis and signaling, which controls root development.

Using procedures known as RNA-directed DNA methylation (RdDM), small interfering RNAs (siRNAs) help promote the growth of roots. They use targeted DNA methylation to silence genes that prevent root expansion. (Jia et al., 2009) talked about the role that siRNAs made from transposable elements and repetitive sequences play in preserving genomic stability, which is essential for healthy root growth programs. For instance, root development in response to stress is impacted by siRNA-mediated silencing of the gene encoding the transcription factor WRKY23 (Knip, 2012).

By interacting with genome-modifying proteins to control gene expression through chromatin reconstruction, long non-coding RNAs (lncRNAs) support root growth. In Arabidopsis, the lncRNA COOLAIR affects flowering and root development, emphasizing its function in plant growth (Yamaguchi and Abe, 2012). Through their associations with chromatin-modifying proteins, lncRNAs like ASCO (At5g09760) and DELAYED FLOWERING 1 (DLF1) are engaged in root-specific processes, including as responses to stress and root development control (Meng et al., 2010; Stauffer and Maizel, 2014). Furthermore, in Arabidopsis, chromatin state and gene expression are influenced by the lncRNA ROOTLESS WITH LADYBIRD1 (RUL1), which controls root growth (Roule et al., 2022). Together with each other, these non-coding RNAs help regulate the gene expression patterns that are important to root growth and environmental flexibility, guaranteeing that roots grow as best they can in response to internal and external stresses.

EPIGENETIC RESPONSES TO ENVIRONMENTAL STRESSES

Types of Environmental Stresses

Epigenetic Responses to Environmental Stresses: Abiotic Stresses

Abiotic stressors that greatly affect plant growth and development include salinity, drought, temperature extremes, and other environmental conditions. These stressors cause a range of epigenetic changes in plants, which aid in their ability to adjust and endure harsh circumstances. These alterations include adjustments to the histone structure, DNA methylation, and non-coding RNAs. Taken together, these modifications control genes that are sensitive to stress and preserve cellular homeostasis as showed in Figure 3.

Figure 3. Epigenetic regulation maintains constant tolerance against abiotic stress

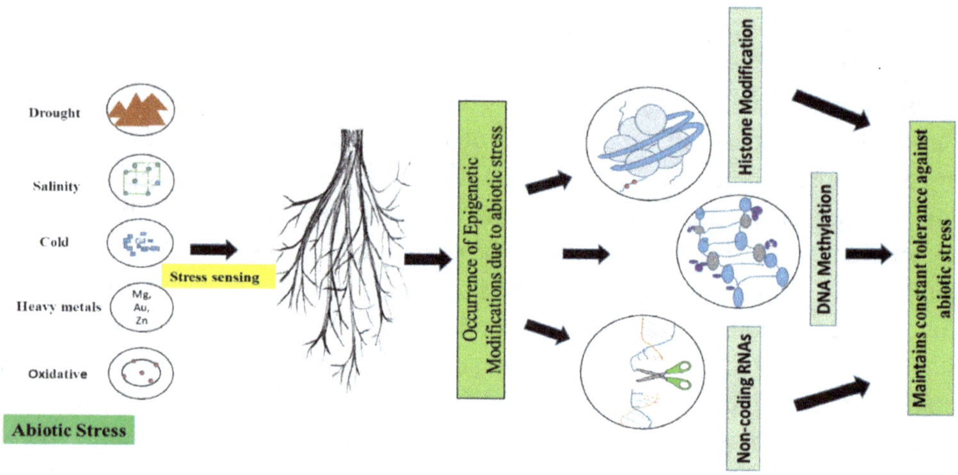

Drought Stress:

Since epigenetic processes in roots under drought stress modify gene expression without altering the DNA sequence, they are important to how plants respond to water-limited conditions. DNA methylation is one of the main epigenetic modifications that can either repress or activate genes that are responsive to stress, especially those that are involved in modifications to root architecture that maximize water absorption. For example, during a drought, demethylation can activate genes involved in the formation of lateral roots, thereby encouraging the development of a deeper and more extensive root system to access water from deeper soil layers, while hypermethylation of specific genomic regions can repress genes which prevent root elongation (Kumari et al., 2022). Drought stress-induced the demethylation process has been shown to activate genes related to abscisic acid (ABA) signaling, a crucial hormone in the drought response, in species such as Arabidopsis thaliana (Ding et al., 2012).

Histone changes like acetylation and methylation, as along with DNA methylation, are important in controlling gene expression during drought stress. Genes involved in water transportation and root development can be expressed due to a more relaxed chromatin structure caused by high histone acetylation at loci responsive to drought. For instance, increased drought tolerance in roots has been linked to histone acetylation at the promoter regions of aquaporin genes, which aid in water absorption (Shanker and Maheswari, 2017). Small RNAs, in particular microRNAs (miRNAs), play a critical role in another epigenetic mechanism by regulating the production of transcription factors and genes linked to stress, thereby fine-tuning root responses to drought. For example, it has been demonstrated that miR169 and miR393 control root development and ABA signaling pathways during drought, improving the plant's capacity to store water and retain root function in stressful situations (Kumari et al., 2022).

Salinity Stress

Plants need epigenetic reactions in their roots under salinity stress in order to adapt and live in high-salinity settings. Stress from salinity generates changes in DNA methylation, histones, and small RNA activity, all of which control the expression of genes linked to stress signaling pathways, root architecture, and ion transport. Under salt stress, methylation of the DNA patterns frequently shifts; hypomethylation activates genes that respond to stress, while hypermethylation can quiet genes that are harmful to stress tolerance. For example, methylation levels of salt-responsive genes in *Arabidopsis thaliana* and rice (Oryza sativa) have been found to be altered. These genes are especially important for ion homeostasis, as sodium/hydrogen antiporters are essential for preserving ionic balance in salinity (Viggiano and de Pinto, 2017). Plants can reach less salinized soil zones for water uptake by elongating their roots due to the demethylation process of genes controlling the development of roots (Karan et al., 2012).

In addition, acetylation and methylation of histones are crucial for the response to salt stress. Stress-responsive genes' promoter areas have more histone acetylation, which reduces the chromatin framework and improves gene expression. Histone acetylation, for instance, has been connected to the activation of genes in Medicago truncatula that are involved in proline synthesis, a crucial osmoprotectant that aids in preserving root cell structure in the face of extreme salinity (Chinnusamy and Zhu, 2009). Additionally, by controlling the expression of genes involved in ion transport and stress signaling, short RNAs—particularly microRNAs, or miRNAs—modify root adaptability. Specifically, miR393 targets genes in the auxin signaling pathway, which is essential for salt tolerance and root development. This improves the plant's capacity to control ionic toxicity and sustain root growth in salinized environments (Zhang and Stellwag, 2008). By modifying the expression of genes necessary for stress tolerance and root adaptation, these epigenetic modifications create a complex regulatory system that enables plants to precisely control the growth and function of their roots and increase their resistance to salinity stress.

Temperature Extremes

High levels of temperature, both hot and cold, alter genes involved in hormone regulation, root organization, and response to stress by creating epigenetic variations that include DNA methylation, histone modifications, and small RNA function. For instance, extreme heat has been shown to cause a hypomethylation of DNA in the plant *Oryza sativa* (rice), which stimulates the genes for heat shock protein (HSP), which prevents root cells against folding errors and assembly of proteins (Fang et al., 2015). Similar to this, Arabidopsis thaliana experiences the demethylation in genes linked to auxin and ethylene signaling pathways in response to cold stress, which encourages the production of lateral roots and root elongation to help the plant acclimate to freezing temperatures (Yu et al., 2019, Ijaz et al., 2023).

With heat stress, histone changes are also important for the epigenetic regulation of roots. Histone acetylation at stress-responsive gene loci is elevated during heat stress, which promotes the activation of antioxidant genes that shield root cells from heat-induced oxidative damage. For instance, histone acetylation in rapeseed (*Brassica napus*) increases the expression of genes that are involved in detoxifying reactive oxygen species, which aids in the roots' ability to withstand heat stress (Wei et al., 2022). On the other hand, cold stress suppresses genes that would prevent root extension by causing histone methylation. This has been shown in *Medicago truncatula*, whose roots are able to grow and obtain water under freezing temperatures due to histone methylation at important loci (Bokszczanin, 2013, Ijaz et al., 2024a).

Root reaction to high temperatures are further influenced by small RNAs, specifically microRNAs (miRNAs). In addition, it has been established that miR156 and miR169 control transcription factors that respond to stress, allowing roots to grow and remain stable in both hot and cold environments. According to (Bokszczanin, 2013), miR156 in Arabidopsis targets SPL transcription factors, promoting root development and stress adaptation during cold stress, while miR169 regulates NF-YA transcription factors under heat stress, improving thermotolerance in roots.

These illustrations show how essential epigenetic systems are for temperature stress change, enable plants to perform well in harsh conditions and enhance root function.

Heavy Metal Stress

Plants that are stressed by heavy metals need epigenetic responses in their roots to regulate the expression of genes and withstand harmful conditions without changing the sequence that makes up their DNA. Heavy metals that produce reactive oxygen species and disrupt with root growth, such as arsenic (As), cadmium (Cd), and lead (Pb), can alter DNA methylation, histone modifications, and small RNA activity. In plants exposed to heavy metals, methylation modifications in the DNA are often found; hypomethylation activates stress-responsive genes that improve detoxification and tolerance, while hypermethylation silences genes that may increase metal sensitivity. According to Wang et al., 2016, cadmium exposure in *Arabidopsis thaliana,* for instance, results in global DNA hypermethylation, which reduces genes involved in cadmium transport and accumulation. This helps shield root cells from dangerous metal levels. Similar to this, lead stress in rice (*Oryza sativa*) has been demonstrated to cause DNA hypomethylation, which in turn promotes genes linked to the synthesis of antioxidants and metal transporters, which aid in the accumulation and detoxification of lead ions by root cells (Feng et al., 2016).

The regulation of the reaction of plants to contaminants is additionally greatly impacted by histone changes, including acetylation and methylation. Genes involved in metal detoxification and the oxidative stress response are activated as a result of elevated histone acetylation that occurs at heat-responsive sites. For example, in *Brassica napus*, stress from cadmium causes histone acetylation at the promoters of genes that code for the enzyme Glutathione-S-transferase (GST), which is important in detoxifying reactive oxygen species (ROS) generated during metal stress (Banerjee et al., 2021). Conversely, the histone methylation process has the ability to either activate or repress particular genes. Histone methylation has been shown to suppress genes linked to arsenic transport in response to arsenic stress. This lowers the amount of arsenic absorbed by roots and shields the crops from arsenic toxicity (Talukdar, 2017).

Smaller RNAs, in particular microRNAs (miRNAs), serve as significant participants in the epigenetic management of root responses to heavy metals, in addition to DNA and histone changes. The expression of genes associated with metal detoxification and stress responses is regulated by miRNAs like miR398 and miR397. To improve Arabidopsis roots' ability to resist metal stress, for instance, miR398 targets superoxide dismutase (SOD), an enzyme that is important in reducing oxidative damage brought on by heavy metals like copper (Cu) (Ding and Zhu, 2009). MiR397 also affects root architecture and improves metal sequestration in root tissues, which decreases metal translocation to aerial parts of the plant (Zhang et al., 2022). By allowing plants to fine-tune their root growth, detoxification procedures, and stress responses under toxic stress, such epigenetic alterations form a dynamic regulatory system that improves plant survival in areas of contamination.

Oxidative Stress

Under circumstances involving oxidative stress, epigenetic changes in roots are essential for coordinating gene expression and helping plants withstand the harmful effects of reactive oxygen species (ROS). ROS builds up in root cells as a result of oxidative stress, which is brought on by environmental variables such as salt, dryness, or exposure to heavy metals. ROS can harm lipids, proteins, and DNA. To prevent this, plants activate epigenetic mechanisms such as DNA methylation, histone changes, and the activity of short RNAs. Stress-responsive genes are frequently activated or silenced as a result of alterations in DNA methylation that are brought on by oxidative stress. For example, in *Arabidopsis thaliana*, oxidative stress-induced ROS exposure causes the demethylation of genes linked to antioxidants, like those encoding catalase and superoxide dismutase (SOD), which allows these enzymes to scavenge ROS and shield root cells from oxidative damage (Miryeganeh, 2021). Demethylation provides a quick reaction of defense and promotes root tolerance to stress.

Important regulators of oxidative stress responses are also histone changes. While histone methylation can either activate or repress genes depending on the site of modification, histone acetylation normally increases gene expression by loosening the structure of the chromatin. Histone acetylation in the promoter regions of genes implicated in ROS elimination under oxidative stress has been detected in rice (Oryza sativa), resulting in the activation of antioxidant genes that lessen ROS damage in root tissues (Hernandez et al., 2022, Ijaz et al., 2024b). On the other hand, stress-sensitive genes can be suppressed by histone methylation at particular loci, avoiding needless cellular reactions that might worsen oxidative damage.

Furthermore, during oxidative stress, tiny RNAs—in particular, microRNAs, or miRNAs—modify gene expression by either targeting mRNAs for translation inhibition or breakdown. For instance, miR398 targets genes that encode the essential enzymes that detoxify reactive oxygen species (ROS) in roots under oxidative stress conditions, copper/zinc superoxide dismutase (CSD1 and CSD2) (Sunkar et al., 2006). Increased production of those antioxidant enzymes during oxidative stress is made possible by downregulating miR398 in Arabidopsis, which shields root cells from an excessive accumulation of reactive oxygen species. Plants are able to precisely regulate gene expression in response to oxidative stress thanks to these epigenetic alterations, which provide effective protection against antioxidant activation and support root survival under adverse situations.

EPIGENETIC RESPONSES TO ENVIRONMENTAL STRESSES: BIOTIC STRESSES

Biotic stresses have an enormous effect on the health and productivity of plants. These pressures include pathogen attack, herbivory, and interactions with other organisms including symbionts and pests. In order to cope with these stressors, plants have developed complex regulatory systems that include non-coding RNAs, histone modifications, and DNA methylation in order to control gene expression and strengthen defense mechanisms Figure 4.

Figure 4. Epigenetic regulation maintains constant tolerance against abiotic stress

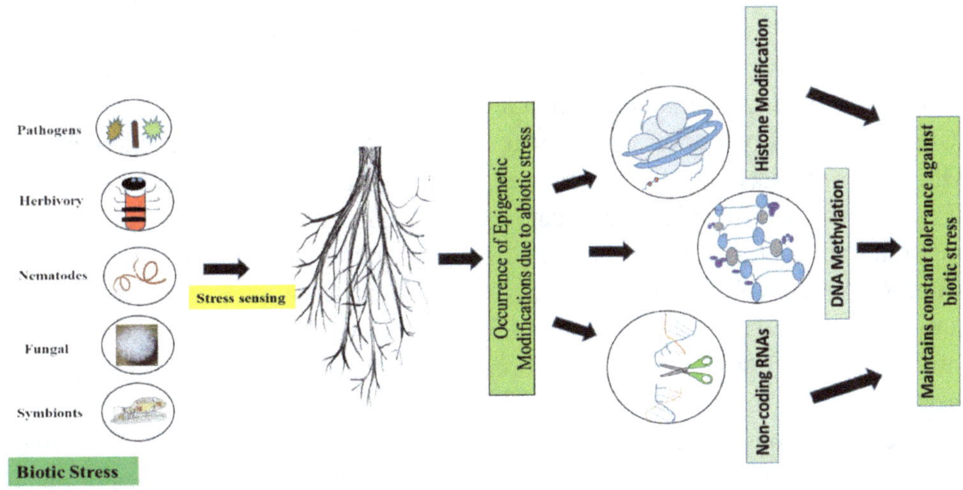

Pathogen Attack

When bacteria, fungi, viruses, or oomycetes attack a plant, a complex epigenetic reaction takes place, which aids the plant in establishing a successful defense. An essential function of DNA methylation is to control genes that respond to pathogens. Variations in DNA methylation caused by pathogens have the ability to mute transposable elements and activate genes linked to defense, both of which have the potential to impair immune responses (Matzke and Mosher, 2014; Alvarez et al.,

2010). By changing the layout of chromatin and allowing the activation of genes like those encoding peptides that protect against bacteria and pathogenesis-related proteins, histone modifications, such as acetylation and methylation, affect pathogen defense (Ding and Wang, 2015; Lee et al., 2016). Non-coding RNAs, or RNAs, have a crucial role in pathogen defense, particularly microRNAs (miRNAs) and small interfering RNAs (siRNAs). While miRNAs control the expression of genes involved in the immune response, siRNAs have the ability to target and destroy viral RNA (Ruiz-Ferrer and Voinnet, 2009; Luo et al., 2024).

Herbivory

When insects or other living things eat plants, a complex stress response is triggered in the plants. This response is controlled by different epigenetic alterations that impact the defensive mechanisms and root development of the plants. The expression of genes involved in the manufacture of jasmonic acid (JA) and secondary metabolites, which are essential for herbivore tolerance, is altered by DNA methylation, especially in the promoters of important genes. For instance, herbivorous organism by *Pieris brassicae* induces hypermethylation in genes associated to root growth in Arabidopsis thaliana, which impacts root elongation (Kellenberger et al., 2018). Additionally, important are histone alters like acetylation and methylation, which alter the access of chromatin and activate genes linked to defense. During herbivore attacks, Nicotiana attenuate exhibits enhanced JA-responsive gene expression due to elevated histone acetylation, as shown by (Mostafa et al., 2022). MicroRNAs (miRNAs) are one type of non-coding RNA (ncRNA) that regulate herbivore-induced responses further. MiRNAs improve defense responses in Oryza sativa by modifying JA signaling and root growth (Kumar et al., 2022). Together, these epigenetic mechanisms allow plants to modify their root development and defenses in response to herbivore pressure. Some of these modifications can even be handed down to subsequent generations, as demonstrated by the case of plants made from Arabidopsis subjected to larvae of the *Spodoptera littoralis* (Verhoeven and Van Gurp, 2012).

Nematode Infection

Nematode diseases result from considerable epigenetic changes which impact both root growth and parasite resistance. Nematode infections are caused by microscopic worms that infect plant roots. During parasite infection, DNA methylation patterns are changed; hypermethylation of particular genes frequently results in their silence, which can aid in the management of damage caused by nematodes. For instance, it has been discovered that hypermethylation of the RAP2.6 genes in

Arabidopsis thaliana suppresses the gene's expression, lowering the plant's vulnerability to infection by the root-knot nematode (*Meloidogyne incognita*) (Hewezi, 2020). Similarly, in Glycine max (soybean), variations in DNA methylation influenced the expression of genes linked to the formation of root galls produced by nematodes, hence reducing damage (Rambani et al., 2020). Nematode resistance is also influenced by histone alterations. According to Mahomed, 2022, found that in *Solanum lycopersicum* (tomato), for example, increasing histone H3 acetylation at defense gene loci improved the expression of genes implicated in resistance to cyst nematodes (*Heterodera schachtii*). In addition, it was discovered that *Medicago truncatula* was affected by histone H3K9 methylation. Demethylation at particular genes triggered root defense mechanisms, enhancing parasite resistance (Wang et al., 2020). By altering gene expression to strengthen root resistance and resilience, these epigenetic reactions assist plants in controlling worm infections.

Fungal Infections

Fungal infections have the capacity to alter a plant's epigenetic makeup significantly, affecting how roots grow and triggering defensive mechanisms. Changes in proteins and DNA methylation both have important functions in controlling the expression of genes related to fungal resistance. For instance, in *Arabidopsis thaliana*, *Fusarium oxysporum* infection causes changes in DNA methylation patterns at the promoters of defense-related genes, which repress genes related to normal root development and activate genes that strengthen the plant's defenses, such as those involved in producing antifungal substances and enhancing cell walls (He et al., 2020). Similarly, in *Oryza sativa* (rice), a *Magnaporthe oryzae* infection causes DNA demethylation at loci relevant to defense, which activates the genes encoding chitinase, which breaks down the fungal cell wall (Arora et al., 2022).

Another significant procedure by which plants control their root' reactions to fungal infections is modification of histone. In *Nicotiana tabacum*, infection by *Rhizoctonia solani* led to increased histone H3 acetylation at the promoters of defense genes, such as those producing PR (pathogenesis-related) proteins, improving resistance (Mierziak and Wojtasik, 2024). Reduced histone H3K9 methylation was discovered in defense-related genes in *Solanum Lycopersicon* (tomato) after *Verticillium dahliae* infection. This resulted in their overexpression and enhanced synthesis of antifungal chemicals such as phytoalexins (Mahomed, 2022).

Furthermore, during fungal infections, non-coding RNAs regulate gene expression. According to Soto-Suarez et al., 2017, miR396 has been demonstrated to target genes in Oryza sativa that are involved in root development during infection by *Fusarium graminearum*. This modulation of auxin signaling balances root growth with the activation of defensive pathways. Plants could continuously regulate their defensive mechanisms and development of roots thanks to these epigenetic alterations, which maintain an adaptive balance between fungal disease resistance and growth.

Interactions with Symbionts

Mycorrhizae and nitrogen-fixing bacteria are examples of symbiotic organisms with which interactions can interact to produce epigenetic changes that enhance nutrient intake and maximize root development. Histone alterations and DNA methylation play a critical role in regulating the expression of the genes included in these advantageous interactions. For instance, in *Medicago truncatula*, phosphate transporter gene promoter DNA demethylation was caused by symbiotic association with *arbuscular mycorrhizal* fungus (AMF), which promoted phosphorus uptake and plant growth (Uhe et al., 2018). Similar alterations in DNA methylation were linked to *Oryza sativa* symbiosis with AMF. These modifications affected genes related to root architecture and nutrient absorption, enhancing the plant's ability to adapt to nutrient-poor soils (Zanetti et al., 2024).

Additionally, histone changes are essential for symbiotic connections. During interaction with Rhizobium bacteria, it was observed that histone H3 acetylation increased at the promoters of nitrogen fixation-related genes in *Lotus japon*icus, allowing root nodulation and improving nitrogen uptake (Asgari, 2014). Furthermore, it has been demonstrated that histone H3K4 trimethylation at particular loci enhances the expression of genes related to mycorrhizal symbiosis in Phaseolus vulgaris (common bean), which enhances root colonization and nutrient absorption (Fiorilli et al., 2024).

Symbiotic relationships are regulated by non-coding RNAs as well. For example, miR171 was demonstrated to regulate NODULATION SIGNALING PATHWAY 1 (NSP1) expression in Medicago truncatula during Rhizobium root colonization, thereby achieving a balance between nodule development and root growth (Han and Zhu, 2022). Plants are able to maximize the benefits of symbiosis as well as root development through this regulation.

APPLICATIONS AND IMPLICATIONS

Agricultural Applications

In the face of environmental stressors, epigenetic changes have shown to be a potent tool for farming, providing creative ways to boost crop resilience and yield. Researchers can create crops that are more resilient to different stressors by modifying these alterations, which will ultimately increase agricultural sustainability and food security.

Enhancing Stress Tolerance Through Epigenetic Modifications

Epigenetic Regulation of Stress Tolerance:

Plant adaptation to environmental challenges depends on epigenetic alterations such as DNA methylation, histone modifications, and non-coding RNA regulation. By changing the way stress-responsive genes are expressed, these alterations help plants cope with challenges such as salt, drought, and severe temperatures. It has been demonstrated, for instance, that DNA methylation controls genes related to drought tolerance by altering gene expression patterns that assist plants in managing osmotic stress and conserving water (Kumar et al., 2016, Ullah et al., 2023a). According to santos et al., 2017, histone changes like acetylation and methylation additionally impact the expression of genes linked to stressful responses, enhancing resistance to illnesses and stressors from the environment. Furthermore, by controlling gene expression at various levels, non-coding RNAs such as lncRNAs and miRNAs alter responses to stress pathways and enhance stress resilience (Huang et al., 2023, Ullah et al., 2023b).

Transgenerational Stress Memory

The potential of epigenetic variations to produce transgenerational stress memory is one of their primary benefits. According to Jablonka and Lamb, 1995, crops that experience stress from the environment might pass on their higher stress sensitivity to their progeny through epigenetic modifications. Plants having this transgenerational memory are better able to withstand repeated shocks and can be used to create crop types that are more resilient over the long run. For instance, it has been demonstrated that stress-induced epigenetic modifications in crops like wheat and rice improve stress tolerance in succeeding generations, offering a useful technique for breeding stress-resistant varieties (Ali et al., 2022, Ullah et al., 2023).

Epigenetic Tools for Crop Improvement

Epigenetic Editing

With the advent of CRISPR/Cas9-based systems and other recent developments in the field, farmers now use precise tools at their disposal to adjust crop attributes and alter epigenetic markers. Researchers can precisely control the expression of genes sensitive to stress by utilizing these technologies to modify DNA methylation patterns and histone modifications (Liu et al., 2022, Haider et al., 2021a, Haider et al., 2021b). By using epigenetic editing, for instance, it is possible to increase a crop's resistance to environmental stressors like drought and saltiness by activating or silencing particular genes involved in stress tolerance. This strategy opens up new possibilities for improving productivity in agriculture and creating crops with customized responses when stressed.

Epigenetic Markers for Breeding

Crop breeding efforts can be greatly improved by identifying and employing epigenetic indicators linked to stress tolerance. Selecting and developing crops with desired characteristics can be done more effectively by breeders by screening for particular epigenetic modifications associated with stress resilience. For example, drought-tolerant crop types such as wheat and rice have been identified and bred using epigenetic characteristics associated with drought tolerance (Samantara et al., 2021, Haider et al., 2022a). These indicators aid in the breeding of crops that are more adapted to a variety of environmental situations by providing insightful information about the genetic and epigenetic underpinnings of stress responses.

CONCLUSION

Plant stress responses and the regulation of root growth depend heavily on epigenetic changes. Plants have a capacity to regulate the expression of genes in accordance with developmental cues and environmental stimuli through a variety of alterations, such as DNA methylation, histone modifications, and the involvement of non-coding RNAs. In contrast to mutations in the genome, epigenetic modifications are dynamic and reversible, giving crops the adaptability they need to deal with changing environmental conditions without affecting their fundamental genetic code. The internal structure of the root system, which directly affects nutrient uptake, water acquisition, and general plant health, depends on this plasticity.

Key processes in the growth of roots, including differentiation of cells, expansion, and pattern creation, are controlled by epigenetic changes. For example, DNA methylation is essential for the temporal and spatial control of root growth genes, which guarantees healthy development at various phases. In contrast, chromatin accessibility is modulated by histone modifications, which regulate the activation or repression of genes related to lateral root development and root elongation. Furthermore, by adjusting the gene expression networks that dictate the form and function of roots, non-coding RNAs—including microRNAs and long non-coding RNAs—act as crucial regulators.

Plants use epigenetic mechanisms to modify gene expression patterns in response to environmental challenges like salt, drought, and nutrient shortage, which helps the plants survive. Stress, which can be for instance, can activate genes that promote root growth, increase the density of root hair, or start root branching to optimize the absorption of water and nutrients by altering DNA methylation and histone changes.

Furthermore, non-coding RNAs play a crucial role in controlling the expression of stress-responsive genes, providing an additional degree of control over the way roots respond to difficult circumstances. Crucially, epigenetic memory enables plants to retain memories of past stress exposure, enabling a stronger reaction in the event that the stressor arises again. The dual function that epigenetic alterations play in plant resilience is highlighted by their capacity to impact both stress responses and developmental processes. Knowing how epigenetics affects root growth is becoming more and more crucial as environmental stresses like soil erosion, water shortages, and climate change continue to endanger world agriculture. Through the identification of distinct epigenetic regulators and their corresponding molecular targets, scientists can devise tactics to bolster the resilience of crops and improve root characteristics for superior performance in harsh conditions. Additionally, future studies must concentrate on understanding the extensive regulatory systems influenced by epigenetic alterations as they relate to stress memory and long-term adaptability. To fully realize the promise of plants in enhancing agricultural sustainability, research must focus on how they integrate different types of epigenetic signals to regulate root development and stress responses. Furthermore, by incorporating epigenetic information into breeding programs, crops with increased stress tolerance, better nutrient efficiency, and higher yield potential in a variety of situations may be developed.

REFERENCES

. Ahmad, Y., Haakim, Z., Iqbal, J., Abbasi, B. A., Mahmood, T., & Kazi, M. (2024b). Technological Innovations for Abiotic Stress Resistance in Horticultural Crops. OMICs-based Techniques for Global Food Security, 233-244.

Ahmad, Y., Haider, S., Iqbal, J., Abbasi, B. A., Yaseen, T., & Mahmood, T. (2022a). The mechanisms of genome editing technologies in crop plants. In *Principles and Practices of OMICS and Genome Editing for Crop Improvement* (pp. 295–313). Springer International Publishing.

Ahmad, Y., Haider, S., Iqbal, J., Naseer, S., Attia, K. A., Mohammed, A. A., & Mahmood, T. (2024a). In-silico analysis and transformation of OsMYB48 transcription factor driven by CaMV35S promoter in model plant–Nicotiana tabacum L. conferring abiotic stress tolerance. *GM Crops and Food: Biotechnology in Agriculture and the Food Chain*, *15*(1), 130–149. PMID: 38551174

Ali, A., Altaf, M. T., Nadeem, M. A., Karaköy, T., Shah, A. N., Azeem, H., & Chung, Y. S. (2022). Recent advancement in OMICS approaches to enhance abiotic stress tolerance in legumes. *Frontiers in Plant Science*, *13*, 952759. PMID: 36247536

Ali, I., Salah, K. B. H., Sher, H., Ali, H., Ullah, Z., Ali, A., & Mahmood, T. (2022). Drought stress enhances the efficiency of floral dip method of Agrobacterium-mediated transformation in Arabidopsis thaliana. *Brazilian Journal of Biology*, *84*, e259326. PMID: 35703626

Alvarez, M. E., Nota, F., & Cambiagno, D. A. (2010). Epigenetic control of plant immunity. *Molecular Plant Pathology*, *11*(4), 563–576. PMID: 20618712

Arora, H., Singh, R. K., Sharma, S., Sharma, N., Panchal, A., Das, T., & Prasad, M. (2022). DNA methylation dynamics in response to abiotic and pathogen stress in plants. *Plant Cell Reports*, *41*(10), 1931–1944. PMID: 35833989

Asgari, S. (2014). Epigenetic modifications underlying symbiont–host interactions. *Advances in Genetics*, *86*, 253–276. PMID: 25172352

Banerjee, S., Mitra, M., Agarwal, P., & Roy, S. (2021). 79Unveiling the Mechanism of Transcriptional and Epigenetic Responses in Plants under Heavy Metal Stress: An Update. In *Heavy Metal Toxicity in Plants* (pp. 79–94). CRC Press.

Bannister, A. J., & Kouzarides, T. (2011). Regulation of chromatin by histone modifications. *Cell Research*, *21*(3), 381–395. PMID: 21321607

Bazin, J., Khan, G. A., Combier, J. P., Bustos-Sanmamed, P., Debernardi, J. M., Rodriguez, R., & Lelandais-Brière, C. (2013). miR396 affects mycorrhization and root meristem activity in the legume M edicago truncatula. *The Plant Journal, 74*(6), 920–934. PMID: 23566016

. Be, B. (2006). A bivalent chromatin structure marks key developmental genes in embryonic stem cells. *cell, 125*, 315-326.

Berger, S. L. (2007). The complex language of chromatin regulation during transcription. *Nature, 447*(7143), 407–412. PMID: 17522673

Bian, H., Xie, Y., Guo, F., Han, N., Ma, S., Zeng, Z., & Zhu, M. (2012). Distinctive expression patterns and roles of the miRNA393/TIR1 homolog module in regulating flag leaf inclination and primary and crown root growth in rice (Oryza sativa). *The New Phytologist, 196*(1), 149–161. PMID: 22846038

Bird, A. (2002). DNA Methylation Patterns and Epigenetic Memory. *Genes & Development, 16*(1), 6–21. PMID: 11782440

Bokszczanin, K. L.Solanaceae Pollen Thermotolerance Initial Training Network (SPOT-ITN) Consortium, & Fragkostefanakis, S. (2013). Perspectives on deciphering mechanisms underlying plant heat stress response and thermotolerance. *Frontiers in Plant Science, 4*, 315. PMID: 23986766

Boyko, A., & Kovalchuk, I. (2008). Epigenetic control of plant stress response. *Environmental and Molecular Mutagenesis, 49*(1), 61–72. PMID: 17948278

Chan, S. W. L., Henderson, I. R., & Jacobsen, S. E. (2005). Gardening the genome: DNA methylation in Arabidopsis thaliana. *Nature Reviews. Genetics, 6*(5), 351–360. PMID: 15861207

Chen, D. H., Huang, Y., Jiang, C., & Si, J. P. (2018). Chromatin-based regulation of plant root development. *Frontiers in Plant Science, 9*, 1509. PMID: 30386363

Chen, X.. (2021). Small RNAs and Their Role in Plant Development. *Nature Reviews. Molecular Cell Biology, 22*(10), 589–603.

Chinnusamy, V., & Zhu, J. K. (2009). Epigenetic regulation of stress responses in plants. *Current Opinion in Plant Biology, 12*(2), 133–139. PMID: 19179104

Ding, B., & Wang, G. L. (2015). Chromatin versus pathogens: The function of epigenetics in plant immunity. *Frontiers in Plant Science, 6*, 675. PMID: 26388882

Ding, H., Gao, J., Qin, C., Ma, H., Huang, H., Song, P., & Zhang, Z. (2014). The dynamics of DNA methylation in maize roots under Pb stress. *International Journal of Molecular Sciences, 15*(12), 23537–23554. PMID: 25526567

Ding, Y., Fromm, M., & Avramova, Z. (2012). Multiple exposures to drought'train'transcriptional responses in Arabidopsis. *Nature Communications*, *3*(1), 740. PMID: 22415831

Ding, Y. F., & Zhu, C. (2009). The role of microRNAs in copper and cadmium homeostasis. *Biochemical and Biophysical Research Communications*, *386*(1), 6–10. PMID: 19501049

Esteller, M. (2011). Non-coding RNAs in human disease. *Nature Reviews. Genetics*, *12*(12), 861–874. PMID: 22094949

Fang, Y., Liao, K., Du, H., Xu, Y., Song, H., Li, X., & Xiong, L. (2015). A stress-responsive NAC transcription factor SNAC3 confers heat and drought tolerance through modulation of reactive oxygen species in rice. *Journal of Experimental Botany*, *66*(21), 6803–6817. PMID: 26261267

Feng, S. J., Liu, X. S., Tao, H., Tan, S. K., Chu, S. S., Oono, Y., & Yang, Z. M. (2016). Variation of DNA methylation patterns associated with gene expression in rice (Oryza sativa) exposed to cadmium. *Plant, Cell & Environment*, *39*(12), 2629–2649. PMID: 27412910

Fiorilli, V., Martínez-Medina, A., Pozo, M. J., & Lanfranco, L. (2024). Plant Immunity Modulation in Arbuscular Mycorrhizal Symbiosis and Its Impact on Pathogens and Pests. *Annual Review of Phytopathology*, *62*(1), 127–156. PMID: 39251211

Guo, Q., Liu, Q. A., Smith, N., Liang, G., & Wang, M. B. (2016). RNA silencing in plants: Mechanisms, technologies and applications in horticultural crops. *Current Genomics*, *17*(6), 476–489. PMID: 28217004

Haider, S., Iqbal, J., Naseer, S., Shaukat, M., Abbasi, B. A., Yaseen, T., & Mahmood, T. (2022b). Unfolding molecular switches in plant heat stress resistance: A comprehensive review. *Plant Cell Reports*, *41*(3), 775–798. PMID: 34401950

Haider, S., Iqbal, J., Naseer, S., Yaseen, T., Shaukat, M., Bibi, H., & Mahmood, T. (2021a). Molecular mechanisms of plant tolerance to heat stress: Current landscape and future perspectives. *Plant Cell Reports*, •••, 1–25. PMID: 33890138

Haider, S., Iqbal, J., Shaukat, M., Naseer, S., & Mahmood, T. (2021b). The epigenetic chromatin-based regulation of somatic heat stress memory in plants. *Plant Gene*, *27*, 100318.

Haider, S., Raza, A., Iqbal, J., Shaukat, M., & Mahmood, T. (2022a). Analyzing the regulatory role of heat shock transcription factors in plant heat stress tolerance: A brief appraisal. *Molecular Biology Reports*, *49*(6), 5771–5785. PMID: 35182323

Han, H., & Zhou, Y. (2022). Function and regulation of microRNA171 in plant stem cell homeostasis and developmental programing. *International Journal of Molecular Sciences*, *23*(5), 2544. PMID: 35269685

Harris, C. J., Scheibe, M., Wongpalee, S. P., Liu, W., Cornett, E. M., Vaughan, R. M., & Jacobsen, S. E. (2018). A DNA methylation reader complex that enhances gene transcription. *Science*, *362*(6419), 1182–1186. PMID: 30523112

He, C., Zhang, Z., Li, B., & Tian, S. (2020). The pattern and function of DNA methylation in fungal plant pathogens. *Microorganisms*, *8*(2), 227. PMID: 32046339

Hewezi, T. (2020). Epigenetic mechanisms in nematode–plant interactions. *Annual Review of Phytopathology*, *58*(1), 119–138. PMID: 32413274

Huang, J., Zhou, W., Zhang, X., & Li, Y. (2023). Roles of long non-coding RNAs in plant immunity. *PLoS Pathogens*, *19*(5), e1011340. PMID: 37167200

Ijaz, S., Iqbal, J., Abbasi, B. A., Tufail, A., Ullah, Z., Sharifi-Rad, J., & Uddin, S. (2023). Biofortification: Lessons from the Past and Strategies for Future Food Security. In *Legumes Biofortification* (pp. 521–545). Springer International Publishing.

. Ijaz, S., Iqbal, J., Abbasi, B. A., Yaseen, T., Rehman, S., Kazi, M., & Mahmood, T. (2024a). Role of OMICS-Based Technologies in Plant Sciences. OMICs-based Techniques for Global Food Security, 45-66.

. Ijaz, S., Iqbal, J., Abbasi, B. A., Yaseen, T., Rehman, S., Kazi, M. & Mahmood, T. (2024b). Role of OMICS-Based Technologies in Plant Sciences. OMICs-based Techniques for Global Food Security, 45-66.

Ikeuchi, M., Iwase, A., & Sugimoto, K. (2015). Control of plant cell differentiation by histone modification and DNA methylation. *Current Opinion in Plant Biology*, *28*, 60–67. PMID: 26454697

Jablonka, E., & Lamb, M. J. (1995). *Epigenetic inheritance and evolution: the Lamarckian dimension*. Oxford University Press.

Jaenisch, R., & Bird, A. (2003). Epigenetic regulation of gene expression: How the genome integrates intrinsic and environmental signals. *Nature Genetics*, *33*(3), 245–254. PMID: 12610534

Jang, I. C., Chung, P. J., Hemmes, H., Jung, C., & Chua, N. H. (2011). Rapid and reversible light-mediated chromatin modifications of Arabidopsis phytochrome A locus. *The Plant Cell*, *23*(2), 459–470. PMID: 21317377

Jia, Y., Lisch, D. R., Ohtsu, K., Scanlon, M. J., Nettleton, D., & Schnable, P. S. (2009). Loss of RNA–dependent RNA polymerase 2 (RDR2) function causes widespread and unexpected changes in the expression of transposons, genes, and 24-nt small RNAs. *PLOS Genetics*, *5*(11), e1000737. PMID: 19936292

Kar, M. M., & Raichaudhuri, A. (2021). Role of microRNAs in mediating biotic and abiotic stress in plants. *Plant Gene*, *26*, 100277.

Karan, R., DeLeon, T., Biradar, H., & Subudhi, P. K. (2012). Salt stress induced variation in DNA methylation pattern and its influence on gene expression in contrasting rice genotypes. *PLoS One*, *7*(6), e40203. PMID: 22761959

Kellenberger, R. T., Schlüter, P. M., & Schiestl, F. P. (2018). Chapter I Herbivore-induced DNA demethylation changes floral signalling and attractiveness to pollinators in Brassica rapa. *Insect-Mediated Microevolution of Flowering Plants, 11*, 45.

Kim, J. M., Sasaki, T., Ueda, M., Sako, K., & Seki, M. (2015). Chromatin changes in response to drought, salinity, heat, and cold stresses in plants. *Frontiers in Plant Science*, *6*, 114. PMID: 25784920

. Knip, M. L. U. (2012). *Daysleeper: from genomic parasite to indispensable gene*

Kouzarides, T. (2007). Chromatin modifications and their function. *Cell*, *128*(4), 693–705. PMID: 17320507

Kumar, K., Mandal, S. N., Neelam, K., & de Los Reyes, B. G. (2022). MicroRNA-mediated host defense mechanisms against pathogens and herbivores in rice: Balancing gains from genetic resistance with trade-offs to productivity potential. *BMC Plant Biology*, *22*(1), 351. PMID: 35850632

Kumar, S., & Singh, A. (2016). Epigenetic regulation of abiotic stress tolerance in plants. *Advances in Plants & Agriculture Research*, *5*(5), 00179.

Kumari, P., Khan, S., Wani, I. A., Gupta, R., Verma, S., Alam, P., & Alaklabi, A. (2022). Unravelling the role of epigenetic modifications in development and reproduction of angiosperms: A critical appraisal. *Frontiers in Genetics*, *13*, 819941. PMID: 35664328

Lauria, M., & Rossi, V. (2011). Epigenetic control of gene regulation in plants. *Biochimica et Biophysica Acta (BBA)-. Gene Regulatory Mechanisms*, *1809*(8), 369–378.

Lee, S., Fu, F., Xu, S., Lee, S. Y., Yun, D. J., & Mengiste, T. (2016). Global regulation of plant immunity by histone lysine methyl transferases. *The Plant Cell*, *28*(7), 1640–1661. PMID: 27354553

Li, H., Yan, S., Zhao, L., Tan, J., Zhang, Q., Gao, F., & Li, L. (2014). Histone acetylation associated up-regulation of the cell wall related genes is involved in salt stress induced maize root swelling. *BMC Plant Biology*, *14*, 1–14. PMID: 24758373

. Li, Y., & Tollefsbol, T. O. (2011). DNA methylation detection: bisulfite genomic sequencing analysis. *Epigenetics protocols*, 11-21.

Liu, H., Chen, W., Li, Y., Sun, L., Chai, Y., Chen, H., & Huang, C. (2022). CRISPR/Cas9 technology and its utility for crop improvement. *International Journal of Molecular Sciences*, *23*(18), 10442. PMID: 36142353

Liu, X., Hao, L., Li, D., Zhu, L., & Hu, S. (2015). Long non-coding RNAs and their biological roles in plants. *Genomics, Proteomics & Bioinformatics*, *13*(3), 137–147. PMID: 25936895

Luo, C., Bashir, N. H., Li, Z., Liu, C., Shi, Y., & Chu, H. (2024). Plant microRNAs regulate the defense response against pathogens. *Frontiers in Microbiology*, *15*, 1434798. PMID: 39282567

Mahomed, N. B. (2022). *Investigation of histone methylation of tomato curly stunt virus (ToCSV) in resistant and susceptible tomato lines* (Doctoral dissertation, University of Johannesburg).

Matzke, M. A., & Mosher, R. A. (2014). RNA-directed DNA methylation: An epigenetic pathway of increasing complexity. *Nature Reviews. Genetics*, *15*(6), 394–408. PMID: 24805120

Meng, Y., Ma, X., Chen, D., Wu, P., & Chen, M. (2010). MicroRNA-mediated signaling involved in plant root development. *Biochemical and Biophysical Research Communications*, *393*(3), 345–349. PMID: 20138828

Mierziak, J., & Wojtasik, W. (2024). Epigenetic weapons of plants against fungal pathogens. *BMC Plant Biology*, *24*(1), 175. PMID: 38443788

Miryeganeh, M. (2021). Plants' epigenetic mechanisms and abiotic stress. *Genes*, *12*(8), 1106. PMID: 34440280

Miura, A., Yonebayashi, S., Watanabe, K., Toyama, T., Shimada, H., & Kakutani, T. (2001). Mobilization of transposons by a mutation abolishing full DNA methylation in Arabidopsis. *Nature*, *411*(6834), 212–214. PMID: 11346800

Mostafa, S., Wang, Y., Zeng, W., & Jin, B. (2022). Plant responses to herbivory, wounding, and infection. *International Journal of Molecular Sciences*, *23*(13), 7031. PMID: 35806046

Nejat, N., & Mantri, N. (2018). Emerging roles of long non-coding RNAs in plant response to biotic and abiotic stresses. *Critical Reviews in Biotechnology*, *38*(1), 93–105. PMID: 28423944

Rambani, A., Pantalone, V., Yang, S., Rice, J. H., Song, Q., Mazarei, M., & Hewezi, T. (2020). Identification of introduced and stably inherited DNA methylation variants in soybean associated with soybean cyst nematode parasitism. *The New Phytologist*, *227*(1), 168–184. PMID: 32112408

Reinhart, B. J., Weinstein, E. G., Rhoades, M. W., Bartel, B., & Bartel, D. P. (2002). MicroRNAs in plants. *Genes & Development*, *16*(13), 1616–1626. PMID: 12101121

Rinn, J. L., & Chang, H. Y. (2012). Genome regulation by long noncoding RNAs. *Annual Review of Biochemistry*, *81*(1), 145–166. PMID: 22663078

Roudier, F., Ahmed, I., Bérard, C., Sarazin, A., Mary-Huard, T., Cortijo, S., & Colot, V. (2011). Integrative epigenomic mapping defines four main chromatin states in Arabidopsis. *The EMBO Journal*, *30*(10), 1928–1938. PMID: 21487388

Roulé, T., Crespi, M., & Blein, T. (2022). Regulatory long non-coding RNAs in root growth and development. *Biochemical Society Transactions*, *50*(1), 403–412. PMID: 34940811

Ruiz-Ferrer, V., & Voinnet, O. (2009). Roles of plant small RNAs in biotic stress responses. *Annual Review of Plant Biology*, *60*(1), 485–510. PMID: 19519217

Samantara, K., Shiv, A., de Sousa, L. L., Sandhu, K. S., Priyadarshini, P., & Mohapatra, S. R. (2021). A comprehensive review on epigenetic mechanisms and application of epigenetic modifications for crop improvement. *Environmental and Experimental Botany*, *188*, 104479.

Santos, A. P., Ferreira, L. J., & Oliveira, M. M. (2017). Concerted flexibility of chromatin structure, methylome, and histone modifications along with plant stress responses. *Biology (Basel)*, *6*(1), 3. PMID: 28275209

Santos-Rosa, H., Schneider, R., Bannister, A. J., Sherriff, J., Bernstein, B. E., Emre, N. T., & Kouzarides, T. (2002). Active genes are tri-methylated at K4 of histone H3. *Nature*, *419*(6905), 407–411. PMID: 12353038

Shahzad, Z., Eaglesfield, R., Carr, C., & Amtmann, A. (2020). Cryptic variation in RNA-directed DNA-methylation controls lateral root development when auxin signalling is perturbed. *Nature Communications*, *11*(1), 218. PMID: 31924796

Shanker, A. K., & Maheswari, M. (2017). Small RNA and drought tolerance in crop plants. *Indian Journal of Plant Physiology / Official Publication of the Indian Society for Plant Physiology*, *22*, 422–433.

Sharma, M., Sidhu, A. K., Samota, M. K., Gupta, M., Koli, P., & Choudhary, M. (2023). Post-Translational Modifications in Histones and Their Role in Abiotic Stress Tolerance in Plants. *Proteomes*, *2023*(11), 38. PMID: 38133152

Shemer, O., Landau, U., Candela, H., Zemach, A., & Williams, L. E. (2015). Competency for shoot regeneration from Arabidopsis root explants is regulated by DNA methylation. *Plant Science*, *238*, 251–261. PMID: 26259192

. Singh, S., Yadav, S., Singh, A., Mahima, M., Singh, A., Gautam, V., & Sarkar, A. K. (2020). Auxin signaling modulates LATERAL ROOT PRIMORDIUM 1 (LRP 1) expression during lateral root deve

Singroha, G., & Sharma, P. (2019). Epigenetic modifications in plants under abiotic stress. *Epigenetics*, ●●●, 10.

Slotkin, R. K., & Martienssen, R. A. (2007). Transposable Elements and the Epigenetic Regulation of the Genome. *Nature Reviews. Genetics*, *8*(4), 272–285. PMID: 17363976

Soto-Suárez, M., Baldrich, P., Weigel, D., Rubio-Somoza, I., & San Segundo, B. (2017). The Arabidopsis miR396 mediates pathogen-associated molecular pattern-triggered immune responses against fungal pathogens. *Scientific Reports*, *7*(1), 44898. PMID: 28332603

Stauffer, E., & Maizel, A. (2014). Post-transcriptional regulation in root development. *Wiley Interdisciplinary Reviews. RNA*, *5*(5), 679–696. PMID: 24827552

Sunkar, R., Kapoor, A., & Zhu, J. K. (2006). Posttranscriptional induction of two Cu/Zn superoxide dismutase genes in Arabidopsis is mediated by downregulation of miR398 and important for oxidative stress tolerance. *The Plant Cell*, *18*(8), 2051–2065. PMID: 16861386

. Talukdar, D. (2017). Epigenetics in arsenic toxicity: mechanistic response, alterations, and regulations. *Arsenic Contamination in the Environment: The Issues and Solutions*, 67-101

Tsilimigka, F., Poulios, S., Mallioura, A., & Vlachonasios, K. (2022). ADA2b and GCN5 affect cytokinin signaling by modulating histone acetylation and gene expression during root growth of Arabidopsis thaliana. *Plants*, *11*(10), 1335. PMID: 35631760

Uhe, M., Hogekamp, C., Hartmann, R. M., Hohnjec, N., & Küster, H. (2018). The mycorrhiza-dependent defensin MtDefMd1 of Medicago truncatula acts during the late restructuring stages of arbuscule-containing cells. *PLoS One*, *13*(1), e0191841. PMID: 29370287

Ullah, R., Ullah, Z., Iqbal, J., Chalgham, W., & Ahmad, A. (2023). Aspartic Acid-Based Nano-Copper Induces Resilience in Zea mays to Applied Lead Stress Via Conserving Photosynthetic Pigments and Triggering the Antioxidant Biosystem. *Sustainability*, *15*(16), 12186.

Ullah, Z., Iqbal, J., Abbasi, B. A., Akhtar, W., Kanwal, S., Ali, I., & Mahmood, T. (2023b). Assessment of Gus Expression Induced by Anti-Sense Os PPO Gene Promoter and Antioxidant Enzymatic Assays in Response to Drought and Heavy Metal Stress in Transgenic Arabidopsis thaliana. *Sustainability*, *15*(17), 12783.

Ullah, Z., Iqbal, J., Abbasi, B. A., Ijaz, S., Sharifi-Rad, J., Yaseen, T., & Mahmood, T. (2023a). How Genome Editing Can Be Helpful in the Biofortification of Legumes. In *Legumes Biofortification* (pp. 207–232). Springer International Publishing.

Van Dijk, K., Ding, Y., Malkaram, S., Riethoven, J. J. M., Liu, R., Yang, J., & Fromm, M. (2010). Dynamic changes in genome-wide histone H3 lysine 4 methylation patterns in response to dehydration stress in Arabidopsis thaliana. *BMC Plant Biology*, *10*, 1–12. PMID: 21050490

Vangheluwe, N., & Beeckman, T. (2021). Lateral root initiation and the analysis of gene function using genome editing with CRISPR in Arabidopsis. *Genes*, *12*(6), 884. PMID: 34201141

. Varriale, A. (2017). DNA methylation in plants and its implications in development, hybrid vigour, and evolution. *Plant epigenetics*, 263-280.

Verhoeven, K. J., & van Gurp, T. P. (2012). Transgenerational effects of stress exposure on offspring phenotypes in apomictic dandelion. *PLoS One*, *7*(6), e38605. PMID: 22723869

. Viggiano, L., & de Pinto, M. C. (2017). Dynamic DNA methylation patterns in stress response. *Plant epigenetics*, 281-302.

Wang, H., He, L., Song, J., Cui, W., Zhang, Y., Jia, C., & Liu, W. (2016). Cadmium-induced genomic instability in Arabidopsis: Molecular toxicological biomarkers for early diagnosis of cadmium stress. *Chemosphere*, *150*, 258–265. PMID: 26907594

Wang, H., Tong, X., Tang, L., Wang, Y., Zhao, J., Li, Z., & Zhang, J. (2022). RLB (RICE LATERAL BRANCH) recruits PRC2-mediated H3K27 tri-methylation on OsCKX4 to regulate lateral branching. *Plant Physiology*, *188*(1), 460–476. PMID: 34730827

Wang, Q., Yung, W. S., Wang, Z., & Lam, H. M. (2020). The histone modification H3K4me3 marks functional genes in soybean nodules. *Genomics*, *112*(6), 5282–5294. PMID: 32987152

Wei, J., Shen, Y., Dong, X., Zhu, Y., Cui, J., Li, H., & Liu, Z. (2022). DNA methylation affects freezing tolerance in winter rapeseed by mediating the expression of genes related to JA and CK pathways. *Frontiers in Genetics*, *13*, 968494. PMID: 36061187

Wiles, E. T., & Selker, E. U. (2017). H3K27 methylation: A promiscuous repressive chromatin mark. *Current Opinion in Genetics & Development*, *43*, 31–37. PMID: 27940208

Wu, J., Yamaguchi, N., & Ito, T. (2019). Histone demethylases control root elongation in response to stress-signaling hormone abscisic acid. *Plant Signaling & Behavior*, *14*(7), 1604019. PMID: 30983495

Yamaguchi, A., & Abe, M. (2012). Regulation of reproductive development by non-coding RNA in Arabidopsis: To flower or not to flower. *Journal of Plant Research*, *125*, 693–704. PMID: 22836383

Yao, X., Feng, H., Yu, Y., Dong, A., & Shen, W. H. (2013). SDG2-mediated H3K4 methylation is required for proper Arabidopsis root growth and development. *PLoS One*, *8*(2), e56537. PMID: 23483879

Yu, J. Y., Zhang, Z. G., Huang, S. Y., Han, X., Wang, X. Y., Pan, W. J., & Qi, Z. M. (2019). Analysis of miRNAs targeted storage regulatory genes during soybean seed development based on transcriptome sequencing. *Genes*, *10*(6), 408. PMID: 31142023

Yun, P., Kaya, C., & Shabala, S. (2024). Hormonal and epigenetic regulation of root responses to salinity stress. *The Crop Journal*.

Zanetti, M. E., Blanco, F., Ferrari, M., Ariel, F., Benoit, M., Niebel, A., & Crespi, M. (2024). Epigenetic control during root development and symbiosis. *Plant Physiology*, *196*(2), 697–710. PMID: 38865442

Zanetti, M. E., Blanco, F., Ferrari, M., Ariel, F., Benoit, M., Niebel, A., & Crespi, M. (2024). Epigenetic control during root development and symbiosis. *Plant Physiology*, *196*(2), 697–710. PMID: 38865442

Zhang, B., Pan, X., & Stellwag, E. J. (2008). Identification of soybean microRNAs and their targets. *Planta*, *229*, 161–182. PMID: 18815805

Zhang, F., Yang, J., Zhang, N., Wu, J., & Si, H. (2022). Roles of microRNAs in abiotic stress response and characteristics regulation of plant. *Frontiers in Plant Science*, *13*, 919243. PMID: 36092392

Zhang, Y., & Reinberg, D. (2001). Transcription Regulation by Histone Methylation: Interplay between Different Types of Histone Marks. *Current Opinion in Genetics & Development*, *11*(2), 151–158.

Zhu, J., Jeong, J. C., Zhu, Y., Sokolchik, I., Miyazaki, S., Zhu, J. K., & Bressan, R. A. (2008). Involvement of Arabidopsis HOS15 in histone deacetylation and cold tolerance. *Proceedings of the National Academy of Sciences of the United States of America*, *105*(12), 4945–4950. PMID: 18356294

Chapter 6
Epigenetic Crosstalk:
Integrating Multiple Stress Signals in Plants

Sadia Zafar
https://orcid.org/0000-0003-0879-459X
University of Education, Faisalabad, Pakistan

Muhammad Arslan Ashraf
Government College University, Faisalabad, Pakistan

Rizwan Rasheed
Government College University, Faisalabad, Pakistan

Maria Bilal
Division of Science and Technology, University of Education, Lahore, Pakistan

ABSTRACT

Plant stress significantly affects plant growth, development, and productivity, creating unfavorable conditions that disrupt physiological functions and survival. Environmental pollution, climate change, and global warming intensify the impact of stressors on growth and yield. Plants adapt through a complex innate immune system. While progress has been made in identifying stress-inducible genes and using advanced sequencing tools, signaling mechanisms under combined stressors remain poorly understood. Studies reveal overlapping but distinct responses to environmental stimuli, intricate interactions between abiotic and biotic stress responses, and improper signaling crosstalk during multiple stresses. Epigenetic crosstalk integrates signals to regulate traits through interconnected pathways. We explore chemical regulators in epigenetic research and propose interdisciplinary strategies for developing new regulators.

DOI: 10.4018/979-8-3693-8648-4.ch006

INTRODUCTION

The world's food security is threatened by biotic and abiotic stressors, which are part of the natural ecosystem and include a wide range of biotic (like bacteria, phytoplasmas, oomycetes, nematodes, biotrophic and necrotrophic fungi, and non-cellular pathogens like viruses and viroids) and abiotic (like heat, cold, drought, salinity, mechanical wounds, high light intensity, freezing, heavy metals, and metalloids) that are constantly and simultaneously present in the natural environments of plants. As sessile organisms, plants have developed numerous intricate immune response pathways that allow them to withstand both individual and multiple stressors (Nejat et al., 2016).

Figure 1.

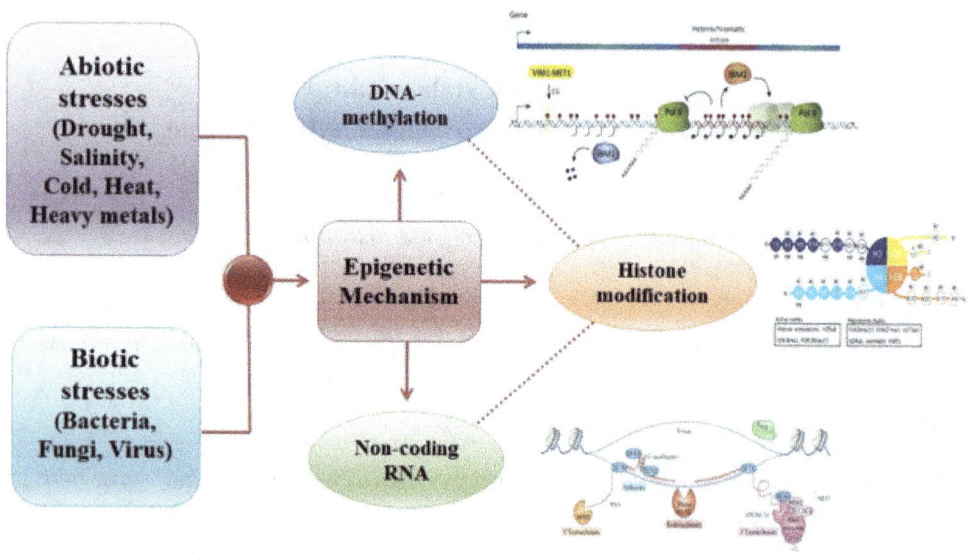

In order to protect plant cells from the difficulties they face, interactions between plant-biotic and/or abiotic stressors set off a variety of defensive processes at both the cellular and molecular levels (Kim et al., 2014). Li et al. (2017) state that phytohormones are tiny, multifunctional compounds produced by plants that, in trace amounts, activate the transmission of signals cascades and are detected by certain proteins. According to Du et al., (2015), there are currently nine primary phytohormones with different chemical structures: strigolactones, jasmonic acid, cytokinins, auxin, ethylene, gibberellins, salicylic acid, abscisic acid, gibberellins, and brassinosteroids.

Throughout a plant's life cycle, phytohormones are essential for a variety of functions, including seed germination and development, flowering, fertilization, and reactions to environmental cues including light and abiotic/biotic stress signals. According to Duan et al. (2018), the production, metabolism, transportation, ligand detection, and transmission of signals pathways of various phytohormones vary. Based on the developmental stage of the plant or in response to external stimuli, the number of transcripts and post-transcriptional modifications of the encoded enzymes, the carriers, receptors, and transducers for signals linked to these methods are handled spatiotemporally to allow for precise programmed development and adaptation to continuously changing circumstances (Song et al., 2019). DNA methylation, histone modifications, chromatin remodeling, and regulation by non-coding RNAs are examples of processes that fall within the broad category of epigenetics, which is the study of heritable control of gene expression without necessitating changes in the genetic code (He et al., 2021).The molecular processes and essential components behind these plant functions have been the subject of in-depth research for many years (Shang & He, 2022). It should be mentioned that depending on the genetic composition, length, and severity of stress, different plant species have unique inherited stress tolerance systems (Song et al., 2021). Plant breeders use phenotypic differences resulting from DNA sequence mutations to improve plant populations because of chromatin state alterations (Sudan et al., 2018). Notably, plants have a strong response to their surroundings due to their capacity for memory and experience-based learning (Bäurle & Trindade, 2020).

It is commonly known that during stressful situations, epigenetic modifications are crucial for maintaining the expression of genes and transcriptome remodeling. "DM, modifications to histone (HM), and non-coding RNA molecules," among other chromatin structural changes, enable these changes. Rather, it entails changes to DNA or related proteins that could affect how genes function (Liu et al., 2023). Plants can convey stress memories to their offspring through these changes, which may be permanent and inherited (Sudan et al., 2018). During development and in responses to environmental stimuli, such as flowering (Luo & He, 2020), the development of fruits (Tang et al., 2020), the development of seeds (Markulin et al., 2021; Ding et al., 2022), and responses to biotic and abiotic stress (Ashapkin et al., 2020), epigenetic mechanisms affect the phenotypic versatility of plants. According to recent studies, different stressors can result in differing DM levels in the same plants (Saeed et al., 2022). It's interesting to note that apomictic Taraxacum officinale plants under abiotic stress showed significant variations in DM irrespective of the type of stress. These suggest that even in the absence of genetic variation across people, epigenetic inheritance may play a significant role in plant adaptation. Stress form, genotype, tissue, and organism all affect stress-induced methylation patterns, which in turn affect stress-responsive gene regulation (González et al., 2016). While it can occur in CHG and CHH (where H is A, C, or T) sequencing scenarios, DM primarily occurs at the cytosine residue in a CG context (CG methylation) in plants (Kawakatsu et al., 2017).

DMs and HMs alter gene expression by blocking transcription factor binding or altering DNA accessibility to regulatory proteins (Suelves et al., 2016). When the immune system detects stress, it starts a number of defense processes to shield plants from different dangers. Receptor proteins, kinase cascades, hormone signaling pathways, heat shock proteins, ROS, transcription factors, and transduction of signals cascades are a few examples of these systems. But according to recent studies, the abnormal signaling crosstalk depicted in Figure 2 causes plants' immune responses to numerous stressors to diverge considerably from those to a single stimulus (Suzuki et al., 2014). This chapter covers the primary factors affecting the responses of signaling pathways to various stresses, their shared and overlapping responses, and the connections between the responses of signaling pathways to various pressures.

Figure 2. Plants show stress recognition, signal transduction and epigenetic variation due to multiple stresses

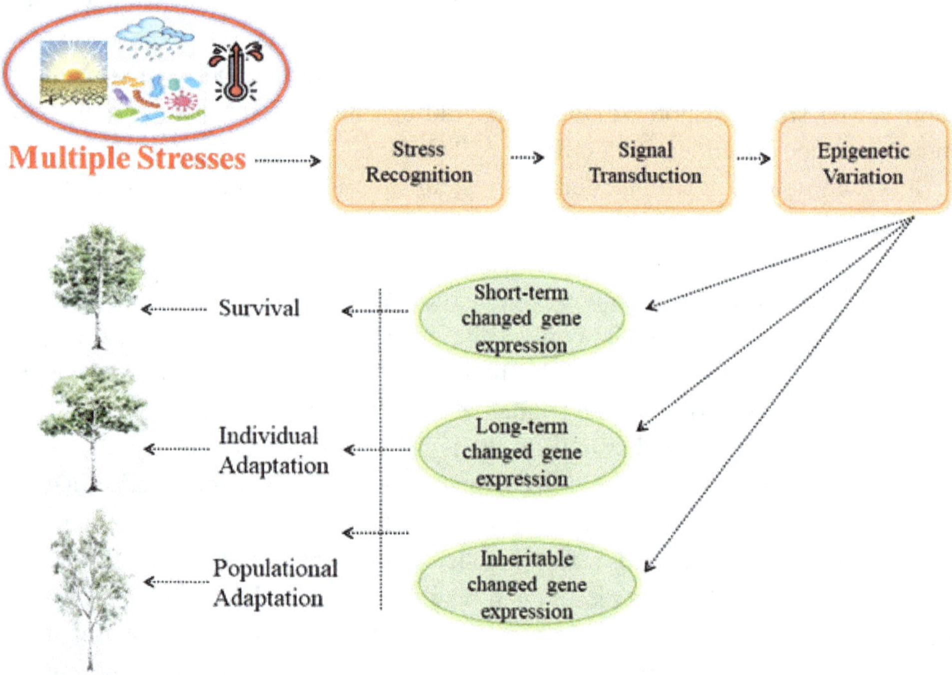

PLANT SIGNALING COMPOUNDS AND CROSSTALK WITH EPIGENETIC MECHANISMS UNDER STRESS REGULATIONS

In order for plants to survive and become more fit, their natural immune system must be engaged when they are subjected to biotic (heavy metals, salt, cold, and heat) and abiotic challenges. Plant survival in harsh conditions depends on compacted signaling networks and their interaction (Smékalová et al., 2014). Once a signal from a particular receptor is received, it sets off signaling pathway activities by stimulating secondary signals and protein phosphorylation cascades such as MAPK signaling. ROS, Ca2, NO, H2S, H2O2, phospholipids, and PGRs are examples of secondary messengers that function as secondary signals in signaling cascades (Suzuki et al., 2012).

Under typical conditions, these signals control abiotic and biotic interactions as well as basic functions such as division and development of cells, differentiation, and programmed death of cells (Sasabe & Machida, 2012). They have several functions in signaling pathways.

Using microarrays, the cellular response of A. thaliana to drought stress and a simultaneous infection by the plant-parasitic worm *Heterodera schachtii* was examined. It was discovered that drought stress increased *A. thaliana's* susceptibility to nematode infection. Additionally, following these stress treatments, induction patterns of genes with different expressions demonstrated that, in addition to a specific response to each stress, the combination of both abiotic and biotic stresses activated a particular response (Atkinson, 2011).

Under typical conditions, these signals control abiotic and biotic interactions as well as basic functions such as division and development of cells, reactive oxygen species (ROS), metabolic stress, and a lack of nutrients are caused by changes in membrane structure brought on by stress caused by salt (Tuteja, 2007). The growth responses to salinity are divided into two stages (Munns, 1993). Hormone signals from the roots slow down development and stop when the plant achieves maturity. The decline in growth can be attributed to a salt-specific effect that often takes weeks to years to show up. The second stage, which elevates levels above a cell's capacity for storing salts into vacuoles, is brought on by reduced supply of water and salt accumulation in transpiring leaves (Munns, 1993). Thus, a variety of factors, including plant tissue species, growth stage, stressors, stress intensity, time period, time of day, geographic origin, etc., influence the degree of crosstalk between various physiological defense pathways and the plant immune response to specific stress and the combination of various environmental stimuli (Pandey et al., 2015). These findings also demonstrate the presence of complex relationships and the regulation of plant responses by complex systems, including overlap and specific signaling pathways.

PHYTOHORMONES SIGNALING AND ITS CROSSTALK WITH EPIGENETIC MECHANISMS

The first phytohormone to be discovered was auxin. According to Darwin's studies on plant tropism movement, auxin is required for phenotypic plasticity in response to environmental cues (Edwards & Moles, 2009). Numerous aspects of plant biology, such as basic developmental processes, abiotic stress responses, programmed death of cells, and dynamic responses to light and gravity, are influenced by auxin (Benjamins & Scheres, 2008). Auxin creates a bell-shaped dose response curve that regulates plant development because it suppresses growth above a threshold and promotes growth at low concentrations (Thimann, 1938).

First discovered for its capacity to stimulate cell division, cytokinin has now been demonstrated to be involved in a variety of plant growth mechanisms and reactions to environmental stimuli (Cortleven et al., 2019). Trans zeatin (tZ) and isopentenyladenine (iP) are the two main side chain forms of all natural cytokinins that are connected to the N6 position of the adenine scaffold.

These side chains are isoprenoid and aromatic. Many studies have been conducted on the synthesis of isoprenoid cytokinins. Methylene erythritol 4 phosphate (MEP) or dimethylallyl diphosphate (DMAPP), which is generated by the mevalonate (MVA) pathway, is converted by isopentenyl transferase (IPT) into iP nucleotides, which cytochrome CYP735A can then utilize to generate tZ nucleotides. Adenosine nucleosidase and 5' ribonucleotide phosphohydrolase transform nucleotide forms of iP into their free base forms (Sakakibara, 2006). Callus development and shoot regeneration depend on both cytokinin signaling and epigenetic processes (Ikeuchi et al., 2019). WUSCHEL (WUS) is one known downstream control of cell fate decisions that may be co-regulated by cytokinin signaling and epigenetic pathways in addition to the sorts of reciprocal regulation previously discussed (Li et al., 2011).

Gibberella fujikuroi is a plant fungal infection that generates the phytohormone gibberellin A (GA), which leads to in lodging and sterility (Hedden & Sponsel, 2015). Fruit growth and development, flowering, germination, growth, and both biotic and abiotic stress responses are all impacted by GAs. GA is produced and metabolized by plants, as is well known. The maintenance of GA homeostasis depends on these oxidases through a reciprocal regulatory mechanism (Hedden & Thomas, 2012).

Among the many functions of the gaseous phytohormone ethylene in plant growth and development are ripening fruit (Liu et al., 2020), flower and leaf senescence (Shibuya, 2018), germination of seed (Corbineau et al., 2014), and reactions to stressors like flooding and salinity (Khan et al., 2020). ACC oxidases (ACOs), 1-aminocyclopropane-1-carboxylic acid (ACC) synthetases (ACSs), and S-adenosylmethionine (SAM) synthetase all gradually mediate the Yang cycle, which is how plants produce endogenous ethylene from methionine (Larsen, 2015).

According to Liu et al. (2020), epigenetics also influences biological processes like fruit ripening and leaf senescence that are connected to ethylene.

A type of sesquiterpene lactones called strigolactones (SLs) are produced from carotenoid and regulate various aspects of plant growth, such as the shape of shoots and roots. Additionally, they aid in plant-to-plant communication with arbuscular fungi and parasitic plants (Xie et al., 2010). A variety of SL structures, both canonical and non-canonical, are present in various plants (Mashiguchi et al., 2021). Genetic studies support a canonical signaling pathway in which the canonical receptor D14 identifies SLs and subsequently facilitates the proteasome-mediated reduction in INHIBITOR OF MAX2 1-LIKEs (SMXLs) via the E3 ubiquitin ligase MAX2, despite the contentious biochemical mechanism of D14's SL perception (Mashiguchi et al., 2021).

EPIGENETIC CROSSTALK REGULATION MECHANISMS OF PLANT

Plants use a variety of epigenetic strategies to alter the expression of certain genes in response to stress. These strategies include DM, short RNA-mediated gene silencing, chromatin remodeling, and histone alterations (Figure 2). Maintaining genomic integrity and controlling gene expression are both crucial processes, and each strategy has unique and significant significance. In the kingdom of plants, the process of DM-induced epigenetic alteration is well understood. The cytosine residues of the DNA, specifically the CpG dinucleotides, receive a methyl group addition during this process.Gene expression and phenotypic plasticity may be greatly impacted by stress-induced alterations in DM patterns (Huang et al., 2022). According to recent research, DM is a dynamic process that reacts to a range of external stressors (Kawakatsu et al., 2017). It helps draw chromatin-modifying proteins and inhibit some transcription factors from attaching to DNA (Sadakierska-Chudy et al., 2015). Additionally, this process identifies patterns of histone modification and aids in the recruitment of repressor complexes containing MBD proteins, DNMTs, and HD7ACs. Although DM will produce complicated DM states in crossbreeding populations, unlike DNA sequence modifications, it can result in novel and appealing phenotypes that are not possible with genetic variation (Huang et al., 2022). Additionally, changes in the methylation of regions that affect gene expression, including cis-elements, may lead to changed gene expression and a unique phenotype because diabetes mellitus is associated with altered gene expression (Tonosaki et al., 2022).

In reaction to stress, plants can employ DM to add a methyl group to their DNA. For example, a gene called ATDM1 in Arabidopsis thaliana methylates specific genes related to drought tolerance, leading the plant to react to stress caused by drought

(Lee et al., 2023). DM changes brought on by pathogen infection might result in the activation or repression of defensive genes. A more compact chromatin shape and greater suppression of gene expression may result from DM's capacity to mute genes and maybe draw proteins that alter histones (Suelves et al., 2016).

Figure 3. Epigenetic mechanism in plants (In response to stress, plants employ a range of epigenetic techniques to change the expression of certain genes. These tactics include histone modifications, chromatin remodeling, DM, and small non-coding RNA

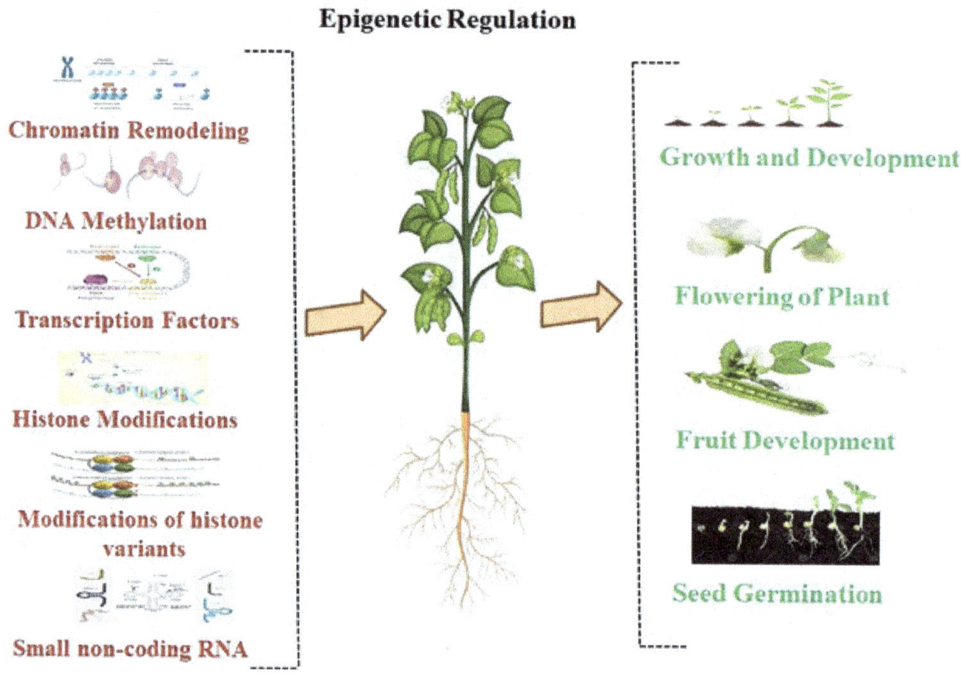

Histone Modification (HM)

Histone modifications (HMs) control chromatin and gene expression; the kind and location of HM can either promote or inhibit gene expression (Nunez-Vazquez et al., 2022). Research has shown that plants' reactions to stress involve ongoing regulation of HMs. According to research, severe drought can alter histone acetylation, which leads to an upregulation of genes that react to stress (Hu et al., 2019). Their study's findings highlight the significance of heavy metals in plants' reactions to stress and provide insight into the underlying regulatory mechanisms. Research indicates that HM significantly affects how plants react to both biotic and abiotic stresses (Nunez-Vazquez et al., 2022). For example, it has been shown that the histone deacetylase inhibitor trichostatin A increases Arabidopsis thaliana's resistance to the fungal disease Botrytis cinerea (Nunez-Vazquez et al., 2022).Moreover, complex signaling pathways in plants are triggered by abiotic stressors such as drought, salinity, severe temperatures, and heavy metals, which results in alterations to histones. The stress-responsive genes of Arabidopsis thaliana are affected by drought stress, which has been shown to increase histone H3 acetylation levels (Wollmann et al., 2017). The ability of the plant to withstand water limitation is directly correlated with changes in its histones, as this acetylation is associated with the production of genes that respond to stress (Zeng et al., 2019). Salt stress has also been demonstrated to change histone methylation patterns, which affects the expression of genes linked to ion homeostasis and osmotic regulation (Zeng et al., 2019). In potatoes (*Solanum tuberosum*), cold stress enhances chromatin accessibility through bivalent histone modifications (H3K4me3 and H3K27me3) of activated genes (Zeng et al., 2019). In fact, a methylation reader protein particular to plants, named SHORT LIFE, was discovered lately. Both repressive (H3K27me3) and active (H3K4me3) markers can be detected by it. The way that excessive temperatures change genes that react to heat stress highlights the ever-changing nature of histone changes (Wu et al., 2019).

Small RNA-Mediated Gene Silencing

Plant stress responses are significantly influenced by remodeling of chromatin complexes and short RNA-mediated gene silencing (Park et al., 2017). siRNAs and miRNAs, which control the expression of genes under stress. The RNA-induced silencing complex (RISC) uses these short, non-coding RNA molecules to decide whether to translate or block mRNAs. The latter is used to make miRNAs, whereas siRNAs are generated from either endogenous long double-stranded RNA precursors or endogenous hairpin-shaped precursors (Xie & Yu, 2015). According to Zhou et al. (2019), both kinds of short RNAs work by matching nucleotides with the targeted mRNAs, which causes translational inhibition or mRNA cleavage.

According to research by Zhou et al. (2019) and Singroha et al. (2021), plants that are stressed by drought generate reactive oxygen species (ROS) in their chloroplasts and peroxisomes.

He et al. (2023) claim that in order to counteract the harmful effects of ROS on cells, plants produce antioxidative enzymes like ascorbate peroxidase, glutathione reductase, peroxidase, catalase, and superoxide dismutase. Remarkably, oxidative stress networks are known to be regulated by certain plant miRNAs, including miR398 and miR528 (Yang et al., 2022).Additionally, recent research shows that stress-responsive miRNAs target genes linked to stress, hence modulating plant stress tolerance. According to new research, stress-induced lncRNAs control genes and pathways that are sensitive to stress (Zhan & Meyers, 2023).These RNAs have the ability to either attach to messenger RNAs (mRNAs) and stop them from being translated into proteins or they can cause certain mRNAs to degrade (Ferdous et al., 2015). It has been demonstrated that miR159 modulates drought tolerance genes in Arabidopsis thaliana. Plants use miRNAs to adapt to abiotic stressors including heat and drought (Ferdous et al., 2015).

Chromatin Remodeling Complexes

The epigenetic regulation of plant gene expression, especially in response to stress, is significantly influenced by chromatin remodeling complexes (Pandey et al., 2016). These complexes alter the structure of the chromatin, which affects DNA's accessibility to regulatory proteins and transcriptional machinery (Pandey et al., 2016). Chromatin remodeling complexes alter the activation of genes in plants that respond to biotic and abiotic stress, influencing the plant's ability to adapt and endure under challenging circumstances (Kim, 2019). By encouraging structural changes that either activate or inhibit genes that responds towards stress, chromatin remodeling complexes control gene expression patterns and enable plants to react to stress.

This optimizes gene expression and guarantees survival (Wang & Qiao, 2020).

Using ATP hydrolysis energy, chromatin remodeling complexes in plants respond to stress slide, evict, or modify the composition of nucleosomes to change gene expression (He et al., 2023). Histone-modifying enzymes can help or impede gene transcription by catalyzing posttranslational changes of histone proteins. DNA cannot be accessed by transcription factor polymerases, and several other nuclear proteins due to the chromatin architecture. This is because stress signals generate particular structural changes in the chromatin that expose DNA (Bhadouriya et al., 2021). Because chromatin remodeling complexes allow for quick and reversible changes in gene expression brought on by plant stress, they are essential for epigenetic control (Pecinka et al., 2020). Plants can react to a variety of stimuli efficiently thanks to this dynamic regulation, which doesn't change the fundamental genetic code. Through their interactions with DM and small RNA-mediated silencing mechanisms, among other epigenetic processes, chromatin remodeling complexes coordinate comprehensive reactions in response to stress. The responsiveness, resilience, and adaptation of the crop are influenced by the interaction of these many epigenetic regulation stages. Furthermore, recent studies have demonstrated a clear connection between stressful situations and epigenetic modifications that impact the expression of genes responsive to stress, suggesting that environmental cues might influence the specificity and functionality of chromatin remodeling complexes (Kim, 2021).

Mitogen-Activated Protein Kinase Signaling (MAPK)

APK is dependent on a serine/threonine kinase that mediates post translational phosphorylation signaling. This kinase phosphorylates MAPKs after reversibly phosphorylating MAPKK (a dual-specificity kinase) (Smékalová et al., 2014). Plant protein kinase, cytoskeleton-associated protein, and transcription factors are all phosphorylated by MAPKs (Nakagami et al., 2005). Several studies have recently determined that the generation of auxin and Ca2 channel activity, which increase NO buildup, are responsible for the commencement of root growth. This NO is responsible for the alteration of MPK cascade enzyme activity and Ca2 channel movement. MAPK activation can also be caused by ROS molecules, which are metabolic byproducts that arise under stressful circumstances (Nakagami et al., 2005). In both normal and stressful situations, plant hormones are a remarkable signaling molecule. Hormone signals that start MAPK cascades include auxin, JA, brassinosteroids (BRs), strigolactones, ABA, SA and ET. These signaling molecules interact with one another and form a distinct signaling network that reacts to both normal and extreme situations (Chini et al., 2016).

Transcription Factors

The following transcription factors (TFs) are identified by Frescatada-Rosa et al., (2015): APETALA2/ETHYLENE-RESPONSE, MYB, NAC, basic-helix-loop helix (bHLH), WRKY, ELEMENT BINDING FACTOR (AP2/ERF), and basic domain leucine zipper (bZIP). At the center of gene regulatory networks are transcription factors (TFs), which can either positively or negatively impact the regulation of genes to precisely regulate the rate of DNA synthesis. regulate both biotic and abiotic factors and various facets of the plant process of growth (Tsuda & Somssich, 2015). A variety of biotic and abiotic stimuli can activate Arabidopsis thaliana activating factor1 (ATAF1), a transcription factor belonging to the NAC family, one of the largest TF families exclusively found in plants. The drought-inducible gene ATAF1 was significantly regulated by both drought and high salinity in an ABA-deficient mutant aba2, indicating that ATAF1 expression is independent of ABA. Wu et al. (2009) state that ATAF1 helps regulate stomata and is highly expressed in stomatal guard cells. In order to enhance the closure of stomatal and plant drought tolerance, ATAF1 may employ both ABA-dependent and ABA-independent pathways of signaling (Jansen et al., 2008).

CONCLUSION

As sessile organisms, plants have developed sophisticated defenses that allow them to withstand both individual and combined pressures. To reduce the negative effects of stress on plant growth, development, and productivity and to extend the lifespan of plants, it is essential to comprehend the intricacy of plant defense mechanisms and stress signaling. Recent research on plant immunological responses to multiple stressors alone and in combination has demonstrated the complexity and high level of signaling epigenetic interaction in defenses against different threats. We still don't fully understand the signaling of epigenetic crosstalk between plant defense pathways under different stress conditions, despite the significant advancements in high throughput sequencing technologies and powerful bioinformatics techniques for transcriptome, proteome, and genome-wide analysis. A recent study discovered that the creation of ROS, MAP kinase cascades, and hormone signaling—all of which are non-specifically induced by a range of environmental stimuli—are essential for the immune system's response to several stressors. Receptor proteins, the first line of defense for the immune system, are crucial modulators of the interplay between multiple stress stimuli. As both beneficial and detrimental regulators of plant defense, transcription factors are also crucial for crosstalk and adjusting reactions to various stressors. It is still necessary to clarify the relationships between TFs that regulate genes that are responsive to stress and hormones.

REFERENCES

Ashapkin, V. V., Kutueva, L. I., Aleksandrushkina, N. I., & Vanyushin, B. F. (2020). Epigenetic mechanisms of plant adaptation to biotic and abiotic stresses. *International Journal of Molecular Sciences*, *21*(20), 7457. PMID: 33050358

Atkinson, N. J. (2011). *Plant molecular response to combined drought and nematode stress*. University of Leeds.

Bäurle, I., & Trindade, I. (2020). Chromatin regulation of somatic abiotic stress memory. *Journal of Experimental Botany*, *71*(17), 5269–5279. PMID: 32076719

Benjamins, R., & Scheres, B. (2008). Auxin: The looping star in plant development. *Annual Review of Plant Biology*, *59*(1), 443–465. PMID: 18444904

Bhadouriya, S. L., Mehrotra, S., Basantani, M. K., Loake, G. J., & Mehrotra, R. (2021). Role of chromatin architecture in plant stress responses: An update. *Frontiers in Plant Science*, *11*, 603380. PMID: 33510748

Chini, A., Gimenez-Ibanez, S., Goossens, A., & Solano, R. (2016). Redundancy and specificity in jasmonate signalling. *Current Opinion in Plant Biology*, *33*, 147–156. PMID: 27490895

Corbineau, F., Xia, Q., Bailly, C., & El-Maarouf-Bouteau, H. (2014). Ethylene, a key factor in the regulation of seed dormancy. *Frontiers in Plant Science*, *5*, 539. PMID: 25346747

Cortleven, A., Leuendorf, J. E., Frank, M., Pezzetta, D., Bolt, S., & Schmülling, T. (2019). Cytokinin action in response to abiotic and biotic stresses in plants. *Plant, Cell & Environment*, *42*(3), 998–1018. PMID: 30488464

Ding, X., Jia, X., Xiang, Y., & Jiang, W. (2022). Histone modification and chromatin remodeling during the seed life cycle. *Frontiers in Plant Science*, *13*, 865361. PMID: 35548305

Du, J., Johnson, L. M., Jacobsen, S. E., & Patel, D. J. (2015). DNA methylation pathways and their crosstalk with histone methylation. *Nature Reviews. Molecular Cell Biology*, *16*(9), 519–532. PMID: 26296162

Duan, C. G., Zhu, J. K., & Cao, X. (2018). Retrospective and perspective of plant epigenetics in China. *Journal of Genetics and Genomics*, *45*(11), 621–638. PMID: 30455036

Edwards, W., & Moles, A. T. (2009). Re-contemplate an entangled bank: The power of movement in plants revisited. *Botanical Journal of the Linnean Society*, *160*(2), 111–118.

Ferdous, J., Hussain, S. S., & Shi, B. J. (2015). Role of micro RNA s in plant drought tolerance. *Plant Biotechnology Journal*, *13*(3), 293–305. PMID: 25583362

Frescatada-Rosa, M., Robatzek, S., & Kuhn, H. (2015). Should I stay or should I go? Traffic control for plant pattern recognition receptors. *Current Opinion in Plant Biology*, *28*, 23–29. PMID: 26344487

González, A. P. R., Chrtek, J., Dobrev, P. I., Dumalasová, V., Fehrer, J., Mráz, P., & Latzel, V. (2016). Stress-induced memory alters growth of clonal offspring of white clover (Trifolium repens). *American Journal of Botany*, *103*(9), 1567–1574. PMID: 27620180

He, K., Cao, X., & Deng, X. (2021). Histone methylation in epigenetic regulation and temperature responses. *Current Opinion in Plant Biology*, *61*, 102001. PMID: 33508540

He, M. Y., Ren, T. X., Jin, Z. D., Deng, L., Liu, H. J., Cheng, Y. Y., & Chang, H. (2023). Precise analysis of potassium isotopic composition in plant materials by multi-collector inductively coupled plasma mass spectrometry. *Spectrochimica Acta. Part B, Atomic Spectroscopy*, *209*, 106781.

Hedden, P., & Sponsel, V. (2015). A century of gibberellin research. *Journal of Plant Growth Regulation*, *34*, 740–760. PMID: 26523085

Hedden, P., & Thomas, S. G. (2012). Gibberellin biosynthesis and its regulation. *The Biochemical Journal*, *444*(1), 11–25. PMID: 22533671

Hu, Y., Lu, Y., Zhao, Y., & Zhou, D. X. (2019). Histone acetylation dynamics integrates metabolic activity to regulate plant response to stress. *Frontiers in Plant Science*, *10*, 1236. PMID: 31636650

Huang, H., Wu, N., Liang, Y., Peng, X., & Shu, J. (2022). SLNL: A novel method for gene selection and phenotype classification. *International Journal of Intelligent Systems*, *37*(9), 6283–6304.

Ikeuchi, M., Favero, D. S., Sakamoto, Y., Iwase, A., Coleman, D., Rymen, B., & Sugimoto, K. (2019). Molecular mechanisms of plant regeneration. *Annual Review of Plant Biology*, *70*(1), 377–406. PMID: 30786238

Kawakatsu, T., Nery, J. R., Castanon, R., & Ecker, J. R. (2017). Dynamic DNA methylation reconfiguration during seed development and germination. *Genome Biology*, *18*, 1–12. PMID: 28911331

Khan, M. I. R., Trivellini, A., Chhillar, H., Chopra, P., Ferrante, A., Khan, N. A., & Ismail, A. M. (2020). The significance and functions of ethylene in flooding stress tolerance in plants. *Environmental and Experimental Botany*, *179*, 104188.

Kim, J. H. (2019). Chromatin remodeling and epigenetic regulation in plant DNA damage repair. *International Journal of Molecular Sciences*, *20*(17), 4093. PMID: 31443358

Kim, J. H. (2021). Multifaceted chromatin structure and transcription changes in plant stress response. *International Journal of Molecular Sciences*, *22*(4), 2013. PMID: 33670556

Kim, Y., Tsuda, K., Igarashi, D., Hillmer, R. A., Sakakibara, H., Myers, C. L., & Katagiri, F. (2014). Mechanisms underlying robustness and tunability in a plant immune signaling network. *Cell Host & Microbe*, *15*(1), 84–94. PMID: 24439900

Larsen, P. B. (2015). Mechanisms of ethylene biosynthesis and response in plants. *Essays in Biochemistry*, *58*, 61–70. PMID: 26374887

Lee, S., Choi, J., Park, J., Hong, C. P., Choi, D., Han, S., & Hwang, I. (2023). DDM1-mediated gene body DNA methylation is associated with inducible activation of defense-related genes in Arabidopsis. *Genome Biology*, *24*(1), 106. PMID: 37147734

Li, J., Li, C., & Smith, S. M. (2017). *Hormone metabolism and signaling in plants*. Academic press.

Li, W., Liu, H., Cheng, Z. J., Su, Y. H., Han, H. N., Zhang, Y., & Zhang, X. S. (2011). DNA methylation and histone modifications regulate de novo shoot regeneration in Arabidopsis by modulating WUSCHEL expression and auxin signaling. *PLOS Genetics*, *7*(8), e1002243. PMID: 21876682

Liu, X., Zhu, K., & Xiao, J. (2023). Recent advances in understanding of the epigenetic regulation of plant regeneration. *aBIOTECH*, *4*(1), 31–46. PMID: 37220541

Liu, Y., Tang, M., Liu, M., Su, D., Chen, J., Gao, Y., & Li, Z. (2020). The molecular regulation of ethylene in fruit ripening. *Small Methods*, *4*(8), 1900485.

Luo, X., & He, Y. (2020). Experiencing winter for spring flowering: A molecular epigenetic perspective on vernalization. *Journal of Integrative Plant Biology*, *62*(1), 104–117. PMID: 31829495

Markulin, L., Škiljaica, A., Tokić, M., Jagić, M., Vuk, T., Bauer, N., & Leljak Levanić, D. (2021). Taking the wheel–De novo DNA methylation as a driving force of plant embryonic development. *Frontiers in Plant Science*, *12*, 764999. PMID: 34777448

Mashiguchi, K., Seto, Y., & Yamaguchi, S. (2021). Strigolactone biosynthesis, transport and perception. *The Plant Journal*, *105*(2), 335–350. PMID: 33118266

Munns, R. (1993). Physiological processes limiting plant growth in saline soils: Some dogmas and hypotheses. *Plant, Cell & Environment*, *16*(1), 15–24.

Nakagami, H., Pitzschke, A., & Hirt, H. (2005). Emerging MAP kinase pathways in plant stress signalling. *Trends in Plant Science*, *10*(7), 339–346. PMID: 15953753

Nejat, N., Rookes, J., Mantri, N., & Cahill, D. (2016). Plant-pathogen interactions: Toward development of next-generation disease-resistant plants. *Critical Reviews in Biotechnology*, *37*(2), 229–237. PMID: 26796880

Nunez-Vazquez, R., Desvoyes, B., & Gutierrez, C. (2022). Histone variants and modifications during abiotic stress response. *Frontiers in Plant Science*, *13*, 984702. PMID: 36589114

Nunez-Vazquez, R., Desvoyes, B., & Gutierrez, C. (2022). Histone variants and modifications during abiotic stress response. *Frontiers in Plant Science*, *13*, 984702. PMID: 36589114

Pandey, P., Ramegowda, V., & Senthil-Kumar, M. (2015). Shared and unique responses of plants to multiple individual stresses and stress combinations: Physiological and molecular mechanisms. *Frontiers in Plant Science*, *6*, 723. PMID: 26442037

Park, J. J., Dempewolf, E., Zhang, W., & Wang, Z. Y. (2017). RNA-guided transcriptional activation via CRISPR/dCas9 mimics overexpression phenotypes in Arabidopsis. *PLoS One*, *12*(6), e0179410. PMID: 28622347

Pecinka, A., Chevalier, C., Colas, I., Kalantidis, K., Varotto, S., Krugman, T., & Pradillo, M. (2020). Chromatin dynamics during interphase and cell division: Similarities and differences between model and crop plants. *Journal of Experimental Botany*, *71*(17), 5205–5222. PMID: 31626285

Sadakierska-Chudy, A., Kostrzewa, R. M., & Filip, M. (2015). A comprehensive view of the epigenetic landscape part I: DNA methylation, passive and active DNA demethylation pathways and histone variants. *Neurotoxicity Research*, *27*, 84–97. PMID: 25362550

Saeed, F., Chaudhry, U. K., Bakhsh, A., Raza, A., Saeed, Y., Bohra, A., & Varshney, R. K. (2022). Moving beyond DNA sequence to improve plant stress responses. *Frontiers in Genetics*, *13*, 874648. PMID: 35518351

Sakakibara, H. (2006). Cytokinins: Activity, biosynthesis, and translocation. *Annual Review of Plant Biology*, *57*(1), 431–449. PMID: 16669769

Sasabe, M., & Machida, Y. (2012). Regulation of organization and function of microtubules by the mitogen-activated protein kinase cascade during plant cytokinesis. *Cytoskeleton*, *69*(11), 913–918. PMID: 23027702

Shang, J. Y., & He, X. J. (2022). Chromatin-remodeling complexes: Conserved and plant-specific subunits in Arabidopsis. *Journal of Integrative Plant Biology*, *64*(2), 499–515. PMID: 34964264

Shibuya, K. (2018). Molecular aspects of flower senescence and strategies to improve flower longevity. *Breeding Science*, *68*(1), 99–108. PMID: 29681752

Singroha, G., Sharma, P., & Sunkur, R. (2021). Current status of microRNA-mediated regulation of drought stress responses in cereals. *Physiologia Plantarum*, *172*(3), 1808–1821. PMID: 33956991

Smékalová, V., Doskočilová, A., Komis, G., & Šamaj, J. (2014). Crosstalk between secondary messengers, hormones and MAPK modules during abiotic stress signalling in plants. *Biotechnology Advances*, *32*(1), 2–11. PMID: 23911976

Song, X., Li, Y., Cao, X., & Qi, Y. (2019). MicroRNAs and their regulatory roles in plant–environment interactions. *Annual Review of Plant Biology*, *70*(1), 489–525. PMID: 30848930

Song, Z. T., Zhang, L. L., Han, J. J., Zhou, M., & Liu, J. X. (2021). Histone H3K4 methyltransferases SDG25 and ATX1 maintain heat-stress gene expression during recovery in Arabidopsis. *The Plant Journal*, *105*(5), 1326–1338. PMID: 33278042

Sudan, J., Raina, M., & Singh, R. (2018). Plant epigenetic mechanisms: role in abiotic stress and their generational heritability. *3 Biotech*, *8*, 1-12

Suelves, M., Carrió, E., Núñez-Álvarez, Y., & Peinado, M. A. (2016). DNA methylation dynamics in cellular commitment and differentiation. *Briefings in Functional Genomics*, *15*(6), 443–453. PMID: 27416614

Suzuki, N., Koussevitzky, S. H. A. I., Mittler, R. O. N., & Miller, G. A. D. (2012). ROS and redox signalling in the response of plants to abiotic stress. *Plant, Cell & Environment*, *35*(2), 259–270. PMID: 21486305

Suzuki, N., Rivero, R. M., Shulaev, V., Blumwald, E., & Mittler, R. (2014). Abiotic and biotic stress combinations. *The New Phytologist, 203*(1), 32–43. PMID: 24720847

Tang, D., Gallusci, P., & Lang, Z. (2020). Fruit development and epigenetic modifications. *The New Phytologist, 228*(3), 839–844. PMID: 32506476

Thimann, K. V. (1938). Hormones and the analysis of growth. *Plant Physiology, 13*(3), 437. PMID: 16653503

Tonosaki, K., Fujimoto, R., Dennis, E. S., Raboy, V., & Osabe, K. (2022). Will epigenetics be a key player in crop breeding? *Frontiers in Plant Science, 13*, 958350. PMID: 36247549

Tsuda, K., & Somssich, I. E. (2015). Transcriptional networks in plant immunity. *The New Phytologist, 206*(3), 932–947. PMID: 25623163

Tuteja, N. (2007). Mechanisms of high salinity tolerance in plants. *Methods in Enzymology, 428*, 419–438. PMID: 17875432

Wang, L., & Qiao, H. (2020). Chromatin regulation in plant hormone and plant stress responses. *Current Opinion in Plant Biology, 57*, 164–170. PMID: 33142261

Wollmann, H., Stroud, H., Yelagandula, R., Tarutani, Y., Jiang, D., Jing, L., & Berger, F. (2017). The histone H3 variant H3. 3 regulates gene body DNA methylation in Arabidopsis thaliana. *Genome Biology, 18*, 1–10. PMID: 28521766

Wu, R., Wang, T., Richardson, A. C., Allan, A. C., Macknight, R. C., & Varkonyi-Gasic, E. (2019). Histone modification and activation by SOC1-like and drought stress-related transcription factors may regulate AcSVP2 expression during kiwifruit winter dormancy. *Plant Science, 281*, 242–250. PMID: 30824057

Wu, Y., Deng, Z., Lai, J., Zhang, Y., Yang, C., Yin, B., & Xie, Q. (2009). Dual function of Arabidopsis ATAF1 in abiotic and biotic stress responses. *Cell Research, 19*(11), 1279–1290. PMID: 19752887

Xie, M., & Yu, B. (2015). siRNA-directed DNA methylation in plants. *Current Genomics, 16*(1), 23–31. PMID: 25937811

Xie, X., Yoneyama, K., & Yoneyama, K. (2010). The strigolactone story. *Annual Review of Phytopathology, 48*(1), 93–117. PMID: 20687831

Yang, Y., Huang, J., Sun, Q., Wang, J., Huang, L., Fu, S., & He, Y. (2022). microRNAs: Key players in plant response to metal toxicity. *International Journal of Molecular Sciences, 23*(15), 8642. PMID: 35955772

Zeng, Z., Zhang, W., Marand, A. P., Zhu, B., Buell, C. R., & Jiang, J. (2019). Cold stress induces enhanced chromatin accessibility and bivalent histone modifications H3K4me3 and H3K27me3 of active genes in potato. *Genome Biology*, *20*, 1–17. PMID: 31208436

Zhan, J., & Meyers, B. C. (2023). Plant small RNAs: Their biogenesis, regulatory roles, and functions. *Annual Review of Plant Biology*, *74*(1), 21–51. PMID: 36854480

Zhou, R., Kong, L., Yu, X., Ottosen, C. O., Zhao, T., Jiang, F., & Wu, Z. (2019). Oxidative damage and antioxidant mechanism in tomatoes responding to drought and heat stress. *Acta Physiologiae Plantarum*, *41*, 1–11.

Chapter 7
Epigenetic Influences on Plant–Microbe Interactions and Stress Adaptation

Rahul Kumar
https://orcid.org/0009-0005-4262-5521
Mahatma Gandhi Central University, India

Shreya Anand
Mizoram University, India

Prakash Kumar Sarangi
Central Agricultural University, India

Garima Khaspuria
https://orcid.org/0009-0007-8927-580X
Vivekananda Global University, India

Akhilesh Kumar Singh
Mahatma Gandhi Central University, India

Saurabh Singh Rathore
https://orcid.org/0000-0002-6997-487X
Mahatma Gandhi Central University, India

ABSTRACT

Plant-microbe interactions are vital for ecological balance and crop growth, with epigenetics playing a key role in regulating these relationships. Epigenetics involves changes in gene expression without altering DNA, enabling plants to respond to stress and interact with microbes effectively. These mechanisms help plants adapt by activating or suppressing genes in response to microbial presence. Dynamic relationships with bacteria, fungi, and viruses significantly influence plant health and growth. Epigenetic modifications enhance plant resilience by regulating gene expression during interactions, optimizing plant function under stress. This chapter

DOI: 10.4018/979-8-3693-8648-4.ch007

summarizes recent findings on the impact of epigenetics on plant-microbe interactions and stress adaptation, highlighting its potential to revolutionize sustainable agriculture and ecosystem management practices.

INTRODUCTION

The interaction between plants and microbes represents a cornerstone of both natural ecosystems and agricultural systems, playing an essential role in sustaining ecosystem balance and enhancing agricultural productivity (Aqeel et al., 2023). These interactions are characterized by a diverse range of relationships, from mutualistic and symbiotic associations to parasitic dynamics, each contributing to nutrient cycling, soil health, and plant growth in unique ways. The mutually beneficial relationships within these interactions are of particular importance, as they facilitate the exchange of critical resources and services that support plant development and soil ecosystem stability (Wiesmann et al., 2023).

A prime example of such symbiotic relationships is the mutualistic association between plants and beneficial microorganisms, such as mycorrhizal fungi and nitrogen-fixing bacteria. These microbes colonize the roots of plants, forming intricate partnerships that optimize resource acquisition and growth (Liu et al., 2023). Mycorrhizal fungi, for instance, extend their hyphal networks into the surrounding soil, significantly expanding the absorptive capacity of plant roots. This increased surface area allows plants to access water and essential nutrients, particularly phosphorus, which is often a limiting factor in soils. In return, plants supply the fungi with carbohydrates produced during photosynthesis, enabling both organisms to thrive (Fall et al., 2022).

This relationship between mycorrhizal fungi and plants is a testament to the efficiency of co-evolution, where each partner benefits from the other's specialized capabilities. Phosphorus acquisition, a key function of these fungi, is particularly crucial in agricultural systems where soil nutrients are often depleted due to intensive farming practices. By leveraging these natural processes, farmers can reduce their reliance on chemical fertilizers, thereby promoting sustainable agriculture and minimizing environmental impacts.

Similarly, nitrogen-fixing bacteria such as Rhizobium species form nodules on the roots of leguminous plants, where they convert atmospheric nitrogen into ammonia through a process known as biological nitrogen fixation (Raza et al., 2020). Ammonia serves as a bioavailable form of nitrogen that plants can readily assimilate, promoting robust growth and productivity. This symbiosis not only enhances the nitrogen content of the soil but also reduces the need for synthetic nitrogen fertilizers, which are energy-intensive to produce and can contribute to environmental issues such as water pollution and greenhouse gas emissions.

The benefits of nitrogen fixation extend beyond individual plants to the broader agroecosystem. Increased soil nitrogen levels promote the growth of subsequent crops, enabling crop rotation systems that maintain soil fertility over time. This natural enrichment of soil health underscores the critical role of microbial partnerships in sustainable farming practices.

In addition to nutrient acquisition, microbes play a vital role in protecting plants from soil-borne pathogens. Beneficial microbes can outcompete harmful organisms by occupying ecological niches within the rhizosphere, thereby preventing the proliferation of pathogens. Furthermore, certain microbes produce natural antibiotics and antifungal compounds that directly inhibit the growth of harmful microbes. These protective mechanisms strengthen plant immunity, enabling plants to withstand various biotic stresses more effectively (Das et al., 2021).

Beyond these well-documented examples, microbial communities associated with plants contribute to ecosystem stability in multifaceted ways. Soil microbes decompose organic matter, breaking down complex organic compounds into simpler molecules that plants can absorb. This decomposition process recycles essential nutrients back into the soil, fostering a self-sustaining system that supports long-term agricultural productivity (Prasad et al., 2021).

These microbial processes are not uniform; instead, they are influenced by numerous factors, including plant species, soil type, and environmental conditions. The microbial community, often referred to as the plant's "second genome," is as unique and dynamic as the plants themselves (Glick et al., 2021). The specificity of these interactions highlights the need for a tailored approach in agricultural management, where understanding the microbial preferences of specific crops can lead to optimized farming practices.

For instance, practices such as crop rotation and cover cropping can significantly influence the microbial populations in the soil. By diversifying plant species within a given area, farmers can encourage a variety of beneficial microbes, reducing the risk of soil degradation and enhancing nutrient availability. Similarly, the application of biofertilizers introduces specific strains of beneficial microbes to the soil, directly improving its fertility and structure. These practices align with the principles of sustainable agriculture, ensuring that farming systems are resilient, productive, and environmentally friendly (Nizamani et al., 2024).

While plant-microbe interactions form the foundation of agricultural ecosystems, another fascinating layer of complexity lies in the field of epigenetics. Epigenetics explores how gene expression is regulated through mechanisms that do not involve changes to the DNA sequence itself (Li, 2021). This emerging field has profound implications for understanding how plants adapt to their environment and interact with microbes.

Epigenetic modifications, such as DNA methylation, histone modifications, and small RNA activity, are critical for controlling gene expression in plants. These mechanisms determine which genes are turned on or off in response to various environmental stimuli, enabling plants to adapt dynamically to changing conditions (Zhang et al., 2021). For example, in the face of environmental stressors such as drought, extreme temperatures, or pathogen attacks, plants can undergo epigenetic changes that enhance their resilience and survival (Yajnik et al., 2024).

One remarkable feature of epigenetics is its potential for transgenerational inheritance. Epigenetic states can sometimes be passed from one generation to the next, allowing plants to inherit beneficial traits without altering their DNA sequence (Ashapkin et al., 2020). This form of inheritance is particularly advantageous in plants, as it facilitates rapid adaptation to environmental changes, often outpacing the slower process of genetic mutation.

The implications of epigenetics for agriculture are immense. By understanding and harnessing these mechanisms, scientists can develop crop varieties that exhibit enhanced resilience to stressors without the need for genetic modification. For instance, crops could be bred to retain favorable epigenetic states that improve drought tolerance, pest resistance, or nutrient efficiency. This approach offers a sustainable alternative to traditional genetic engineering, aligning with consumer preferences for non-GMO crops and reducing the environmental impact of agriculture (Abdulraheem et al., 2024).

Epigenetics also plays a significant role in shaping plant-microbe interactions. When plants encounter beneficial or harmful microbes, epigenetic mechanisms such as DNA methylation and histone modifications come into play to regulate the plant's response (Miryeganeh, 2021). These modifications can enhance the plant's ability to recruit beneficial microbes or strengthen its defenses against pathogens. For example, plants may alter their root exudates—a mixture of sugars, amino acids, and secondary metabolites released into the soil—to attract specific microbial communities that support growth or immunity.

Understanding the molecular basis of these interactions opens new avenues for agricultural innovation. By leveraging epigenetic mechanisms, researchers can develop strategies to manipulate plant-microbe interactions in favor of enhanced productivity and sustainability. For instance, biofertilizers could be designed to not only supply beneficial microbes but also trigger epigenetic changes in plants that optimize their use of these microbes.

The study of epigenetics and plant-microbe interactions is supported by advanced research methodologies, such as genome-wide association studies, transcriptomics, and epigenomic mapping. These techniques allow scientists to uncover the precise molecular pathways involved in these processes, providing a detailed understanding of how plants regulate gene expression in response to microbial signals and environmental cues. This knowledge has the potential to revolutionize agricultural practices, enabling farmers to optimize crop performance and resilience while minimizing inputs and environmental impact.

The intricate relationship between plants and microbes serves as the backbone of both natural ecosystems and managed agricultural systems. These interactions not only drive fundamental processes such as nutrient cycling and soil fertility but also offer sustainable solutions to modern agricultural challenges. By fostering beneficial microbial communities and understanding the role of epigenetics in regulating plant responses, we can build resilient farming systems that are better equipped to handle environmental stressors and reduce reliance on chemical inputs. The integration of these insights into agricultural practices promises a future where productivity and sustainability go hand in hand.

KEY EPIGENETIC MECHANISMS IN PLANT-MICROBE INTERACTIONS.

Epigenetic processes play a crucial role for plants in coping with stresses from other organisms and their surroundings. They assist plants in managing their interactions with microbes and enhancing their resilience. These processes modify gene function without changing the DNA itself. They primarily achieve this through DNA methylation, alterations to histones, and the involvement of nc- RNAs, especially in interactions between plants and microbes (Shah et al., 2022).

Methylation in DNA sequence can alter the expression of genes related to defense, enabling plants to react swiftly to attacks from pathogens (Wilkinson, 2020). For instance, hypomethylation of specific promoters can activate defense genes, resulting in improved resistance to pathogenic microbes. Conversely, hypermethylation can silence these genes, making plants more susceptible to infection (Qin et al., 2021). Modifications of histones, including acetylation, methylation, and phosphorylation, are vital components of this regulatory network (Bure et al., 2022). Alterations in chromatin configuration determine the accessibility of DNA for the mechanisms that activate or deactivate genes. In response to microbial invasion, plants can modify the histone marks to either enhance or reduce the functionality of genes that aid in their defense mechanisms. Additionally, plants produce various forms of non-coding RNAs, such as siRNAs (small interfering RNAs) and microRNAs (miRNAs), which are crucial for controlling gene expression post-transcription (Mostafa et al., 2022). Non-coding RNAs can guide messenger RNAs (mRNAs) for degradation or inhibit their translation, thus altering the expression of genes involved in immune responses and stress adaptation (Guillemin et al., 2022). Plants engage with beneficial microorganisms such as mycorrhizal fungi and rhizobia. This interaction may lead to alterations in the plant's genetic expression, referred to as epigenetic changes. These helpful microbes can enhance nutrient absorption in plants, promote more efficient growth, and improve their ability to cope with stress (Kaya et al., 2023). For example, arbuscular mycorrhizal fungi have been found to affect DNA methylation in host plants, which in turn improves their defiance to both biotic stresses like pathogen attacks and abiotic stresses such as salinity and drought (Byregowda et al., 2022). Rhizobia can cause changes in a plant's genes that help improve nitrogen fixation and boost plant health, especially in tough conditions. These changes can be reversible, which means Plants are capable of modifying its gene expression in response to climatic conditions. This ability helps plants stay balanced and resilient (Aranda et al., 2022; Bhanu et al., 2022). These epigenetic alteration can be handed down through generations, meaning that stress adaptations acquired via epigenetic modifications may enhance plant populations' long-term survival and fitness (Ashapkin et al., 2020). This transgenerational inheritance of stress tolerance traits offers an encouraging path for enhancing crops and fostering sustainable farming, especially given the impacts of climate change and growing environmental issues (Abdulraheem et al., 2024). Understanding how plants and microbes interact at the genetic level can help us develop new ways to strengthen plants. By optimizing the relationships between plants and helpful microbes, we can also reduce the effects of diseases and environmental stresses on crop productivity.

METHYLATION IN PLANT DNA

DNA methylation is very important in plants (Figure 1). It involves adding methyl groups to the DNA, particularly to the cytosine bases. This process is crucial because it helps control gene expression, maintains genome stability, and shuts down transposable elements. In plants, methylation occurs in three different sequences: CHH, CG, and CHG ("H" stands for A, T, or C) (Kumar et al., 2021). There are enzymes called DNA methyltransferases and demethylases that set up and maintain these methylation patterns by adding and removing the methyl groups.

In plants, methylation in DNA plays a crucial role in controlling how genes are expressed. It is like a molecular switch that can turn genes on or off by adding methyl groups to specific parts of the plant's genetic code (Kumar et al., 2021). This process is really important for making sure that genes are activated at a particular time and in the right sequence. For example, when a seed is developing, the patterns of DNA methylation change to control the genes responsible for making the seed grow and mature. Similarly, in the plant's other parts, DNA methylation helps to manage the genes that control growth and development, like those involved in hormone signalling (Dhar et al., 2021).

DNA methylation is essential for maintaining genome stability. It silences transposable DNA sequences capable of moving around the genome and potentially causing mutations or disrupting gene function if left unchecked. Methylation prevents these elements from being transcribed and mobilized (Moreira et al., 2021). This process is particularly significant in plants, which possess complex genomes with an increased proportion of transposable components. By regulating these elements, DNA methylation preserves the integrity of the genome and ensures the proper functioning of genes (Mushtaq et al., 2021)

Environmental factors can greatly influence the DNA methylation patterns found in plants. Stressful situations such as drought, high salinity, and attacks from pathogens can result in variations in methylation levels, which in turn can modify gene expression. These alterations can assist plants in adjusting to their surroundings by turning on genes that respond to stress and shutting down those that are not needed during difficult conditions (Arora et al., 2022). For example, Genes that play a role in conserving water and tolerating external pressure may undergo demethylation and activation when faced with drought conditions, helping the plant manage reduced water availability. In contrast, genes that are not critical for survival in stressful situations may become methylated and suppressed to save energy and resources (Arora et al., 2022; Praveen et al., 2023).

Besides its function in stress reflex, Methylation also act a vital role in reproduction. It governs the expression of genes that manage the formation of reproductive organs., such as flowers, and the formation of gametes (sperm and egg cells) (He & Feng, 2022). Methylation in DNA patterns can disturb the timing of flowering, seed development, and the viability of seeds. For example, methylation in certain regions of the genome is necessary for the activation of genes that promote flowering. This means transplants flower at the appropriate time, which is crucial for successful reproduction and seed production (Zhou et al., 2020). Methylation in DNA is a versatile and dynamic epigenetic change that is central to plant development, growth, and adaptation to environmental changes. Its capability to regulate gene expression, maintain genome stability, and stimuli to stress makes it a key player in plant biology. Understanding the mechanisms and functions of DNA methylation in plants provides insights into fundamental biological processes and has practical implications for agriculture and crop improvement. By manipulating methylation patterns, it may be possible to enhance plant resilience to stress, improve yield, and develop crops better suited to changing environmental conditions.

Figure 1. DNA methylation

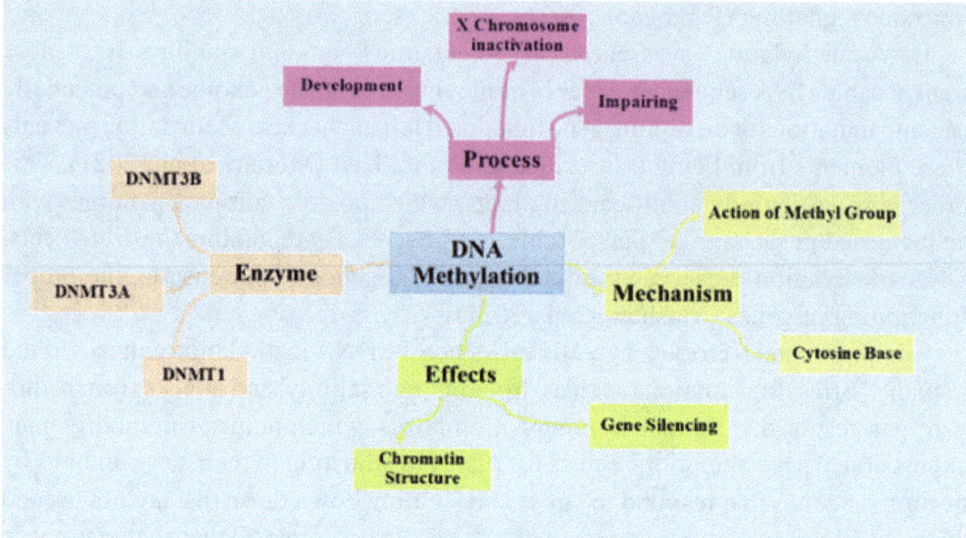

KINDS OF DNA METHYLATION

DNA-methylation serves as an essential epigenetic shifts in plants, occurring in three main contexts: CHG, CHH, and CG (with H signifying any nucleotide apart from G). The most prevalent form is CG methylation, which is observed in both gene bodies and transposable elements (Kumar et al., 2021). Methylation is important for controlling gene activity and keeping the genome stable. CHG and CHH methylation mainly occur in transposable elements and are crucial for silencing these elements. This helps prevent instability in the genome (Liang et al., 2022). Establishing and maintaining these methylation patterns involves a complex interplay of enzymes. DNA methyltransferases add methyl groups to cytosine residues, while demethylases remove these groups, allowing for dynamic regulation of methylation states (Kumar et al., 2021). MET1 is the main enzyme that keeps CG methylation in check. CMT3 is important for methylating CHG, while DRM2 is essential for CHH methylation. The RdDM (RNA-directed DNA methylation) mechanism is crucial in establishing CHH methylation (Wang et al., 2022). This process involves small interfering RNAs (siRNAs) that help add chemical tags (methylation) to specific parts of the DNA. This tagging can change based on the environment and developmental signals, which helps plants adapt. As a result, DNA methylation affects how plants grow, reproduce, and respond to stress (Sun et al., 2022). Variations in DNA methylation motif can affect the activation or inhibition of genes associated with flowering time, seed formation, and defense strategies. In general, the complex regulation of DNA methylation in plants highlights its crucial role in preserving genomic stability and facilitating adaptive responses to environmental shifts (Kumar et al., 2021).

DNA METHYLATION FUNCTIONS

DNA Methylation in Plants Serves Several Critical Functions

Gene regulation: Gene regulation encompasses the methylation in promoter site, which products in the suppress of genes. This mechanism is vital for managing gene expression throughout development and in reaction to environmental signals, enabling the accurate control of gene activity in particular stages of development (Clark & Rager, 2020).

Transposable elements (TEs): They comprise a considerable part of plant genomes. If not adequately regulated, they have the potential to induce mutations and lead to genomic instability. DNA methylation, particularly through CHG and CHH methylation, plays a decisive role in silencing these elements. This helps prevent their transposition and ensures genomic integrity (Liang et al., 2022).

Genome Stability: DNA methylation plays a decisive role in preserving the stability to the genome by inhibiting transposable elements and ensuring the integrity of heterochromatin. This is particularly important in plants, given that a large part of their genome is made up of repetitive sequences that are susceptible to rearrangements (Ramakrishnan et al., 2021).

Epigenetic Inheritance: DNA methylation can be inherited, indicating that patterns of methylation can be transmitted from one generation to the next. This form of inheritance permits the passing of information that goes beyond the DNA sequence, enabling plants to adjust to environmental changes over several generations (Sugden et al., 2020).

DNA methylation is a flexible and strong epigenetic process in plants that act a crucial role in regulating genes. It also helps maintain genome stability and suppress transposable elements. Its capacity to be inherited and adjusted in reaction to environmental elements highlights its importance in plant growth, adaptation, and evolution. Gaining insight into the mechanisms and roles of DNA methylation in plants enhances our understanding of their complex regulatory systems and potential uses in agriculture, including crop enhancement and increased resilience to stress.

MECHANISMS OF DNA METHYLATION IN PLANTS

Methylation mention the addition of a methyl group to the cytosine base of DNA, resulting in the formation of a compound known as 5-methylcytosine (5-mC). This process is performed by enzymes called DNMTs (DNA methyltransferases). In plants Methylation in DNA occurs through three patterns: CG, CHG, and CHH, where H may represent either A, T, or C (Kumar et al., 2021). The enzymes involved include Domain Rearranged Methyltransferases (DRM) for de novo methylation, Methyltransferase 1 (MET1) for maintaining CG methylation, Chromomethylase 3 (CMT3) for CHG methylation, and DRM2 or CMT2 for CHH methylation (Leichter, 2022). In situations of stress like drought, elevated salinity, or attacks from pathogens, plants can undergo differential DNA methylation, leading to either hypermethylation or hypomethylation of specific genomic regions, which in turn affects gene expression (Arora et al., 2022; Praveen et al., 2023).

Hyper-methylation typically diminishes gene expression, whereas hypomethylation can activate genes that react to stress. In plants, a distinctive mechanism called RNA-assisted DNA methylation (RdDM) utilizes small interfering RNAs (siRNAs) to direct the methylation machinery to specific DNA regions. This process results in new methylation and is vital for regulating mobile genetic elements and stress-responsive genes (Tang, 2020). Its changes can be temporary or permanent. Some stress-induced patterns can be inherited, giving plants a form of stress memory that enhances their ability to cope with future stress, improving their survival and adaptability (Sharma et al., 2022). The interaction among DNA methylation and other epigenetic changes, such as histone modifications and non-coding RNAs, further fine-tunes how plants respond to stress. Histone adjustement can affect how accessible DNA is to methylation activities, while non-coding RNAs can regulate the expression of genes that play a role in stress responses (Bure et al., 2022). The operation of CRISPR/Cas-based epigenome amelioration has made it possible to accurately alter DNA methylation in plants, revealing its significance in stress responses. Gaining insights into these processes allows scientists to formulate methods for boosting crop resilience against environmental challenges, which in turn supports sustainable agriculture and food security. Merging sophisticated genomic technologies with conventional breeding techniques presents significant opportunities for enhancing plant tolerance to stress and improving overall productivity.

DNA METHYLATION IN STRESS RESPONSES

Plants continuously face different types of environmental challenges, along with drought, high salinity, extreme temperatures, and attacks by pathogens (Kumar et al., 2021). These pressures can significantly affect the development and yield of plants. Over time, plants have developed intricate strategies to endure and prosper in challenging conditions, allowing them to sense, react to, and retain information about environmental stresses. One such mechanism, DNA methylation, aids plants in adjusting gene expression in reaction to stress.

Regulation of Stress-Responsive Genes: DNA methylation is essential in modulating genes that respond to stress. In typical circumstances, certain stress-responsive genes can be silenced by methylation within the DNA (Liu et al., 2023). When a plant faces stress, certain areas of its genome might experience demethylation, resulting in the activation of those genes. This flexible regulation allows plants to quickly adapt to changes in their environment. For instance, under drought conditions, the demethylation of promoter regions for drought-responsive genes can trigger their expression, allowing the plant to start adaptive mechanisms like closing stomata and producing Osmo protectants.

Memory of Stress Events: DNA methylation also contributes to the "memory" of plant stress events. When a plant experiences stress, it can "remember" this encounter through epigenetic marks like DNA methylation. This memory allows the plant to respond more effectively to subsequent stresses (Sharma et al., 2022). For instance, after an initial exposure to a stressor, plants may retain specific DNA methylation patterns that prime stress-responsive genes for quicker activation upon re-exposure. This phenomenon, known as "stress memory" or "priming," is particularly important for perennial plants that may experience recurring stress conditions over their lifespan.

Transposable Elements and Genome Stability: Under stressful conditions, transposable elements (TEs)—which are mobile genetic components—can become activated, potentially interfering with normal gene functioning if not properly regulated (Colonna et al., 2022). It is crucial in silencing these TEs and maintaining genome stability under stress. By keeping TEs repressed, DNA methylation prevents them from causing harmful mutations that could compromise the plant's capability to cope with stress.

DNA methylation does not act in isolation: Epigenetic modifications often work together with other changes, like histone modifications and nc RNAs. This communication between different epigenetic processes helps modulate gene expression when the cell is under stress. For example, histone changes can affect DNA methylation patterns, and the reverse is also true. This interaction creates a flexible and responsive epigenetic landscape (Blecua et al., 2020).

To sum up, DNA methylation serves as a crucial epigenetic process that enables plants to adjust to environmental challenges. It aids in the adjustement of genes that respond to stress and preserves genome stability, which is vital for plants to thrive under harsh circumstances. As we deepen our comprehension of plant epigenetics, we can create more resilient crops, which is essential in addressing the challenges posed by universal climate change.

HISTONE MODIFICATION IN PLANTS

Conversion of histones in plants perform a vital role as an epigenetic mechanism that controls gene expression, affecting growth, development, and reactions to environmental changes. These alterations impact the structure and accessibility of chromatin, potentially increasing or decreasing gene expression (Acharjee et al., 2023). In plants, modifications to histones, including methylation, acetylation, ubiquitination, phosphorylation, and sumoylation, play a role in regulating genes. Among these, histone methylation—especially at lysine sites—is one of the most extensively researched modifications and has the potential to either activate or silence genes, depending on the particular residue involved and the quantity of methyl groups added (Acharjee et al., 2023). For example, trimethylation at histone H3 lysine 4 (H3K4me3) is attached to active gene transcription, whereas trimethylation at lysine 27 (H3K27me3) signifies gene repression. The equilibrium between these modifications is crucial for proper gene expression during plant development. In Arabidopsis, Polycomb Group (PcG) proteins preserve H3K27me3 to silence developmental genes until they are required, ensuring that processes such as flowering occur at the right time in response to environmental factors (Prall et al., 2019).

Histone acetylation act a crucial role in plants by facilitating gene activation. When lysine residues undergo acetylation, they become neutral due to the loss of their positive charge. This alteration reduces the bond between histones and the negatively charged DNA, allowing transcription factors easier access to the chromatin (Vanzan et al., 2023). In Arabidopsis, two categories of enzymes known as histone acetyltransferases (HATs) and histone deacetylases (HDACs) counteract one another to modify the acetylation status of histones. This change helps control gene expression based on environmental factors like light, temperature, and stress. For example, when plants face drought, they adjust the acetylation of certain histones to turn on genes that help them manage water shortage (Abdulraheem et al., 2024). This process is part of the broader plant stress response, histone modifications are essential for plants to quickly adjust to shifting environmental circumstances.

Histone modifications in plants are crucial not only for responding to external stimuli but also for essential developmental processes like seed germination, flowering, and root development. For instance, during seed dormancy, specific modifications such as H3K27me3 repress growth-related genes until conditions are suitable for germination (Zhao et al., 2019). When a plant gets the right signals, like water and light, it can remove the markings that stop it from growing. This allows the plant to turn on genes needed for germination. In the same way, changes to proteins called histones help control when the plant flowers. This guarantees that blooming occurs at the appropriate time and within favourable conditions (Shi et al., 2022). In rice, changes to histones regulate the expression of genes essential for the transition from vegetative to reproductive growth, affecting traits such as yield and stress tolerance (Gupta et al., 2022).

In addition, histone changes serve significant functions in plats progress, stress response, and immunity. Plants have developed complex systems to protect themselves from pathogens, and these modifications help control genes related to their defense (Huang et al., 2022). During a pathogen attack, specific histone modifications can regulate genes in immune responses, enabling plants to adapt their defense mechanisms to different pathogens. Overall, histone modifications in plants form a dynamic system that integrates developmental cues and environmental signals to control gene expression, ensuring survival and adaptation (Cheng et al., 2024).

NON-CODING RNA$_S$ STRESS RESPONSE IN PLANTS

Non-coding RNAs (ncRNAs) play a requisite role in how plants manage stress. They help regulate gene expression, enabling plants to adapt to different environmental conditions. The main types of ncRNAs that participate in stress responses are microRNAs (miRNAs), small interfering RNAs (siRNAs), and long non-coding RNAs (lncRNAs). Each category functions in distinct yet interrelated ways to support plants in reacting to stress (Zhang et al., 2022; Chen et al., 2020). For instance, MicroRNAs (miRNAs) help control gene activity after genes are made by pairing with specific messenger RNAs (mRNAs). This pairing can lead to the breakdown of the mRNA or stop the production of proteins. During stress from drought, high salt, or temperature changes, miRNAs often adjust the expression of genes related to hormone signals, protection against damage, and regulating moisture levels. For example, miR398 manages the activity of superoxide dismutase (SOD), an important enzyme that helps the body respond to oxidative stress, by breaking down its mRNA when under stress (Kumar et al., 2021). In a similar manner, miR156 plays a role in responding to heat and drought stress by regulating SQUAMOSA Promoter Binding Protein-Like genes, which affect multiple growth processes such as flowering and root growth (Pagano et al., 2021). On the other hand, siRNAs participate in the RNA interference pathway (RNAi), guiding the silencing complex to complementary mRNA sequences, leading to transcriptional silencing or mRNA cleavage (Sioud, 2021). They play a essential role in protecting against viral infections and help maintain genome stability during stressful conditions. Long non-coding RNAs, which exceed 200 nucleotides in length, are increasingly acknowledged as important regulators of plant responses to stress. They function in various ways, such as providing structural support, serving as distractions, or directing enzymes to specific DNA regions, thereby influencing gene expression. Recently, circular RNAs (circRNAs), a category of long non-coding RNA (lncRNA), was discovered to be involved in stress responses. Their primary function is to bind to microRNAs (miRNAs), thereby preventing these miRNAs from inhibiting the expression of its target genes (Budak et al., 2020).

Plants use non-coding RNAs (ncRNAs) to change their gene expression when faced with stress. This gives them a flexible and layered way to regulate their responses. Different types of ncRNAs work together, adding complexity to this process. This interaction helps plants remember past stresses and respond quickly and effectively. In short, ncRNAs are crucial for adjusting plant gene expression under stress, providing extra control that boosts their resilience and adaptability to challenges (Prabhakaran, 2024).

TECHNOLOGICAL ADVANCES IN EPIGENETICS RESEARCH

CRISPR/Cas-Driven Epigenome Alteration: Progress, Uses, and Clinical Value

The application of CRISPR/Cas for editing the epigenome marks a major breakthrough in plant biotechnology, allowing for accurate alterations of the epigenetic environment to improve plant characteristics (Miglani et al., 2020). Table 1 summarizes the various technological advancements in in epigenetics research. Developing the dead Cas9 (dCas9) protein has been important in genetic research. dCas9 does not cut DNA, so researchers can target specific parts of the genome without causing any breaks. By attaching dCas9 to different tools that affect gene activity, scientists can add or remove chemical marks on DNA. This enables them to regulate gene expression with accuracy (Policarpi et al., 2021). This technology has been crucial in helping us understand how plants respond to stress and grow. By specifically targeting DNA methylation and demethylation, we can turn genes on or off that help plants tolerate stress. This allows plants to better deal with difficult conditions like drought, salt, and extreme temperatures. Additionally, we use CRISPR/Cas technology to change histones, which are proteins that DNA wraps around. This change affects how easily DNA can be accessed for gene expression. With this control over gene expression, we can enhance plant growth and development, which could lead to higher crop yields and better adaptation to climate changes (Shakirova et al., 2020). This technology can improve disease resistance in plants by increasing the activity of genes that help defend against pathogens. It gives plants stronger ways to fight infections. CRISPR/Cas-based epigenome editing is also a valuable tool for researchers. It helps them understand how epigenetic marks affect gene expression. This knowledge can lead to new methods for improving crops and making agriculture more sustainable. For instance, creating crops that can endure environmental challenges and diseases can decrease the dependence on chemical inputs such as fertilizers and pesticides, which is advantageous for the ecosystem. Furthermore, the precise editing of the epigenome enables the enhancement of the nutritional value of crops, which is essential for ensuring food security worldwide. In addition, the ability to transmit epigenetic modifications to subsequent generations presents intriguing opportunities for breeding initiatives. Traits acquired through epigenome editing can be passed down, accelerating the development of new plant varieties that offer higher yields, improved nutrition, and increased resilience to stress (Singh et al., 2020). CRISPR/Cas technology for epigenome editing is a robust method with significant promise for enhancing plant science and agriculture. It enables researchers to perform accurate modifications to the epigenetic framework, aiding in the regulation of gene expression. This technology opens new ways to enhance plant traits and tackle global issues related to food security and environmental sustainability. As research continues in this area, we can expect even more innovative uses for this technology, helping plants meet the needs of a growing population in a changing world.

Table 1. Technological advances in epigenetics research

Technological Tool/ Method	**Description**	**Application in Plant Biotechnology**	**Examples**	**Reference**
Bisulfite Sequencing (BS-Seq)	Transforms nonmethylated cytosines into uracil, enabling the identification of methylated cytosines via sequencing.	- Complete analysis of DNA methylation across the genome - Detecting methylation patterns as a reaction to stress.	- Detection of drought-induced methylation changes in rice.	(Akhter et al., 2021)
Chromatin Immunoprecipitation Sequencing (ChIP-Seq)	Uses antibodies to target specific histone modifications or DNA-binding proteins, followed by sequencing.	- Mapping histone modifications - Identifying binding sites of transcription factors in response to stress.	- Identifying H3K4me3 histone marks associated with gene activation in *Arabidopsis*.	(Paul et al., 2020)
RNA Sequencing (RNA-Seq)	High-throughput sequencing to analyze the transcriptome, including coding and non-coding RNAs.	- Profiling gene expression changes during plant-microbe interactions - Identifying stress-responsive miRNAs.	- Discovery of miRNAs involved in phosphate starvation response in plants.	(Chen et al., 2016)
Test for Chromatin Sequencing Accessible to Transposase (ATAC-Seq).	Analyzes open chromatin regions by using transposase enzymes to tag accessible DNA sites for sequencing.	- Mapping chromatin accessibility - Studying regulatory elements involved in stress response.	- Identifying open chromatin regions associated with drought tolerance in maize.	(Shashikant & Ettensohn, 2019).
CRISPR-based Epigenome Editing	Uses dCas9 fused with epigenetic modifiers (e.g., methyltransferases, demethylases) to edit specific epigenetic marks.	- Targeted DNA methylation or demethylation to control gene expression - Functional validation of regulatory elements.	- Inducing demethylation at stress-responsive genes in *Arabidopsis* to enhance drought tolerance.	(Dinkar et al., 2024)

continued on following page

Table 1. Continued

Technological Tool/ Method	Description	Application in Plant Biotechnology	Examples	Reference
Whole Genome Bisulfite Sequencing	A thorough approach for examining DNA methylation throughout the whole genome.	- Detailed methylation profiling for examining epigenetic alterations triggered by stress.	- Genome-wide methylation analysis of heat stress responses in *Brassica rapa*.	(Akhter et al., 2021)
Methylated DNA Immunoprecipitation Sequencing (MeDIP-Seq)	Enriches methylated DNA by specific antibodies, followed by sequencing.	- Identifying areas with elevated DNA methylation levels - Investigating alterations in methylation throughout plant growth.	- Identifying hypermethylated regions in drought-tolerant vs. sensitive plant varieties.	(Zhao et al., 2014)
Hi-C and Chromatin Conformation Capture (3C/4C/5C)	Analyzes the 3D organization of chromatin to study interactions between distant genomic regions.	- Grasping the function of chromatin structure in maintaining gene expression in response to stress.	- Investigating how chromatin loops influence gene expression in Arabidopsis during stress conditions.	(Grob & Cavalli, 2018)
Single Cell Epigenomics	Methods such as single cell RNA sequencing and single cell ATAC sequencing are utilized to examine epigenetic conditions at the individual cellular level.	- Investigating cell type-particular epigenetic modification in response to environmental stress.	- Analysing the single cell epigenomic changes in root cells exposed to salt stress.	(Clark et al., 2016)
High-throughput DNA Methylation Arrays	Detects methylation at specific CpG sites across the genome using array-based technology.	- Screening for methylation changes in different plant genotypes or stress conditions.	- Large-scale screening of methylation patterns in different cultivars of wheat.	(Zhou et al., 2022)
Small RNA Sequencing	Analyzes the population of small RNAs (miRNAs, siRNAs) involved in gene silencing and epigenetic regulation.	- Recognizing small RNAs that respond to stress - Investigating RNA-directed DNA methylation (RdDM) pathways.	- Analysis of miRNA expression in tomato plants during infection by pathogens.	(Kumar et al., 2018)

Epigenome Editing in Plants

The use of CRISPR/Cas technology for ameliorating the epigenome is a precise and effective method for altering gene expression in plants without altering the DNA sequence, (Shin et al., 2022). This technique takes advantage of the CRISPR system's ability to target specific parts of the genome. A guide RNA (gRNA) helps direct a protein called Cas to these locations. Instead of making breaks in the DNA, a modified version of the Cas protein, known as dCas9, is used. This version does not cut DNA. Instead, dCas9 is combined with tools that can change gene activity or add chemical marks to DNA. These changes can affect how genes are expressed (Karlson et al., 2021).

The process starts with creating a specific gRNA that matches a DNA sequence at the target site. This gRNA assists in regulating the dCas9 protein to the precise spot within the plant genome. The dCas9 protein is modified so it can no longer cut DNA. Instead, it serves as a platform for adding or removing chemical tags on the DNA, using enzymes like histone acetyltransferases, deacetylases, methyltransferases, or demethylases (Syding et al., 2020). These epigenetic factors can alter histones or DNA in a precise way to either enhance or inhibit gene expression. For instance, the histones acetylation of is typically related to the activation of transcription, whereas the methylation of histones or DNA is corelated with the repression of genes (Albini et al., 2019). CRISPR/Cas-mediated epigenome editing can accurately regulate gene expression in plants. It achieves this by focusing on key regulatory regions, which makes it an important resource for both research and agriculture. In practice, CRISPR-based epigenetic editing can activate helpful genes that improve stress resistance, disease tolerance, or growth. It can also turn off unwanted genes that can lower crop yield or quality (Karmakar et al., 2022). This technique has clear benefits compared to traditional genetic modification methods. It does not create permanent changes to the DNA, which is important for getting regulatory approval and gaining public trust in genetically modified organisms (GMOs). Additionally, it can target multiple sites at the same time using multiplexed gRNAs. This allows for the regulation of complex traits that involve several genes (Vats & Deshmukh, 2019).

A notable illustration of CRISPR based epigenome editing in plants is the application of dCas9 linked to the histone acetyltransferase p300, which has been shown to enhance gene expression at targeted locations (Gardiner et al., 2022). This innovative method alters the way in which plants express their genes, leading to significant changes in a variety of plant characteristics such as growth rate, yield, disease resistance, and stress tolerance. A notable improvement in this technique is its ability to facilitate heritable modifications; this means that the changes made can be passed down to subsequent generations of plants. This feature is particularly advantageous for breeding programs aimed at developing crops with enhanced traits—such as improved nutritional content or greater resilience to environmental challenges—without the necessity of making permanent changes to the plant's underlying DNA sequence. By utilizing CRISPR/Cas technology for the purpose of editing the epigenome, researchers gain a potent tool within the realm of plant biotechnology. This technique enables precise and targeted regulation of gene expression through temporary modifications to epigenetic markers, thereby preserving the integrity of the original DNA sequence. Such a capability opens up a plethora of opportunities for enhancing the overall performance of crops. Ultimately, the ability to fine-tune gene expression through these temporary adjustments could lead to increased agricultural productivity, improved sustainability practices, and enhanced resilience of crops in the face of various environmental stressors, such as drought, pests, and diseases. This innovative approach could revolutionize the way we cultivate and manage crops, offering solutions to some of the most pressing challenges in agriculture today.

CONCLUSION

Gaining a deeper understanding of how plants respond to their surrounding environments necessitates a thorough examination of the intricate interplay between epigenetic modifications and the symbiotic relationships that plants form with various microbial communities. Epigenetic mechanisms, which encompass a range of processes including DNA methylation, histone modifications, and the regulatory roles of non-coding RNAs, are pivotal in the control of gene expression. These modifications enable plants to not only perceive and respond to a wide array of stressors but also to adapt their growth and development accordingly. Biotic stressors, such as attacks from pests or infections by pathogens, trigger specific epigenetic changes that can enhance a plant's defensive capabilities. For instance, certain epigenetic alterations can activate genes responsible for synthesizing protective compounds, thereby fortifying the plant against potential threats. Conversely, abiotic stress factors—including extreme climatic conditions like drought, temperature fluctuations, and nutrient-deficient soils—also invoke significant epigenetic responses. In these cases, plants may engage in epigenetic reprogramming that leads to adjustments in metabolic pathways, improving their resilience and survival rates under harsh conditions.

The intricate relationship between epigenetic processes and microbial interactions reveals a profound complexity in how plants adapt to various environmental challenges. This interplay offers valuable insights into the evolutionary strategies that enable plants to flourish across diverse ecosystems. One particularly fascinating aspect of these epigenetic modifications is their capacity to persist for prolonged periods, sometimes extending across multiple generations. This transgenerational inheritance allows plants to develop a form of "memory," storing information about previous stressors they have encountered. As a result, plants become better equipped to withstand and navigate similar challenges in the future. For example, when faced with stressors such as drought, nutrient deficiency, or pathogen attacks, plants may undergo epigenetic changes that alter their gene expression without changing the underlying DNA sequence. These changes can lead to enhanced resilience, allowing them to respond more effectively to recurring stressors. This adaptive mechanism not only illustrates the remarkable resilience inherent in plant biology but also highlights the sophisticated and dynamic processes that govern their interactions with a constantly changing environment. Understanding these interactions further emphasizes the importance of preserving biodiversity and ecosystem health, as each plant species contributes uniquely to the larger tapestry of life on Earth.

In their intricate interactions with a diverse array of microbes, plants utilize epigenetic changes to forge and sustain beneficial relationships, particularly with helpful partners like mycorrhizal fungi and nitrogen-fixing bacteria. These symbiotic partnerships play a crucial role in enhancing nutrient uptake, promoting vigorous plant growth, and bolstering resilience against environmental stresses. On the other hand, the same epigenetic regulation is essential for equipping plants with the means to fend off harmful microbes, highlighting the delicate balance that underpins plant health and vigour. This equilibrium is vital, enabling plants not only to thrive but also to safeguard themselves from threats in their environment. Furthermore, plants possess a remarkable capacity to adapt their growth and physiological functions in response to the ever-changing conditions around them, a trait that is imperative for their survival. Recent breakthroughs in CRISPR/Cas technology present exciting opportunities for scientists, allowing for precise modifications of epigenetic marks. This innovative approach can lead to the creation of crop varieties that are better equipped to endure stressors and maximize food production. Importantly, this method enables the alteration of specific epigenetic features without making changes to the plant's underlying genetic structure, paving the way for enhanced agricultural resilience and sustainability.

Non-coding RNAs play a vital role in how plants manage and respond to various stressors, such as drought, salinity, and pathogen attacks. They do this by modulating gene expression after the initial transcription of genes, which significantly influences a broad spectrum of stress responses, developmental processes, and overall plant growth dynamics. Research into the interplay between epigenetics and plant-microbe interactions is particularly important, as it highlights the intricate mechanisms by which plants adapt to their environments. Epigenetic regulation involves chemical modifications that alter gene activity without changing the underlying DNA sequence, and understanding this complex regulatory network is essential for grasping plant biology comprehensively. By delving deeper into the epigenetic modifications that occur in response to environmental stresses, scientists can devise innovative strategies aimed at enhancing crop performance and ensuring agricultural sustainability. This understanding is increasingly necessary to tackle contemporary challenges in agriculture, such as climate change, population growth, and the demand for food security. The potential to leverage epigenetic mechanisms to improve crop resilience is significant. For instance, identifying specific non-coding RNAs that regulate stress responses could lead to the development of crop varieties that are better suited to withstand adverse conditions. Continued research in this area promises to enrich our knowledge of plant biology and mechanisms of stress response, paving the way for revolutionary approaches to boosting the resilience and productivity of crops. Such advancements will be crucial for fostering a sustainable agricultural future, ensuring that we can meet the nutritional needs of a growing global population while minimizing our environmental impact.

REFERENCES

Abdulraheem, M. I., Xiong, Y., Moshood, A. Y., Cadenas-Pliego, G., Zhang, H., & Hu, J. (2024). Mechanisms of plant epigenetic regulation in response to plant stress: Recent discoveries and implications. *Plants*, *13*(2), 163. PMID: 38256717

Acharjee, S., Chauhan, S., Pal, R., & Tomar, R. S. (2023). Mechanisms of DNA methylation and histone modifications. *Progress in Molecular Biology and Translational Science*, *197*, 51–92. PMID: 37019597

Akhter, Z., Bi, Z., Ali, K., Sun, C., Fiaz, S., Haider, F. U., & Bai, J. (2021). In response to abiotic stress, DNA methylation confers epigenetic changes in plants. *Plants*, *10*(6), 1096. PMID: 34070712

Albini, S., Zakharova, V., & Ait-Si-Ali, S. (2019). Histone modifications. In *Epigenetics and regeneration* (pp. 47–72). Academic Press.

Aqeel, M., Ran, J., Hu, W., Irshad, M. K., Dong, L., Akram, M. A., & Deng, J. (2023). Plant-soil-microbe interactions in maintaining ecosystem stability and coordinated turnover under changing environmental conditions. *Chemosphere*, *318*, 137924. PMID: 36682633

Aranda Lastra, M. (2022). Epigenetics and Acquired Tolerance to Environmental Stress. In *Coral Reef Conservation and Restoration in the Omics Age* (pp. 135–150). Springer International Publishing.

Arora, H., Singh, R. K., Sharma, S., Sharma, N., Panchal, A., Das, T., & Prasad, M. (2022). DNA methylation dynamics in response to abiotic and pathogen stress in plants. *Plant Cell Reports*, *41*(10), 1931–1944. PMID: 35833989

Ashapkin, V. V., Kutueva, L. I., Aleksandrushkina, N. I., & Vanyushin, B. F. (2020). Epigenetic mechanisms of plant adaptation to biotic and abiotic stresses. *International Journal of Molecular Sciences*, *21*(20), 7457. PMID: 33050358

Bhanu, B. D., Alluri, A., Shanker, A. K., & Ulaganathan, K. (2022). DNA methylation in plants and its role in abiotic stress tolerance. *Climate Change and Crop Stress*, 539-564.

Blecua, P., Martinez-Verbo, L., & Esteller, M. (2020). The DNA methylation landscape of hematological malignancies: An update. *Molecular Oncology*, *14*(8), 1616–1639. PMID: 32526054

Budak, H., Kaya, S. B., & Cagirici, H. B. (2020). Long non-coding RNA in plants in the era of reference sequences. *Frontiers in Plant Science*, *11*, 276. PMID: 32226437

Bure, I. V., Nemtsova, M. V., & Kuznetsova, E. B. (2022). Histone modifications and non-coding RNAs: Mutual epigenetic regulation and role in pathogenesis. *International Journal of Molecular Sciences*, *23*(10), 5801. PMID: 35628612

Byregowda, R., Prasad, S. R., Oelmüller, R., Nataraja, K. N., & Prasanna Kumar, M. K. (2022). Is endophytic colonization of host plants a method of alleviating drought stress? Conceptualizing the hidden world of endophytes. *International Journal of Molecular Sciences*, *23*(16), 9194. PMID: 36012460

Chen, H., Arsovski, A. A., Yu, K., & Wang, A. (2016). Genome-wide investigation using sRNA-Seq, degradome-Seq and transcriptome-Seq reveals regulatory networks of microRNAs and their target genes in soybean during soybean mosaic virus infection. *PLoS One*, *11*(3), e0150582. PMID: 26963095

Chen, L., Zhu, Q. H., & Kaufmann, K. (2020). Long non-coding RNAs in plants: Emerging modulators of gene activity in development and stress responses. *Planta*, *252*(5), 92. PMID: 33099688

Cheng, Y. J., Wang, J. W., & Ye, R. (2024). Histone dynamics responding to internal and external cues underlying plant development. *Plant Physiology*, *194*(4), 1980–1997. PMID: 38124490

Clark, J., & Rager, J. E. (2020). Epigenetics: An overview of CpG methylation, chromatin remodeling, and regulatory/noncoding RNAs. Environmental epigenetics in toxicology and public health, 3-32.

Clark, S. J., Lee, H. J., Smallwood, S. A., Kelsey, G., & Reik, W. (2016). Single-cell epigenomics: Powerful new methods for understanding gene regulation and cell identity. *Genome Biology*, *17*, 1–10. PMID: 27091476

Colonna Romano, N., & Fanti, L. (2022). Transposable elements: Major players in shaping genomic and evolutionary patterns. *Cells*, *11*(6), 1048. PMID: 35326499

Das, K. A., Chosdon, T., Nawale, R., Koder, S. B., Bhosale, R. K., Rana, M., & Srivastava, S. (2021). Role of Rhizospheric Microbiome against Soil-Borne Pathogens. Biopesticides International, 17(1).

Dhar, G. A., Saha, S., Mitra, P., & Nag Chaudhuri, R. (2021). DNA methylation and regulation of gene expression: Guardian of our health. *Nucleus (Austin, Tex.)*, *64*(3), 259–270. PMID: 34421129

Dinkar, V., Pandey, S., Kumar, A., Shiv, A., Lal, D., Bharati, A., & Chandra, A. K. (2024). Epigenetic regulations under plant stress: A cereals perspective. *Environmental and Experimental Botany*, •••, 105688.

Fall, A. F., Nakabonge, G., Ssekandi, J., Founoune-Mboup, H., Apori, S. O., Ndiaye, A., & Ngom, K. (2022). Roles of arbuscular mycorrhizal fungi on soil fertility: Contribution in the improvement of physical, chemical, and biological properties of the soil. *Frontiers in Fungal Biology*, *3*, 723892. PMID: 37746193

Gardiner, J., Ghoshal, B., Wang, M., & Jacobsen, S. E. (2022). CRISPR–Cas-mediated transcriptional control and epi-mutagenesis. *Plant Physiology*, *188*(4), 1811–1824. PMID: 35134247

Glick, B. R., & Gamalero, E. (2021). Recent developments in the study of plant microbiomes. Microorganisms, 9(7), 1533., S., Malav, L. C., Choudhary, J., Kannojiya, S., Kundu, M., Kumar, S., & Yadav, A. N. (2021). Soil microbiomes for healthy nutrient recycling. Current trends in microbial biotechnology for sustainable agriculture, 1-21.

Grob, S., & Cavalli, G. (2018). Technical review: a Hitchhiker's guide to chromosome conformation capture. *Plant chromatin dynamics: methods and protocols*, 233-246.

Guillemin, A., Kumar, A., Wencker, M., & Ricci, E. P. (2022). Shaping the innate immune response through post-transcriptional regulation of gene expression mediated by RNA-binding proteins. *Frontiers in Immunology*, *12*, 796012. PMID: 35087521

Gupta, C., & Salgotra, R. K. (2022). Epigenetics and its role in effecting agronomical traits. *Frontiers in Plant Science*, *13*, 925688. PMID: 36046583

He, S., & Feng, X. (2022). DNA methylation dynamics during germline development. *Journal of Integrative Plant Biology*, *64*(12), 2240–2251. PMID: 36478632

Huang, C. Y., & Jin, H. (2022). Coordinated epigenetic regulation in plants: A potent managerial tool to conquer biotic stress. *Frontiers in Plant Science*, *12*, 795274. PMID: 35046981

Karlson, C. K. S., Mohd-Noor, S. N., Nolte, N., & Tan, B. C. (2021). CRISPR/dCas9-based systems: Mechanisms and applications in plant sciences. *Plants*, *10*(10), 2055. PMID: 34685863

Karmakar, A., Taufiqa, S., Baig, M. J., & Molla, K. A. (2022). Increasing disease resistance in host plants through genome editing. *Proceedings of the Indian National Science Academy*, *88*(3), 417–429.

Kaya, C., Uğurlar, F., & Adamakis, I. D. S. (2023). Epigenetic and Hormonal Modulation in Plant–Plant Growth-Promoting Microorganism Symbiosis for Drought-Resilient Agriculture. *International Journal of Molecular Sciences*, *24*(22), 16064. PMID: 38003254

Kumar, A., Kaur, R., & Rajam, M. V. (2021). Regulatory role of micro-RNAs in plants under challenging environmental conditions with special focus on drought and salinity. *Harsh Environment and Plant Resilience: Molecular and Functional Aspects*, 121-140.

Kumar, V., Khare, T., Shriram, V., & Wani, S. H. (2018). Plant small RNAs: The essential epigenetic regulators of gene expression for salt-stress responses and tolerance. *Plant Cell Reports*, *37*, 61–75. PMID: 28951953

Leichter, S. M. (2022). *Dissecting the Molecular Basis and Function of Non-CG Methylation in Plants*. The University of Wisconsin-Madison.

Li, Y. (2021). Modern epigenetics methods in biological research. *Methods (San Diego, Calif.)*, *187*, 104–113. PMID: 32645449

Liang, W., Li, J., Sun, L., Liu, Y., Lan, Z., & Qian, W. (2022). Deciphering the synergistic and redundant roles of CG and non-CG DNA methylation in plant development and transposable element silencing. *The New Phytologist*, *233*(2), 722–737. PMID: 34655488

Liu, X. Q., Xie, M. M., Hashem, A., Abd-Allah, E. F., & Wu, Q. S. (2023). Arbuscular mycorrhizal fungi and rhizobia synergistically promote root colonization, plant growth, and nitrogen acquisition. *Plant Growth Regulation*, *100*(3), 691–701.

Miglani, G. S., Kaur, A., & Kaur, L. (2020). Plant gene expression control using genome-and epigenome-editing technologies. *Journal of Crop Improvement*, *34*(1), 1–63.

Miryeganeh, M. (2021). Plants' epigenetic mechanisms and abiotic stress. *Genes*, *12*(8), 1106. PMID: 34440280

Moreira, D. A., de Lanna, C. A., da Cruz, J. G. V., & Boroni, M. (2021). To Build or To Break: The Dual Impact of Interspersed Transposable Elements in Cancer. In *Human Genome Structure, Function and Clinical Considerations* (pp. 245–273). Springer International Publishing.

Mostafa, S., Wang, Y., Zeng, W., & Jin, B. (2022). Plant responses to herbivory, wounding, and infection. *International Journal of Molecular Sciences*, *23*(13), 7031. PMID: 35806046

Mushtaq, A., Mir, U. S., Hunt, C. R., Pandita, S., Tantray, W. W., Bhat, A., & Pandita, T. K. (2021). Role of histone methylation in maintenance of genome integrity. *Genes*, *12*(7), 1000. PMID: 34209979

Nizamani, M. M., Hughes, A. C., Qureshi, S., Zhang, Q., Tarafder, E., Das, D., & Zhang, Z. G. (2024). Microbial biodiversity and plant functional trait interactions in multifunctional ecosystems. *Applied Soil Ecology*, *201*, 105515.

Pagano, L., Rossi, R., Paesano, L., Marmiroli, N., & Marmiroli, M. (2021). miRNA regulation and stress adaptation in plants. *Environmental and Experimental Botany*, *184*, 104369.

Paul, A., & Riddle, N. C. (2020). ChIP in series: Sequential ChIP. *Epigenetics Methods*, 297-321.

Policarpi, C., Dabin, J., & Hackett, J. A. (2021). Epigenetic editing: Dissecting chromatin function in context. *BioEssays*, *43*(5), 2000316. PMID: 33724509

Prabhakaran, S. (2024). RNA in Plant Nutrient Availability. Noncoding RNA: The Dark Matter of the Genome, 165.

Prall, W., Sharma, B., & Gregory, B. D. (2019). Transcription is just the beginning of gene expression regulation: The functional significance of RNA-binding proteins to post-transcriptional processes in plants. *Plant & Cell Physiology*, *60*(9), 1939–1952. PMID: 31155676

Prasad, S., Malav, L. C., Choudhary, J., Kannojiya, S., Kundu, M., Kumar, S., & Yadav, A. N. (2021). Soil microbiomes for healthy nutrient recycling. Current trends in microbial biotechnology for sustainable agriculture, 1-21.

Praveen, A., Dubey, S., Singh, S., & Sharma, V. K. (2023). Abiotic stress tolerance in plants: A fascinating action of defense mechanisms. 3. *Biotech*, *13*(3), 102. PMID: 36866326

Qin, W., Scicluna, B. P., & van der Poll, T. (2021). The role of host cell DNA methylation in the immune response to bacterial infection. *Frontiers in Immunology*, *12*, 696280. PMID: 34394088

Ramakrishnan, M., Satish, L., Kalendar, R., Narayanan, M., Kandasamy, S., Sharma, A., & Zhou, M. (2021). The dynamism of transposon methylation for plant development and stress adaptation. *International Journal of Molecular Sciences*, *22*(21), 11387. PMID: 34768817

Raza, A., Zahra, N., Hafeez, M. B., Ahmad, M., Iqbal, S., Shaukat, K., & Ahmad, G. (2020). Nitrogen fixation of legumes: Biology and Physiology. *The plant family Fabaceae: biology and physiological responses to environmental stresses*, 43-74.

Shah, I. M., Dar, M. A., Bhat, K. A., Dar, T. A., Ahmad, F., & Ahmad, S. M. (2022). Long non-coding RNAs: biogenesis, mechanism of action and role in different biological and pathological processes. In *Recent Advances in Noncoding RNAs*. IntechOpen.

Shakirova, K. M., Ovchinnikova, V. Y., & Dashinimaev, E. B. (2020). Cell reprogramming with CRISPR/Cas9 based transcriptional regulation systems. *Frontiers in Bioengineering and Biotechnology*, *8*, 882. PMID: 32850737

Sharma, M., Kumar, P., Verma, V., Sharma, R., Bhargava, B., & Irfan, M. (2022). Understanding plant stress memory response for abiotic stress resilience: Molecular insights and prospects. *Plant Physiology and Biochemistry*, *179*, 10–24. PMID: 35305363

Shashikant, T., & Ettensohn, C. A. (2019). Genome-wide analysis of chromatin accessibility using ATAC-seq. *Methods in Cell Biology*, *151*, 219–235. PMID: 30948010

Shi, M., Wang, C., Wang, P., Zhang, M., & Liao, W. (2022). Methylation in DNA, histone, and RNA during flowering under stress condition: A review. *Plant Science*, *324*, 111431. PMID: 36028071

Shin, H., Choi, W. L., Lim, J. Y., & Huh, J. H. (2022). Epigenome editing: Targeted manipulation of epigenetic modifications in plants. *Genes & Genomics*, •••, 1–9. PMID: 35000141

Singh, R. K., Prasad, A., Muthamilarasan, M., Parida, S. K., & Prasad, M. (2020). Breeding and biotechnological interventions for trait improvement: Status and prospects. *Planta*, *252*, 1–18. PMID: 32948920

Sioud, M. (2021). RNA interference: story and mechanisms. Design and Delivery of SiRNA Therapeutics, 1-15.

Sugden, K., Hannon, E. J., Arseneault, L., Belsky, D. W., Corcoran, D. L., Fisher, H. L., & Caspi, A. (2020). Patterns of reliability: Assessing the reproducibility and integrity of DNA methylation measurement. *Patterns (New York, N.Y.)*, *1*(2). PMID: 32885222

Sun, M., Yang, Z., Liu, L., & Duan, L. (2022). DNA methylation in plant responses and adaption to abiotic stresses. *International Journal of Molecular Sciences*, *23*(13), 6910. PMID: 35805917

Syding, L. A., Nickl, P., Kasparek, P., & Sedlacek, R. (2020). CRISPR/Cas9 epigenome editing potential for rare imprinting diseases: A review. *Cells*, *9*(4), 993. PMID: 32316223

Tang, K. (2020). Small RNA and DNA methylation in plants. Plant Small RNA, 353-376.

Vanzan, L., Sklias, A., Boskovic, M., Herceg, Z., Murr, R., & Suter, D. M. (2023). Mechanisms of histone modifications. In *Handbook of epigenetics* (pp. 27–54). Academic Press.

Vats, S., Kumawat, S., Kumar, V., Patil, G. B., Joshi, T., Sonah, H., & Deshmukh, R. (2019). Genome editing in plants: Exploration of technological advancements and challenges. *Cells*, *8*(11), 1386. PMID: 31689989

Wang, L., Zheng, K., Zeng, L., Xu, D., Zhu, T., Yin, Y., & Yang, D. L. (2022). Reinforcement of CHH methylation through RNA-directed DNA methylation ensures sexual reproduction in rice. *Plant Physiology*, *188*(2), 1189–1209. PMID: 34791444

Wiesmann, C. L., Wang, N. R., Zhang, Y., Liu, Z., & Haney, C. H. (2023). Origins of symbiosis: Shared mechanisms underlying microbial pathogenesis, commensalism and mutualism of plants and animals. *FEMS Microbiology Reviews*, *47*(6), fuac048. PMID: 36521845

Wilkinson, S. W. (2020). Investigating the mechanisms of long-lasting within generation jasmonate induced resistance (Doctoral dissertation, University of Sheffield).

Yajnik, K. N., Singh, I. K., & Singh, A. (2024). lncRNAs and epigenetics regulate plant's resilience against biotic stresses. *Plant Physiology and Biochemistry*, •••, 108892. PMID: 38964086

Zhang, Y., Sun, Z., Jia, J., Du, T., Zhang, N., Tang, Y., & Fang, D. (2021). Overview of histone modification. Histone Mutations and Cancer, 1-16., V. V., Kutueva, L. I., Aleksandrushkina, N. I., & Vanyushin, B. F. (2020). Epigenetic mechanisms of plant adaptation to biotic and abiotic stresses. *International Journal of Molecular Sciences*, *21*(20), 7457.

Zhang, Y., Zhou, Y., Zhu, W., Liu, J., & Cheng, F. (2022). Non-coding RNAs finetune the balance between plant growth and abiotic stress tolerance. *Frontiers in Plant Science*, *13*, 965745. PMID: 36311129

Zhao, M. T., Whyte, J. J., Hopkins, G. M., Kirk, M. D., & Prather, R. S. (2014). Methylated DNA immunoprecipitation and high-throughput sequencing (MeDIP-seq) using low amounts of genomic DNA. [Formerly Cloning and Stem Cells]. *Cellular Reprogramming*, *16*(3), 175–184. PMID: 24773292

Zhao, T., Zhan, Z., & Jiang, D. (2019). Histone modifications and their regulatory roles in plant development and environmental memory. *Journal of Genetics and Genomics*, *46*(10), 467–476. PMID: 31813758

Zhou, P., Zhang, X., Fatima, M., Ma, X., Fang, H., Yan, H., & Ming, R. (2020). DNA methylome and transcriptome landscapes revealed differential characteristics of dioecious flowers in papaya. *Horticulture Research*, •••, 7. PMID: 32528693

Zhou, W., Hinoue, T., Barnes, B., Mitchell, O., Iqbal, W., Lee, S. M., & Laird, P. W. (2022). DNA methylation dynamics and dysregulation delineated by high-throughput profiling in the mouse. *Cell Genomics*, *2*(7). PMID: 35873672

Chapter 8
Epigenetic Regulation of Plant Immune Responses to Pathogens

Fariba Shohani
Urmia University, Iran

Arash Fazeli
Ilam University, Iran

Seyede Roghie Ghadirnezhad Shiade
Sari Agricultural Science and Natural Resources University, Iran

Amin Fathi
https://orcid.org/0000-0002-7539-0053
Islamic Azad University, Iran

Javed Iqbal
https://orcid.org/0000-0002-9032-2622
Bacha Khan University, Charsadda, Pakistan

ABSTRACT

Plants as immobile organisms are perpetually subjected to environmental changes and various biotic stresses. Pathogens such as viruses, bacteria and fungi can severely hinder plant growth and development thus presenting significant challenges to global food security. The relationship between plants and pathogens has led to evolution of intrinsic system such as gene expression networks, which enable plants to protect themselves from pathogenic threats. These recognition mechanisms allow plants to identify invading microbial pathogens and trigger subsequent defense responses. Thus, plants have evolved advanced and highly regulated strategy to

DOI: 10.4018/979-8-3693-8648-4.ch008

address pathogen threats and to relay information to subsequent generation ultimately creation of stress memory. Recent studies reveal involvement of regulated epigenetic mechanisms; DNA methylation, histone modifications, chromatin remodeling complexes and non-coding RNAs in the regulation of defense-related genes. In this chapter, we explored these epigenetic mechanisms and defensive pathways employed by plants in response to biotic stresses.

INTRODUCTION

Plants are regularly vulnerable to diverse attack from harmful pathogens. The entry of viruses and bacteria into plants typically occur through natural openings like stomata or through area of damaged tissue, whereas oomycetes and fungi can directly invade the surfaces of their host plants (Fan, 2022). Plant tissues, characterized by their rich nutrient content, present an appealing target for pathogens. These pathogens can infiltrate the plants nutrient supply by entering through stomatal opening or by breaching physical barriers to access the underlying tissues. Plants regulate physical barriers such as the cuticle, lignin, and silica to suppress pathogen invasion (**Figure 1**). However, specific pathogens can successfully breach these protective barriers. simultaneously, activation of defense mechanisms may negatively have an impact on plant growth, necessitating precise regulation (Fan, 2022).

Figure 1. Suppression of pathogens mediated by physical barriers

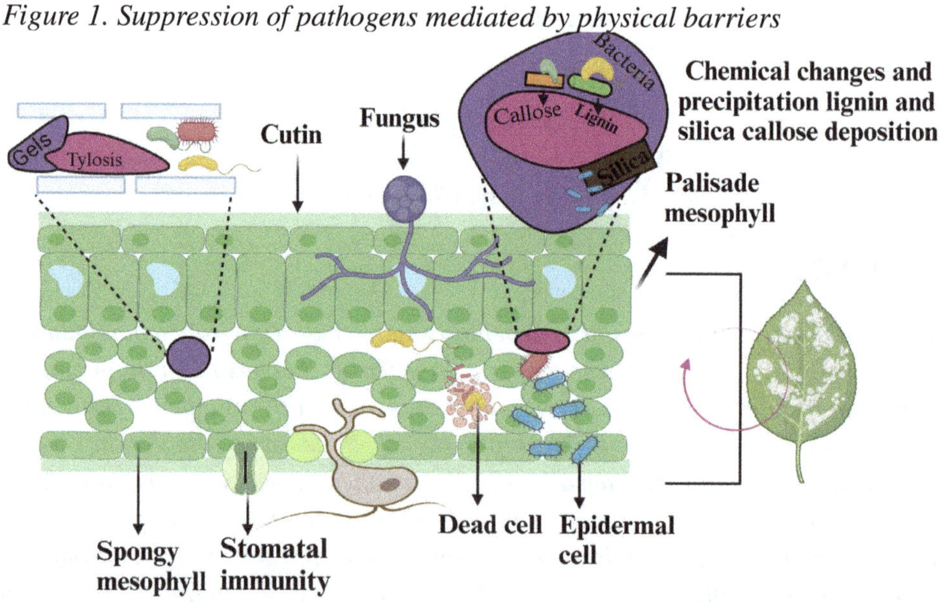

Plants prevent the entry of pathogens by closing stomata and producing a cuticle layer on the epidermis. When pathogens succeed in evading these physical defenses, plants respond by reforming their structural features, including the generation of lignin and silica, to combat pathogen invasion (Jian et al., 2024).

The immune systems of plants play crucial role in protect themselves against pests and diseases. The immune system of plant consisted of an innate component that is genetically determined, and an adaptive component influenced by environmental conditions. in contrast to animals, where innate and adaptive immunity are controlled by distinct mechanisms and cells, these components in plants are more intertwined, relying on overlapping pathways and mechanisms (Mauch-Mani et al., 2017). The cornerstone of the plant innate immune system lies in the recognition of danger signals and subsequent activation of chemical and mechanical defenses. These immune signals are typically molecules derived from the invading organism or those generated due to cellular damage (Hannan Parker, Wilkinson, & Ton, 2022).

Plants, as nutrient-rich hosts, are frequently targeted by a wide range of pathogens. Host innate immunity is triggered by recognizing pathogen-associated molecular patterns (PAMPs) through specific pattern-recognition receptors (PRRs) (Cao et al., 2022; Yuan et al., 2021). PAMP-triggered immunity (PTI) activates diverse cellular defense pathways, such as signaling cascades, defense-related gene expression, and hypersensitive responses (Arora et al., 2022). Pathogen attacks lead to extensive gene expression changes in plants, often regulated through epigenetic modifications. In host plants, pathogens encounter ubiquitous defense mechanisms, including: a) the recognition of pathogen-associated factors by corresponding NLR genes, and

b) RNA silencing mechanisms, also referred to as RNA interference (RNAi).

Besides, RNA silencing includes two separate branches; I) post-transcriptional gene silencing (PTGS), which suppresses gene expression by inhibiting translation or initiating mRNA degradation. II) transcriptional gene silencing (TGS), which inhibits gene transcription by promoting chromatin modifications. Epigenetic processes such as chromatin remodeling, DNA methylation, and histone modifications are critical for regulating plant defense responses to pathogen infections. The relationship between DNA methylation and gene expression is highly complex and depends on factors such as tissue type, sequence context, transposon activity, and genomic location (Tirnaz & Batley, 2019).

This chapter explores the latest research regarding the influence of epigenetic factors on the interactions between plants and pathogens.

FUNDAMENTAL MECHANISMS OF EPIGENETIC MODIFICATIONS IN GENE TRANSCRIPTION REGULATION

DNA Methylation

DNA methylation is a continuous process that occurs throughout a plant's life and plays critical roles in stress adaptation, genome management, and growth regulation (Xiao et al., 2021). This common epigenetic modification typically takes place at accessible and stable cytosine residues. Changes in DNA methylation in response to stimuli can be long-lasting and potentially heritable across generations (Chen et al., 2022, Ahmad et al., 2022a).

DNA methylation is catalyzed by specific DNA methyltransferases at the 5' position of the cytosine base, resulting in 5-methylcytosine (5-mC). In plants, methylation can happen across three different sequence contexts: CG, CHG, and CHH, where H represents A, T, or C. Three distinct types of DNA methyltransferases (DNA Methyltransferase 1 (MET1), Domains Rearranged Methyltransferase (DRM), and Chromomethylase (CMT)) have been identified in plants. Thease are De novo methylation, which is mediated by DRMs. Besides, maintenance of hemimethylated DNA occurs in specific contexts such as CG methylation is maintained by MET1; CHG methylation is maintained by CMT3 and, to a lesser extent, by CMT2; CHH methylation is maintained by DRM2 or CMT2 (Arora et al., 2022).

Plants employ RNA-directed DNA methylation (RdDM) as a defense mechanism to combat viral infections. RdDM facilitates de novo methylation of viral coding and promoter regions, suppressing transcription and replication of the viral minichromosome (Arora et al., 2022). Analysis of the methylome have demonstrated how DNA methylation in regulation plant gene expression in response to viral infections. For example, the infection of *Nicotiana tabacum* with cucumber mosaic virus (CMV; family *Bromoviridae*) leads to decreased, methylation levels in the promoter regions of critical genes—including Argonaute and defense-related genes such as *Phenylalanine Ammonia-Lyase (PAL)* and PR-like proteins—decrease. This reduction underscores the role of DNA demethylation in resistance signaling pathways (Wang et al., 2018)

Histone Modifications

The regulation of gene expression occurs through changes in the structure of chromatin (Mierziak and Wojtasik, 2024). The eukaryotic genome is organized and compacted into nucleosomes, consisting of 146 base pairs of DNA encased around the histone octamer. This structure allows long DNA strands to fit precisely within the nucleus (Yunqing Jian, Shim, and Ma, 2021). a histone octamer consists of two copies of each molecule of the histones H2A, H2B, H3, and H4. histone tails that emerge from the nucleosome core are modified by several post-translational modifications (PTMs), such as methylation (me), acetylation (ac), ubiquitination (ub), and others (Fan, 2022). These PTMs can modify histone-histone or histone-DNA interactions, thereby influencing transcriptional processes (Krajewski, 2022).

Overall, the trimethylation of histone H3 lysine 4 (H3K4me3) and H3K36me2/3 is associated with actively transcribed regions, whereas H3K9me2 and H3K27me3 are linked to transcriptionally repressed regions. H3K4me is catalyzed by a conserved protein complex (the COMPASS-like complex) and is predominantly located in chromatin. Several H3K4 methyltransferases play a role in regulating plant immunity. For instance, ATX1 has been identified as a "master regulator" that activates the transcription factor *WRKY70* by promoting H3K4me3 deposition (Xie and Duan, 2023). Histone acetylation dynamics represent one of the most extensively studied PTM mechanisms. Similar to other PTMs, acetylation is reversible and is dynamically modulated by the actions of histone acetyltransferases (HATs) and histone deacetylases (HDACs). Generally, HATs facilitate gene activation by adding acetyl groups (CH_3COO^-) to lysine residues on the N-terminal histone tails (**Figure 2**) (Kumar, Thakur, and Prasad, 2021).

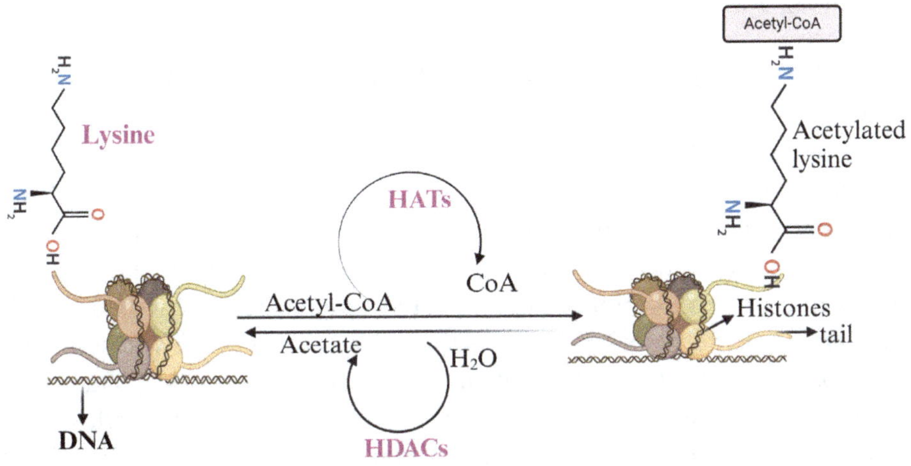

Chromatin Modifications

The DNA strands located in the nucleus are closely coiled and compressed, leading to the development of dense chromatin structures. The dynamic regulation of chromatin facilities the precis timing, spatial organization and sequential coordination of DNA-related cellular prosses (Yunqing Jian, Shim, and Ma, 2021). the regulation of chromatin dynamics plays a crucial role in eukaryotic gene, incorporating covalent modifications of histones, changes in DNA structure and ATP-dependent chromatin remodeling complexes. Chromatin remodeling complexes use the energy from ATP hydrolysis to move, destabilize, evict, or remodel nucleosomes, thereby modulating the access of transcription machinery to DNA. ATP-dependent chromatin remodeling complexes can be classified into four subfamilies: SWI/SNF, ISWI, CHD, and INO80. These subfamilies share a conserved ATPase domain and regulate discrete biological processes, such as transcriptional regulation, DNA replication, DNA repair, homologous recombination, and chromosome segregation (Yunqing Jian, Shim, and Ma, 2021). Among these four subfamilies, the SWI/SNF complex is evolutionarily conserved and serves as the most powerful chromatin-access regulator. initially recognized in *Saccharomyces cerevisiae* through genetic screening and biochemical purification techniques, this substance has also been reported in various other organisms, including *S. cerevisiae* and *Arabidopsis* (Yunqing Jian, Shim, and Ma, 2021).

Role of the SWI/SNF Complex in Plant-Pathogen Interactions

Plants, as non-motile organisms, need to adeptly modulate their gene expression to adapt to variable environments, a process that is predominantly shaped by the ever-changing structure of chromatin (Chang et al., 2020; Z. T. Song, Liu, and Han, 2021). Increasing evidence has shown that the SWI/SNF complex plays a critical role in plant responses to abiotic stresses, including temperature, drought, salinity, osmotic stress, and hormonal signaling pathways, as well as biotic stresses (Chang et al., 2020; Han et al., 2021, Ahmad et al., 2022b). Unlike animals, plants lack an adaptive immune system that produces antibodies and do not have mobile cells for pathogen detection, prevention, or infection reduction. Instead, plants have developed an innate immune system to defend against microbial attacks. Phytohormones, including salicylic acid (SA), jasmonic acid (JA), ethylene (ET), abscisic acid (ABA), auxins, gibberellins (GA), cytokinins, and brassinosteroids, have emerged as key players in plant immune responses. Accumulating evidence suggests that chromatin structures play a fundamental role in regulating plant defense responses by physically interacting with components or regulators of phytohormone signaling pathways (Yunqing Jian, Shim, and Ma, 2021).

In addition to regulating plant hormone signaling pathways, the SWI/SNF complex plays an important role in modulating the transcription of resistance genes. In *Arabidopsis*, SWP73A directly binds to the promoters of NLR (NOD-like receptor) genes to suppress their expression. Moreover, it may act as a reader of H3K9me2 to enhance this transcriptional repression (Qin et al., 2021, Haider et al., 2021a). Beyond the subunits of the SWI/SNF complex, another *Arabidopsis* ATPase, SYD, plays a crucial role in plant defense against specific biotic stresses. Genetic, biochemical, and biological evidence has shown that SYD is involved in the activation of genes that control JA- and ethylene (ET)-dependent defense responses, positively regulating their expression levels following *Botrytis cinerea* infection. Furthermore, SYD is required for resistance against the necrotrophic pathogen *B. cinerea*, but not the biotrophic pathogen *Pseudomonas syringae*, suggesting that chromatin remodeling is selectively involved in resistance to different pathogens (Yunqing Jian, Shim, and Ma, 2021, Haider et al., 2021b).

Non-Coding RNAs (ncRNAs)

Non-Coding RNAs (ncRNAs) are RNA molecules that do not code for proteins but play a crucial role in regulating gene expression via multiple mechanisms. ncRNAs can influence chromatin leading to gene silencing through direct interactions with epigenetic factors. Additionally, they can engage with transcription factors to either inhibit or promote the expression of target genes (S. Kumar et al., 2020, Ahmad et al., 2024a). ncRNAs are categorized primarily by their length in to three groups, small ncRNAs (sRNAs) (18-30 nt), medium-sized ncRNAs (31-200 nt), and long non-coding RNAs (lncRNAs) (> 200 nt) (Song et al., 2021, Haider et al., 2024a).

MicroRNAs (miRNAs) are small ncRNAs, ranging from 20 to 24 nucleotides length, encoded by MIRNA genes. These molecules can regulate complex biological processes in plants. MIRNA genes are typically transcribed by RNA polymerase II (pol II), producing a primary miRNA transcript (pri-miRNA) that includes a 7-methylguanosine cap (7MGpppG) and a polyadenylated tail (Yang et al., 2021). Beyond their involvement in the modification of miRNA function, plant also regulate their responses to pathogens through indirect pathways. key plant hormones, such as auxins, salicylic acid (SA), jasmonic acid (JA), and ethylene are essential for orchestrating defense responses when faced with pathogen challenges. The involvement of miRNAs in these responses is noteworthy, as their function in suppressing plant hormone signaling pathways to influence defense reactions was first reported by Yang et al. (2021). Additionally, miR160 has been reported to regulate potato immunity against *Phytophthora infestans* by targeting AUXIN RESPONSE FACTOR 10 (StARF10) and StARF16 genes (Natarajan et al., 2023, Haider et al., 2021b).

Small interfering RNAs (siRNAs) are double-stranded molecules that act as suppressors of specific gene expression. In the past, RNA interference (RNAi) in plants was mainly used to enhance resistance against diseases by silencing susceptibility genes (genes that negatively regulate plant defense responses). However, in the past decade, RNAi has been more widely exploited for generating plants with "pathogen-induced resistance," where resistance is conferred through siRNAs that can silence genes essential for pathogen infection or lifecycle (Natarajan et al., 2023).

EPIGENETIC REGULATION IN PLANT IMMUNITY

DNA Methylation

Recent reports have highlighted changes in DNA methylation in plants in response to biotic stress factors. Biotic stress factors include the effects of bacteria, fungi, and viruses, which impact the normal growth and development of plants, ultimately reducing plant growth and yield. DNA methylation has been found to play a significant role in plant resistance to biotic stress (Hewezi et al., 2017). For example, DNA methylation levels in tobacco were altered in response to infection by the Tobacco Mosaic Virus (TMV), with these changes closely associated with the activation of stress-responsive genes (Xiao et al., 2021, Ahmad et al., 2024b). Gene silencing plays a crucial role in plant-pathogen interactions. Both post-transcriptional gene silencing (PTGS) and transcriptional gene silencing (TGS) act as innate defense responses against viral attacks (Jin, Zhao, and Guo, 2020).

The involvement of epigenetic modifications in antiviral defense is evidenced by the resistance genes Ty1 and Ty3, which share similarities with RNA polymerase III-dependent RNA (RDR3) and are responsible for resistance to the Tomato Yellow Leaf Curl Virus (TYLCV; Geminiviridae family) in tomatoes. DNA methylation levels of the viral genome increase in response to viral infections (Arora et al., 2022; Prasad et al., 2019). Early studies have shown that the replication of the Tomato Yellow Mosaic Virus in tobacco protoplasts was significantly reduced when their DNA was methylated under laboratory conditions (Mulaudzi, Koorsen, and Mwaba, 2023). Furthermore, Arabidopsis plants deficient in methylation, including cytosine methyltransferases, methyl cycle enzymes, and Dicer-like proteins, exhibited heightened sensitivity to Cabbage Leaf Curl Virus (CaLCuV) and Beet Curly Top Virus (BCTV) (Mierziak and Wojtasik, 2024). The role of viroids in triggering the DNA methylation machinery has been studied in tomato, where Potato spindle tuber viroid (PSTVd; Pospiviroidae family) was shown to increase the expression of genes involved in DNA methylation, such as DRM2, CMT3, MET1, AGO4, and HEN1. This results in DNA hypermethylation, which inhibits the replication and transcription of Sardinia Virus (TYLCSV; Geminiviridae family) (Arora et al., 2022).

Histone Modifications

Histone methylation is another important epigenetic mark that controls gene expression and various biological processes. Post-translational chemical modifications of histone tails provide epigenetic information that aids in the regulation of the genome. These changes in the histone tail directly influence the nucleosome structure, leading to transcriptional alterations that either activate or silence genes. Generally, euchromatic regions of the genome are highly enriched with acetylated histones. For example, acetylation at H3 lysine 27 (H3K27ac) is associated with transcriptional activation, while trimethylation at H3 lysine 27 (H3K27me3) is linked to transcriptional silencing (Zhang et al., 2020).

Histone methylation is also connected to gene activation and repression in *Magnaporthe oryzae*. In this fungus, a cellulase gene is activated in response to a substrate, with histone methylation accelerating this activation. The positive regulation of the cellulase gene MoCel7C occurs upon contact with the host leaf (Zhou et al., 2024). In rice, 15 JmjC genes are transcribed following infection by the bacterial pathogen *Xanthomonas oryzae* pv *oryzae* (Xoo). Among these, JMJ704 represses the transcription of defense-related genes such as NRR, OsWRKY62, and Os-11 N3 by reducing H3K4me2/3 levels, thereby positively regulating rice's defensive responses (Fan, 2022).

Overexpression of BcRPD3 in *Botrytis cinerea* leads to a significant reduction in histone H3 and H4 acetylation levels, which delays fungal vegetative growth and significantly weakens the formation of infectious structures and pathogenicity (Zhang et al., 2020). Additionally, histone methylation at H3K4me2 has been reported to influence the transcription of stress-responsive genes, including resistance (R) proteins, detoxifying enzymes, and genes involved in ion transport and cell death, in *Phaseolus vulgaris* infected with the rust pathogen (*Uromyces appendiculatus*) (Ramos-Cruz, Troyee, and Becker, 2021, Ali et al., 2022).

Chromatin Remodeling

Chromatin structure is essential for organizing the plant genome and is dynamically regulated by histones and a variety of other modifiers. Increasing evidence suggests that Chromatin Remodeling Complexes (CRCs) play significant roles in regulating biological processes, including plant immunity (Zhi and Chang, 2021). For example, the SWI/SNF chromatin remodeling complex in *Arabidopsis* regulates plant defense responses to biotic stress. The SWI2/SNF2 class chromatin remodeler *SPLAYED* (SYD) transiently induces mechanical wounding and positively regulates genes involved in jasmonic acid (JA) and ethylene (ET) signaling in response to the necrotrophic fungus *Botrytis cinerea* (Parker and Wilkinson, 2022). Another chromatin remodeling complex, SWR1, enhances resistance to the white mold fungus *Sclerotinia sclerotiorum* in *A. thaliana* through a signaling process mediated by the receptor kinase ERECTA (ER) (Mierziak and Wojtasik, 2024). *SPLAYED* (SYD) regulates the expression of several defense-related genes within the JA/ET signaling pathways, playing a crucial role in resistance to *B. cinerea*. In normal growth conditions, *CHR19* suppresses multiple SA/JA-related genes, regulating the plant's resistance to various pathogens. Loss of *CHR19* results in increased sensitivity to *B. cinerea* (Tao et al., 2023).

Non-Coding RNAs

Long non-coding RNAs (lncRNAs) are found in a wide range of organisms, including plants, fungi, prokaryotes, and even viruses. In plants, lncRNAs are key regulators of various biological processes, including gene expression (Huang et al., 2023). In *Solanum lycopersicum* (tomato), over 600 differentially expressed lncRNAs have been identified in response to *Phytophthora infestans* infection, and lncRNA16397 and lncRNA33732 modulate plant defense by regulating reactive oxygen species (ROS) accumulation (Huang et al., 2023). A study of interactions between tomato and *Ralstonia solanacearum* revealed 23 differentially expressed lincRNAs, suggesting their involvement in bacterial wilt infection by modulating JA/ethylene signaling pathways or regulating AGO protein expression (Cao et al., 2022).

One of the key mechanisms of gene regulation involves RNA silencing. This process includes small RNAs (sRNAs), which are divided into two main categories: miRNAs and siRNAs. miRNAs typically range from 21 to 24 nucleotides in length, while siRNAs are derived from double-stranded RNA (dsRNA) and may require RNA-dependent RNA polymerases. One of the first reported miRNAs involved in plant resistance is *miR393* from *A. thaliana*. It is induced by PAMP receptors like flg22 and auxin, leading to the downregulation of the auxin signaling pathway and activation of pattern-triggered immunity (PTI) (Mierziak and Wojtasik, 2024). The interplay between the synthesis and signaling of ET, SA, JA, auxin, and polyamines is critical in managing biotic stress in rice (Li et al., n.d.; Val-Torregrosa and Bund, 2021, Ijaz et al., 2023). Several miRNAs, such as *miR1846*, regulate ACC oxidase, controlling genes responsive to ethylene like *ERF43*, *ERF33*, *EREBP56*, *ERF90*, *ERF73*, and *EREBP124* (Sabrina and Abdul, 2022). Additionally, specific miRNAs regulate the conversion of S-adenosyl-L-methionine (SAM) precursors required for the synthesis of methyl jasmonates, methyl salicylates, and polyamines, linking the ethylene pathway with SA and JA pathways. Genes controlling SAM and JA in rice are regulated by *miR5512a/b* and *miR1428b/c/d*, highlighting SAM levels and distribution as central to defense signaling. Auxin signaling is also modulated by specific miRNAs (Sabrina and Abdul, 2022, Ijaz et al., 2024).

CONCLUSION

At present, the world is dealing with pressing challenges associated with food security and environmental degradation, emphasizing the vital contribution of biology to security food resources on a global level. Throughout their evolutionary history, plants have employed genetic and epigenetic modifications to adapt to various biotic and abiotic stresses. Numerous studies have demonstrated that plant resistance to diseases can be significantly influenced by epigenetic markers. Therefore, epigenetic regulation can be considered an effective strategy for improving resistance to plant diseases.

Increasing integration of epigenetic mechanisms such as DNA methylation, histone modifications, chromatin remodeling, and non-coding RNA, significantly strengthens plant defense responses to pathogens, ultimately leading to improved agricultural yield potential. Over time, these investigations have enriched our knowledge regarding the mediation of plant defense responses to pathogens highlighted the role of epigenetic markers in this process. Successful plant defense mechanisms and pathogen infection are influenced by the dynamic control of epigenetic markers t which are vital for either triggering or silencing the expression of genes associated with defense. Although epigenetic regulators have been identified for plant immunity against pathogens, the mechanisms reported primarily focus on how specific regulators affect key defense genes. However, it remains unclear how genetic factors are recruited by the plant defense systems. Therefore, future studies should investigate these regulatory mechanisms in greater detail.

DECLARATION

The chapter has been edited with the help of ChatGPT, an AI-powered tool, to improve the clarity and flow of the content.

REFERENCES

Ahmad, Y., Haakim, Z., Iqbal, J., Abbasi, B. A., Mahmood, T., & Kazi, M. (2024b). Technological Innovations for Abiotic Stress Resistance in Horticultural Crops. OMICs-based Techniques for Global Food Security, 233-244.

Ahmad, Y., Haider, S., Iqbal, J., Abbasi, B. A., Yaseen, T., & Mahmood, T. (2022a). The mechanisms of genome editing technologies in crop plants. In *Principles and Practices of OMICS and Genome Editing for Crop Improvement* (pp. 295–313). Springer International Publishing.

Ahmad, Y., Haider, S., Iqbal, J., Naseer, S., Attia, K. A., Mohammed, A. A., & Mahmood, T. (2024a). In-silico analysis and transformation of OsMYB48 transcription factor driven by CaMV35S promoter in model plant–Nicotiana tabacum L. conferring abiotic stress tolerance. *GM Crops and Food: Biotechnology in Agriculture and the Food Chain*, *15*(1), 130–149. PMID: 38551174

Ali, I., Salah, K. B. H., Sher, H., Ali, H., Ullah, Z., Ali, A., & Mahmood, T. (2022). Drought stress enhances the efficiency of floral dip method of Agrobacterium-mediated transformation in Arabidopsis thaliana. *Brazilian Journal of Biology*, *84*, e259326. PMID: 35703626

Arora, H.. (2022). DNA Methylation Dynamics in Response to Abiotic and Pathogen Stress in Plants. *Plant Cell Reports*, *41*(10), 1931–1944. DOI: 10.1007/s00299-022-02901-x PMID: 35833989

Aslam, H. M. U., Chikh-Ali, M., Zhou, X. G., Zhang, S., Harris, S., Chanda, A. K., & Killiny, N. (2024). Epigenetic modulation of fungal pathogens: A focus on Magnaporthe oryzae. *Frontiers in Microbiology*, *15*, 1463987.

Cao, P.. (2022). Genome-Wide Identification of Long Intergenic Non-Coding RNAs for Ralstonia Solanacearum Resistance in Tomato (Solanum Lycopersicum). *Frontiers in Plant Science*, *13*(September), 1–14. PMID: 36186038

Chang, Y. N.. (2020). Epigenetic Regulation in Plant Abiotic Stress Responses. *Journal of Integrative Plant Biology*, *62*(5), 563–580. PMID: 31872527

Chen, C.. (2022). Long-Term Effect of Epigenetic Modification in Plant – Microbe Interactions : Modification of DNA Methylation Induced by Plant Growth- Promoting Bacteria Mediates Promotion Process. *Microbiome*, •••, 1–19. DOI: 10.1186/s40168-022-01236-9 PMID: 35209943

Fan, T. (2022)... *Histone Modification and Chromatin Remodeling in Plant Response to Pathogens.*, (October), 1–14.

Haider, S., Iqbal, J., Naseer, S., Yaseen, T., Shaukat, M., Bibi, H., & Mahmood, T. (2021a). Molecular mechanisms of plant tolerance to heat stress: Current landscape and future perspectives. *Plant Cell Reports*, •••, 1–25. PMID: 33890138

Haider, S., Iqbal, J., Shaukat, M., Naseer, S., & Mahmood, T. (2021b). The epigenetic chromatin-based regulation of somatic heat stress memory in plants. *Plant Gene*, *27*, 100318.

Hannan Parker, A., Wilkinson, S. W., & Ton, J. (2022). Epigenetics: A catalyst of plant immunity against pathogens. *The New Phytologist*, *233*(1), 66–83.

Hewezi, T., Lane, T., Piya, S., Rambani, A., Rice, J. H., & Staton, M. (2017). Cyst nematode parasitism induces dynamic changes in the root epigenome. *Plant Physiology*, *174*(1), 405–420.

Huang, Juan, Wenling Zhou, Xiaoming Zhang Id, and Yi Li Id. 2023. "Roles of Long Non-Coding RNAs in Plant Immunity.": 1–30. http://dx.doi.org/.DOI: 10.1371/journal.ppat.1011340

Ijaz, S., Iqbal, J., Abbasi, B. A., Tufail, A., Ullah, Z., Sharifi-Rad, J., & Uddin, S. (2023). Biofortification: Lessons from the Past and Strategies for Future Food Security. In *Legumes Biofortification* (pp. 521–545). Springer International Publishing.

Ijaz, S., Iqbal, J., Abbasi, B. A., Yaseen, T., Rehman, S., Kazi, M., & Mahmood, T. (2024). Role of OMICS-Based Technologies in Plant Sciences. *OMICs-based Techniques for Global Food Sec*urity, 45-66.

Jian, Y.. (2024). How Plants Manage Pathogen Infection. *EMBO Reports*, *25*(1), 31–44. PMID: 38177909

Jian, Y., Shim, W. B., & Ma, Z. (2021). Multiple functions of SWI/SNF chromatin remodeling complex in plant-pathogen interactions. *Stress Biology*, *1*(1), 18.

Jin, Y., Zhao, J., & Guo, H. (2020). ScienceDirect Recent Advances in Understanding Plant Antiviral RNAi and Viral Suppressors of RNAi. *Current Opinion in Virology*, *46*, 65–72. DOI: 10.1016/j.coviro.2020.12.001 PMID: 33360834

Krajewski, W. A. (2022). Histone Modi Fi Cations, Internucleosome Dynamics, and DNA Stresses : How They Cooperate to ' Functionalize '. *Nucleosomes.*, *13*(April), 1–11.

Kumar, S., Gonzalez, E. A., Rameshwar, P., & Etchegaray, J. P. (2020). Non-Coding RNAs as Mediators of Epigenetic Changes in Malignancies. *Cancers (Basel)*, *12*(12), 1–32. PMID: 33291485

Kumar, V., Thakur, J. K., & Prasad, M. (2021). Histone Acetylation Dynamics Regulating Plant Development and Stress Responses. *Cellular and Molecular Life Sciences : CMLS*, *78*(10), 4467–4486. DOI: 10.1007/s00018-021-03794-x PMID: 33638653

Li, N. et al. "Signaling Crosstalk between Salicylic Acid and Ethylene / Jasmonate in Plant Defense : Do We Understand What They Are Whispering ?"

Mauch-Mani, B., Baccelli, I., Luna, E., & Flors, V. (2017). Defense Priming: An Adaptive Part of Induced Resistance. *Annual Review of Plant Biology*, *68*(April), 485–512. PMID: 28226238

Mierziak, J., & Wojtasik, W. (2024). Epigenetic weapons of plants against fungal pathogens. *BMC Plant Biology*, *24*(1), 175.

Mulaudzi, P. E., Koorsen, G., & Mwaba, I. (2023).. . *The Identi Fi Cation of the Methylation Patterns of Tomato Curly Stunt Virus in Resistant and Susceptible Tomato Lines.*, (June), 1–11.

Natarajan, B., Kalsi, H. S., Godbole, P., Malankar, N., Thiagarayaselvam, A., Siddappa, S., & Banerjee, A. K. (2018). MiRNA160 is associated with local defense and systemic acquired resistance against Phytophthora infestans infection in potato. *Journal of Experimental Botany*, *69*(8), 2023–2036.

Prasad, A.. (2019). Critical Reviews in Biotechnology Recent Advances in Small RNA Mediated Plant- Virus Interactions. *Critical Reviews in Biotechnology*, *39*(4), 587–601. DOI: 10.1080/07388551.2019.1597830 PMID: 30947560

Qin, X. et al. (2021). "Article The Chromatin-Remodeling Protein BAF60 / SWP73A Regulates the Plant Immune Receptor NLRs Ll Article The Chromatin-Remodeling Protein BAF60 / SWP73A Regulates the Plant Immune Receptor NLRs.": 425–34.

Ramos-Cruz, D., Troyee, A. N., & Becker, C. (2021). Epigenetics in plant organismic interactions. *Current Opinion in Plant Biology*, *61*, 102060.

Sabrina N., & Abdul, N. (2022). "The Role of Non-Coding RNA in Rice Immunity."

Song, L.. (2021). Role of Non-Coding RNAs in Plant Immunity. *Plant Communications, 2*(3), 100180. DOI: 10.1016/j.xplc.2021.100180 PMID: 34027394

Song, Z. T., Liu, J. X., & Han, J. J. (2021). Chromatin remodeling factors regulate environmental stress responses in plants. *Journal of Integrative Plant Biology, 63*(3), 438–450.

Tao, Z., Yan, F., Hahn, M., & Ma, Z. (2023). Regulatory Roles of Epigenetic Modifications in Plant - Phytopathogen Interactions. *Crop Health*, •••, 1–20. DOI: 10.1007/s44297-023-00003-y

Tirnaz, S., & Batley, J. (2019). DNA methylation: Toward crop disease resistance improvement. *Trends in Plant Science, 24*(12), 1137–1150.

Val-Torregrosa, B., Bundó, M., & San Segundo, B. (2021). Crosstalk between nutrient signalling pathways and immune responses in rice. *Agriculture, 11*(8), 747.

Wang, B., Yang, X., Wang, Y., & Xie, Y. (2018).. . *Crossm., 92*(18), 1–16.

Xiao, D.. (2021).. . *Crosstalk of DNA Methylation Triggered by Pathogen in Poplars With Different Resistances., 12*(December), 1–15.

Xie, S. S., & Duan, C. G. (2023). Epigenetic regulation of plant immunity: From chromatin codes to plant disease resistance. *aBIOTECH, 4*(2), 124–139.

Yang, X.. (2021). Mirna Mediated Regulation and Interaction between Plants and Pathogens. *International Journal of Molecular Sciences, 22*(6), 1–13. PMID: 33805611

Yuan, M.. (2021). ScienceDirect PTI-ETI Crosstalk : An Integrative View of Plant Immunity. *Current Opinion in Plant Biology, 62*, 102030. DOI: 10.1016/j.pbi.2021.102030 PMID: 33684883

Zhang, N., Yang, Z., Zhang, Z., & Liang, W. (2020). BcRPD3-mediated histone deacetylation is involved in growth and pathogenicity of Botrytis cinerea. *Frontiers in Microbiology, 11*, 1832.

Zhi, P., & Chang, C. (2021).. . *Exploiting Epigenetic Variations for Crop Disease Resistance Improvement., 12*(June), 1–13.

Chapter 9
Epigenetic Response Underlying Drought and Salt Stress Tolerance in Plants

Muhammad Talha Aslam
https://orcid.org/0009-0000-0630-1961
Jiangxi Agricultural University, Nanchang, China

Saira Ghafoor
Jiangxi Agricultural University, Nanchang, China

Ali Raza
Northwest A&F University, China

Imran Khan
University of Agriculture, Faisalabad, Pakistan

Muhammad Umer Chattha
University of Agriculture, Faisalabad, Pakistan

Muhammad Zia-ul Haq
https://orcid.org/0000-0001-9914-3367
University of Agriculture, Faisalabad, Pakistan

Ziming Wu
Jiangxi Agricultural University, Nanchang, China

ABSTRACT

Global food security is seriously threatened by the harsh effects of salinity and drought on crop production. Plants detect environmental stimuli and modify defensive mechanisms through complex regulatory networks to cope with various abiotic stress challenges. In plants, genetic and epigenetic mechanisms are helpful to explain drought and salinity stress tolerances. Plant-epigenetic-response involves the stable modification in gene expression and "marking" of DNA/proteins. Moreover, many epigenetic pathways allow plants to fine-tune gene expression, critical for phenotypic

DOI: 10.4018/979-8-3693-8648-4.ch009

plasticity and adaptation. These include chromatin remodeling, epi-transcriptomics, DNA methylation, histone changes, and small RNA-mediated gene suppression. We summarized the latest research findings on epigenetic responses to drought and salt stress signals in different crops in this chapter. Furthermore, it is helpful in improving the understanding of epigenetics, helping breeders better incorporate genetics into crop development initiatives that target adaptability to abiotic stress.

INTRODUCTION

Drought and salinity are the key environmental factors affecting plant growth and normal functioning. Similar to the negative consequences of salinity, drought stress Drought and salinity are the key environmental factors affecting plant growth and normal functioning. Osmotic stress from drought and salinity dehydrates plant cells (Aslam et al., 2023). Both ABA-independent and -dependent pathways can cause changes to constitutively produced TFs in plants under osmotic stress (Zhu, 2002; Ji et al., 2023). Osmotic stress and specific ion toxicity are the two main impairments that plants undergo with increasing salt concentrations. These impairments caused nutrient deficiency and ionic imbalance. These consequences lead to a decrease in plant development by interfering with physiological and metabolic processes, such as decreased absorption of water and nutrients, membrane malfunction, and impairments of vital processes like respiration, protein synthesis, and photosynthesis (Arzani et al., 2023). Numerous studies show that plants exhibit a type of "stress memory" and are unable to disregard the onset of difficult environmental conditions (Gallusci et al., 2023; Hoseini and Arzani, 2023). By rearranging their genetic and metabolic pathways, they direct or oversee their adaptation to recurrent, persistent, and coupled abiotic stressors (Walter et al., 2013). Numerous complex regulatory mechanisms, both pre- and post-transcriptional, are essential for cellular adaptability and response to environmental stress. High-affinity potassium transporters (HKTs), pyrroline-5-carboxylate synthetase (P5CS), inward rectifying K+ channel (AKT1), and ABA biosynthesis genes (i.e., NCED1, NCED3, ABA1, ABA2, ABA3, ABF2, ABF4, ABI3, ABI4, and ABI5) can all be induced by various epigenetic mechanisms in response to drought and salinity stimuli (Arzani and Ashraf, 2016).

The term 'epigenetics' has multiple definitions in literature. However, a commonly used definition relevant to the role of epigenetics in plant defense responses posits that epigenetics is the mechanism that facilitates enduring changes in gene expression. Two basic epigenetic changes exist: the covalent alteration of DNA via methylation and post-translational modifications of histones that modulate chromatin structure. These flexible pathways regulate the ability of DNA and chromatin to facilitate gene transcription. They involve the physical marking of DNA or associated histones, enabling genotypically identical cells to exhibit different phenotypes (Felsenfeld, 2014). Moreover, various noncoding RNAs, including micro RNAs and long non-coding RNAs, are modified, affecting numerous biochemical reactions. Recent advancements have broadened our viewpoint about plant epigenetic mechanisms in environmental stresses (Mladenov et al., 2021). Therefore, heritable chemical changes in covalent DNA and histones govern gene function without changing the nucleotide sequence, establishing a set of regulators that affect plant responses and stress resilience (Li et al., 2021b; Singroha et al., 2022).

DNA Methylation

A key mechanism in controlling gene expression and a plant's ability to respond to abiotic stress is DNA methylation (Sun et al., 2022). The main mechanism of DNA methylation in plants is the addition of a methyl group at either the N6 position of adenine, which results in N6-methyladenine (6mA), or the 5′ position of cytosine, which results in 5-methylcytosine (5mC) (Li et al., 2020). DNA methylation predominantly occurs on cytosines that are adjacent to guanine nucleobases (CpG) and is widely recognized as an indicator of transcriptional repression in genes. It disrupts the binding of transcriptional machinery on regulatory sites and recruits transcriptional repressors to promote the formation of a heterochromatin state. This repression is catalyzed by DNA methyltransferases (DMTs) that is an enzyme that transfer the methyl groups to the 5-position of cytosine bases, a process critical for initiating heterochromatin formation (Quadrana and Colot, 2016; Dong et al., 2022).

The genomic loci of DNA methylation, including gene bodies, regulatory elements, transcriptional start sites, transposable elements, and repetitive sequences, have different functional implications in addition to the sequence and amount of methylation. Gene body methylation (GBM) in plants refers to the presence of CG methylation within the coding regions of genes, unlike the lack of methylated cytosine at the transcriptional start and termination sites (Bewick and Schmitz, 2017). Numerous activities, such as transcription elongation, alternative splicing, and differential promoter usage, may be impacted by DNA methylation within gene bodies (Jones, 2012). Genes associated with GBM, in contrast to other epigenetic modifications that result in transcriptional repression, are fundamentally expressed and are associated with constitutive genes that demonstrate moderate to high expression levels (Han et al., 2022; Hoseini and Arzani, 2023).

Thus, alterations in methylation patterns might cause notable plastic responses to abiotic stress in addition to promoting phenotypic diversity (Quadrana and Colot, 2016; Noshay and Springer, 2021). DNA methylation is essential to plant stress responses, as DNA methylation enzyme disruptions affect plant life (Kumar et al., 2024). Thus, DNA methylation provides a wide range of distinct of gene expressions depending on the sequence and chromosomal position (Niederhuth and Schmitz, 2017). Unlike highly variable gene expression, genic region DNA methylation appears to be less impacted by environmental changes. Different tissues, species, and environmental factors affects the differential gene expressions. Pigeonpea (*Cajanus cajan* L. Huth) genes with high gene body methylation (GBM) had the lowest expression, whereas, those with intermediate GBM had the highest (Junaid et al., 2022). GBM loss lowered gene expression in Eutrema salsugineum (Muyle and Gaut, 2019), Arabidopsis thaliana (Horvath et al., 2019), and Populus euphratica (Su et al., 2018). However, drought stress-induced gene body DNA methylation inversely related to gene expression in maize (Rehman et al., 2021).

Researchers noticed that GBM plays a major part in controlling alternative splicing (AS) that increased the transcriptome and proteome complexity (Hu et al., 2014). A single gene produces multiple mRNA isoforms, which enhances phenotypic diversity (Reddy et al., 2013). Genes that encode AS factors are generally down-regulated during transcription but frequently experience alternative splicing, a process affected by alterations in DNA methylation. RNA- and BS-seq analysis of isogenic lines of rice seedlings (identical tissue) showed that AS alterations in plants cause cytosine methylation reduction in gene body areas (Hu et al., 2014). In maize, DNA methylation within the gene body shows a positive relationship with exon splicing events during drought stress (Wang et al., 2021).

In DNA, adenine nucleotides are also methylated by adding a CH3 group at the N1 or N6 position, though at much lower rates than cytosine, add complexity through the methylation of the purine ring's first position (N1) and sixth position (C6) forms N1-methyladenine and N6-methyladenine (6-mA) (Sedgwick et al., 2007). 6-mA affects transcriptional inhibition and activation, transgenerational chromatin control, and stress intensity depending on genomic position. Liang et al. (2018) found 6mA in the gene body of Arabidopsis thaliana and rice genes as active. DNA N6-methyladenine, unlike 5-methylcytosine in the gene promoter, does not always repress but modulates gene expression differently. A 6 mA density increase in rice increases salt, heat, and cold resistance (Hoseini and Arzani, 2023).

DNA methylation has a major impact on how plants regulate gene expression during stress (Akhter et al., 2021). Rice cultivars, and specifically the drought-sensitive IR20, have shown evidence of a link between DNA methylation and tolerance to drought stress. According to Gayacharan and Joel (2013), IR20 shows hypomethylation when drought stress is applied, whereas tolerant cultivars show hypermethylation. DNA methylation in response to drought stress is anticipated to vary based on genotype and growth stage, as well as being dependent on specific organs or tissues. Drought causes a 12.1% alteration in methylation in *O. sativa* L., affecting multiple genotypes, tissues, and growth stages. Leaf DNA methylation was higher than root DNA methylation throughout identical developmental stages, suggesting roots are crucial to water deprivation response (Suji and Joel, 2010). Salinity stress affects 5mC methylation or demethylation in transcription regulation in diverse plant species, altering downstream salinity-related gene expression. In soybean and rice, 5mC demethylation at the promoter of specific transporters increases salt tolerance by increasing transporter gene expression (Zhang et al., 2020). In contrast, increased 5mC turnover rates caused by salinity stress may impact ion transporter gene expressions (HKT2;1 and HKT2;3) or microRNA, leading to improved salinity tolerance (Kord et al., 2019).

This suggests the possibility of stress memory, since the changes in methylation/demethylation are quite persistent even after the plant recovers. Numerous research have examined the biochemical, physiological, and epigenome mechanisms of stress memory in crops to improve stress tolerance. According to Feng et al. (2016), the gene implicated in proline synthetase, light-induced P5CS1, may be modulated at the transcriptional level to generate stress memory in Arabidopsis following salt stress treatment.

HISTONE POST-TRANSCRIPTIONAL MODIFICATIONS

Chromosome structure and chromatin alteration regulate gene accessibility to transcription factors (TFs), which strengthen the plant's ability to respond to drought and salt stress (Nguyen et al., 2019; Pierats-Llobet et al., 2016). Epigenetic inheritance pathways allow for the transmission of stress-related memories from one generation to another (Sun et al., 2021). An octamer of histones, consisting of two copies of each of histones H2A, H2B, H3, and H4, makes up the nucleosome, the basic functional unit of chromatin. The DNA molecule that contains this octamer is surrounded by 147 base pairs. Abiotic stresses, such drought and salt stress, cause plants to show transcriptional activation in certain genes. The quick and dynamic elimination of the H2A.Z variation from stress-related genes is responsible for this reaction. Nonetheless, H2A.Z nucleosomes within gene bodies protect genes that are sensitive to drought and salt stress under normal conditions (Nguyen and Cheong, 2018).

Zhang et al. (2020) noted that histones can undergo post-translational covalent modifications that change their structure. This changes the accessibility of cis-regulatory regions to TFs. Among the various histone post-translational modifications (HPTMs) identified (Zhu et al., 2015), methylation, acetylation, phosphorylation, and ubiquitination have received increased focus regarding plant stress responses. The modifications on histones result from the activity of histone "writer" complexes, which include histone methyltransferases (HMT), acetyltransferases (HAT), ubiquitinases, and kinases. Conversely, these modifications are removed by "eraser" enzymes such as demethylases (HDM), deacetylases (HDA), de-ubiquitinases, and phosphatases (Maeji and Nishimura, 2018).

Histone Acetylation

In most eukaryotes, histone acetylation and deacetylation of lysine regulate chromatin shape and function, affecting many cellular activities (Ma et al., 2013). Acetylation of lysine residues (K) on histones H3 and H4, particularly K9, K14, K18, K23, K27, and K36 for H3 and K5, K8, K12, K16, and K20 for H4, reduces histone-DNA binding strength. This loosens chromatin by neutralizing histones' positive charge. Deacetylation compacts chromatin and blocks TF access to DNA (Berger, 2007). The antagonism between HATs and HDACs, which write and remove the histone acetylation mark, dynamically regulates chromatin structure (Boycheva et al., 2014). Histone acetylation changes the surface of nucleosomes, affecting their ability to bind gene transcription proteins (Berger, 2007). Histone acetylation regulates chromatin folding, which affects the homeostatic gene program and gene expression in stressed plants. This regulation depends on the level and/or distribution of acetylated histones (Hu et al., 2019). Histone and nucleosome density changes in promoters drive gene activation during drought and salinity, but ORF changes are rare. Drought stress altered histones such H3K9ac and H3K4me3 in Rd20 and Rd29A compared to growth conditions (Hoseini and Arzani, 2023). In rice, Chinese cabbage, and Brachypodium distachyon, moisture stress increases HAT gene expression (Hou et al., 2021; Hoseini and Arzani, 2023). A wheat cultivar with increased drought tolerance upregulated HAT genes (TaHAC2, TaHAG2, and TaHAG3) under drought stress, but not other drought-sensitive cultivars (Li et al., 2021a). H3K9/14 acetylation regulates MYB54, PGX3, and CTL1 expression in Arabidopsis, promoting salt tolerance (Wang et al., 2019). H3K9 hyperacetylation in the promoter region promotes cell wall-associated gene expression in maize during salt stress (Mao et al., 2015). Rice with salt stress upregulates ABA-responsive genes like OsLEA3, OsbZIP72, OsABI5, and OsNHX1. Salt stress and ABA therapy increase histone H4 acetylation in promoter areas, which upregulate (Ullah et al., 2021). HMT significantly upregulated gene 9-cis-epoxy carotenoid dioxygenase, which synthesizes ABA (Ding et al., 2011).

Histone Methylation

Histone modification is a dynamic process in which different writer enzymes, eraser enzymes, and reader proteins, respectively, add, remove, recognize, and interpret histones. The hydrophobicity of histone side chains is altered by methylation at single, double, or triple sites on histone tails, particularly at lysine (K) or arginine (R). This alteration affects how well the histones bind to reader proteins and transcriptional machinery. The precise amino acid location of the alteration determines how K and R methylation alters gene expression and chromatin architecture (Lämke and Bäurle, 2017). Studies show that DNA methylation—hyper or hypo—affects gene functioning in challenging conditions. Additionally, alterations in DNA methylation (hyper or hypo) affect gene functioning. Methylation of gene bodies and promoters modulates organ-specific gene expression during salt stress. Numerous studies on histone methylation dynamics and the response to effective HMT/HDMT mutations underline its importance in growth and prptection against environmental stresses (Miryeganeh, 2021).

Sun et al. (2019) reported the stress-related gene expression requires the removal of restrictive methylation marks and active mark addition. Furthermore, salinity stress reduces repressive histone marks like H3K27me and 3H3K9me2 and increases functional methylation marks like H3K4me3 on salt-tolerance genes. Sun et al. (2019) found that in soybean, when plants are exposed to salt, their genes become inactive. This inactivation is accompanied with the development of H3K27me3 in coding regions or promoters absent in control plants. In addition, H3K27me3 deficiency in drought-responsive TF loci is linked to improved stress tolerance in Arabidopsis (Ramirez-Prado et al., 2019). H3K4me3 alteration of 4837 rice genes under moisture stress affected the activation of a limited number of drought-responsive genes (Zong et al., 2013). Reduced H4R3sme2 (histone4 arginine3 symmetric dimethylation) promotes the flowering gene FLC and stress-responsive genes in Arabidopsis during salt stress (Zhang et al., 2011). H4R3sme2-type histone methylation in the ANAC055 promoter region increases P5CS1 expression and proline levels in wild-type Arabidopsis under drought stress (Fu et al., 2017).

Histones Ubiquitylation

Ubiquitination is a type of post-translational protein modification that selectively degrades proteins in cells by adding ubiquitin after translation, making them more susceptible to protease destruction (March and Farrona, 2018). This process primarily regulates protein stability but can also influence subcellular localization, transcriptional output, and other functional properties of proteins (March and Farrona, 2018; Chen et al., 2020; Hoseini, & Arzani, 2023). Target proteins that are multi-ubiquitinated are often degraded by the 26S proteasome, but proteins that are mono-ubiquitinated are able to modulate their function and localization apart from proteolysis (Chen et al., 2020). Moreover, March and Farrona (2018) reported mono-ubiquitination is an important form of HPTM that happens at lysine 121 (K121) for histone H2A and lysine 143 (K143) for histone H2B. Furthermore, mono-ubiquitination of histone lysine H2B (H2Bub1) is the active sites for seed dormancy, and the control of salinity and drought stresses (Chen et al., 2020). MPK4 and MYB42 downregulated in salt stress and responsible for the reduction of H3K4me3 and H2Bub1 in hypersensitive mutants of A. thaliana (Sun et al., 2020). Ubiquitination modifications of H2A.Z have been shown to regulate stress-responsive genes in Arabidopsis during drought stress (Sura et al., 2017).

Histone Phosphorylation

Histone H3 is capable of undergoing phosphorylation at the arginine and threonine positions (Hoseini, & Arzani, 2023). The relevance of this epigenetic modification in the cell cycle has been extensively studied (Houben et al., 2007), but its behavior in plants against various environmental stressors is still unknown. In addition to facilitating a speedy reaction to stressors, histone post-translational modifications (HPTMs) like methylation serve as an environmental memory that helps plants adjust to new conditions. This memory mechanism allows previously stressed plants to cope better with the same stress during successive exposures (Lämke and Bäurle, 2017). Table 1 summarizes studies examining epigenetic changes in crops under drought stress.

Table 1. Histone, DNA and RNA base epigenetic adjustment in different crops under drought stress

Crop	Epigenetic adjustment			References
	Histone	DNA	RNA	
Rice		Promoter Hypomethylation (CLT1 and PSBP gene)		(Ding et al., 2011)
Maize		Hypermethylation		(Wang et al., 2021)
Rice		-do-		(Kou et al., 2022)
Wheat		-do-		Duan et al., 2020)
Maize	H3K4me3, H3K9ac			(Forestan et al., 2020)
tton	H2Bub, H3K4me3			(Chen et al., 2019)
Rice	H2Bub1			(Ma et al., 2013)
Tobacco	HDAC			(Ma et al., 2017)
Barley	H3K9ac, H3K14ac, H3K27ac			(Papaefthimiou et al., 2010)

continued on following page

Table 1. Continued

Crop	Epigenetic adjustment			References
	Histone	DNA	RNA	
Rice			Non-coding RNA (mir162b), Non-coding RNA (mir164)	(Tian et al., 2015), (Feng et al., 2016)
Soybean			Non-coding RNA (miR169g)	(Ni et al., 2013)

'Blank cells indicate that the specific epigenetic adjustment was not reported in the cited study, '-do-' reveals similar adjustment as given in the cell above.'

CHROMATIN REMODELING COMPLEXES (CRC)

CRC play a key role in plants to control gene functions through epigenetics under stress situations (Tonosaki et al., 2022). These complexes change chromatin shape, affecting DNA accessibility to regulatory proteins and transcriptional machinery (Pandey et al., 2016). CRC regulate the genes responsible for stress tolerance, improves the plant survival in stress (Kim, 2019). This is due to the sliding, evicting, or altering the composition of nucleosomes or by using energy from ATP hydrolysis (He et al., 2023). Histone modification enzymes limit or stimulate gene transcription by changing histone proteins posttranslationally. Highly compacted chromatin blocks polymerase access to DNA as well as TFs and other nuclear proteins. When chromatin is exposed to stress signals, it changes structurally, making DNA accessible (Bhadouriya et al., 2021).

NONCODING RNAs

Despite lacking an open reading frame for translation, noncoding RNAs (ncRNAs) play a crucial role in different cellular activities. This classification includes SiRNAs, miRNAs, lncRNAs, and circular RNAs. A large percentage of noncoding RNAs, including lncRNAs and miRNAs, regulate DNA replication, transcription, translation, and stability (Mattick et al., 2023). Noncoding RNAs, though not epigenetic elements, play a major role in epigenetic alteration. Gene regulation relies on 21–24-nucleotide single-stranded RNAs called miRNAs and siRNAs. Small RNAs (sRNA) or RNA interference (RNAi) are abundant in plant and animal genomes. Growth, cellular behavior, gene expression, genome protection, and adaptation to biotic and abiotic challenges depend on them. Small RNAs have many functions, including helping plants adapt to drought (Ferdous et al., 2015). These short non-coding RNAs regulate genomic activities such chromatin shape, transcription, and translation (Borges and Martienssen, 2015). Processing by RNA-dependent RNA polymerases (RDRs) produces siRNAs from double-stranded RNA intermediates (Song et al., 2019). Small RNAs, notably miRNA against drought, cold, salinity, and oxidative stresses (Banerjee et al., 2017), highlighting the importance of intrinsic endogenic pathways in alleviating these difficulties. Modern crop development efforts benefit from transgenic methods' resilience and productivity under environmental challenges. RNA interference helps crops overcome abiotic stressors. Younis et al. (2014) found that RACK1-modified RNAi rice was more drought-tolerant than wild varieties. Table 2 summarizes studies examining epigenetic changes in crops under salt stress.

Table 2. Histone, DNA and RNA base epigenetic adjustment in different crops under salt stress

Crop	Epigenetic adjustment			References
	Histone	DNA	RNA	
Alfalfa	H3K9Ac			(Deng et al., 2018)
Soybean	HDAC		miR482bd-5	(Cadavid et al., 2020)
Cotton			LncRNA973	(Zhang et al., 2019)
Tomato			miR156, miR398	(Çakır et al., 2021)
Sweet potato			nta-miR156a_R + 3, farmiR159_L + 2_1ss22T, mes-MIR319ep5_2ss12GC19 GA	(Yang et al., 2020)
Rice			miR171b, miR167f	(Parmar et al., 2020)
Wheat			TaemiR408	(Bai et al., 2018)
Maize		Hypomethylation		Shams et al., 2020)
Chili pepper		-do-		(Shams et al., 2020)
Sugar beet		-do-		(Skorupa et al., 2021)

continued on following page

Table 2. Continued

Crop	Epigenetic adjustment			References
	Histone	DNA	RNA	
Wheat		-do-		(Hosseinpour et al., 2022)
Wheat		Coding region hypermethylation (TaHKT2;1 & TaHKT2;3)		(Kumar et al., 2018)

'Blank cells indicate that the specific epigenetic adjustment was not reported in the cited study, '-do-' reveals similar adjustment as given in the cell above'

EPITRANSCRIPTOMICS

Post-transcriptional modification alters epigenetics besides DNA methylation and histone changes. This modifications in mRNA, tRNA, rRNA, and lncRNA improve plant regulatory systems (Rajkumar et al., 2020). However, the most common modifications are N5-methylcytidine (m5C) and N6-methyladenosine (m6A), with other mRNA abnormalities. Numerous studies show that covalent RNA changes regulate eukaryotic transcript abundance and quality. The modifications have an impact on the secondary structure of RNA molecules, their stability, translatability, degradation, and utilization (Hoseini and Arzani, 2023). Research has shown that these epitranscriptomic signals significantly influence the plant's response to salinity stress by regulating RNA fate (Hu et al., 2022; Shoaib et al., 2022). In drought stress, m6A levels within the 5'UTR region increased, promoting the translation of transcripts associated with resistance (Shoaib et al., 2022). As a result, stress leads to a reconfiguration of m6A throughout the transcriptome, causing an increased abundance of mRNAs featuring 5′ UTR m6A.

The results indicate that the distribution of m6A on mRNAs is stable across different organisms, regardless of whether they are under stress or in normal conditions. The transport of RNA information molecules over long distances, such as siRNAs, miRNAs, tRNAs, rRNAs, and mRNAs, which play systemic signaling roles in vascular tissues, aids in the adaptation of plants to abiotic stress (Thieme et al., 2015; Zhang et al., 2018).

CONCLUSION

Crop production losses from drought and salinity stressors are a major problem for staple food crops, and the future climate change scenario is predicted to make the problem worse. Epigenetic technology can improve stress tolerance in agricultural cultivars by creating new epialleles. Epigenetic changes can modify gene transcription, improving abiotic stress adaption. These alterations also operate as "plant stress memory," helping plants adapt to environmental pressures and giving future generations resilience. The complex interaction between epigenetic regulatory levels shapes stress responses, resilience, and adaptability in plants. Understanding the mechanism behind epigenetic regulation in plants during stress conditions has improved the fundamental difficulties in different breeding approaches. This entails improving technology, data workflow systems, plant species knowledge across all epigenetic levels, and epigenomic data integration with other "omics" datasets.

REFERENCES

Akhter, Z., Bi, Z., Ali, K., Sun, C., Fiaz, S., Haider, F. U., & Bai, J. (2021). In response to abiotic stress, DNA methylation confers epigenetic changes in plants. *Plants*, *10*(6), 1096. PMID: 34070712

Arzani, A., & Ashraf, M. (2016). Smart engineering of genetic resources for enhanced salinity tolerance in crop plants. *Critical Reviews in Plant Sciences*, *35*(3), 146–189.

Arzani, A., Kumar, S., & Mansour, M. M. F. (2023). Salt tolerance in plants: Molecular and functional adaptations. *Frontiers in Plant Science*, *14*, 1280788. PMID: 37731972

Aslam, M. T., Imran, K. H. A. N., Chattha, M. U., Maqbool, R., Ziaulhaq, M., Lihong, W., & Arshad, M. (2023). The critical role of nitrogen in plants facing the salinity stress: Review and future prospective. *Notulae Botanicae Horti Agrobotanici Cluj-Napoca*, *51*(3), 13347–13347.

Bai, Q., Wang, X., Chen, X., Shi, G., Liu, Z., Guo, C., & Xiao, K. (2018). Wheat miRNA TaemiR408 acts as an essential mediator in plant tolerance to Pi deprivation and salt stress via modulating stress-associated physiological processes. *Frontiers in Plant Science*, *9*, 499. PMID: 29720988

Banerjee, S., Sirohi, A., Ansari, A. A., & Gill, S. S. (2017). Role of small RNAs in abiotic stress responses in plants. *Plant Gene*, *11*, 180–189.

Berger, S. L. (2007). The complex language of chromatin regulation during transcription. *Nature*, *447*(7143), 407–412. PMID: 17522673

Bewick, A. J., & Schmitz, R. J. (2017). Gene body DNA methylation in plants. *Frontiers in Plant Science*, *36*, 103–110. PMID: 28258985

Bhadouriya, S. L., Mehrotra, S., Basantani, M. K., Loake, G. J., & Mehrotra, R. (2021). Role of chromatin architecture in plant stress responses: An update. *Frontiers in Plant Science*, *11*, 603380. PMID: 33510748

Borges, F., & Martienssen, R. A. (2015). The expanding world of small RNAs in plants. *Nature Reviews. Molecular Cell Biology*, *16*(12), 727–741. PMID: 26530390

Boycheva, I., Vassileva, V., & Iantcheva, A. (2014). Histone acetyltransferases in plant development and plasticity. *Current Genomics*, *15*(1), 28. PMID: 24653661

Cadavid, I. C., da Fonseca, G. C., & Margis, R. (2020). HDAC inhibitor affects soybean miRNA482bd expression under salt and osmotic stress. *Journal of Plant Physiology*, *253*, 153261. PMID: 32947244

Çakır, Ö., Arıkan, B., Karpuz, B., & Turgut-Kara, N. (2021). Expression analysis of miRNAs and their targets related to salt stress in *Solanum lycopersicum* H-2274. *Biotechnology, Biotechnological Equipment*, *35*(1), 275–282.

Chen, H., Feng, H., Zhang, X., Zhang, C., Wang, T., & Dong, J. (2019). An *Arabidopsis* E3 ligase HUB 2 increases histone H2B monoubiquitination and enhances drought tolerance in transgenic cotton. *Plant Biotechnology Journal*, *17*(3), 556–568. PMID: 30117653

Chen, K., Tang, W. S., Zhou, Y. B., Xu, Z. S., Chen, J., Ma, Y. Z., & Li, H. Y. (2020). Overexpression of GmUBC9 gene enhances plant drought resistance and affects flowering time via histone H2B monoubiquitination. *Frontiers in Plant Science*, *11*, 555794. PMID: 33013972

Deng, F., Zhang, X., Wang, W., Yuan, R., & Shen, F. (2018). Identification of *Gossypium hirsutum* long non-coding RNAs (lncRNAs) under salt stress. *BMC Plant Biology*, *18*, 1–14. PMID: 29370759

Ding, Y., Avramova, Z., & Fromm, M. (2011). The Arabidopsis trithorax-like factor ATX1 functions in dehydration stress responses via ABA-dependent and ABA-independent pathways. *The Plant Journal*, *66*(5), 735–744. PMID: 21309869

Dong, T., Song, S., Wang, Y., Yang, R., Chen, P., Su, J., & Duan, H. (2022). Effects of 5-azaC on Iridoid Glycoside Accumulation and DNA Methylation in *Rehmannia glutinosa. Frontiers in Plant Science*, *13*, 913717. PMID: 35812974

Duan, H., Li, J., Zhu, Y., Jia, W., Wang, H., Jiang, L., & Zhou, Y. (2020). Responsive changes of DNA methylation in wheat (*Triticum aestivum*) under water deficit. *Scientific Reports*, *10*(1), 7938. PMID: 32404881

Felsenfeld, G. (2014). A brief history of epigenetics. *Cold Spring Harbor Perspectives in Biology*, *6*(1), a018200. PMID: 24384572

Feng, X. J., Li, J. R., Qi, S. L., Lin, Q. F., Jin, J. B., & Hua, X. J. (2016). Light affects salt stress-induced transcriptional memory of P5CS1 in Arabidopsis. *Proceedings of the National Academy of Sciences of the United States of America*, *113*(51), E8335–E8343. PMID: 27930298

Ferdous, J., Hussain, S. S., & Shi, B. J. (2015). Role of microRNAs in plant drought tolerance. *Plant Biotechnology Journal*, *13*(3), 293–305. PMID: 25583362

Forestan, C., Farinati, S., Zambelli, F., Pavesi, G., Rossi, V., & Varotto, S. (2020). Epigenetic signatures of stress adaptation and flowering regulation in response to extended drought and recovery in *Zea mays. Plant, Cell & Environment*, *43*(1), 55–75. PMID: 31677283

Fu, R., Zhang, M., Zhao, Y., He, X., Ding, C., Wang, S., Feng, Y., Song, X., Li, P., & Wang, B. (2017). Identification of salt tolerance-related microRNAs and their targets in maize (*Zea mays L.*) using high-throughput sequencing and degradome analysis. *Frontiers in Plant Science*, *8*, 864. PMID: 28603532

Gallusci, P., Agius, D. R., Moschou, P. N., Dobránszki, J., Kaiserli, E., & Martinelli, F. (2023). Deep inside the epigenetic memories of stressed plants. *Trends in Plant Science*, *28*(2), 142–153. PMID: 36404175

Gayacharan, & Joel, A. J. (2013). Epigenetic responses to drought stress in rice (*Oryza sativa L.*). *Physiology and Molecular Biology of Plants, 19*, 379-387.

Han, B., Wu, D., Zhang, Y., Li, D. Z., Xu, W., & Liu, A. (2022). Epigenetic regulation of seed-specific gene expression by DNA methylation valleys in castor bean. *BMC Biology*, *20*(1), 57. PMID: 35227267

He, M. Y., Ren, T. X., Jin, Z. D., Deng, L., Liu, H. J., Cheng, Y. Y., & Chang, H. (2023). Precise analysis of potassium isotopic composition in plant materials by multi-collector inductively coupled plasma mass spectrometry. *Spectrochimica Acta. Part B, Atomic Spectroscopy*, *209*, 106781.

Horvath, R., Laenen, B., Takuno, S., & Slotte, T. (2019). Single-cell expression noise and gene-body methylation in *Arabidopsis thaliana. Heredity*, *123*(2), 81–91. PMID: 30651589

Hoseini, M., & Arzani, A. (2023). Epigenetic adaptation to drought and salinity in crop plants. *Journal of Plant Molecular Breeding*, *11*(2), 1–16.

Hosseinpour, A., Ilhan, E., Özkan, G., Öztürk, H. İ., Haliloglu, K., & Cinisli, K. T. (2022). Plant growth-promoting bacteria (PGPBs) and copper (II) oxide (CuO) nanoparticle ameliorates DNA damage and DNA methylation in wheat (*Triticum aestivum* L.) exposed to NaCl stress. *Journal of Plant Biochemistry and Biotechnology*, *31*(4), 751–764.

Hou, J., Ren, R., Xiao, H., Chen, Z., Yu, J., Zhang, H., Shi, Q., Hou, H., He, S., & Li, L. (2021). Characteristic and evolution of HAT and HDAC genes in Gramineae genomes and their expression analysis under diverse stress in *Oryza sativa. Planta*, *253*, 1–22. PMID: 33606144

Houben, A., Demidov, D., Caperta, A. D., Karimi, R., Agueci, F., & Vlasenko, L. (2007). Phosphorylation of histone H3 in plants—A dynamic affair. *Biochimica et Biophysica Acta*, *1769*(5-6), 308–315. PMID: 17320987

Hu, J., Cai, J., Xu, T., & Kang, H. (2022). Epitranscriptomic mRNA modifications governing plant stress responses: Underlying mechanism and potential application. *Plant Biotechnology Journal*, *20*(12), 2245–2257. PMID: 36002976

Hu, L., Li, N., Xu, C., Zhong, S., Lin, X., Yang, J., Zhou, T., Yuliang, A., Wu, Y., & Chen, Y.-R. (2014). Mutation of a major CG methylase in rice causes genome-wide hypomethylation, dysregulated genome expression, and seedling lethality. *Proceedings of the National Academy of Sciences of the United States of America*, *111*(29), 10642–10647. PMID: 25002488

Hu, Y., Lu, Y., Zhao, Y., & Zhou, D.-X. (2019). Histone acetylation dynamics integrates metabolic activity to regulate plant response to stress. *Frontiers in Plant Science*, *10*, 1236. PMID: 31636650

Ji, W., Yu, H., Shangguan, Y., Cao, J., Chen, X., Zhao, L., & Xu, Z. (2023). Transcriptome profiling of Gossypium Anomalum Seedlings reveals key regulators and metabolic pathways in response to Drought stress. *Plants*, *12*(2), 312. PMID: 36679025

Jones, P. A. (2012). Functions of DNA methylation: Islands, start sites, gene bodies and beyond. *Nature Reviews. Genetics*, *13*(7), 484–492. PMID: 22641018

Junaid, A., Singh, N. K., & Gaikwad, K. (2022). Evolutionary fates of gene-body methylation and its divergent association with gene expression in pigeonpea. *The Plant Genome*, *15*(3), e20207. PMID: 35790083

Kim, J. H. (2019). Chromatin remodeling and epigenetic regulation in plant DNA damage repair. *International Journal of Molecular Sciences*, *20*(17), 4093. PMID: 31443358

Kord, H., Fakheri, B., Ghabooli, M., Solouki, M., Emamjomeh, A., Khatabi, B., & Ghaffari, M. R. (2019). Salinity-associated microRNAs and their potential roles in mediating salt tolerance in rice colonized by the endophytic root fungus *Piriformospora indica*. *Functional & Integrative Genomics*, *19*, 659–672. PMID: 30903405

Kou, S., Gu, Q., Duan, L., Liu, G., Yuan, P., Li, H., Wu, Z., Liu, W., Huang, P., & Liu, L. (2022). Genome-wide bisulphite sequencing uncovered the contribution of DNA methylation to rice short-term drought memory formation. *Journal of Plant Growth Regulation*, *41*(7), 2903–2917.

Kumar, S., Chinnusamy, V., & Mohapatra, T. (2018). Epigenetics of modified DNA bases: 5-methylcytosine and beyond. *Frontiers in Genetics*, *9*, 640. PMID: 30619465

Kumar, S., Seem, K., Kumar, S., Singh, A., Krishnan, S. G., & Mohapatra, T. (2024). DNA methylome analysis provides insights into gene regulatory mechanisms for better performance of rice under fluctuating environmental conditions: Epigenomics of adaptive plasticity. *Planta*, *259*(1), 4. PMID: 37993704

Lämke, J., & Bäurle, I. (2017). Epigenetic and chromatin-based mechanisms in environmental stress adaptation and stress memory in plants. *Genome Biology*, *18*(1), 1–11. PMID: 28655328

Li, H., Liu, H., Pei, X., Chen, H., Li, X., Wang, J., & Wang, C. (2021a). Comparative genome-wide analysis and expression profiling of histone acetyltransferases and histone deacetylases involved in the response to drought in wheat. *Journal of Plant Growth Regulation*, •••, 1–14.

Li, J., Jia, W., Wang, H., Zhu, Y., Duan, Z., Jiang, L., Zhou, Y., & Duan, H. (2020). Morpho-physiological adaptation and DNA methylation of wheat seedlings under osmotic stress. *Crop & Pasture Science*, *71*(4), 349–355.

Li, S., He, X., Gao, Y., Zhou, C., Chiang, V. L., & Li, W. (2021b). Histone acetylation changes in plant response to drought stress. *Genes*, *12*(9), 1409. PMID: 34573391

Liang, Z., Shen, L., Cui, X., Bao, S., Geng, Y., Yu, G., Liang, F., Xie, S., Lu, T., & Gu, X. (2018). DNA N6-adenine methylation in *Arabidopsis thaliana. Developmental Cell, 45*(3), 406-416. e403.

Ma, X., Lv, S., Zhang, C., & Yang, C. (2013). Histone deacetylases and their functions in plants. *Plant Cell Reports*, *32*, 465–478. PMID: 23408190

Ma, X., Zhang, B., Liu, C., Tong, B., Guan, T., & Xia, D. (2017). Expression of a *Populus* histone deacetylase gene 84KHDA903 in tobacco enhances drought tolerance. *Plant Science*, *265*, 1–11. PMID: 29223330

Maeji, H., & Nishimura, T. (2018). Epigenetic mechanisms in plants. [). Academic Press.]. *Advances in Botanical Research*, *88*, 21–47.

Mao, H., Wang, H., Liu, S., Li, Z., Yang, X., Yan, J., & Qin, F. (2015). A transposable element in a NAC gene is associated with drought tolerance in maize seedlings. *Nature Communications*, *6*(1), 8326. PMID: 26387805

March, E., & Farrona, S. (2018). Plant deubiquitinases and their role in the control of gene expression through modification of histones. *Frontiers in Plant Science*, *8*, 2274. PMID: 29387079

Mattick, J. S., Amaral, P. P., Carninci, P., Carpenter, S., Chang, H. Y., Chen, L. L., & Wu, M. (2023). Long non-coding RNAs: Definitions, functions, challenges and recommendations. *Nature Reviews. Molecular Cell Biology*, *24*(6), 430–447. PMID: 36596869

Miryeganeh, M. (2021). Plants' epigenetic mechanisms and abiotic stress. *Genes*, *12*(8), 1106. PMID: 34440280

Mladenov, V., Fotopoulos, V., Kaiserli, E., Karalija, E., Maury, S., Baranek, M., Segal, N. A., Testillano, P. S., Vassileva, V., & Pinto, G. (2021). Deciphering the epigenetic alphabet involved in transgenerational stress memory in crops. *International Journal of Molecular Sciences*, *22*(13), 7118. PMID: 34281171

Muyle, A., & Gaut, B. S. (2019). Loss of gene body methylation in *Eutrema salsugineum* is associated with reduced gene expression. *Molecular Biology and Evolution*, *36*(1), 155–158. PMID: 30398664

Nguyen, N. H., & Cheong, J.-J. (2018). H2A.Z-containing nucleosomes are evicted to activate AtMYB44 transcription in response to salt stress. *Biochemical and Biophysical Research Communications*, *499*(4), 1039–1043. PMID: 29649476

Nguyen, N. H., Jung, C., & Cheong, J.-J. (2019). Chromatin remodeling for the transcription of type 2C protein phosphatase genes in response to salt stress. *Plant Physiology and Biochemistry*, *141*, 325–331. PMID: 31207493

Ni, Z., Hu, Z., Jiang, Q., & Zhang, H. (2013). GmNFYA3, a target gene of miR169, is a positive regulator of plant tolerance to drought stress. *Plant Molecular Biology*, *82*, 113–129. PMID: 23483290

Niederhuth, C. E., & Schmitz, R. J. (2017). Putting DNA methylation in context: From genomes to gene expression in plants. *Biochimica et Biophysica Acta*, *1860*(1), 149–156. PMID: 27590871

Noshay, J. M., & Springer, N. M. (2021). Stories that can't be told by SNPs; DNA methylation variation in plant populations. *Current Opinion in Plant Biology*, *61*, 101989. PMID: 33445144

Pandey, G., Sharma, N., Sahu, P. P., & Prasad, M. (2016). Chromatin-based epigenetic regulation of plant abiotic stress response. *Current Genomics*, *17*(6), 490–498. PMID: 28217005

Papaefthimiou, D., Likotrafiti, E., Kapazoglou, A., Bladenopoulos, K., & Tsaftaris, A. (2010). Epigenetic chromatin modifiers in barley: III. Isolation and characterization of the barley GNAT-MYST family of histone acetyltransferases and responses to exogenous ABA. *Plant Physiology and Biochemistry*, *48*(2-3), 98–107. PMID: 20117010

Parmar, S., Gharat, S. A., Tagirasa, R., Chandra, T., Behera, L., Dash, S. K., & Shaw, B. P. (2020). Identification and expression analysis of miRNAs and elucidation of their role in salt tolerance in rice varieties susceptible and tolerant to salinity. *PLoS One*, *15*(4), e0230958. PMID: 32294092

Peirats-Llobet, M., Han, S.-K., Gonzalez-Guzman, M., Jeong, C. W., Rodriguez, L., Belda-Palazon, B., Wagner, D., & Rodriguez, P. L. (2016). A direct link between abscisic acid sensing and the chromatin-remodeling ATPase BRAHMA via core ABA signaling pathway components. *Molecular Plant*, *9*(1), 136–147. PMID: 26499068

Quadrana, L., & Colot, V. (2016). Plant transgenerational epigenetics. *Annual Review of Genetics*, *50*, 467–491. PMID: 27732791

Rajkumar, M. S., Shankar, R., Garg, R., & Jain, M. (2020). Bisulphite sequencing reveals dynamic DNA methylation under desiccation and salinity stresses in rice cultivars. *Genomics*, *112*(5), 3537–3548. PMID: 32278023

Ramirez-Prado, J. S., Latrasse, D., Rodriguez-Granados, N. Y., Huang, Y., Manza-Mianza, D., Brik-Chaouche, R., Jaouannet, M., Citerne, S., Bendahmane, A., & Hirt, H. (2019). The Polycomb protein LHP1 regulates *Arabidopsis thaliana* stress responses through the repression of the MYC2-dependent branch of immunity. *The Plant Journal*, *100*(6), 1118–1131. PMID: 31437321

Reddy, A. S., Marquez, Y., Kalyna, M., & Barta, A. (2013). Complexity of the alternative splicing landscape in plants. *The Plant Cell*, *25*(10), 3657–3683. PMID: 24179125

Rehman, Z., Baig, A., Shabir, G., Aslam, K., Shahzad, M., Rehman, S., ... & Khan, A. R. (2021). Drought stress induces differential DNA methylation shift at symmetric and asymmetric cytosine sites in the promoter region of *ZmEXPB2* gene in maize.

Sedgwick, B., Bates, P. A., Paik, J., Jacobs, S. C., & Lindahl, T. (2007). Repair of alkylated DNA: Recent advances. *DNA Repair*, *6*(4), 429–442. PMID: 17112791

Shams, M., Yildirim, E., Arslan, E., & Agar, G. (2020). Salinity induced alteration in DNA methylation pattern, enzyme activity, nutrient uptake and H2O2 content in pepper (*Capsicum annuum* L.) cultivars. *Acta Physiologiae Plantarum*, *42*, 1–12.

Shoaib, Y., Usman, B., Kang, H., & Jung, K. H. (2022). Epitranscriptomics: An additional regulatory layer in plants' development and stress response. *Plants*, *11*(8), 1033. PMID: 35448761

Singroha, G., Kumar, S., Gupta, O. P., Singh, G. P., & Sharma, P. (2022). Uncovering the epigenetic marks involved in mediating salt stress tolerance in plants. *Frontiers in Genetics*, *13*, 811732. PMID: 35495170

Skorupa, M., Szczepanek, J., Mazur, J., Domagalski, K., Tretyn, A., & Tyburski, J. (2021). Salt stress and salt shock differently affect DNA methylation in salt-responsive genes in sugar beet and its wild, halophytic ancestor. *PLoS One*, *16*(5), e0251675. PMID: 34043649

Song, X., Li, Y., Cao, X., & Qi, Y. (2019). MicroRNAs and their regulatory roles in plant–environment interactions. *Annual Review of Plant Biology*, *70*, 489–525. PMID: 30848930

Su, Y., Bai, X., Yang, W., Wang, W., Chen, Z., Ma, J., & Ma, T. (2018). Single-base-resolution methylomes of *Populus euphratica* reveal the association between DNA methylation and salt stress. *Tree Genetics & Genomes*, *14*, 1–11.

Suji, K., & Joel, A. J. (2010). An epigenetic change in rice cultivars under water stress conditions. *Electronic Journal of Plant Breeding*, *1*(4), 1142–1143.

Sun, C., Ali, K., Yan, K., Fiaz, S., Dormatey, R., Bi, Z., & Bai, J. (2021). Exploration of epigenetics for improvement of drought and other stress resistance in crops: A review. *Plants*, *10*(6), 1226. PMID: 34208642

Sun, L., Song, G., Guo, W., Wang, W., Zhao, H., Gao, T., Lv, Q., Yang, X., Xu, F., & Dong, Y. (2019). Dynamic changes in genome-wide histone3 lysine27 trimethylation and gene expression of soybean roots in response to salt stress. *Frontiers in Plant Science*, *10*, 1031. PMID: 31552061

Sun, M., Yang, Z., Liu, L., & Duan, L. (2022). DNA methylation in plant responses and adaption to abiotic stresses. *International Journal of Molecular Sciences*, *23*(13), 6910. PMID: 35805917

Sun, Y., Zhao, J., Li, X., & Li, Y. (2020). E2 conjugases UBC1 and UBC2 regulate MYB42-mediated SOS pathway in response to salt stress in *Arabidopsis*. *The New Phytologist*, *227*(2), 455–472. PMID: 32167578

Sura, W., Kabza, M., Karlowski, W. M., Bieluszewski, T., Kus-Slowinska, M., Pawełoszek, Ł., Sadowski, J., & Ziolkowski, P. A. (2017). Dual role of the histone variant H2A.Z in transcriptional regulation of stress-response genes. *The Plant Cell*, *29*(4), 791–807. PMID: 28258158

Thieme, C. J., Rojas-Triana, M., Stecyk, E., Schudoma, C., Zhang, W., Yang, L., Miñambres, M., Walther, D., Schulze, W. X., & Paz-Ares, J. (2015). Endogenous *Arabidopsis* messenger RNAs transported to distant tissues. *Nature Plants*, *1*(4), 1–9. PMID: 27247031

Tian, C., Zuo, Z., & Qiu, J.-L. (2015). Identification and characterization of ABA-responsive microRNAs in rice. *Genes & Genomics*, *42*(7), 393–402. PMID: 26233894

Tonosaki, K., Fujimoto, R., Dennis, E. S., Raboy, V., & Osabe, K. (2022). Will epigenetics be a key player in crop breeding? *Frontiers in Plant Science*, *13*, 958350. PMID: 36247549

Ullah, F., Xu, Q., Zhao, Y., & Zhou, D. X. (2021). Histone deacetylase HDA710 controls salt tolerance by regulating ABA signaling in rice. *Journal of Integrative Plant Biology*, *63*(3), 451–467. PMID: 33289304

Walter, J., Jentsch, A., Beierkuhnlein, C., & Kreyling, J. (2013). Ecological stress memory and cross stress tolerance in plants in the face of climate extremes. *Environmental and Experimental Botany*, *94*, 3–8.

Wang, Q., Xu, J., Pu, X., Lv, H., Liu, Y., Ma, H., Wu, F., Wang, Q., Feng, X., & Liu, T. (2021). Maize DNA methylation in response to drought stress is involved in target gene expression and alternative splicing. *International Journal of Molecular Sciences*, *22*(15), 8285. PMID: 34361051

Wang, T., Xing, J., Liu, Z., Zheng, M., Yao, Y., Hu, Z., Peng, H., Xin, M., Zhou, D., & Ni, Z. (2019). Histone acetyltransferase GCN5-mediated regulation of long non-coding RNA At4 contributes to phosphate starvation response in *Arabidopsis*. *Journal of Experimental Botany*, *70*(21), 6337–6348. PMID: 31401648

Yang, Z., Zhu, P., Kang, H., Liu, L., Cao, Q., Sun, J., Dong, T., Zhu, M., Li, Z., & Xu, T. (2020). High-throughput deep sequencing reveals the important role that microRNAs play in the salt response in sweet potato (*Ipomoea batatas* L.). *BMC Genomics*, *21*, 1–16. PMID: 32066373

Younis, A., Siddique, M. I., Kim, C.-K., & Lim, K.-B. (2014). RNA interference (RNAi) induced gene silencing: A promising approach of hi-tech plant breeding. *International Journal of Biological Sciences*, *10*(10), 1150. PMID: 25332689

Zhang, W., Wang, N., Yang, J., Guo, H., Liu, Z., Zheng, X., Li, S., & Xiang, F. (2020). The salt-induced transcription factor GmMYB84 confers salinity tolerance in soybean. *Plant Science*, *291*, 110326. PMID: 31928665

Zhang, X., Dong, J., Deng, F., Wang, W., Cheng, Y., Song, L., Hu, M., Shen, J., Xu, Q., & Shen, F. (2019). The long non-coding RNA lncRNA973 is involved in cotton response to salt stress. *BMC Plant Biology*, *19*, 1–16. PMID: 31666019

Zhang, Z., Zhang, S., Zhang, Y., Wang, X., Li, D., Li, Q., Yue, M., Li, Q., Zhang, Y.-E., & Xu, Y. (2011). *Arabidopsis* floral initiator SKB1 confers high salt tolerance by regulating transcription and pre-mRNA splicing through altering histone H4R3 and small nuclear ribonucleoprotein LSM4 methylation. *The Plant Cell*, *23*(1), 396–411. PMID: 21258002

Zhu, J. K. (2002). Salt and drought stress signal transduction in plants. *Annual Review of Plant Biology*, *53*(1), 247–273. PMID: 12221975

Zhu, N., Cheng, S., Liu, X., Du, H., Dai, M., Zhou, D.-X., Yang, W., & Zhao, Y. (2015). The R2R3-type MYB gene OsMYB91 has a function in coordinating plant growth and salt stress tolerance in rice. *Plant Science*, *236*, 146–156. PMID: 26025528

Zong, W., Zhong, X., You, J., & Xiong, L. (2013). Genome-wide profiling of histone H3K4-tri-methylation and gene expression in rice under drought stress. *Plant Molecular Biology*, *81*, 175–188. PMID: 23192746

Chapter 10
Thermal Stress and Epigenetic Regulation in Plants

Fariba Shohani
Urmia University, Iran

Arash Fazeli
Ilam University, Iran

Nicolas L. Taylor
https://orcid.org/0000-0003-2004-5217
The University of Western Australia, Australia

Seyede Roghie Ghadirnezhad Shiade
Sari Agricultural Science and Natural Resources University, Iran

Ammara Saleem
University of the Punjab, Pakistan

Amin Fathi
https://orcid.org/0000-0002-7539-0053
Islamic Azad University, Iran

ABSTRACT

Today, climate change presents a significant challenge that impacts the growth and performance of plants, leading to adverse environmental conditions that disrupt their metabolic and physiological processes, ultimately threatening their survival. Plants have developed crucial genetic and epigenetic regulatory mechanisms, such as DNA methylation and histone modifications, in addition to managing stress-responsive genes to withstand thermal stress. Chromatin regulators are vital in sustaining this gene network in highly stressful conditions. While many of these modifications are often temporary, specific changes may be enduring and can be inherited by the next generation, contributing to plant stress memory. Understanding genetic mechanisms and the characteristics of chromatin regulators and histone modifications reveals the complexity of plant stress memory.

DOI: 10.4018/979-8-3693-8648-4.ch010

INTRODUCTION

Plants are subjected to diverse environmental challenges, classified as abiotic stresses, encompassing heat, cold, drought, salinity, ultraviolet (UV) radiation, and heavy metal toxicity. This vulnerability arises from their immobile nature (Xie et al., 2019). In recent years, numerous studies have reported the detrimental effects of extreme temperature fluctuations (heat and cold stress) on plant growth and development. Although plants have developed complex strategies to cope with heat stress (Sato et al., 2024, Ahmad et al., 2022a), the ongoing macroclimatic changes have elevated temperatures to critical levels, transforming ecological and environmental conditions. Consequently, researchers are striving to elucidate the epigenetic mechanisms through which plants mitigate the threats of severe heat stress. Similarly, cold stress can lead to reduced growth and development, resulting in decreased crop yields. For instance, extreme cold and freezing can disrupt plasma membrane permeability, induce programmed cell apoptosis, and cause chlorosis and wilting. This damage and its impact on photosynthesis can have irreparable effects on plant physiology (Verma et al., 2021, Ahmad et al., 2022b). Fundamental changes at the molecular level are essential for enhancing plant resilience against temperature stress. In recent years, epigenetic regulation has emerged as an effective and efficient control mechanism (Lämke and Bäurle, 2017). Epigenetic modifications, such as DNA methylation, histone modifications, and RNA interference (RNAi) mediated by non-coding RNA (ncRNA), play crucial roles in plant survival under abiotic stress conditions. Furthermore, epigenetic mechanisms can establish stress memory that can be passed on to subsequent generations of plants when encountering similar stresses (Friedrich et al., 2019). Therefore, deciphering these epigenetic mechanisms is vital for developing stress-resistant crops. In recent years, extensive research has been conducted to identify molecular targets, creating resilient lines through genetic manipulation that can better tolerate extreme temperature fluctuations. This chapter summarizes the responses of various regulatory epigenetic mechanisms to extreme temperature changes, such as heat and cold, in plants.

HEAT, COLD STRESS AND EPIGENETIC REGULATORY MECHANISMS

Epigenetically controlled mechanisms, such as DNA methylation, non-coding RNA modifications, and chromatin remodelling, play a vital role in plant survival under unfavorable environmental conditions. These epigenetic mechanisms can establish stress memory within the plant, which may be transmitted to subsequent generations (Figure 1).

Figure 1. Diverse epigenetic regulatory mechanisms in response to extreme temperature stress (heat and cold). SWI/SNF, SWItching/Sucrose Non-Fermenting; HDA6 (Histone Deactylases-6); lncRNA (Long-Non-Coding RNA, miRNA (Micro RNA, H3K4 (HistoneH3 LysineK4)

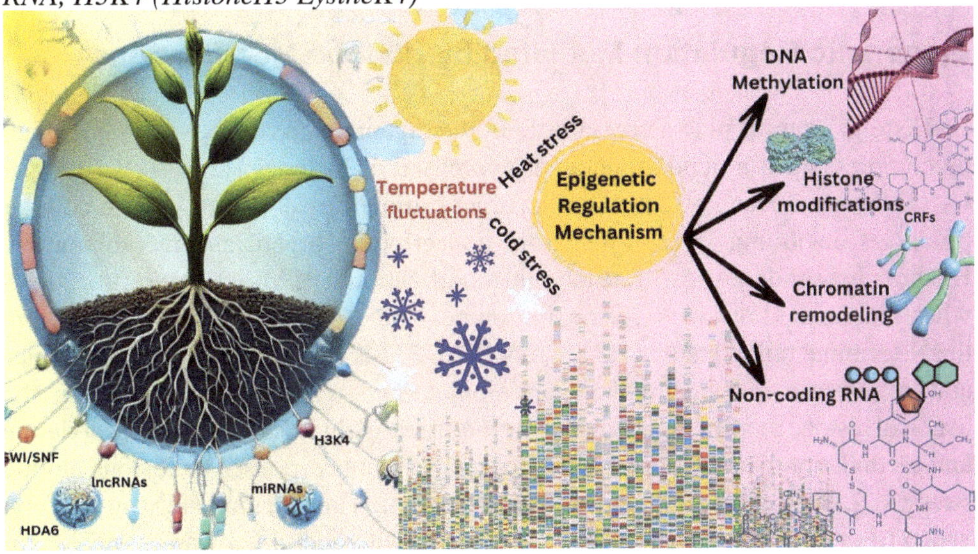

DNA Methylation

DNA methylation is recognized as one of the earliest discoveries of epigenetic regulatory control mechanisms, which is considered a relatively stable, heritable, and transgenerational mark. DNA methylation generally refers to the transfer of a methyl group to the C5 position of cytosine, resulting in the formation of 5-methylcytosine (5mC). In mammals, the primary type of DNA methylation is CG methylation, while non-CG methylation is limited to specific tissues, such as embryonic stem cells. In contrast, both CG and non-CG methylation, including DNA methylation in CHG (symmetrical) and CHH (asymmetrical) contexts (where H = A, C, or T), are extensively identified in higher plants (Sun et al., 2022). DNA methylation, a heritable epigenetic mark, influences gene expression and genome stability, thereby regulating many biological processes. Furthermore, DNA methylation changes can lead to plant growth abnormalities (Ramakrishnan et al., 2022). Within the plant genome, methylation is predominantly distributed in heterochromatin and euchromatin. However, in plants such as *Arabidopsis* and rice, it can also occur in promoter regions, intergenic regions, transposable elements (TEs), and even in exon regions (Verma et al., 2021). Promoter methylation is required for the repression of gene expression, while exon regions seem to be associated with gene activation. Additionally, TE methylation plays a critical role in maintaining genome integrity, with methylation and demethylation regulated by methyltransferases and demethylases, respectively (Verma et al., 2021).

Epigenetic Regulation Mediated by Histone Modifications

Histone proteins play a crucial role in the organization of DNA within the cell by compressing it around an octamer composed of histones H2A, H2B, H3, and H4 (Talbert and Henikoff, 2021). This organization is essential for various cellular processes, including DNA replication, transcription, repair, and recombination. Histones may undergo post-translational modifications (PTMs) (Verma et al., 2021), where amino acid residues (serine, threonine, lysine, and arginine) primarily from the N-terminal tail of histone H3 are modified. PTMs can alter chromatin dynamics and architecture, influencing the expression of stress-responsive genes through the action of histone modifiers. These histone modifiers include acetylation/deacetylation and methylation/demethylation, contributing to plant responses to various abiotic stresses (S. Kumar and Mohapatra, 2024). Acetylation involves transferring an acetyl group to the ε-amino group of lysine residues, with acetyl-CoA as a cofactor for histone acetyltransferases (HATs) that catalyze this process. Acetylation is regulated by two key enzymes: histone acetyltransferases (HATs) and histone deacetylases (HDACs) (Ahmad et al., 2024a, Kumar and Mohapatra, 2024). This modification produces an "open" chromatin state, representing DNA's most transcriptionally active regions. HATs and HDACs are enzymes responsible for histone modifications through acetylation or deacetylation, which are associated with plant responses to abiotic stress. In soybean, at least 14 HATs belonging to the CBP, GNAT/MYST, and TF-II250 families have been identified. In plants, the interaction of HATs with transcription factors (TFs) activates stress-related genes in response to abiotic stress (Cadavid, Balbinott, and Margis, 2023). Histone deacetylation, performed by histone deacetylases (HDACs), leads to a "closed" chromatin state necessary for the repression of gene expression. In *Arabidopsis*, approximately eighteen HDACs have been identified, belonging to the RPD3/HDA1 superfamily/family, HD2, and SIR2 (Verma et al., 2021). Histone deacetylases act as "erasers," removing acetyl groups from lysine residues of histones that HATs have acetylated. Overall, deacetylation results in gene repression, thereby regulating multiple biological processes. Similar to HATs, the number of HDACs varies among plant species; for instance, 14 have been identified in tomato, 13 in grape, and 28 in soybean (Kumar, Thakur, and Prasad, 2021). Furthermore, Verma et al. (2022) reported that in *Arabidopsis*, histone deacetylase 19 (HDA19) functions as a global repressor, while HDA6 directs the silencing of essential genes through RNA-directed DNA methylation (RdDM). Several reports have demonstrated the role of histone acetylation in plant development and acclimation during various stresses by regulating a set of genes for specific targets (Fig 2, Tables 1 and 2). Histone acetylation also contributes to the maintenance and regulation of genome stability. Histone methylation status (HMT) is dynamically regulated by histone methyltransferases and histone demethylases, which add or remove methyl marks. They primarily occur at lysine methylation sites (H3K4, H3K9, H3K27, H3K36, and H4K20) and two arginine methylation sites (H3R17 and H4R3), which can undergo mono-, di-, or trimethylation (Shi, Cui, and Shen, 2023). Histone methylation can influence chromatin structure and regulate gene transcription. Generally, H3K4me3 and H3K36me3 are associated with transcriptional activation, while H3K9me2 and H3K27me3 are linked to transcriptional repression (Shi, Cui, and Shen, 2023, Ahmad et al., 2024b). A distinguishing feature of HMTs is the presence of SET domains, including Su(var), E(z), and Trithorax. Histone methyltransferases can be classified based on their methylation residue specificity into lysine methyltransferases (HKMTs) and arginine methyltransferases (PRMTs) (Shi, Cui, and Shen, 2023). In *Arabidopsis* and rice, 31 and 25 genes encoding SET domain-containing proteins have been identified, respectively (Verma et al., 2021). In *Arabidopsis*, lysine methylation predominantly occurs at K4, K9, K27, and K36 of histone H3 (Lopez et al., 2022). Demethylation events are carried out by histone demethylases (HDMs), which act as "erasers," removing methyl marks. In *Arabidopsis*, Jumonji (JMJ) proteins serve as histone demethylases, performing demethylation of H3K27. Approximately 21 JMJ proteins have been identified, belonging to five subfamilies: the JMJC domain, which includes the groups JMJD6, Kdm4/Jhdm3, Kdm5/Jarid1, and Kdm3/Jhdm2 (Verma et al., 2021). JMJC domain-containing proteins, functioning as histone demethylases, play important roles in stress and light responses by modifying a series of associated transcription factors or specific phenotypic genes (Wang et al., 2022). Jumonji proteins are crucial for flower development and flowering time control. Lysine-specific demethylase 1 (LSD1) represents another family of HDMs that removes mono- and dimethyl groups from lysine residues (Verma et al., 2021, Ali et al., 2022). In *Arabidopsis*, four LSD1 homologs (FLD, LDL1, LDL2, and LDL3) have been identified, playing essential roles in regulating flowering (Wang et al., 2022).

Figure 2. Showing the role of epigenetic regulation through histone modifications

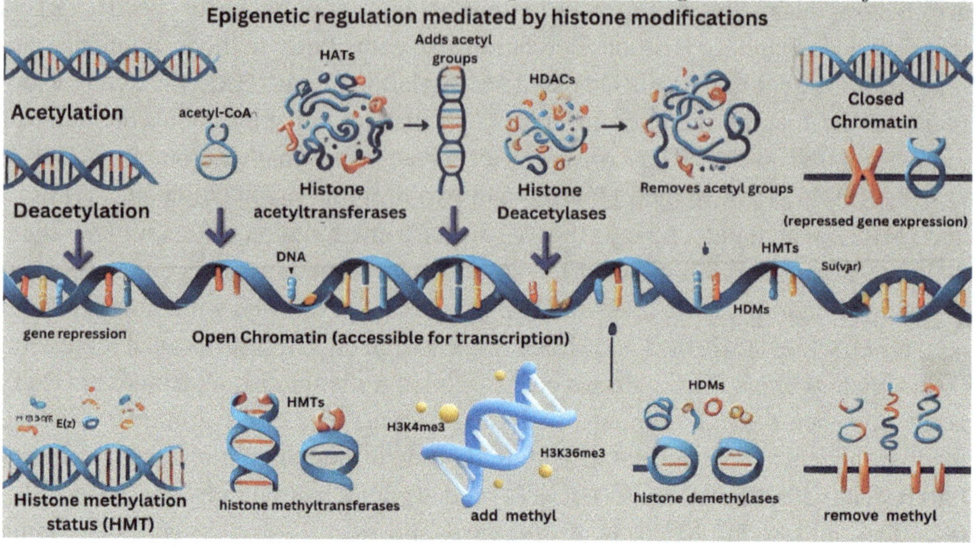

Adenosine Triphosphate (ATP)-Dependent Chromatin Remodelling

Plants interpret and transmit a range of environmental signalling pathways that lead to transcriptional reprogramming. The function of these stress signals involves altering chromatin structure, which facilitates the accessibility of DNA to various chromatin remodelling factors (CRFs). This may include the removal of histones and their subsequent post-translational modifications to regulate gene expression (Bhadouriya et al. 2021). The chromatin structure is modified by destabilizing nucleosomes (histone octamers) in an ATP-dependent energy-derived process mediated by CRFs. Different classes of chromatin remodelers, including SWItching/Sucrose Non-Fermenting (SWI/SNF), imitation switch (ISWI), inositol 80 (INO80), and chromodomain (CHD) factors, play crucial roles in the epigenetic regulation of various abiotic stressors. The identified ATP-dependent remodelling factors facilitate the transition of chromatin from a transcriptionally inactive state to an active state (Bhadouriya et al. 2021; Verma et al. 2022, Ijaz et al., 2023). The SWI/SNF subfamily belongs to the translocase and helicase superfamily, composed of 8 to 14 subunits, which include the helicase-SANT domain, terminal C bromodomain, and post-catalytic ATPase domain (Bhadouriya et al. 2021). The SWI/SNF complex contains four types of SNF2-ATPases: Brahma (BRM), SPLAYED (SYD), CHR12, and CHR23, which have been identified in *Arabidopsis*. BRM and SYD regulate vegetative and reproductive growth (Verma et al., 2021, Ijaz et al., 2024a).

Non-Coding RNA

Non-coding RNAs (ncRNAs) are transcribed from the genome but are not translated into proteins. Some ncRNAs play critical roles in plant growth and development, metabolism, and tolerance to abiotic stress. Current research on plant non-coding RNAs primarily focuses on two areas: small RNAs with lengths less than 50 nucleotides (nt), including microRNAs (miRNAs) and small interfering RNAs (siRNAs), and long non-coding RNAs (lncRNAs), which typically exceed 200 nt in length, including long intergenic non-coding RNAs (lincRNAs) (Liao et al. 2018). Plant miRNAs (20-24 nt) are transcribed from MIR genes by RNA polymerase II (RNA pol II) (Dong, Hu, and Zhang 2022). These miRNAs have been reported to participate in gene expression regulation through DNA methylation (Verma et al. 2022). Increasing evidence indicates that miRNAs play pivotal roles in organ development and stress responses. lncRNAs constitute a class of endogenous ncRNAs longer than 200 nt lacking coding potential (Liao et al. 2018). Numerous studies have shown that lncRNAs may serve as precursors to miRNAs, and emerging evidence suggests that many lncRNAs are important in various biological processes, such as flowering time regulation, gene silencing, root organogenesis, seedling photomorphogenesis, and reproduction (Baruah et al. 2021; Liao et al. 2018; Varshney et al. 2019, Ijaz et al., 2024b). lncRNAs likely play both shared and distinct roles in plant stress responses (Xiao, Wang, and Xu 2022). For example, COOLAIR regulates histone methylation (H3K36me3 and H3K27me3) and demethylation (H3K4me2) in response to cold exposure, modulating flowering locus C (FLC) in *Arabidopsis* (Verma et al. 2022).

HEAT STRESS AND EPIGENETIC REGULATION

Global climate change extensively affects the natural environment and human civilization. The increase in average temperatures, coupled with recurrent heat waves, negatively impacts agricultural crop yields and wild species within natural ecosystems. Heat stress disrupts physiological processes such as photosynthesis, ultimately reducing plant growth and productivity. To mitigate the effects of heat stress, plants epigenetically modify chromatin structure through covalent changes, including DNA methylation and histone modifications (Table 1).

Table 1. Epigenetic regulation of heat stress in plants

Species	Epigenetic regulators	Epigenetic marks	References
Arabidopsis thaliana	CMT2	DNA Methylation	(Jin et al. 2024)
Gossypium hirsutum	DRM2/CMT2	DNA Methylation	(Ma et al. 2018)
Populus qiongdaoensis	lncRNAs and miRNAs	ncRNA mediated regulation	(Xu et al. 2020)
Triticum aestivum	HD-ZIP	microRNAs	(Begum 2022)
Arabidopsis thaliana	At3g57230 and At3g09220	microRNAs	(F. Zhang et al. 2022a)
Musa sapientum	SPL	microRNAs	(Zhu et al. 2019)
Arabidopsis thaliana	DMRs	DNA Methylation	(Malabarba et al. 2021)
Arabidopsis thaliana	SWR1	Histone	(Jin et al. 2024)
Arabidopsis thaliana	SDG25 and ATX1	Histone	(Z. T. Song et al. 2021)
Arabidopsis thaliana	JMJ	Histone methylation	(Yamaguchi et al. 2021)
Populus qiongdaoensis	lncRNAs and miRNAs	ncRNA mediated regulation	(Verma et al. 2022)
Arabidopsis thaliana	DRM2, NRPD1, and NRPE1	DNA methylation	(Malik 2022)
Arabidopsis thaliana	ARF10, ARF16, and ARF17	microRNAs	(Lin et al. 2018)
Arabidopsis thaliana	HSFA2	Histone	(Rehman et al. 2023)
Arabidopsis thaliana	BRU1	Histone	(Brzezinka, Altmann, and Bäurle 2019)
Brassica juncea	SPL2	microRNAs	(Ahmed et al. 2019)
Arabidopsis thaliana	SWR1 (ARP6)	Chromatin remodeling	(Verma et al. 2022)
Arabidopsis thaliana	PIF4 and YUCCA8	Chromatin	(Zioutopoulou et al. 2021)

DNA Methylation and Demethylation

Mutants of *Arabidopsis* that exhibit deficiencies in various epigenetic regulatory pathways were investigated for their epigenetic responses to heat stress. Two mutants, nrpd2 and hda6, demonstrated sensitivity to heat stress. It has been suggested that transcriptional responses to heat stress partially depend on the RNA-directed DNA methylation (RdDM) pathway (Verma et al. 2022). In *Arabidopsis*, severe heat stress can impact seed morphology, seed physiology, DNA methylation levels, and seed expression. Differentially methylated regions (DMRs) were observed in the promoter and gene sequences of germinating seeds during periods of severe heat stress (Malabarba et al. 2021). In *Arabidopsis*, gene expression significantly alters during seed development under severe heat stress, enhancing the mechanisms involved in the heat stress response. It was found that DNA demethylation mediated by the ROS1 gene can disrupt seed germination by influencing the expression of germination-related genes. Conversely, under severe heat stress, most DMRs were located in the promoter regions of genes associated with germination (Sun et al. 2022, Ullah et al., 2023a). Changes in DNA methylation dynamics in response to heat stress were examined in *Arabidopsis*. The results were assessed under heat stress conditions, control, and recovery periods following the removal of stress, revealing a reduction in DNA methylation levels during the recovery phase compared to heat stress conditions (Korotko et al. 2021).

In another study, the dynamics of DNA methylation in response to short-term and long-term heat stress were analysed in *Gossypium hirsutum*. Heat stress significantly inhibited growth and was associated with alterations in the abundance of transcripts from numerous heat stress-related genes, particularly H3K4me2 and H4K5ac. Furthermore, changes in the abundance of H3K4me2 and H4K5ac were correlated with the expression of heat stress-responsive genes, including GhHSFA1a (Heat Shock Factor A1a), GhHSFA2, GhHSP3 (Heat Shock Protein 3), GhRBCS (small subunit of Rubisco), and GhHSFA1a (EthylERF1A). The findings indicated that H3K4me2 transcripts increased under short-term heat stress, while H4K5ac transcripts were elevated throughout heat stress, especially following prolonged heat exposure. Overall, it was observed that DNA methylation, H3K4me2, and H4K5ac could modulate the expression of heat stress-related genes, contributing to the regulation of heat stress in *Gossypium hirsutum* (K. He, Cao, and Deng 2021, Ullah et al., 2023b).

Similarly, the responses of two lines (84021 and H05) of *Gossypium hirsutum* to heat stress (normal and extreme temperatures) were evaluated. Analysis revealed that heat stress resulted in a deficiency of S-adenosyl-L-homocysteine hydrolase 1 and inhibited methyltransferases, leading to hypomethylation in line H05. In contrast, the 84021 line exhibited increased expression of histone genes, resulting in enhanced DNA stability (Verma et al. 2022). The expression patterns of DNA methyltransferases (MET1, CMT3, and DRM2) were examined during heat stress. They also assessed the effects of elevated temperatures on protein-coding genes (At3g50770 and At5g43260). In single- and multicellular structures of soybean (*Glycine max* L.), an increase in the expression of At3g50770 was observed, suggesting that CHH tissues are more susceptible to hypomethylation under heat stress (Hossain et al. 2017, Haider et al., 2021a).

Histone Modifications

The regulation of growth and development in plants is closely linked to gene expression markers. Chromatin structure is crucial in determining physiological patterns for plant transcription (Bhadouriya et al. 2021). The unmodified protein 5 (GCN5) acts as a histone acetyltransferase, activating heat stress-responsive genes in *Arabidopsis*. Deficiencies in GCN5 result in ecotypes tolerant to heat stress and with disrupted responses to heat stress. Under thermal stress conditions, the expression of various transcriptional regulators such as HSF, MBF1c, and UVH6 was reduced in GCN5 mutants compared to wild type. During heat stress, GCN5 facilitates the activation of these regulators through acetylation of H3K9 and H3K14 (Hu et al. 2015). Ueda et al. (2018) reported an increased tolerance to various stresses, including drought, salinity, and heat stress, in the hda19 mutant compared to the wild type. They also suggested that HDA19 could be a promising target for enhancing resilience against various abiotic stresses. In banana fruit, MaMYB4 represses the transcription of a histone deacetylase (ω-3 MaFADs), thus increasing histone acetylation levels of H3 and H4 under cold stress (Song et al. 2019, Haider et al., 2021b). Prolonged heat stress induces excessive expression of memory genes, such as HSP18.2, HSP21, and HSP22.0, correlated with increased H3K4me3 and H3K4me2 levels due to substantial accumulation (Rehman et al., 2023). It has been reported that histone methylation such as H3K4me2 and H3K4me3 are associated with heat stress (Rehman et al. 2023). In one study, Shekhawat et al. (2020) emphasized the importance of endophytic bacteria (Enterobacter sp. SA187) in the heat stress response of *Arabidopsis*, proposing that priming of heat stress memory genes could be induced by acclimation. However, transient changes in histone methylation levels (H3K4me3) at memory gene loci under heat stress might induce constructive priming (Shekhawat et al. 2020). For plant growth, proteins containing the SET domain are essential and significantly minimize responses to various stresses by altering histone methylation status. In *Arabidopsis*, both the Arabidopsis trithorax-like factor (ATX1) and histone methyltransferases H3K4 (SDG25) are important for heat stress tolerance. It has been suggested that SDG25 and ATX1 help maintain the expression of stress-responsive genes during recovery periods after stress. Moreover, it was concluded that a decrease in histone methylation levels (H3K4me3) and an increase in cytosine DNA methylation levels by mutations in SDG25 and ATX1 might contribute to this process (Song et al. 2020; Song et al. 2021, Haider et al., 2022a).

Chromatin Remodelling and Heat Stress Regulation

Plants respond to heat by employing various ATP-dependent chromatin remodelling factors (CRFs) that play a critical role in temperature sensing. For example, in *Arabidopsis*, the expression of the CHR12 gene increases during drought and heat stress, and it is believed to play a role in halting primary stem growth and flower bud development (Verma et al., 2021). In response to low temperatures, mutations in swi3c alter the expression of several involved genes. The binding of SWI/SNF complexes containing SWI3C to target genes ICE1, MYB15, and CBF1 is temperature-dependent. Thus, it can be concluded that the swi3c mutation affects the transcription of several cold-responsive genes (COR). These findings, along with genetic analyses of the double mutant swi3c/ice1 and enhanced freezing tolerance, indicate that SWI/SNF CRCs significantly contribute to the fine-tuning of plant growth responses to varying temperature regimes (Gratkowska-Zmuda et al. 2020). The interaction of HD2C (histone deacetylase) with the BRM-SWI/SNF remodelling complex regulates the level of tH4K16ac, decreasing heat-activated genes in hd2c mutants in *Arabidopsis*. This study demonstrated that HD2C epigenetically controls heat stress responses in *Arabidopsis* (Buszewicz et al. 2016). In another study, the presence of heat-responsive elements (HREs) in several transposons, including ONSEN, COPIA37, TERESTRA, and ROMANIAT5, was reported in Arabidopsis *thaliana* and *Arabidopsis lyrata*, suggesting that HREs likely play a fundamental role in chromatin remodelling within transposable elements regulated by BRM (Kovalchuk 2023). ATPases are involved in many growth processes that are both chromatin-dependent (transcription regulation and stress memory maintenance) and independent (such as pri-miRNA processing). The analysis showed that many proteins involved in responses to abiotic and biotic stress, such as the HEAT-STRESS-ASSOCIATED 32-kDa protein (HSA32) and BRM, interact with each other (Kovalchuk 2023).

Heat Stress Mediated by Non-Coding RNA Regulation

Heat stress has been reported to mediate the differential expression of long non-coding RNAs (lncRNAs) in *Brassica juncea* in response to heat and drought stress. The identification and expression of lncRNAs associated with various abiotic stresses were found to be linked to antioxidant pathways (Bhatia et al., 2020). Exposure of cucumber plants (*Cucumis sativus L.*) to severe temperature stress can induce the differential expression of non-coding RNAs, including microRNAs (miRNAs), lncRNAs, and circular RNAs (circRNAs). Furthermore, it has been suggested that non-coding RNAs, particularly lncRNAs, may play a regulatory role in cucumber plants under heat stress (He et al., 2020, Haider et al., 2022b). They observed the differential expression of miRNAs and lncRNAs under heat stress and demonstrated the involvement of various phytohormones in heat tolerance (Verma et al., 2021). In heat stress conditions, the differential expression of miRNAs and lncRNAs highlights the roles of various phytohormones in modulating heat stress. In another study, two rice genotypes, heat-tolerant Nagina and heat-sensitive Vandana, were subjected to both short- and long-term heat stress conditions, revealing that osa-miR1436, osa-miR5076, osa-miR5161, and osa-miR6253 were differentially expressed in both genotypes, indicating their essential roles in heat stress responses (Mangrauthia et al., 2017). The first identified miRNA that positively modulated heat stress was miR156, which is significantly induced and regulates the transcription factor gene target (SPL) at the transcriptional level following heat stress (F. Zhang et al., 2022b). It has been reported that in response to heat stress and to mitigate its harmful effects, miRNAs regulate the abundance of transcription factors and proteins associated with reactive oxygen species (ROS) metabolism, as well as participate in the signalling of several mitogen-activated protein kinases (MAPK) pathways (Ahmed et al., 2019; Zhang et al., 2019).

EPIGENETIC MODIFICATION IN RESPONSE TO COLD STRESS

Cold stress, among various environmental stresses, can significantly adversely impact the physiological activities, biochemical metabolism, growth, development, and productivity of plants. The challenges posed by cold stress, particularly spring frost, present a significant cost to growers annually across the globe. Some plants may acquire adaptations to cope with such unfavorable conditions; however, many important crops still lack this phenomenon. To withstand such conditions, plants can remodel chromatin structure through epigenetic markers (DNA methylation, histone modifications, and chromatin remodelling factors) to regulate the expression of specific cold stress-related genes (Table 2).

Table 2. Epigenetic regulation of cold stress in plants

Species	Associated genes / epigenetic regulators	Epigenetic marks	References
Arabidopsis thaliana	COR15A and ATGOLS3	Histone	(Ramakrishnan et al. 2022)
Arabidopsis thaliana	Aux/IAA14	miRNAs	(F. Zhang et al. 2022b)
Oryza sativa	AP2/ERF TF OsERF096	miRNAs	(Sun et al. 2022)

continued on following page

Table 2. Continued

Species	Associated genes / epigenetic regulators	Epigenetic marks	References
Fagopyrum tataricum	CuAO, RPB1, and DHE1	DNA methylation	(Song et al. 2020)
Arabidopsis thaliana	DREB1 and CBFs	DNA methylation	(Kidokoro et al. 2020)
Zea mays	ADA313	microRNAs	(Aydinoglu 2020)
Arabidopsis thaliana	COR15A	Histone	(Banerjee, Wani, and Roychoudhury 2017)
Arabidopsis thaliana	PKL	chromatin remodeler	(Miryeganeh 2021)
Zea mays	ZmMI1	DNA methylation	(Miryeganeh 2021)
Arabidopsis thaliana	FLC	Histone	(Ramakrishnan et al. 2022)
Arabidopsis thaliana	COR15A and GOLS3	chromatin remodeler	(Rehman et al. 2023)
Arabidopsis thaliana	PWR-HOS15-HD2C	Histone acetylation	(Lim et al. 2020)
Arabidopsis thaliana	SWI/SNF	Chromatin remodeling	(Gratkowska-Zmuda et al. 2020)
Brassica rapa ssp	bra-miR398	ncRNA mediated regulation	(Verma et al. 2022)

DNA Methylation and Demethylation

Two rice genotypes (IR64 and LTH) were studied for DNA methylation patterns in response to cold stress at the seedling stage. A significant increase in DNA methylation was observed in the roots of the LTH genotype (Verma et al., 2021). *Hevea brasiliensis* cultivars exposed to short-term cold can induce the expression of DNA methylation (HbMET1) and genes associated with cold stress, such as HbICE1 and HbCBF2. It has also been suggested that cold tolerance in *H. brasiliensis* is controlled by epigenetic changes (Tang, Wang, and Huang, 2018). Both methylation and demethylation occur during adaptation to cold stress. In total, 1,562 differentially methylated genes were identified in response to cold adaptation in *Brassica*, including BrammDH1, BraKAT2, BraSHM4, and Bra4CL2, where their promoters were demethylated, leading to increased transcriptional activity (Sun et al., 2022). In *Populus L*, reduced genomic DNA methylation in meristematic tissues, induced by the cold-dependent DNA demethylase DEMETER-LIKE 10, was associated with breaking bud dormancy (Conde et al., 2017). It has been reported that global methylation varies according to stress conditions in *Malus domestica* Borkh, showing a decrease under strong cold stress. At the same time, no significant reduction was observed under mild cold stress conditions (Sun et al., 2022). Key transcription factors (TFs) such as DREB1/CBF are essential for plant adaptation under cold stress, and they interact with the cold-inducible CBFExpression 1 (ICE1) to mediate cold stress responses in plants (Kidokoro et al., 2020). It has been proposed that epigenetic markers, such as DNA methylation, may improve crop performance by alleviating the effects of cold, drought, salinity, and heat stresses (Akhter et al., 2021). Cold stress induces changes in methylation patterns in cold-tolerant rice seedlings (P427) by stimulating demethylation in the promoter region of the gene Os03g0610900, resulting in increased expression. Altered methylation patterns may be associated with enhanced enzymatic antioxidant activity and abscisic acid accumulation. The results indicated that the ICE-CBF-COR pathway is critical for cold stress tolerance in the rice genotype P427 (Guo et al., 2019).

Histone Modifications

In *Arabidopsis thaliana*, the gene encoding the osmotic stress-responsive protein HOS15, a WD40 repeat protein, exhibits increased expression. This protein represses genes associated with biotic stress by mediating the acetylation of histone H4. A high acetylation level was observed in mutants compared to wild-type plants, resulting in heightened sensitivity to cold stress (Verma et al., 2021). Another study reported that in plants subjected to cold stress, the interaction between HOS15 and histone deacetylase 2C (HD2C) plays a significant role in regulating the transcription of cold-responsive genes, namely COR15A and COR47. The transcription of COR genes under cold stress is supported by elevated levels of histone H3 acetylation and the interaction between transcription factors HOS15 and CBF (Park et al., 2018). Although limited information exists regarding the precise regulatory mechanisms governed by histone acetyltransferases (HATs) and histone deacetylases (HDACs), the dynamic alterations in overall histone acetylation and deacetylation highlight their role in reprogramming gene expression in response to cold stress (Hyun, 2022). Several studies have summarized the responses of plants to cold stress and elucidated the molecular mechanisms underlying cold stress tolerance (Lim et al., 2020). Research has established that epigenetic regulation, particularly histone modifications and DNA methylation, plays a prominent role in vernalisation and the plant response to cold stress (Rehman et al., 2023). Under cold stress conditions, the high expression of histone deacetylase AtHD2D delays morphological and physiological stages in *Arabidopsis*. In contrast, gene-edited (transgenic) plants exhibited better cold tolerance (Han et al., 2016). It has been reported that HDA6 plays a crucial role in freezing tolerance in plants under cold stress (Kumar, Thakur, and Prasad, 2021). HD2C interacts with the WD40 repeat protein HOS15 and CBFs (C-repeat binding factors) at normal temperatures to repress the expression of cold-responsive genes COR15A and COR47. During cold stress, HOS15 facilitates the ubiquitin-mediated degradation of HD2C, which leads to the recruitment of CBFs to the promoter regions of COR genes, enhancing their expression through the hyperacetylation of histones (Zheng et al., 2016).

Cold Stress and Chromatin Remodelling Mediated Regulation

Cold stress can be considered one of the most fundamental abiotic stresses. In plants subjected to cold stress, vegetative and reproductive growth is regulated by the expression of several cold-responsive genes. In response to cold stress, genes exhibit transcriptional changes. For instance, in *Arabidopsis*, the cold-responsive chromatin gene interacts with Argonaute 1 (AGO1). Also, it interacts with siRNA and chromatin remodelling complexes such as SWI/SNF, thereby enhancing cold stress response and the response to plant hormones such as jasmonic acid (Verma et al., 2021). The interaction between the proteins PWR and HOS15 is essential for regulating the transcription of COR genes. This interaction results in the formation of a HOS15/PWR chromatin remodelling complex, which attracts many CBF transcription factors to the promoter regions of COR genes, assisting plants under cold stress (Lim et al., 2021). The PICKLE (PKL) protein, through its interaction with CBF3, plays a significant role as a chromatin remodelling factor in alleviating cold stress in *Arabidopsis* and facilitates the transcription of cold-responsive genes (D29A, COR15A, and COR15B) (Yang et al., 2019).

Cold Stress and Non-Coding RNA-Mediated Regulation

The role of miRNAs in response to cold stress was first identified in *Arabidopsis*, where miR319c, miR393, and miR398a were differentially regulated following cold stress exposure (Aydinoglu, 2020). During cold stress, ta-siRNA contributes to stress modulation in cassava by regulating auxin signaling pathways (Verma et al., 2021). Another study examined the correlation between non-coding RNAs (ncRNAs) and the flowering locus C (FLC) role in chromatin regulation in *Arabidopsis*. It was proposed that the transcription of the antisense FLC may be induced under cold stress and could regulate FLC repression. Additionally, it was determined that prolonged exposure of *Arabidopsis* to cold stress leads to the expression of the lncRNA COOLAIR associated with the FLC locus (Liu et al., 2022). A specific mutation in lncRNA, such as SVALKA, has been reported to interfere with the expression of CBF genes, significantly impacting freezing tolerance in plants under cold stress conditions (Kindgren et al., 2018). In plants with excessive expression of CBF1, increased tolerance to freezing temperatures is associated with a marked reduction in biomass (Kindgren et al., 2018).

DECLARATION OF GENERATIVE AI IN SCIENTIFIC WRITING

While preparing this work, the author(s) used ChatGPT to improve and edit the text. After using this tool, the authors reviewed and edited the content as needed and take full responsibility for the publication's content.

REFERENCES

Ahmad, Y., Haakim, Z., Iqbal, J., Abbasi, B. A., Mahmood, T., & Kazi, M. (2024b). Technological Innovations for Abiotic Stress Resistance in Horticultural Crops. OMICs-based Techniques for Global Food Security, 233-244.

Ahmad, Y., Haider, S., Iqbal, J., Abbasi, B. A., Yaseen, T., & Mahmood, T. (2022a). The mechanisms of genome editing technologies in crop plants. In *Principles and Practices of OMICS and Genome Editing for Crop Improvement* (pp. 295–313). Springer International Publishing.

Ahmad, Y., Haider, S., Iqbal, J., Abbasi, B. A., Yaseen, T., & Mahmood, T. (2022b). The mechanisms of genome editing technologies in crop plants. In *Principles and Practices of OMICS and Genome Editing for Crop Improvement* (pp. 295–313). Springer International Publishing.

Ahmad, Y., Haider, S., Iqbal, J., Naseer, S., Attia, K. A., Mohammed, A. A., & Mahmood, T. (2024a). In-silico analysis and transformation of OsMYB48 transcription factor driven by CaMV35S promoter in model plant–Nicotiana tabacum L. conferring abiotic stress tolerance. *GM Crops and Food: Biotechnology in Agriculture and the Food Chain, 15*(1), 130–149. PMID: 38551174

Ahmed, W.. (2019). Identification of Conserved and Novel MiRNAs Responsive to Heat Stress in Flowering Chinese Cabbage Using High-Throughput Sequencing. *Scientific Reports, 9*(1), 1–11. PMID: 31624298

Akhter, Z.. (2021). In Response to Abiotic Stress, Dna Methylation Confers Epigenetic Changes in Plants. *Plants, 10*(6), 1–21. PMID: 34070712

Ali, I., Salah, K. B. H., Sher, H., Ali, H., Ullah, Z., Ali, A., & Mahmood, T. (2022). Drought stress enhances the efficiency of floral dip method of Agrobacterium-mediated transformation in Arabidopsis thaliana. *Brazilian Journal of Biology, 84*, e259326. PMID: 35703626

Aydinoglu, F. (2020). Elucidating the Regulatory Roles of MicroRNAs in Maize (Zea Mays L.) Leaf Growth Response to Chilling Stress. *Planta, 251*(2), 1–15. DOI: 10.1007/s00425-019-03331-y PMID: 31907623

Banerjee, A., Wani, S. H., & Roychoudhury, A. (2017). Epigenetic Control of Plant Cold Responses. *Frontiers in Plant Science, 8*(September), 1–5. PMID: 28983309

Baruah, P. M.. (2021). Identification and Functional Analysis of Drought Responsive LncRNAs in Tea Plant. *Plant Gene, 27*(March), 100311. DOI: 10.1016/j.plgene.2021.100311

Begum, Y. (2022). Regulatory Role of MicroRNAs (MiRNAs) in the Recent Development of Abiotic Stress Tolerance of Plants. *Gene, 821*(January), 146283. DOI: 10.1016/j.gene.2022.146283 PMID: 35143944

Bhadouriya, S. L.. (2021). Role of Chromatin Architecture in Plant Stress Responses: An Update. *Frontiers in Plant Science, 11*(January). PMID: 33510748

Bhatia, G.. (2020). Genome-Wide Investigation of Regulatory Roles of LncRNAs in Response to Heat and Drought Stress in Brassica Juncea (Indian Mustard). *Environmental and Experimental Botany, 171*, 103922. DOI: 10.1016/j.envexpbot.2019.103922

Brzezinka, K., Altmann, S., & Bäurle, I. (2019). BRUSHY1/TONSOKU/MGOUN3 Is Required for Heat Stress Memory. *Plant, Cell & Environment, 42*(3), 771–781. PMID: 29884991

Buszewicz, D.. (2016). HD2C Histone Deacetylase and a SWI/SNF Chromatin Remodelling Complex Interact and Both Are Involved in Mediating the Heat Stress Response in Arabidopsis. *Plant, Cell & Environment, 39*(10), 2108–2122. PMID: 27083783

Cadavid, I. C., Balbinott, N., & Margis, R. (2023). Beyond Transcription Factors: More Regulatory Layers Affecting Soybean Gene Expression under Abiotic Stress. *Genetics and Molecular Biology, 46*(1), 1–16. PMID: 36706026

Dong, Q., Hu, B., & Zhang, C. (2022). MicroRNAs and Their Roles in Plant Development. *Frontiers in Plant Science, 13*(February). PMID: 35251094

Friedrich, T., Faivre, L., Bäurle, I., & Schubert, D. (2019). Chromatin-Based Mechanisms of Temperature Memory in Plants. *Plant, Cell & Environment, 42*(3), 762–770. PMID: 29920687

Gratkowska-Zmuda, D. M.. (2020). The SWI/SNF ATP-Dependent Chromatin Remodeling Complex in Arabidopsis Responds to Environmental Changes in Temperature-Dependent Manner. *International Journal of Molecular Sciences, 21*(3), 1–19. PMID: 31979421

Guo, H.. (2019). The Methylation Patterns and Transcriptional Responses to Chilling Stress at the Seedling Stage in Rice. *International Journal of Molecular Sciences, 20*(20). PMID: 31615063

Haider, S., Iqbal, J., Naseer, S., Shaukat, M., Abbasi, B. A., Yaseen, T., & Mahmood, T. (2022b). Unfolding molecular switches in plant heat stress resistance: A comprehensive review. *Plant Cell Reports, 41*(3), 775–798. PMID: 34401950

Haider, S., Iqbal, J., Naseer, S., Yaseen, T., Shaukat, M., Bibi, H., & Mahmood, T. (2021a). Molecular mechanisms of plant tolerance to heat stress: Current landscape and future perspectives. *Plant Cell Reports*, •••, 1–25. PMID: 33890138

Haider, S., Iqbal, J., Shaukat, M., Naseer, S., & Mahmood, T. (2021b). The epigenetic chromatin-based regulation of somatic heat stress memory in plants. *Plant Gene, 27*, 100318.

Haider, S., Raza, A., Iqbal, J., Shaukat, M., & Mahmood, T. (2022a). Analyzing the regulatory role of heat shock transcription factors in plant heat stress tolerance: A brief appraisal. *Molecular Biology Reports, 49*(6), 5771–5785. PMID: 35182323

Han, Zhaofen et al. 2016. "AtHD2D Gene Plays a Role in Plant Growth, Development, and Response to Abiotic Stresses in Arabidopsis Thaliana." *Frontiers in Plant Science* 7(MAR2016): 1–13.

He, K., Cao, X., & Deng, X. (2021). Histone Methylation in Epigenetic Regulation and Temperature Responses. *Current Opinion in Plant Biology, 61*, 102001. DOI: 10.1016/j.pbi.2021.102001 PMID: 33508540

He, Xueying et al. 2020. "Systematic Identification and Analysis of Heat-Stress-Responsive LncRNAs, CircRNAs and MiRNAs with Associated Co-Expression and CeRNA Networks in Cucumber (Cucumis Sativus L .).": 736–54.

Hossain, M. S.. (2017). Divergent Cytosine DNA Methylation Patterns in Single-Cell, Soybean Root Hairs. *The New Phytologist, 214*(2), 808–819. PMID: 28106918

Hu, Z.. (2015). Histone Acetyltransferase GCN5 Is Essential for Heat Stress-Responsive Gene Activation and Thermotolerance in Arabidopsis. *The Plant Journal, 84*(6), 1178–1191. PMID: 26576681

Hyun, T. K. (2022). Plant Histone Modifications in Response to Cold Stress. *Botanica Serbica, 46*(1), 1–6.

Ijaz, S., Iqbal, J., Abbasi, B. A., Tufail, A., Ullah, Z., Sharifi-Rad, J., & Uddin, S. (2023). Biofortification: Lessons from the Past and Strategies for Future Food Security. In *Legumes Biofortification* (pp. 521–545). Springer International Publishing.

Ijaz, S., Iqbal, J., Abbasi, B. A., Yaseen, T., Rehman, S., Kazi, M., & Mahmood, T. (2024a). Role of OMICS-Based Technologies in Plant Sciences. OMICs-based Techniques for Global Food Security, 45-66.

Ijaz, S., Iqbal, J., Abbasi, B. A., Yaseen, T., Rehman, S., Kazi, M. & Mahmood, T. (2024b). Role of OMICS-Based Technologies in Plant Sciences. OMICs-based Techniques for Global Food Security, 45-66.

Jin, Q.. (2024). Epigenetic Regulation for Heat Stress Adaptation in Plants: New Horizons for Crop Improvement under Climate Change. *Agronomy (Basel), 14*(9), 1–21.

Kidokoro, S.. (2020). DREB1A/CBF3 Is Repressed by Transgene-Induced DNA Methylation in the Arabidopsis Ice1-1 Mutant. *The Plant Cell, 32*(4), 1035–1048. PMID: 32034036

Kindgren, P., Ard, R., Ivanov, M., & Marquardt, S. (2018). Transcriptional Read-through of the Long Non-Coding RNA SVALKA Governs Plant Cold Acclimation. *Nature Communications, 9*(1). Advance online publication. DOI: 10.1038/s41467-018-07010-6 PMID: 30385760

Korotko, U., Chwiałkowska, K., Sańko-Sawczenko, I., & Kwasniewski, M. (2021). Dna Demethylation in Response to Heat Stress in Arabidopsis Thaliana. *International Journal of Molecular Sciences, 22*(4), 1–20. PMID: 33557095

Kovalchuk, I. (2023). Role of Epigenetic Factors in Response to Stress and Establishment of Somatic Memory of Stress Exposure in Plants. *Plants, 12*(21). PMID: 37960024

Kumar, Suresh, and Trilochan Mohapatra. 2024. "Epigenetic Modifications in Genome Help Remembering the Stress Tolerance Strategy Adopted by the Plant." *Frontiers in Bioscience - Landmark* 29(3).

Kumar, V., Thakur, J. K., & Prasad, M. (2021). Histone Acetylation Dynamics Regulating Plant Development and Stress Responses. *Cellular and Molecular Life Sciences : CMLS, 78*(10), 4467–4486. DOI: 10.1007/s00018-021-03794-x PMID: 33638653

Lämke, J., & Bäurle, I. (2017). Epigenetic and Chromatin-Based Mechanisms in Environmental Stress Adaptation and Stress Memory in Plants. *Genome Biology, 18*(1), 1–11. PMID: 28655328

Liao, P., Li, S., Cui, X., & Zheng, Y. (2018). A Comprehensive Review of Web-Based Resources of Non-Coding RNAs for Plant Science Research. *International Journal of Biological Sciences, 14*(8), 819–832. PMID: 29989090

Lim, C. J.. (2020). The Histone-Modifying Complex PWR/HOS15/HD2C Epigenetically Regulates Cold Tolerance1 [OPEN]. *Plant Physiology, 184*(2), 1097–1111. PMID: 32732349

Lim, C. J.. (2021). HOS15-PWR Chromatin Remodeling Complex Positively Regulates Cold Stress in Arabidopsis. *Plant Signaling & Behavior, 16*(5). Advance online publication. DOI: 10.1080/15592324.2021.1893978 PMID: 33641608

Lin, J. S.. (2018). MicroRNA160 Modulates Plant Development and Heat Shock Protein Gene Expression to Mediate Heat Tolerance in Arabidopsis. *Frontiers in Plant Science, 9*(February), 1–16. PMID: 29449855

Liu, G., Liu, F., Wang, Y., & Liu, X. (2022). A Novel Long Noncoding RNA CIL1 Enhances Cold Stress Tolerance in Arabidopsis. *Plant Science, 323*(July), 111370. DOI: 10.1016/j.plantsci.2022.111370 PMID: 35788028

Lopez, L.. (2022). Genome-Wide Identification of Histone Modification Gene Families in the Model Legume Medicago Truncatula and Their Expression Analysis in Nodules. *Plants, 11*(3). PMID: 35161303

Ma, Y.. (2018). Disrupted Genome Methylation in Response to High Temperature Has Distinct Affects on Microspore Abortion and Anther Indehiscence. *The Plant Cell, 30*(7), 1387–1403. PMID: 29866646

Malabarba, J., Windels, D., Xu, W., & Verdier, J. (2021). Regulation of DNA (de) Methylation Positively Impacts Seed. *Genes, 12*(457), 1–21. PMID: 33807066

Malik, S. (2022). Epigenetic Regulation of Heat Stress in Plant Male Reproduction. *Frontiers in Plant Science,* •••, 13. PMID: 35222484

Mangrauthia, S. K.. (2017). Genome-Wide Changes in MicroRNA Expression during Short and Prolonged Heat Stress and Recovery in Contrasting Rice Cultivars. *Journal of Experimental Botany, 68*(9), 2399–2412. PMID: 28407080

Miryeganeh, M. (2021). Plants' Epigenetic Mechanisms and Abiotic Stress. *Genes, 12*(8). PMID: 34440280

Park, J.. (2018). Epigenetic Switch from Repressive to Permissive Chromatin in Response to Cold Stress. *Proceedings of the National Academy of Sciences of the United States of America, 115*(23). PMID: 29784800

Ramakrishnan, M.. (2022). Epigenetic Stress Memory: A New Approach to Study Cold and Heat Stress Responses in Plants. *Frontiers in Plant Science, 13*(December), 1–17. PMID: 36570899

Rehman, S.. (2023). Regulation of Plant Epigenetic Memory in Response to Cold and Heat Stress: Towards Climate Resilient Agriculture. *Functional & Integrative Genomics, 23*(4), 1–24. DOI: 10.1007/s10142-023-01219-5 PMID: 37700098

Shekhawat, K.. 2020. "Enterobacter Sp. SA187 Mediates Plant Thermotolerance by Chromatin Modification of Heat Stress Genes." *bioRxiv*: 2020.01.16.908756. https://doi.org/DOI: 10.1101/2020.01.16.908756

Shi, L., Cui, X., & Shen, Y. (2023, November). 2024. "The Roles of Histone Methylation in the Regulation of Abiotic Stress Responses in Plants.". *Plant Stress, 11*, 100303. DOI: 10.1016/j.stress.2023.100303

Song, C.. (2019). MaMYB4 Recruits Histone Deacetylase MaHDA2 and Modulates the Expression of ω-3 Fatty Acid Desaturase Genes during Cold Stress Response in Banana Fruit. *Plant & Cell Physiology, 60*(11), 2410–2422. PMID: 31340013

Song, Y.. (2020). Roles of DNA Methylation in Cold Priming in Tartary Buckwheat. *Frontiers in Plant Science, 11*(December). PMID: 33365044

Song, Z. T.. (2021). Histone H3K4 Methyltransferases SDG25 and ATX1 Maintain Heat-Stress Gene Expression during Recovery in Arabidopsis. *The Plant Journal, 105*(5), 1326–1338. PMID: 33278042

Sun, M., Yang, Z., Liu, L., & Duan, L. (2022). DNA Methylation in Plant Responses and Adaption to Abiotic Stresses. *International Journal of Molecular Sciences, 23*(13). PMID: 35805917

Talbert, P. B., & Henikoff, S. (2021). Histone Variants at a Glance. *Journal of Cell Science, 134*(6), 1–10. PMID: 33771851

Tang, X., Wang, Q., & Huang, X. (2018). Chilling-Induced DNA Demethylation Is Associated with the Cold Tolerance of Hevea Brasiliensis. *BMC Plant Biology, 18*(1), 1–16. PMID: 29685105

Ueda, M.. (2018). Versatility of HDA19-Deficiency in Increasing the Tolerance of Arabidopsis to Different Environmental Stresses. *Plant Signaling & Behavior, 13*(8), e1475808. DOI: 10.1080/15592324.2018.1475808 PMID: 30047814

Ullah, Z., Iqbal, J., Abbasi, B. A., Akhtar, W., Kanwal, S., Ali, I., & Mahmood, T. (2023b). Assessment of Gus Expression Induced by Anti-Sense Os PPO Gene Promoter and Antioxidant Enzymatic Assays in Response to Drought and Heavy Metal Stress in Transgenic Arabidopsis thaliana. *Sustainability*, *15*(17), 12783.

Ullah, Z., Iqbal, J., Abbasi, B. A., Ijaz, S., Sharifi-Rad, J., Yaseen, T., & Mahmood, T. (2023a). How Genome Editing Can Be Helpful in the Biofortification of Legumes. In *Legumes Biofortification* (pp. 207–232). Springer International Publishing.

Varshney, D.. (2019). Tissue Specific Long Non-Coding RNAs Are Involved in Aroma Formation of Black Tea. *Industrial Crops and Products*, *133*(January), 79–89. DOI: 10.1016/j.indcrop.2019.03.020

Verma, N.. (2021, November). 2022. "Epigenetic Regulation of Heat and Cold Stress Responses in Crop Plants.". *Plant Gene*, *29*, 100351. DOI: 10.1016/j.plgene.2022.100351

Wang, X.. (2022). Genome-Wide Identification of the Jumonji C Domain-Containing Histone Demethylase Gene Family in Wheat and Their Expression Analysis under Drought Stress. *Frontiers in Plant Science*, *13*(August), 1–12. PMID: 36092409

Xiao, M., Wang, J., & Xu, F. (2022). Methylation Hallmarks on the Histone Tail as a Linker of Osmotic Stress and Gene Transcription. *Frontiers in Plant Science*, *13*(August), 1–14. PMID: 36035677

Xu, J.. (2020). Third-Generation Sequencing Found LncRNA Associated with Heat Shock Protein Response to Heat Stress in Populus Qiongdaoensis Seedlings. *BMC Genomics*, *21*(1), 1–14. PMID: 32831021

Yamaguchi, N.. (2021). H3K27me3 Demethylases Alter HSP22 and HSP17.6C Expression in Response to Recurring Heat in Arabidopsis. *Nature Communications*, *12*(1), 1–16. DOI: 10.1038/s41467-021-23766-w PMID: 34108473

Yang, R.. (2019). A Role for PICKLE in the Regulation of Cold and Salt Stress Tolerance in Arabidopsis. *Frontiers in Plant Science*, *10*(July), 1–11. PMID: 31354770

Zhang, F.. (2022a). Roles of MicroRNAs in Abiotic Stress Response and Characteristics Regulation of Plant. *Frontiers in Plant Science*, *•••*, 13. PMID: 36092392

Zhang, F.. (2022b). Roles of MicroRNAs in Abiotic Stress Response and Characteristics Regulation of Plant. *Frontiers in Plant Science*, *13*, 919243. PMID: 36092392

Zhang, M.. (2019). The MiRNA-Mediated Post-Transcriptional Regulation of Maize in Response to High Temperature. *International Journal of Molecular Sciences*, *20*(7). PMID: 30970661

Zheng, Yu.. (2016). Histone Deacetylase HDA9 Negatively Regulates Salt and Drought Stress Responsiveness in Arabidopsis. *Journal of Experimental Botany*, *67*(6), 1703–1713. PMID: 26733691

Zhu, H.. (2019). Banana SRNAome and Degradome Identify MicroRNAs Functioning in Differential Responses to Temperature Stress. *BMC Genomics*, *20*(1), 1–16. PMID: 30630418

Zioutopoulou, A., Patitaki, E., Xu, T., & Kaiserli, E. (2021). The Epigenetic Mechanisms Underlying Thermomorphogenesis and Heat Stress Responses in Arabidopsis. *Plants*, *10*(11). PMID: 34834802

Chapter 11

Epigenetic Regulation and Plant Adaptation to Environmental Stress: Towards New Strategies for Sustainable Agriculture

Yassine Mouniane
https://orcid.org/0000-0002-6037-5301
Faculty of Sciences, Ibn Tofail University, Kenitra, Morocco

Issam El-Khadir
https://orcid.org/0000-0002-3970-2865
Faculty of Sciences, Ibn Tofail University, Kenitra, Morocco

Ahmed Chriqui
Faculty of Sciences, Ibn Tofail University, Kenitra, Morocco

Brahim El Ouardi
https://orcid.org/0009-0000-4559-6141
Faculty of Sciences, Ibn Tofail University, Kenitra, Morocco

Imane Aitouhanni
https://orcid.org/0009-0004-1400-8027
ENSIAS, SSLAB, Mohammed V University, Rabat, Morocco

Mahmoud Oudghiri
Faculty of Sciences, Ibn Tofail University, Kenitra, Morocco

Abdelaati Soufiani
https://orcid.org/0000-0002-5164-7707
Faculty of Sciences, Ibn Tofail University, Kenitra, Morocco

Driss Hmouni
https://orcid.org/0000-0001-7598-6204
Faculty of Sciences, Ibn Tofail University, Kenitra, Morocco

ABSTRACT

Plants are constantly exposed to various environmental stress, such as drought, salinity, extreme temperatures and pathogen attacks, which have a negative impact on their growth and development. In response to these stresses, highly complex

DOI: 10.4018/979-8-3693-8648-4.ch011

molecular processes are activated, epigenetic mechanisms being among the most important. These mechanisms, which include DNA methylation, histone modifications and non-coding RNAs, regulate gene expression and modify the plant's response to abiotic stresses. In this chapter, we highlight progress on the epigenetic mechanisms associated with the evolution of plants to adapt to abiotic stress. Combined with knowledge gleaned from other omics approaches, these strategies will facilitate the identification of regulatory networks and their dynamics under stress, enabling the development of strategies for sustainable agriculture that will ensure yield security in a context of climate change.

INTRODUCTION

Plants are subject to constant fluctuations in their local environment. Recurrent and unpredictable alterations in climate make conditions more unfavorable, subjecting plants to abiotic stresses such as drought, salinity, temperature extremes, nutrient deficiencies and heavy metal toxicity (Gaude et al., 2024). Today, we are confronted with global stress conditions, and these stresses affect plant physiological response, biomass and agricultural yield worldwide. In order to cope with these abiotic stresses, plants have developed a variety of adaptation strategies encompassing several physiological and biochemical changes (Cheong, 2021; Yajnik et al., 2024). One of these common mechanisms, epigenetic regulation, has been shown to modify the expression of stress-resistant genes. Epigenetic modifications are non-black-letter environmental factors that can affect the way genes act while preserving their DNA sequence. These epigenetic modifications enable plants to respond rapidly to changing environments by facilitating gene expression on demand. Epigenetic mechanisms consist mainly of DNA methylation, histone modification and RNA interference (Gallusci et al., 2023). DNA methylation is a type of chemical modification that regulates gene expression in response to the environment. Similarly, histone modifications have a major impact on chromatin structure by making the DNA portions of aggregates more or less accessible to proteins, notably through phosphorylation, ubiquitination, methylation/demethylation and acetylation/deacetylation. In addition to these mechanisms, epigenetic regulation also encompasses non-coding RNAs and RNA-based gene silencing processes that serve to protect plants against transposable elements and pathogens (Kappel et al., 2023).

An interesting aspect of these epigenetic modifications is their "stress memory" function. Some of these modifications, when they become permanent after exposure to stress, may persist after the stress has disappeared, giving a better capacity for defense against similar future exposures. This somatic memory can manifest itself within a generation and, in some cases, heritable epigenetic changes are passed on to offspring, defining transgenerational memory (Dinkar et al., 2024). The ability to remember past stresses gives plants an adaptive advantage by helping them to cope better with growth, which is particularly important in changing environments. Enhanced plant stress resistance through utilising epigenetic regulation will lead to important progress towards sustainable agriculture. This allows for the development of more resilient crops to extreme climate conditions, by first identifying and then applying these stress-resistance promoting epigenetic modifications in breeding programs. It is, therefore, not only a mechanism to cope with the environmental stress a plant senses at this moment but also leads us to hope that epigenetics might help humanity dealing with their food security in the future (Sajjanar et al., 2024).

The aim of this chapter is to describe a little of the epigenetic mechanisms involved in plant response and adaptation to abiotic stresses, and to discuss some possibilities for translating our knowledge into sustainable agricultural practices.

EPIGENETIC MECHANISMS IN PLANTS

Epigenetic mechanisms involve alterations in DNA, RNA and histone proteins and changes on chromatin structures in plants. These changes affect gene expression and are not changes in the DNA sequence which allows plants to tolerate environmental factors. All of these factors work synergistically to allow plants to adapt genetic expression in response to a changing environment without manipulation of the genome.

DNA Methylation

DNA methylation, primarily as 5-methylcytosine (5mC), is generally transcriptionally repressive (Wu et al., 2019). Such DNA modification occurs in CG, CHG and CHH sequences), and is within periphery controlled by different enzymes (MET1, CMT2). In contrast, demethylation could be either active (with involvement of DNA glycosylases) or passive (due to lack of methylation during replication) (Xiao et al., 2019; Zhang and Zhu, 2024).

Histone Modifications

Histones are the proteins that package the DNA into nucleosomes and can be modified (acetylated, methylated etc.) to regulate gene expression. Transcription is supported by acetylation and inhibition or activation can result from various methylations. COMPASSES and PRC2, for instance, are complexes that modulate pivotal histone marks (Morgan and Shilatifard, 2020; Millán-Zambrano et al., 2022).

Histone Variations

They are involved in regulating gene expression and plant responses to environmental cues. In addition, these variants alter the chromatin structure at specific regions of the genome compared to conventional histones and subsequently have a profound effect on the accessibility of DNA to transcription factors and other regulatory proteins. One example is the enrichment of H2A.Z at regulatory regions of active genes. Gene transcription leading to its presence is associated with an open chromatin structure (Xu et al., 2021; Dinkar et al., 2024).

In response to unfavorable environmental conditions, H2A.Z can translocate to specific genes and upregulate their expression. This histone variant therefore plays an important role in the adaptability of plants to respond rapidly to stimuli such as heat or cold stress. On the other hand, H3. 3 is also involved in the dynamic control of gene expression, particularly in proliferating or activated cells. H3. Importantly, H3.3 is often integrated into chromatin without the need for DNA replication and can thus rapidly alter the transcript state of WT genes (Shen et al., 2015; Cho et al., 2018; Zhou et al., 2020). Under stress, H3. 3 can replace canonical H3 histones in specific regions, altering chromatin structure to help activate genes that facilitate stress tolerance. The action of these histone variants and post-translational modifications (e.g. acetylation and methylation) forms a complex regulatory network for gene expression. This network plays a central role in controlling gene expression and plant adaptation to the environment (Pratx et al., 2023). Increased insight into the function of histone variants can be translated into approaches to improve plant response and adaptation to environmental stressors by manipulating epigenetic control (Gahlaut et al., 2020; Song et al., 2021).

Chromatin Structure

Chromatin architecture is one of the key factors that dictates how genes are expressed and how plants respond to environmental challenges. Chromatin accessibility is specified by fundamental regulators, such as ATP-dependent chromatin remodelers that mediate chromatin reorganization. These remodelers are able to use ATP energy to move or change histones so that DNA may be more or less accessible for transcription factors and regulatory complexes. Functionally relevant is the three-dimensional organization of chromatin (Long et al., 2023).

For instance, as shown in (D, E), chromatin is often divided into distinguishable compartments like compartment A and B. Compartment A is open, transcription-competent, anchored to the nuclear membrane by active locus such as gene hubs or nuclear pore complexes whereas compartment B is condensed and associated with repressed genes. Inside the nucleus of a cell, these two types of compartments will interact with one another and generate chromatin loops, which bring together in close proximity distant genes and regulatory elements. The interaction between genomic regions is essential for coordinating gene expression in response to external signals (Haider et al., 2021).

The dynamicity of this 3D chromatin structure, however, has been proved to be affected by many environmental factors including abiotic stresses like cold or drought. For example, low-temperature exposure can alter chromatin structure, allowing access to particular stress response-associated genes. This chromatin-remodeling mechanism allows plants to respond rapidly by shifting their transcriptional profile to changing conditions, an essential trait for plant survival (Klemm et al., 2019; Shang et al., 2022).

RNA Modifications

Adenosine (m6A) and cytosine (m5C) mRNA methylation are post-transcriptional RNA modifications that affect post-transcriptional regulation of stability, localization, translation and therefore plant growth and stress response. These changes are controlled in a reversible manner, and their importance for plant adaptation is remarkable (Cappannini et al. 2023; Shen et al., 2023).

Figure 1. Epigenetic mechanisms in plants under various abiotic stress

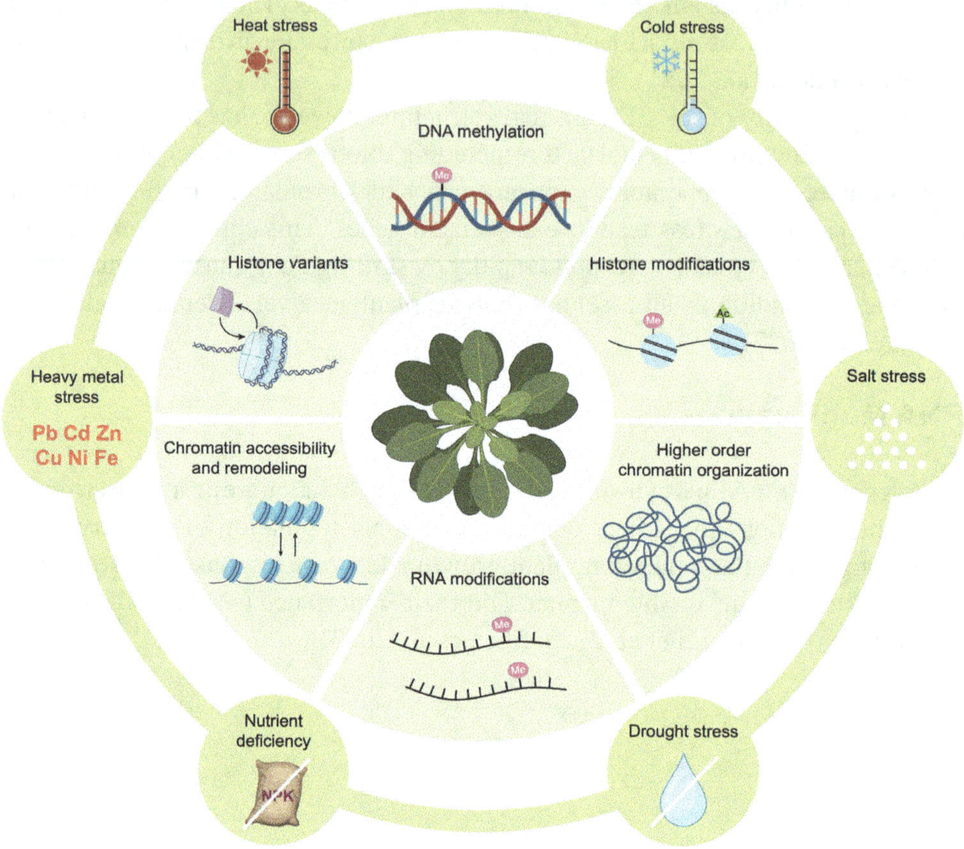

(Ma et al., 2024)

EPIGENETIC REGULATION OF PLANT RESPONSE TO ABIOTIC STRESS

Extreme Temperatures

Plants modulate their development upon exposure to different temperatures in a context- and temperature range-specific manner, from sub-optimal conditions to extremes. Thermomorphogenesis is adaptive responses in species like Arabidopsis that is induced by mild temperature elevation (up to 30 °C); in contrast, extreme temperatures trigger stress responses. This adaptation is mediated by epigenetic mechanisms, especially histone modifications, variant exchange and chromatin remodeling that modulate gene expression changes to thermal fluctuations. Consequently, these epigenetic processes including histone dynamics and chromatin reorganization underline flexible regulation that promotes thermotolerance in plants (Xue et al., 2021; Perrella et al., 2022; Quint et al., 2023).

Heat Stress

Heat stress leads to decondensation of Arabidopsis heterochromatin, which is followed by alteration of 3D chromatin architecture. Histone deacetylases HD2B and HD2C are involved in the regulation of this process through inhibiting histone hyperacetylation. The same has also been true in other crops like rice and tomato where transcription factors, like HSFA1a help regulate gene expression of heat-stress-sensitive genes (Sun et al., 2020).

Heat stress evokes a genome-wide reactivation of transposable elements (TEs), exemplified by the induction of the ONSEN retrotransposon in Arabidopsis, which is dependent on the transcription factors HSFA1 and HSFA2. GCN5 is an essential heat stress tolerance-related histone acetyltransferase that induces histone acetylation on the promoters of heat-sensitive genes. Post-translational regulation is also emphasized by the RNA demethylase AtALKBH9B, which releases ONSEN RNA from stress granules and triggers its transcription and transposition. These dynamic outlines exemplify how plant responses to heat stress are regulated through epigenetic and post-transcriptional mechanisms whereby plants adapt to increasingly hostile environments (Wang et al., 2023; Yang et al., 2023).

Figure 2. Epigenetic regulation of plant response to extreme temperatures

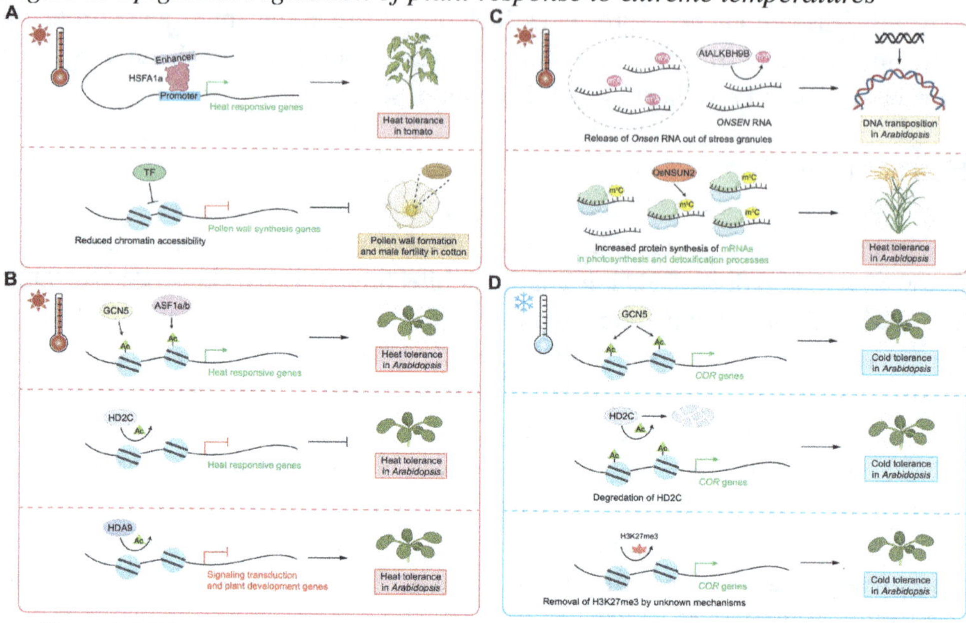

(Ma et al., 2024)

Cold Stress

Cold stress, including chilling temperature (0–15°C) and freezing (≤0°C), represents damaging factor in plant growth, which can cause permanent cell damage due to ice crystals formation. In response to this stress, the CBF-COR signaling pathway directs plant response through the activation of COR genes by CBF transcription factors and facilitates frost resistance (Banerjee et al., 2017; Adhikari et al., 2022). DNA methylation overrides this process: studies in indispensable gene-improving species have found an association between enhanced freezing tolerance and reduced ICE1- locus methylation (Ashapkin et al., 2020; Antoniou-Kourounioti et al., 2021). In addition, the balances of histone acetylation and deacetylation is essential for COR expression upon cold. These mechanisms, shown (D), involve chromatin modifications such as the methylation of histone H3K27me3 and the activity of chromatin remodelers (e.g., PICKLE [PKL]), which modulates gene expression under heat stress.

Figure 3. Different pathways and genes involved in cold stress tolerance in plants

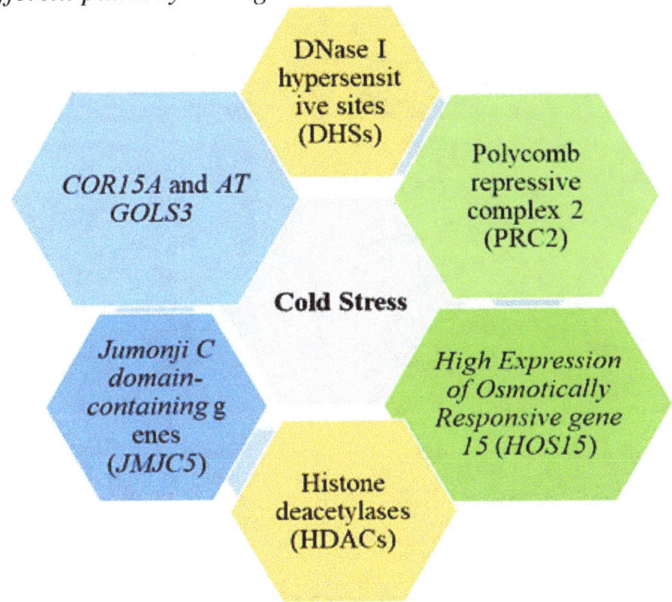

(Shilpa et al., 2024)

Water Stress

Plants use different epigenetic strategies to modulate tolerance to water stress (early and late changes in DNA and histone methylation or in RNA transcription and stability). Stable resistance genes include processes such as DNA methylation, while the expression of stress-sensitive genes is modulated by histone deacetylation. In addition, specific enzymes and transcription factors, such as HDA6 and MYB96, regulate physiological responses to the ABA required for stomatal closure. RNA modifications, including m6A methylation, have also been shown to regulate the translation of resistance genes. These dynamic processes facilitate adaptive gene regulation in plants, which is fundamental for improved drought resistance (Ma et al., 2024; Shilpa et al., 2024).

Figure 4. Epigenetic regulation of plant response to water stress

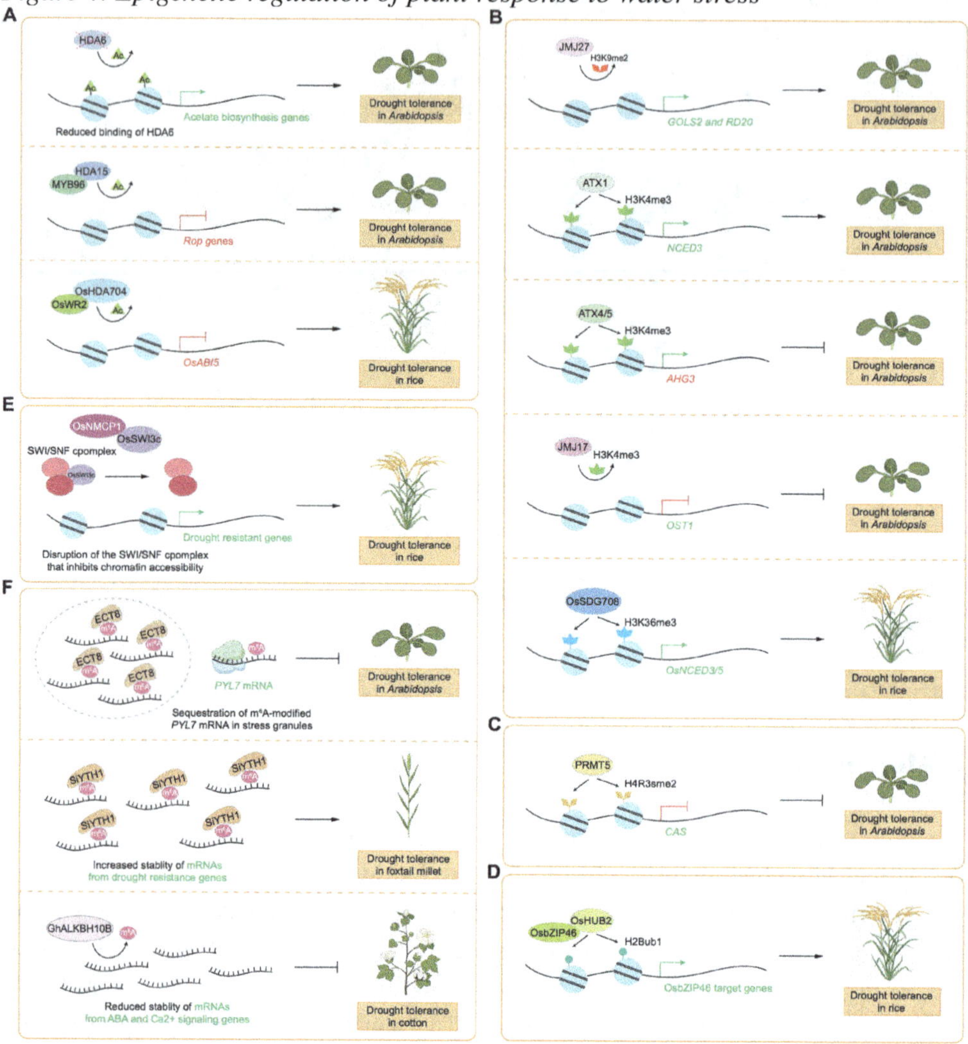

(Ma et al., 2024)

Salt Stress

High salinity induces ionic and osmotic stress in most plants, especially crops that are salt-sensitive and activates the SOS and ABA signaling pathways (Miryeganeh et al., 2022). Salt stress causes DNA and histone methylation at the epigenetic level that induces resistance gene expression (Gupta and Garg, 2023). Salt stress triggers the aforementioned GCN5 gene for cell wall reinforcement in Arabidopsis and other alterations delineating salt tolerance in rice, wheat and soyabean (Wang et al., 2020). Histone acetylation and deacetylation events regulate the expression of essential genes for germination and abiotic-stresses resistance (Ullah et al., 2021).

Figure 5. Epigenetic regulation of plant response to salt stress

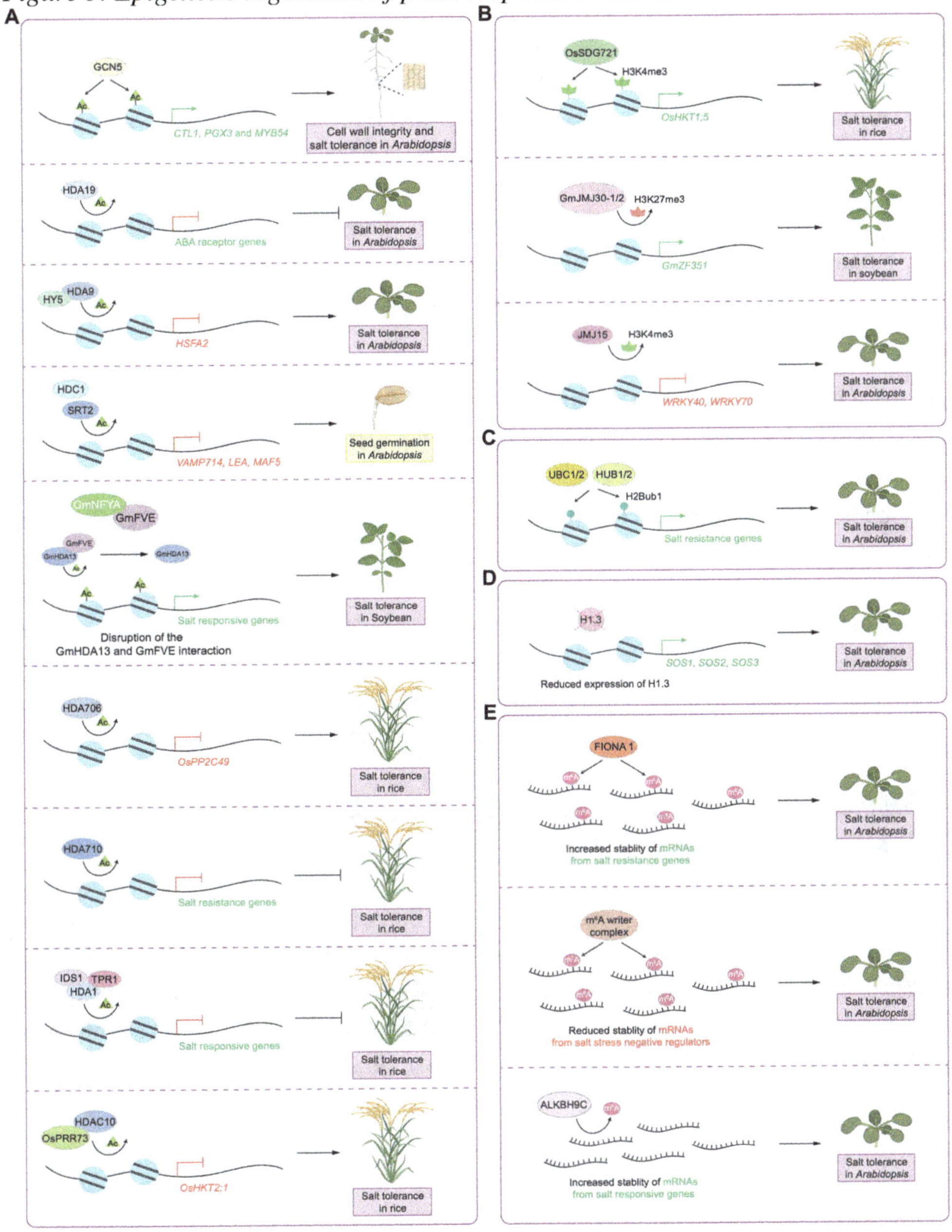

(Ma et al., 2024)

Heavy Metals

While many heavy metals are essential for plant growth (iodine, iron, copper, manganese, molybdenum, nickel and zinc), they can be toxic if they accumulate excessively in the environment as well as other harmful elements (e.g. chromium, cadmium and lead) (Feng et al., 2023). Those metals interfere with plant metabolism (e.g. enzyme activity, protein structure and membrane integrity) (Fan et al., 2020). Increases in DNA methylation in rice and Arabidopsis under cadmium stress, but repression of demethylase genes, probably enhances cadmium tolerance (Cheng et al., 2021). Metal detoxification transporters and DNA demethylation are highly induced in resistant wheat after heavy metal treatments, providing evidence that demethylation is involved in resistance mechanisms. Cadmium is also known to cause demethylation of the metal transporter OsZIP1 and m6A RNA, but the role of these events in stress response has not been studied (Han et al., 2023).

EXAMPLES OF PLANTS IN WHICH EPIGENETIC MODIFICATIONS HAVE BEEN OBSERVED

Generally speaking, according to recent studies on epigenetic modifications, rice (*Oryza sativa*) (Çelik et al., 2019), maize (*Zea mays*) (Qian et al., 2019), wheat (*Triticum aestivum*) (Zhang et al., 2023) and barley (*Hordeum vulgare*) (Azarin et al., 2024) show epigenetic responses under stress conditions, enabling adaptations and resilient development in the face of changing environments, because their life cycle and genome influence the retention and transmission of epigenetic marks.

COMPUTER LANGUAGES FOR PROCESSING EPIGENETIC DATABASES

Python and R are the most commonly used languages for processing and analyzing epigenetic data, thanks to their robust scientific and statistical libraries (pandas, scikit-learn for Python, and Bioconductor for R). Other languages such as SQL are also used to manage large databases. Big data technologies such as Apache Spark and Hadoop can also be mentioned for large-scale data processing (Farooq et al., 2020; Ascensión and Araúzo-Bravo, 2020).

MACHINE LEARNING MODELS AND FUTURE PERSPECTIVES

The use of machine learning algorithms to interpret massive epigenetic data, such as pattern recognition in DNA methylation and histone modification profiles. A recent study showed that machine learning models, in particular Random Forest, were optimized and demonstrated superior accuracy in predicting stress-responsive signatures (Panahi and Golkari, 2024).

Future research in epigenetics aims to extend studies beyond model plants to include crops with complex genomes such as wheat and maize. In addition, the creation of recombinant epigenetic lines (epiRILs) could improve specific traits, such as yield and stress tolerance. More integration of eQTN and eQTM data will help to reveal causal relationships between genetic variation, methylation and gene expression. It is also vital to gain a better understanding of the role of chromatin remodelling complexes in regulating gene expression under stress. Finally, studying the interactions between epigenetic modifications and responses to environmental stresses will enable crops to be adapted to climate change, using advanced technologies including bioinformatics and gene editing with CRISPR/Cas (Lu et al., 2020; Pedro et al., 2021).

CONCLUSION

Molecular insights into plant reactions to thermic and freezing shock is imperative for devising crop augmentation and biodiversity conservation approaches because of climatic-induced challenges. Thus, heat shock proteins (HSPs) and transcription factors (HSFs) are involved in thermotolerance, and epigenetic modifications including DNA methylation and histone modifications regulate cold stress responses. This fine functional connectivity among these mechanisms helps planteadapting to the environment fluctuation and with positive growth. Para: Understanding these processes more deeply will allow us to better predict the effects of climate change on ecosystems and create new ways to increase crop resilience and protect our food security.

REFERENCES

Adhikari, L., Baral, R., Paudel, D., Min, D., Makaju, S. O., Poudel, H. P., Acharya, J. P., & Missaoui, A. M. (2022). Cold stress in plants: Strategies to improve cold tolerance in forage species. *Plant Stress, 4*, 100081. DOI: 10.1016/j.stress.2022.100081

Antoniou-Kourounioti, R. L., Zhao, Y., Dean, C., & Howard, M. (2021). Feeling every bit of winter – distributed temperature sensitivity in vernalization. *Frontiers in Plant Science, 12*, 628726. DOI: 10.3389/fpls.2021.628726 PMID: 33584778

Ascensión, A. M., & Araúzo-Bravo, M. J. (2020). BigMPI4py: Python module for parallelization of big data objects discloses germ layer specific DNA demethylation motifs. *IEEE/ACM Transactions on Computational Biology and Bioinformatics, 19*(3), 1507–1522. DOI: 10.1109/TCBB.2020.3043979 PMID: 33301409

Ashapkin, V. V., Kutueva, L. I., Aleksandrushkina, N. I., & Vanyushin, B. F. (2020). Epigenetic mechanisms of plant adaptation to biotic andabiotic stresses. *International Journal of Molecular Sciences, 21*(20), 7457. DOI: 10.3390/ijms21207457 PMID: 33050358

Azarin, K., Usatov, A., Minkina, T., Duplii, N., Fedorenko, A., Plotnikov, A., Mandzhieva, S., Kumar, R., Yong, J. W. H., Sehar, S., & Rajput, V. D. (2024). Evaluating the phytotoxicological effects of bulk and nano forms of zinc oxide on cellular respiration-related indices and differential gene expression in *Hordeum vulgare* L. *Ecotoxicology and Environmental Safety, 282*, 116670. DOI: 10.1016/j.ecoenv.2024.116670 PMID: 38981388

Banerjee, A., Wani, S. H., & Roychoudhury, A. (2017). Epigenetic control of plant cold responses. *Frontiers in Plant Science, 8*, 1643. DOI: 10.3389/fpls.2017.01643 PMID: 28983309

Cappannini, A., Ray, A., Purta, E., Mukherjee, S., Boccaletto, P., Moafinejad, S.N., Lechner, A., Barchet, C., Klaholz Bruno, P., Stefaniak, F., Bujnicki, J.M. MODOMICS: a database of RNA modifications and related information. 2023 update. Nucleic Acids Research. 52, D239–D244. https://doi.org/DOI: 10.1093/nar/gkad1083

Çelik, Ö., Meriç, S., Ayan, A., & Atak, Ç. (2019). Epigenetic analysis of WRKY transcription factor genes in salt stressed rice (*Oryza sativa* L.) plants. *Environmental and Experimental Botany, 159*, 121–131. DOI: 10.1016/j.envexpbot.2018.12.015

Cheng, Q., Wang, P., Wu, G., Wang, Y., Tan, J., Li, C., Zhang, X., Liu, S., Huang, S., Huang, T., Yang, M., He, H., & Bian, J. (2021). Coordination of m6A mRNA methylation and gene transcriptome in rice response to cadmium stress. *Rice (New York, N.Y.), 14*, 62. DOI: 10.1186/s12284-021-00502-y PMID: 34224034

Cheong, J. J. (2021). Genetic and epigenetic changes in plants in response to abiotic stress. *Genes*, *12*(10), 1603. DOI: 10.3390/genes12101603 PMID: 34680997

Cho, L. H., Yoon, J., Wai, A. H., & An, G. (2018). Histone deacetylase 701 (HD-T7induces flowering in rice by modulating expression of OsIDS1. *Molecules and Cells*, *41*(7), 665–675. DOI: 10.14348/molcells.2018.0148 PMID: 29991671

Dinkar, V., Pandey, S., Kumar, A., Shiv, A., Lal, D., Bharati, A., Joshi, A., Adhikari, S., Singh, A., Pandey, P. K., & Chandra, A. K. (2024). Epigenetic regulations under plant stress: A cereals perspective. *Environmental and Experimental Botany*, *220*, 105688. DOI: 10.1016/j.envexpbot.2024.105688

Fan, S. K., Ye, J. Y., Zhang, L. L., Chen, H. S., Zhang, H. H., Zhu, Y. X., Liu, X. X., & Jin, C. W. (2020). Inhibition of DNA demethylation enhances plant tolerance to cadmium toxicity by improving iron nutrition. *Plant, Cell & Environment*, *43*(1), 275–291. DOI: 10.1111/pce.13670 PMID: 31703150

Farooq, A., Grønmyr, S., Ali, O., Rognes, T., Scheffler, K., Bjørås, M., & Wang, J. (2020). HMST-Seq-Analyzer: A new python tool for differential methylation and hydroxymethylation analysis in various DNA methylation sequencing data. *Computational and Structural Biotechnology Journal*, *18*, 2877–2889. DOI: 10.1016/j.csbj.2020.09.038 PMID: 33163148

Feng, D., Wang, R., Sun, X., Liu, L., Liu, P., Tang, J., Zhang, C., & Liu, H. (2023). Heavy metal stress in plants: Ways to alleviate with exogenous substances. *The Science of the Total Environment*, *897*, 165397. DOI: 10.1016/j.scitotenv.2023.165397 PMID: 37429478

Gahlaut, V., Samtani, H., & Khurana, P. (2020). Genome-wide identification and expression profiling of cytosine-5 DNA methyltransferases during drought and heat stress in wheat (*Triticum aestivum*). *Genomics*, *112*(6), 4796–4807. DOI: 10.1016/j.ygeno.2020.08.031 PMID: 32890700

Gallusci, P., Agius, D. R., Moschou, P. N., Dobránszki, J., Kaiserli, E., & Martinelli, F. (2023). Deep inside the epigenetic memories of stressed plants. *Trends in Plant Science*, *28*(2), 142–153. DOI: 10.1016/j.tplants.2022.09.004 PMID: 36404175

Gaude, A.A., Siqueira, R.H., Botelho, S.B., Jalmi, S.K. 2024. Epigenetic arsenal for stress mitigation in plants. Biochimica et Biophysica Acta (BBA)-General Subjects. 1868 (7), 130620. https://doi.org/DOI: 10.1016/j.bbagen.2024.130620

Gupta, K., & Garg, R. (2023). Unravelling differential DNA methylation patterns in genotype dependent manner under salinity stress response in chickpea. *International Journal of Molecular Sciences, 24*(3), 1863. DOI: 10.3390/ijms24031863 PMID: 36768187

Haider, S., Iqbal, J., Shaukat, M., Naseer, S., Mahmood, T. 2021. The epigenetic chromatin-based regulation of somatic heat stress memory in plants. 27, 100318. https://doi.org/DOI: 10.1016/j.plgene.2021.100318

Han, X., Wang, J., Zhang, Y., Kong, Y., Dong, H., Feng, X., Li, T., Zhou, C., Yu, J., Xin, D., Chen, Q., & Qi, Z. (2023). Changes in the m6A RNA methylome accompany the promotion of soybean root growth by rhizobia under cadmium stress. *Journal of Hazardous Materials, 441*, 129843. DOI: 10.1016/j.jhazmat.2022.129843 PMID: 36113351

Kappel, C., Friedrich, T., Oberkofler, V., Jiang, L., Crawford, T., Lenhard, M., & Bäurle, I. (2023). Genomic and epigenomic determinants of heat stress-induced transcriptional memory in Arabidopsis. *Genome Biology, 24*, 129. DOI: 10.1186/s13059-023-02970-5 PMID: 37254211

Klemm, S. L., Shipony, Z., & Greenleaf, W. J. (2019). Chromatin accessibility and the regulatory epigenome. *Nature Reviews. Genetics, 20*, 207–220. DOI: 10.1038/s41576-018-0089-8 PMID: 30675018

Long, Y., Wendel, J. F., Zhang, X., & Wang, M. (2023). Evolutionary insights into the organization of chromatin structure and landscape of transcriptional regulation in plants. *Trends in Plant Science, 29*(6), 638–649. DOI: 10.1016/j.tplants.2023.11.009 PMID: 38061928

Lu, W., Xiao, L., Quan, M., Wang, Q., El-Kassaby, Y. A., Du, Q., & Zhang, D. (2020). Linkage-linkage disequilibrium dissection of the epigenetic quantitative trait loci (epiQTLs) underlying growth and wood properties in Populus. *The New Phytologist, 225*(3), 1218–1233. DOI: 10.1111/nph.16220 PMID: 31560799

Ma, L., Xing, L., Li, Z., & Jiang, D. (2024). Epigenetic control of plant abiotic stress responses. *Journal of Genetics and Genomics*. Advance online publication. DOI: 10.1016/j.jgg.2024.09.008 PMID: 39322116

Millán-Zambrano, G., Burton, A., Bannister, A. J., & Schneider, R. (2022). Histone post-translational modifications — Cause and consequence of genome function. *Nature Reviews. Genetics, 23*, 563–580. DOI: 10.1038/s41576-022-00468-7 PMID: 35338361

Miryeganeh, M., Marlétaz, F., Gavriouchkina, D., & Saze, H. (2022). De novo genome assembly and *in natura* epigenomics reveal salinity-induced DNA methylation in the mangrove tree *Bruguiera gymnorhiza. The New Phytologist, 233*(5), 2094–2110. DOI: 10.1111/nph.17738 PMID: 34532854

Morgan, M. A. J., & Shilatifard, A. (2020). Reevaluating the roles of histone-modifying enzymes and their associated chromatin modifications in transcriptional regulation. *Nature Genetics, 52*, 1271–1281. DOI: 10.1038/s41588-020-00736-4 PMID: 33257899

Panahi, B., & Golkari, S. (2024). Identification of gene expression signature for drought stress response in barley (*Hordeum vulgare* L.) using machine learning approach. *Current Plant Biology, 100416*. Advance online publication. DOI: 10.1016/j.cpb.2024.100416

Pedro, D. L. F., Amorim, T. S., Varani, A., Guyot, R., Domingues, D. S., & Paschoal, A. R. (2021). An atlas of plant transposable elements. *F1000 Research, 10*, e1194. DOI: 10.12688/f1000research.74524.1 PMID: 35035898

Perrella, G., Bäurle, I., & van Zanten, M. (2022). Epigenetic regulation of thermomorphogenesis and heat stress tolerance. *The New Phytologist, 234*(4), 1144–1160. DOI: 10.1111/nph.17970 PMID: 35037247

Pratx, L., Wendering, P., Kappel, C., Nikoloski, Z., & Bäurle, I. (2023). Histone retention preserves epigenetic marks during heat stress-induced transcriptional memory in plants. *The EMBO Journal, 42*, e113595. DOI: 10.15252/embj.2023113595 PMID: 37937667

Qian, Y., Hu, W., Liao, J., Zhang, J., & Ren, Q. (2019). The Dynamics of DNA methylation in the maize (*Zea mays* L.) inbred line B73 response to heat stress at the seedling stage. *Biochemical and Biophysical Research Communications, 512*(4), 742–749. DOI: 10.1016/j.bbrc.2019.03.150 PMID: 30926168

Quint, M., Delker, C., Balasubramanian, S., Balcerowicz, M., & Casal, J. J.. (2023). 25 Years of thermomorphogenesis research: Milestones and perspectives. *Trends in Plant Science, 28*(10), 1098–1100. DOI: 10.1016/j.tplants.2023.07.001 PMID: 37574427

Sajjanar, B., Aalam, M. T., Khan, O., Dhara, S. K., Ghosh, J., Gandham, R. K., Gupta, P. K., Chaudhuri, P., Dutt, T., Singh, G., & Mishra, B. P. (2024). Genome-wide DNA methylation profiles regulate distinct heat stress response in zebu (*Bos indicus*) and crossbred (*Bos indicus* × *Bos taurus*) cattle. *Cell Stress & Chaperones, 29*(4), 603–614. DOI: 10.1016/j.cstres.2024.06.005 PMID: 38936463

Shang, J. Y., & He, X. J. (2022). Chromatin-remodeling complexes: Conserved and plant-specific subunits in Arabidopsis. *Journal of Integrative Plant Biology*, *64*(2), 499–515. DOI: 10.1111/jipb.13208 PMID: 34964264

Shen, L., Ma, J., Li, P., Wu, Y., & Yu, H. (2023). Recent advances in the plant epitranscriptome. *Genome Biology*, *24*, 43. DOI: 10.1186/s13059-023-02872-6 PMID: 36882788

Shen, Y., Wei, W., & Zhou, D. X. (2015). Histone acetylation enzymes coordinate metabolism and gene expression. *Trends in Plant Science*, *20*(10), 614–621. DOI: 10.1016/j.tplants.2015.07.005 PMID: 26440431

Shilpa., Thakur, R., Prasad, P. (2024). Epigenetic regulation of abiotic stress responses in plants. Biochimica et Biophysica Acta (BBA) - General Subjects. 1868 (9), 130661. https://doi.org/DOI: 10.1016/j.bbagen.2024.130661

Song, Z. T., Zhang, L. L., Han, J. J., Zhou, M., & Liu, J. X. (2021). Histone H3K4 methyltransferases SDG25 and ATX1 maintain heat-stress gene expression during recovery in Arabidopsis. *The Plant Journal*, *105*(5), 1326–1338. DOI: 10.1111/tpj.15114 PMID: 33278042

Sun, L., Jing, Y., Liu, X., Li, Q., Xue, Z., Cheng, Z., Wang, D., He, H., & Qian, W. (2020). Heat stress-induced transposon activation correlates with 3D chromatin organization rearrangement in Arabidopsis. *Nature Communications*, *11*, 1886. DOI: 10.1038/s41467-020-15809-5 PMID: 32312999

Ullah, F., Xu, Q., Zhao, Y., & Zhou, D. X. (2021). Histone deacetylase HDA710 controls salt tolerance by regulating ABA signaling in rice. *Journal of Integrative Plant Biology*, *63*(3), 451–467. DOI: 10.1111/jipb.13042 PMID: 33289304

Wang, J., Nan, N., Li, N., Liu, Y., Wang, T. J., Hwang, I., Liu, B., & Xu, Z. Y. (2020). A DNA methylation reader–chaperone regulator–transcription factor complex activates OsHKT1;5 expression during salinity stress. *The Plant Cell*, *32*(11), 3535–3558. DOI: 10.1105/tpc.20.00301 PMID: 32938753

Wang, N., Wang, Z., Tzourtzou, S., Wang, X., Bi, X., Leimeister, J., Xu, L., Sakamoto, T., Matsunaga, S., & Schaller, A.. (2023). The plant nuclear lamina disassembles to regulate genome folding in stress conditions. *Nature Plants*, *9*, 1081–1093. DOI: 10.1038/s41477-023-01457-2 PMID: 37400513

Wu, J., Ichihashi, Y., Suzuki, T., Shibata, A., Shirasu, K., Yamaguchi, N., & Ito, T. (2019). Abscisic acid-dependent histone demethylation during postgermination growth arrest in Arabidopsis. *Plant, Cell & Environment*, *42*(7), 2198–2214. DOI: 10.1111/pce.13547 PMID: 30859592

Xiao, X., Zhang, J., Li, T., Fu, X., Satheesh, V., Niu, Q., Lang, Z., Zhu, J. K., & Lei, M. (2019). A group of SUVH methyl-DNA binding proteins regulate expression of the DNA demethylase ROS1 in Arabidopsis. *Journal of Integrative Plant Biology*, *61*(2), 110–119. DOI: 10.1111/jipb.12768 PMID: 30589237

Xu, Q., Liu, Q., Chen, Z., Yue, Y., Liu, Y., Zhao, Y., & Zhou, D. X. (2021). Histone deacetylases control lysine acetylation of ribosomal proteins in rice. *Nucleic Acids Research*, *49*(8), 4613–4628. DOI: 10.1093/nar/gkab244 PMID: 33836077

Xue, M., Zhang, H., Zhao, F., Zhao, T., Li, H., & Jiang, D. (2021). The INO80 chromatin remodeling complex promotes thermomorphogenesis by connecting H2A. Z eviction and active transcription in Arabidopsis. *Molecular Plant*, *14*(11), 1799–1813. DOI: 10.1016/j.molp.2021.07.001 PMID: 34242850

Yajnik, K. N., Singh, I. K., & Singh, A. (2024). lncRNAs and epigenetics regulate plant's resilience against biotic stresses. *Plant Physiology and Biochemistry*, *214*, 108892. DOI: 10.1016/j.plaphy.2024.108892 PMID: 38964086

Yang, F., Sun, Y., Du, X., Chu, Z., Zhong, X., & Chen, X. (2023). Plant-specific histone deacetylases associate with ARGONAUTE4 to promote heterochromatin stabilization and plant heat tolerance. *The New Phytologist*, *238*(1), 252–269. DOI: 10.1111/nph.18729 PMID: 36631970

Zhang, H., & Zhu, J. K. (2024). Epigenetic gene regulation in plants and its potential applications in crop improvement. *Nature Reviews. Molecular Cell Biology*. Advance online publication. DOI: 10.1038/s41580-024-00769-1 PMID: 39192154

Zhang, S., Yan, C., Lu, T., Fan, Y., Ren, Y., Zhao, J., Shan, X., Guan, Y., Song, P., Li, D., & Hu, H. (2023). New insights into molecular features of the genome-wide AOX family and their responses to various stresses in common wheat (*Triticum aestivum* L.). *Gene*, *888*, 147756. DOI: 10.1016/j.gene.2023.147756 PMID: 37659597

Zhou, H., Liu, Y., Liang, Y., Zhou, D., Li, S., Lin, S., Dong, H., & Huang, L. (2020). The function of histone lysine methylation related SET domain group proteins in plants. *Protein Science : a Publication of the Protein Society*, *29*(5), 1120–1137. DOI: 10.1002/pro.3849 PMID: 32134523

Chapter 12
Harnessing Epigenetics for the Breeding of Stress-Tolerant Crops

Sadia Zafar
https://orcid.org/0000-0003-0879-459X
University of Education, Faisalabad, Pakistan

Rizwan Rasheed
Government College University, Faisalabad, Pakistan

Muhammad Arslan Ashraf
Government College University, Faisalabad, Pakistan

Inam Mehdi Khan
Division of Science and Technology, University of Education, Lahore, Pakistan

ABSTRACT

The transcriptome's activity depends on both the genomic profile and epigenetic regulation. Epigenetic changes are essential for phenotypic variation and stress resistance triggered by environmental stressors. Stress signals regulate genes via histone modifications, DNA methylation, and miRNA. The epigenetics field has advanced rapidly over two decades. While modern plant breeding is efficient, it falls short of traditional crop breeding demands. Epigenetic methods, such as novel epiallele generation and RNAi techniques, can enhance crop yield and nutritional value. For instance, RNAi has improved barley, wheat, and soybean by reducing anti-nutritional factors and enhancing taste, such as silencing the Omega-3 fatty acid desaturase gene to modify linoleic acid content. HvTX1 gene expression in barley

DOI: 10.4018/979-8-3693-8648-4.ch012

under drought stress boosts tolerance, while HAT genes in rice activate under water and heat stress. These insights aid in developing stress-tolerant crops and deepen our understanding of plant adaptation mechanisms.

INTRODUCTION

Epigenetics is an interesting branch of genetics which provides a theoretical shift in the concept of hereditary material (DNA) and the phenotype, mystery to scientists for many years. Based on Arthur Riggs and his colleagues, epigenetics means "the study of the heritable alterations in gene expression that take place through mitosis and meiosis without change in the DNA sequence" (Miryeganeh 2021). Despite they are distinct from conventional inheritance modes, DNA methylation, histone alterations, chromatin restructuring, and sRNAs are inheritable (Fallet 2024). If we compare plants with mammals, there are certain advantages of using epigenetics in plants. Since the plants grow from somatic tissues they do not require a specific germ tissue as is found in mammals. Further, during gametogenesis, fertilization, and some phases of embryo development, plants possess efficient systems that ensure epigenetic information transfer and stability (Jaligot and Rival, 2015). It is also possible to provide examples of the epigenetic states' instability. Hence, it is crucial to identify such epigenetic states that could be stably maintained in order to control and regulate stable plant development across generations (Gallusci et al., 2017). Plant response to abiotic stresses is always negative in terms of growth and development, and this leads to huge reductions in yields and income. Since the majority of epigenetic studies has been performed on model plants, especially on Arabidopsis (Provart et al., 2016), the accrued knowledge about the epigenetic regulation of environmental responses has stimulated interest in epigenetic regulation of crop resistance to abiotic stress (Figure 1). In order to improve crop adaptability to environmental changes and climate-smart crop breeding, we discuss current research on epigenetic processes, specifically as they relate to agricultural stress response.

Figure 1. Importance of epigenetic processes to key plant processes

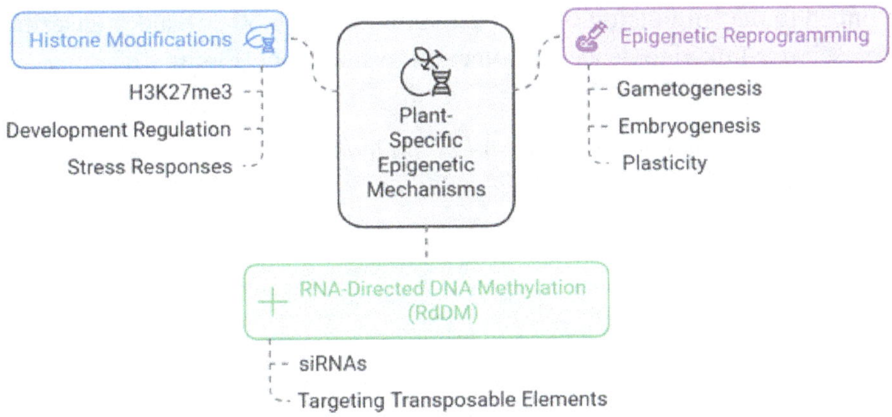

The figure emphasizes how important epigenetic processes are in controlling important agronomic characteristics of crops. DNA methylation, histone modifications, and short RNAs in gene expression affect several plant processes, including germination, growth, blooming, and fruit development. These factors in turn affect agricultural productivity, yields, and quality.

EPIGENETIC MECHANISMS

Chromatin shape regulate gene expression since it determines how open DNA sequences, incorporating coding and regulatory zones, are to transcriptional components (Carlberg 2024). Location of nucleosome is important in determining the ability of the DNA to be transcribed. An octamer of histones, including two copies of each of the histones H2A, H2B, H3, and H4, makes up the individual core nucleosome. It is surrounded by two loops of around 146 base pairs of DNA. The spatial dimensions of the linker DNAs, to which the linker histone H1 is bound, change with the level of chromatin compaction; 33,34. Additional modifications of histone tails, which are covalently attached to the amino terminal of the histone globular domain and project from the nucleosome core particle, lead to changes in the interaction of histones with DNA. Combined, these modifications are called the histone code and are critical regulatory mechanisms in DNA accessibility and chromatin structure (Foroozani et al., 2022). Reversible histone changes, such as histone acetylation and deacetylation and histone methylation and demethylation, have been shown to occur in plants. Acetyl groups, such as histone 3 lysine 14 acetylation (H3K14ac), histone 3 lysine 9 acetylation (H3K9ac), and histone 3 lysine 36 acetylation (H3K36ac), are linked to the stimulation of gene activation. On the other hand, transcriptional suppression is linked to deacetylation at these same loci (Figure 2).

Figure 2. Epigenetic mechanisms in plant include DNA methylation, histone modification, and RNA-directed DNA methylation

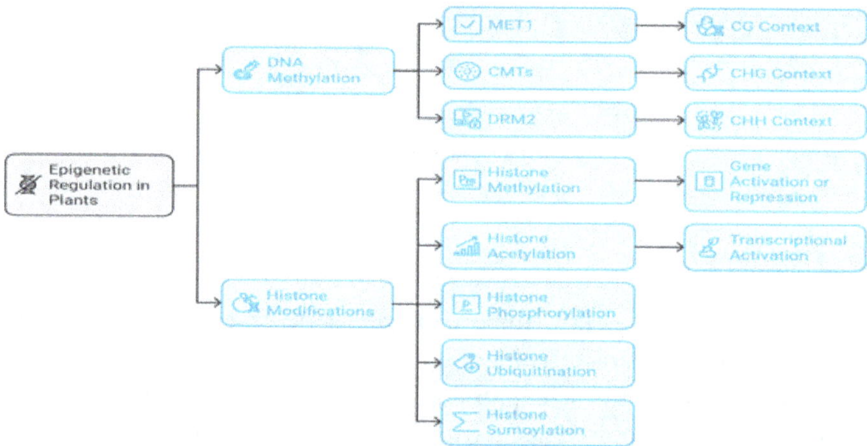

The location of the alteration determines whether histone methylation increases or decreases gene expression. Histone H3 methylation at lysine 4 (H3K4) and lysine 36 (H3K36) is linked to gene activation, while pre-occupation of histones H3 at lysine 9 (H3K9) and lysine 27 (H3K27) has been linked to gene suppression and heterochromatin formation. H3K9me1 and H3K9me2 in plants mark regions, which consist of repetitive sequences and TE and maintain transcriptional quiescence. In addition, while its trimethylation form H3K27me3 is repressive in euchromatin region, monomethylation of H3 lysine 27 (H3K27me1) is frequent in heterochromatin. For instance, H3K27me3 is generally localized over the whole length of transcribed region of developmentally inactive genes (Montgomery et al., 2020). It is possible that methylation that occurs in CHG, CG and CHH sites through various mechanisms influences chromatin at the DNA level in plants. The authors Shen et al. (2021) also concluded that the enzymes CHROMOMYTHYLASE 3 (CMT3) and METHYLTRANSFERASE 1 (MET1) are necessary for the maintenance of either CHG and CG methylation respectively. CHH methylation is initiated from scratch through two possible pathways. The RDDM pathway requires 24-nucleotide small interfering RNA (siRNA) which targets specific genomic sites with AGO proteins that subsequently methylates by DRM2. The second mechanism involves CMT2, which has been shown to associate with DDM1 at chromatin domains containing H1 histone (Akinmusola et al., 2023). Additionally, a class of bifunctional enzymes known as methyl cytosine glycosylases apurinic/apyrimidinic lyases that use the base excision repair mechanism may actively remove DNA methylation marks (Sanders 2018). In particular, DNA methylation might both activate transposable elements and transposable element associated genes depending on environmental signals, regulate imprinting, and modulate gene expression (Lucibelli et al., 2022). New findings on stress induced changes regulated by sRNAs have advanced understanding of epigenetic regulation of transcription in model plants (Kumar & Mohapatra, 2021). Several short ncRNA types have been identified as significant players in de novo DNA methylation, gene silencing and other epigenetic events. Plants generate sRNAs in different ways (Liang et al., 2020). Besides, transcripts longer than 200 nucleotides are referred to as long non-coding RNAs (lncRNAs) together with short RNAs (sRNAs). These transcripts may bind with proteins and nucleic acids and may act as a scaffold by which the nucleus forms specific functional structures.

THE EPIGENETICS ROLE IN GENE EXPRESSION UNDER ABIOTIC STRESS

Stresses arising from the surroundings include water and nutrient deficiencies; temperature and light intensity; and soil characteristics such as salinity and heavy metal content. Genetic as well as epigenetic stress tolerance mechanisms are found in plants against single or multiple stress factors and their combinations due to climate change and increased unpredictability of the climatic events (Miryeganeh, 2021). To comprehend this complexity in cropping systems, it is necessary to understand the genetic and epigenetic underpinnings of the crops' reactions to environmental changes. In one such study, Vuković et al. (2024), the authors used heat stress priming as the experimental paradigm to examine how Arabidopsis remembered environmental stresses. This was done in order to find genes that are solely involved in memory for heat stress and not in the first reaction to heat. The researchers then discovered that once they identified the FORGETTER1 (FGT1) gene, this group of heat-inducible genes were directly binding to the FGT1 protein. This interaction ensures that genes with such functions are constantly accessible and active since their chromatin structures are modified (Figure 3).

Their work may provide the basis for approaches to new approaches in crop breeding programs aimed at increasing resistance to abiotic stress. Breeding applications require the understanding of the regulations governing epigenetics as well as the stability and heritability of the epigenetic signs (Varotto et al., 2020). That is why the aforementioned research provides a solid foundation for additional development of knowledge concerning various processes affecting crop and plant production variability.

Figure 3. Epigenetic regulation of Abiotic stresses in crop plants

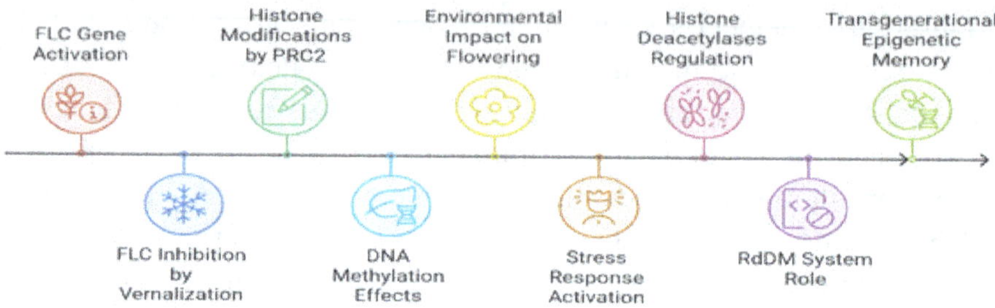

The three mechanisms of epigenetic control are DNA methylation, histone modification and RNA interference.

DNA METHYLATION

The process of attaching a methyl group in the fifth position to the pyrimidine ring in cytosine residues is referred to as DNA methylation, and is mainly carried out by cytosine methyltransferases (Akhter et al., 2021). Often this mechanism is site-specific and is prominent in CpG dinucleotides. If a methyl group is incorporated into the DNA sequence it forms a binding site for several protein complexes. Thereafter, these complexes work with the DNA's Histone scaffolds packaging to modulate the activity of the respective gene. Histones undergo multiplicity of post-translational modifications and exist in multiple forms that are expressed in a regulated manner depending on the situation. For instance, as previously established, drought stress in tomatoes generates the linker histone variation H1-S (Scippa et al., 2004). As for histones, they are known and susceptible to changes like acetylation, methylation, sumoylation, ubiquitination, phosphorylation and there are many variants of histones. Acetylation and phosphorylation are mostly associated with the activation of genes while sumoylation and biotinylation are associated with the inactivation of the genes (Hashiguchi & Komatsu, 2016).

RNA interferences (RNAi) is a specific sequence of gene regulation process that is evolutionarily conserved. RNA interference (RNAi) was to perform its role mainly in shielding the body in contrast to viruses and other pathogens, and that its main function was to have strings control on genes. The two basic RNAi types widely existing in plants are siRNA and miRNA (Tiwari & Rajam, 2022). Both are synthesized from double-stranded RNA precursors as short, single-stranded RNA molecules of 20 – 30 nucleotides length. Epigenetic processes are not only shown to have individual effects, but are also interrelated processes with multiple patterns of combined interactions. For instance, siRNA and DNA methylation are related in a certain manner. SUP32/UBP26 belongs to Arabidopsis deubiquitination enzymes and its mutation leads to reduced siRNA-directed DNA methylation and dimethylation of H3K9; many siRNA clusters are highly methylated. As it has been demonstrated that the plant salt stress-inducible protein GmPHD5 can bind to the DNA-binding protein GmGNAT1 to modulate histone acetylation and methylation crosstalk (Miryeganeh, 2021).

Additionally, it has been found that the met1 mutant of *Arabidopsis thaliana* exhibited high levels of H3K9me2 methylation across a number of genes suggesting these changes may be related (Zhao et al., 2022). Innovations on the epiallele and transgenic RNA interferences paving way to epigenetics would make it possible to amplify the favorable plant characteristics concerned with stress protection, superior nutritional values, and yield (Figure 1).

Figure 4. Epigenetic and gene expression through the method of DNA methylation and acetylation in plants

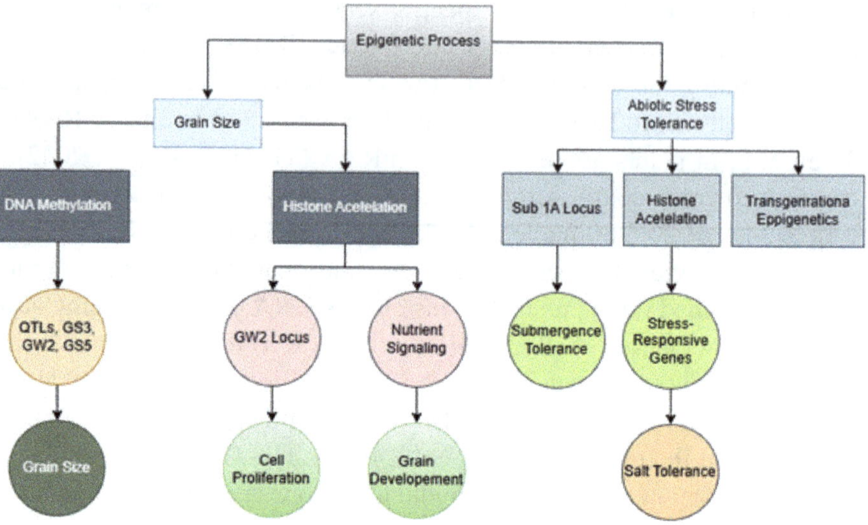

Methylation patterns in DNA evolve when the embryo's cells and tissues get specialized (Akhter et al., 2021). The process is facilitated by a class of DNA methyltransferase enzymes found in eukaryotic organisms and which translocates methyl groups from the methyl donor SAMe to cytosine residues. Gene silence alsow often emerges from the repressive activity of the subsequent 5-methylcytosine (5 mC). Cytosine methylation is not widely distributed across the plant genome including chickpeas; it is preferentially targeted to the repetitive regions present in the form of centromeric region, silent 45S or 5S rRNA gene repeat cluster and in the regions containing transposable elements. In addition, it is also involved in the protein coding regions of developmentally important genes expressed in chickpeas and through a group of divergently regulated enhancers (Mishra et al., 2023). The last pattern of gene body methylation is associated with pre-mRNA splicing and is evolutionarily conserved (Miryeganeh, 2021). As described in the subsequent sections of this article, DNA methylation is deeply involved with apomixis, restriction enzymes, plant immunology, genomic imprinting and gene silence among other functions.

Long-Term Gene Silencing

Patterns of DNA methylation through the process of seed development reprogramming has been linked to long lasting gene silencing, which has phenotypic impact on seed weight and size in a large seed of chickpea line, JGK 3. The understanding of DNA methylation in the regulation of seed size, development and weight was further enhanced with the identification of candidate genes for seed size and weight in relation to hyper-methylation of the CG context within the gene body with differential expression in JGK 3. The buildup of CHH-related DNA methylation in TEs, the subsequent increase in H3K27me3 levels, and the enhanced synthesis of short RNAs in the hyper-methylated TEs during seed development provided evidence of the involvement of the RNA-dependent DNA methylation process (Akhter et al., 2021).

Imprinting of Genes

Expression may be parent of origin specific due to an epigenetic occurrence known as imprinting. This technique of inheritance does not alter the primary DNA sequence; however it includes DNA methylation and histone changes and has no relation to Mendelian genetics. There is for instance genomic imprinting in Arabidopsis, because the quick growth of the endosperm is a positive characteristic (Ashapkin et al., 2020). If this technique is also applied to commercial crops and hybrids, then regeneration may be enhanced and may even transcend the current plant breeding limitations and at the same time secure the longevity of hybrid vigor in successive generations (Akhter et al., 2021).

The levels of Demeter DNA glycosylase domain increases in central cell of the female gametophyte and results in the decrease in MET1 methyltransferase that lead to demethylation of the transposons DNA. Transcripts of the transposons contain a small interfering RNA of 24 nucleotides. This siRNA is then translocated to the egg cell where it triggers some genes to imprint. Likewise, the up-regulation of Demeter molecule results in the DNA of transposons to become demethylated in the male fertilization cell of the vegetative shoot. A 21-nucleotide short interfering RNA is transcribed from these transposons and transported to a sperm cell to induce gene imprinting (Miryeganeh, 2021).

The creation of heritable gene silence may be facilitated by DNA methylation, according to research that examined five legumes: The crops include lotus, chickpeas, soybean, alfalfa and pigeon pea. The study also identified likely DNMTs that might be involved in this process (Garg et al., 2014). Legumes' MTases have been divided into four subfamilies based on their domain structure: DNA, MET, DRM, CMT and nucleotide methyltransferases. The other three MTases are protein of DNA methyltransferase group while DNMT2 is tRNA methyltransferase. Some of these MTases' biological functions have been described with aid of structural genomics that compares plants and animals (Rehman & Tanti, 2020). The comprehensive gene expression profile analysis of methyltransferases (MTases) have been targeted in several research in legumes, elucidating their roles during various developmental stages. MTases are required for the growth and development of plants at all their developmental stages and under different abiotic stresses (Czajka et al., 2021).

Histone Modification and Crop Improvement

Present mostly in eukaryotes and some archaea, histones are a set of proteins that coil around the DNA to form a compact structure called the nucleosome. Arginine and lysine are the common amino acids found in the tail, the N terminal domain of histones (Figure 3) diverse environmental signals lead to diverse covalent modification of histones such as methylation, acetylation, ubiquitination, phosphorylation and biotinylation. These alterations into chromatin structure either increase DNA accessibility for transcription or condense the structure and hinder the attachment of transcriptional apparatus to DNA (Singroha & Sharma, 2019).

Acetylation/Deacetylation

Histone protein, a transcription factor, initiates transcription by acetylating the DNA sequence at the N terminal lysine (Cheong 2021). The attractive force between negatively charged DNA and positively charged histones reduces because when an acetyl group has been added to lysine, histone protein is less positively charged. This in turn leads to the relaxation of the DNA supercoiled structure and transcriptional activation is elicited (Akhter et al., 2021). HATs (Histone acetyl transferases) are those that put on the acetyl group. HATs are classified into five classes in eukaryotes: These have been determined to include the following: The nuclear hormone-related HATs family, TFII250, TATA, binding protein associated factors, MYST, encoding MOZ, Ybf2/Sas3, Sas2, and Tip60, and CBP, CREB binding protein (Figure 4). HATs, however, are classified into four classes in rice and Arabidopsis thaliana: GNAT, CBP, MYST, and TFII250. It has been reported that the number of HATs is not the same in different plants; to date, eight different HATs from plant belonging to the four classes mentioned above have been identified. Elongator is a histone HAT complex with significant sequence homology across the eukaryotic domain. It is formed of six subdomains, ELP1 to ELP6. The elongator HAT complex has been related to the drought responses in Arabidopsis (Miryeganeh, 2021). Not all of the lysine at the alpha-amino group's N-terminal can be acetylated. In various histone proteins, the locations of amino acids shift. For instance, it has been shown that acetylation alterations take place at the lysine residues of histone H3 (K9, K14, K18, K23 and K27) and histone H4 (K5, K8, K12, K16 and K20) in Arabidopsis. Under stress, plants undergo many types of histone acetylation. In A. thaliana under drought stress, H3K9ac is elevated in the promoter regions of the stress-inducible genes RD29A, RD29B, RD20, and RAP2.4 (Rehman & Tanti, 2020).

ABO1 has been shown to be an Arabidopsis homolog of the yeast elongator component ELP1. Supplied evidence was drought-resistant phenotype of abo1 mutant. Further, the authors found that ABA-induced stomata closure was enhanced by the abo1 mutation. Sagor et al. (2016) argue that such findings suggest that the HAT elongator complex is essential for oxidative stress and drought stress response in Arabidopsis. Research shows that nitric oxide affects histone deacetylase within Arabidopsis thaliana, thus enhancing the total histone acetylation. By the functional analysis of the genes, it has been identified that most of the genes are involved in abiotic stress responses only (Ashapkin et al., 2020). In order to determine the interaction of the degree of histone acetylation in cell cycle genes under stress conditions, the maize sample was analyzed and using the bar chart, the total root acetylation levels of maize seedlings was found to be higher than the control group which showed that several genes are activated under stress condition. Interestingly, this increase was more pronounced during heat stress (Wang et al., 2015).

Additionally, the nucleus consistently displayed positive histone modification marks linked to gene activation, such as H3K9ac and H4k5ac, even though heterochromatin knobs and nucleoli only little changed between the control and treatment conditions (Cheong 2021). The stress-responsive gene ZmDREB2A was upregulated in maize seedlings under osmotic stress. Stress produces fast hypermethylation of the promotor region's H3K9 and H4K5, indicating rapid gene activation, according to a ZmDREB2A real-time PCR study (Zhao et al., 2014). It has been shown that the histone deacetylase proteins HD2C and HD6 control the ABA-sensitive genes ABI1 and ABI2 in response to ABA stress by changing the acetylation of histones H3K9 and H3K14 in the promoter region (Saraswat et al., 2017). Zheng et al. (2016), found that Arabidopsis histone deacetylase, or HDA9, regulates the transcription of 60 genes under drought. The ABA sensitive genes ABI1, ABI2, and RD29B showed down-regulated expression in the HDA6 mutant under ABA stress, according to microarray research, suggesting that HDA6 linked to activation of genes (Chen et al., 2010).

HAD9 nonfunctional mutants had a higher fold induction of these genes than the wild type and that 40 of the 60 genes tested had a higher expression level while the rest had a lower expression level these observations may suggest that HDA9 is involved in the deacetylation of the genes. H3K9ac is elevated in had9 mutants, indicating that the promotors in these genes are least acetylated, according to ChIP study of 14 upregulated genes (Zheng et al., 2016). According to a genome-wide analysis of rice's H3K4 trimethylation, 4837 genes have variable methylation, with 910 showing decreased total transcript and 3927 showing increased expression under drought conditions (Saraswat et al., 2017).

Non-Histone Proteins and Their Roles

Plants have non-histone chromosomal proteins, like HMG proteins, that may aid in the regulation of epigenetic genes, much like other eukaryotes. The HMGB family of plant proteins is the most researched and varied grouping (Zhang et al., 2024). This family's members differ in their location, style, degree of expression, and interactions with DNA and other proteins. Mutation and aberrant expression lead to a partial sub-functionalization of the functions of each family members in developmental stages and responses to various stresses (Kumar et al., 2021). The "indirect" demethylation of DNA sequences matching to the identified genes in the central cell of the female gametophyte was attributed to the structure-specific recognition protein (Chandana et al., 2022). Hence, the structure, assembly, and rejection of cohesion are possibly sufficiently shielded (Yan et al., 2013). Some functions can be invested in a few members of the plant family. In addition, the link between the defective meristem silencing 3 or de novo was established as required for the transcription of the DNA dependent RNA polymerase V (Lindermayr et al., 2020).

This is crucial for the production of RdDM proteins, replication factor C1 and replication protein A2 (RPA2) are two of the several mutant screens linked to epigenetic regulators (Chandana et al., 2022). This was also evident when the chromatin mutants with increased phenotypes and the MGO homolog functional loss mutants were combined and the repair of stem cell and meristem occurred. This suggested that a significant number of other nonhistone proteins associated with DNA would also function as either direct or indirect epigenetic modifiers (Pikaard & Scheid, 2014).

Methylation

Two groups of amino acids that are prone to methylation in histones include arginine and lysine. It is possible to methylate various arginine and lysine residues, and those residues may be mono, di, or tri methylated. Acetylation, on the other hand, shows a range of patterns (Akhter et al., 2021). Different types of genes have different responses to methylation; some cause the gene to become active while others lead to the gene's inactivation. For instance, K9 and K27 of H3 is methylated being a repressor while the trimethylation of K4 of H3 enhances transcription. Similar to acetylation, methylation adds methyl groups to histone proteins, altering their hydrophobicity but leaving the net charge unchanged. Histone-DNA interactions may change as a result, or other proteins may develop binding sites that restrict transcription machinery binding and prevent transcription. There are histone lysine methyl transferases (HKMT) that methylate arginine and protein arginine methyl transferases (PRMT) that methylate lysine (Miryeganeh, 2021).

The H3K4me3 methyltransferase AtATX1/SDG27 is a TRITHORAX1 homolog in Arabidopsis. Wider stomatal apertures, improved transpiration, and reduced resilience to drought stress were all seen in AtATX1 mutants. The plant's response to drought and the salicylic acid (SA) pathway are both influenced by AtATX1 (Xu et al., 2020). Moreover, it has been shown that drought stress increases the transcriptional activity of HvTX1, an H3K4 methyltransferase that resembles barley TRX (Shafiq & Khan, 2015). In maize study, 18 genes containing the SET domain were discovered; their expression levels varied in response to drought and salt stress, suggesting that they could be involved in both stress responses (Liu et al., 2021).

DNA Methylation and Crop Improvement

DNA methylation is a specific chemical change caused by the addition of a methyl group to the nitrogenous base in the DNA strand, especially in a sequence-specific way (Figure 5). These nitrogenous bases are mostly cytosine, although they may also be adenine. 5-methyl cytosine is produced when cytosine is methylated. Because it is sequence specific, it may be found at three different positions: CG, CHG, and CHH, where H can be any base except G (Sun et al., 2021). Methylation occurs at all three sequences in plants, while it is mostly seen at CG sequences in mammals (He et al., 2011). It was also found that the overall DNA methylation level in Arabidopsis at different locations is 24% for CG, 6.7% for CHG and 1.7% for CHH. The principal classification of DNA methylation consists of symmetrical and asymmetrical methylation depending on the target sequence. CHH methylation is referred to be asymmetrical, whereas CG and CHG methylation are referred to as symmetrical. DNA methylation exhibits varying behaviors depending on where it occurs, although it is often associated with gene suppression. Whereas gene body methylation has been demonstrated to be associated with gene activation, methylation in TEs and the promoter region of a gene causes gene repression (Kumar, 2019).

An increase in genomic DNA methylation may reduce the forces that drive transcriptome, thus decelerating the metabolic process in the plant. By so doing it is able to accumulate energy when faced with a biotic or abiotic stress and is thus able to overcome the short-term problem. The decrease in methylation of genes linked to resistance, on the other hand, promotes chromatin activation and the production of new genes that provide permanent or long-term stress resistance (Kakoulidou et al., 2021). In order to decrease water loss and boost stress tolerance, CAM plants transition from the C3-photosynthetic cycle to the CAM pathway when they experience a water shortage. The enhancement of satellite DNA hypermethylation and genomic methylation occurs concurrently with this shift to the CAM pathway (Shi et al., 2021). For the change to CAM under water stress, satellite DNA hypermethylation must therefore be accompanied by the formation of a chromatin structure that can control the expression of multiple genes simultaneously. In conditions of water deficit, hypermethylation is also identified in the DNA of the root tips of a pea (*Pisum sativum* L.). Comparing to the data obtained by MSAP, the level of methylation both cytosine residues in CCGG motifs has risen, especially the inner one (Yu et al., 2022).

Research on tomatoes revealed alterations in the Asr gene's DNA methylation pattern. CG, CHG and CHH methylation was detected in high levels in the Asr2,27 regulatory region, 968 bp upstream from the transcription start site, of the tomato plant in response to water deficit conditions. In total, CHH methylation was reduced but significantly after stress across all time points (Mercé et al., 2020). Also, cytosine methylation in Asr2 gene body was examined. Even under 'normal' as well as under 'water deficiency' environmental conditions methylation was found to occur only at the few CG sites that are present at the moment. On the other hand, regardless of whether the plants were under normal conditions, a very small proportion of the numerous CHG and CHH sites were methylated. Notably, two different clones seemed to be highly methylated in every situation after the water stress, which, when thoroughly examined, resulted in an overall rise in CHG and CHH methylation. Research on how cold affects maize seedlings found that although cold stress causes genome-wide DNA methylation in root tissues, genome screening uncovered a 1.8-kb section known as ZmMI1 that was demethylated after chilling. Under typical circumstances, the same section displays a methylation percentage of almost 38%, which is nearly twice as high as other locations. Saraswat et al. (Saraswat et al., 2017) opine that this gene segment is a part of a possible protein coding gene that shows functionality in chilling tolerance. In other study done to determine the DNA methylation pattern in cold grown maize, 28 effectively produced differentially amplified fragments were generated. In silico study revealed that these fragments are involved in functions such as photosynthesis, hormonal regulation, transposon, and cold response (Singroha & Sharma, 2019). OsCMT2 plays a crucial function in rice, according to research on the expression levels of several DNA methyltransferases in rice. Under drought conditions, the expression level of OsCMT2 coding mRNA was shown to have decreased six to four times. This research also shown how OsCMT3 contributes to drought. Transcript accumulation in rice seedlings was shown to be reduced four to six times during drought stress (Yang et al., 2019). Drought significantly affects the overall DNA methylation pattern, according to a research on rice. When taking into consideration variations in tissue, genotype, and developmental stage, drought was able to cause an average total of 12.1% methylation changes. Demethylation tends to increase in stressful situations like heat, drought, and heavy metal stress. Additionally, it was shown that DNA methylation exhibits tissue selectivity. At the same embryonic stage, roots had a lower total DNA methylation level than leaves, suggesting that roots play a more important role in drought than leaves. According to Saraswat et al. (Saraswat et al., 2017), 29% of these DNA methylation/de-methylation regions were shown to remain abnormal even after recovery. Another research on heat stress in rice revealed that OsFIE1 (fertilization independent endosperm) controls the decline in seed size. Analysis revealed that DNA methylation and H3K9me2 methylation are the two variables that regulate OsFIE1 expression. Subsequent investigation indicated that both histone and DNA methylation decreased under heat stress (Gao et al., 2014). Popuplus tricocarpa under drought stress has a higher methylation level (10.04%) according to global methylation study than under well-watered circumstances (7.75%). Additionally, CG and CHG methylations were greater than CHH methylation, according to single base methylome analysis. Furthermore, there was also the comparison of UTRs, coding areas, and repetitive sequences which concluded that coding sequences had higher methylation than UTR sequences and repetitive sequences had higher methylation rates than coding sequences (Zheng et al., 2017). Concerning tobacco BY-2 cells development pattern, it was established that DNA methylation controls the growth and expansion of cells under heat shock. NtCycA13, NtCyc29, and CDKB1-1 transcript levels decreased and NtCycD3-1 levels increased when exposed to mild heat (35°C), which also inhibited 70% of development. NtCycD3-1 genes had hypomethylation, whereas NtEXPA-5 genes had hypermethylation, according to a DNA methylation study based on the McrBC test. NtcycA13 and CDKB1-1 genes were unaffected. In both normal and heat-stressed methylation patterns (Sun et al., 2022).

MicroRNAs and Crop Improvement

A type of endogenously produced, tiny RNA molecules with a length of around 25 nucleotides (nts) that are often genetically conserved are known as microRNAs (miRNAs). Some plants contain many isoforms, which are thought to have arisen as a result of gene duplication and divergence. Numerous processes, including as immunity, cell death, proliferation, and regulation of development of leaves and flowers, as well as abiotic and biotic stressors, are known to be influenced by miRNA (Raza et al., 2023). Targeting complementary mRNA regions for expression or degradation suppression allows plants to regulate gene expression and carry out all of these functions. In contrast to plants, which do not have miRNAs grouped in clusters, animals have miRNAs that are mostly assigned to intergenic regions (Zhang 2015).

The identification of lin-14, the substrate of lin-4, later demonstrated the significance of miRNA in gene regulation (Bracht et al., 2010). After 10 years the discovery of animal miRNA, the first miRNA in plants was found in Arabidopsis thaliana (Hajdarpašić & Ruggenthaler, 2012). Numerous plants have now been shown to have thousands of miRNAs (Saiyed et al., 2022). Many miRNA sequences have been identified by cloning and size fractionated RNA, recent technologies and High Through Put Sequencing techniques along with computing instruments for the identification of miRNA and probable target sequences (Noman et al., 2017).

Synthesis of miRNA in Plants

Although they may also be produced in conjunction with the host gene, the majority of miRNAs are produced from their own genes with separate transcription units. Only primary miRNA (pri-miRNA) is produced by RNA polymerase II. The conversion of pri-miRNA into mature miRNA is carried out by a variety of enzymes (Kumar 2014). Members of the dsRNA binding protein family in Arabidopsis, DCL1, SE, and HYL1, first produce a pre-miRNA to initiate the two-step processing process. This pre-miRNA is further processed by HYL1 and DCL1 to create the miRNA/miRNA duplex (Wani et al., 2020). When HEN1 methylates the miRNA/miRNA duplex, HASTY, a plant homolog of Exportin-5, moves the duplex from the nucleus to the cytoplasm (Raza et al., 2023). Following transit, the miRNA duplex is uncoils by an unidentified helicase, and one strand—referred to as the guide strand—is broken down with the aid of the AGO protein, leaving the mature miRNA remnants (Figure 5). Organic longevity, increasing the yield of agricultural products through the modification of plant height, branching, and size, the growth and development of seedless fruits and increasing the nutritive value are some of the goals that might be addressed by miRNA technology. Through RNA interference, a new tomato cultivar that has higher carotenoids and flavonoids both of which are exceptionally good for health was created. This method has also created a high market value seedless tomato type by silencing a chalcone synthase gene responsible for biosynthesis of flavonoids in tomatoes (Kowsalya et al., 2024).

RNAi technology has been successfully used to develop crops that can withstand stress. Farnesyl transferase expression was downregulated in canola, which increases resistance to seed abortion during blooming brought on by water scarcity. Silencing the activated C-kinase1 receptor resulted in transgenic rice plants that could withstand drought. Drought tolerance is increased in rice when the RINGfinger E3 ligase gene OsDSG1 is knocked down. According to a recent research, bentgrass that overexpresses the rice miR319 gene has higher leaf wax content, which increases its resistance to drought (Dhaliwal et al., 2022). Research revealed that under cold conditions, rice and sugarcane's expression of miR319 changed significantly (Yang et al., 2019). miRNA technology has become a potent molecular engineering tool for crop enhancement in recent years (Mujtaba et al., 2021).

Figure 5. Abiotic stresses tolerance through different mode of epigenetic in plants

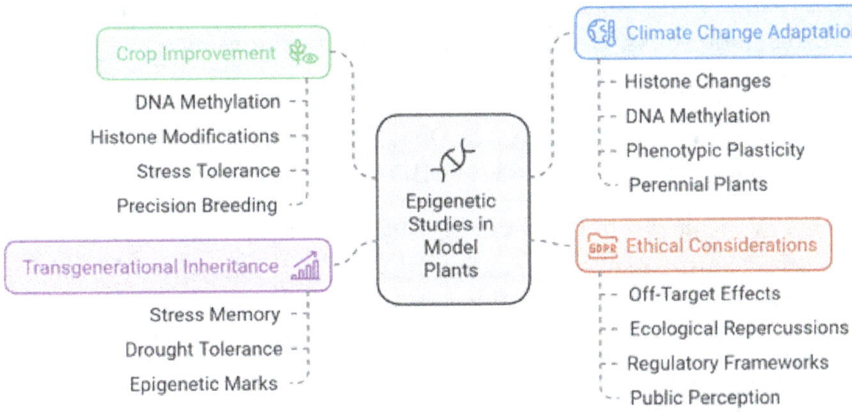

CONCLUSION

Plant growth under stress has been shown to be regulated by factors such as DNA methylation, histone N-tail modifications, and histone variations, which together affect the differential expression of many genes. The precise genes and areas in charge of adjusting to stimuli like heat and drought may be identified by looking at the kinds and processes of these changes. This information is essential for identifying potential targets to increase crop resilience. A possible method for improving crop quality is RNA interference (RNAi) technology, which modifies gene expression to change metabolic pathways and improve agronomic parameters. Furthermore, by lowering the synthesis of inherently harmful substances in edible plant sections, RNAi may improve food safety and more efficiently satisfy the world's nutritional demands.

REFERENCES

Akhter, Z., Bi, Z., Ali, K., Sun, C., Fiaz, S., Haider, F. U., & Bai, J. (2021). In response to abiotic stress, DNA methylation confers epigenetic changes in plants. *Plants*, *10*(6), 1096. PMID: 34070712

Akinmusola, R. Y., Wilkins, C. A., & Doughty, J. (2023). DDM1-Mediated TE silencing in plants. *Plants*, *12*(3), 437. PMID: 36771522

Ashapkin, V. V., Kutueva, L. I., Aleksandrushkina, N. I., & Vanyushin, B. F. (2020). Epigenetic mechanisms of plant adaptation to biotic and abiotic stresses. *International Journal of Molecular Sciences*, *21*(20), 7457. PMID: 33050358

Bracht, J. R., Van Wynsberghe, P. M., Mondol, V., & Pasquinelli, A. E. (2010). Regulation of lin-4 miRNA expression, organismal growth and development by a conserved RNA binding protein in C. elegans. *Developmental Biology*, *348*(2), 210–221. PMID: 20937268

Carlberg, C. (2024). Gene expression and chromatin. In *Gene Regulation and Epigenetics: How Science Works* (pp. 13–29). Springer Nature Switzerland.

Chandana, B. S., Mahto, R. K., Singh, R. K., Ford, R., Vaghefi, N., Gupta, S. K., & Kumar, R. (2022). Epigenomics as potential tools for enhancing magnitude of breeding approaches for developing climate resilient chickpea. *Frontiers in Genetics*, *13*, 900253. PMID: 35937986

Chen, L. T., Luo, M., Wang, Y. Y., & Wu, K. (2010). Involvement of Arabidopsi s histone deacetylase HDA6 in ABA and salt stress response. *Journal of Experimental Botany*, *61*(12), 3345–3353. PMID: 20519338

Cheong, J. J. (2021). Genetic and epigenetic changes in plants in response to abiotic stress. *Genes*, *12*(10), 1603. PMID: 34680997

Czajka, K., Mehes-Smith, M., & Nkongolo, K. (2021). DNA methylation and histone modifications induced by abiotic stressors in plants. *Genes & Genomics*, •••, 1–19. PMID: 34837631

Dhaliwal, I., Rialch, I., Rana, K., Kaur, J., & Kaur, G. (2022). Designing the Rapeseed Genome for Abiotic Stress Tolerance. In *Genomic Designing for Abiotic Stress Resistant Oilseed Crops* (pp. 75–99). Springer International Publishing.

Fallet, M. (2024). Epigenetic Inheritance. In *Epigenetics in Biological Communication* (pp. 87–130). Springer Nature Switzerland.

Foroozani, M., Holder, D. H., & Deal, R. B. (2022). Histone variants in the specialization of plant chromatin. *Annual Review of Plant Biology*, *73*(1), 149–172. PMID: 35167758

Gallusci, P., Dai, Z., Génard, M., Gauffretau, A., Leblanc-Fournier, N., Richard-Molard, C., & Brunel-Muguet, S. (2017). Epigenetics for plant improvement: Current knowledge and modeling avenues. *Trends in Plant Science*, *22*(7), 610–623. PMID: 28587758

Gao, G., Li, J., Li, H., Li, F., Xu, K., Yan, G., & Wu, X. (2014). Comparison of the heat stress induced variations in DNA methylation between heat-tolerant and heat-sensitive rapeseed seedlings. *Breeding Science*, *64*(2), 125–133. PMID: 24987298

Garg, R., Kumari, R., Tiwari, S., & Goyal, S. (2014). Genomic survey, gene expression analysis and structural modeling suggest diverse roles of DNA methyltransferases in legumes. *PLoS One*, *9*(2), e88947. PMID: 24586452

Hajdarpašić, A., & Ruggenthaler, P. (2012). Analysis of miRNA expression under stress in Arabidopsis thaliana. *Bosnian Journal of Basic Medical Sciences*, *12*(3), 169. PMID: 22938544

Hashiguchi, A., & Komatsu, S. (2016). Impact of post-translational modifications of crop proteins under abiotic stress. *Proteomes*, *4*(4), 42. PMID: 28248251

He, X. J., Chen, T., & Zhu, J. K. (2011). Regulation and function of DNA methylation in plants and animals. *Cell Research*, *21*(3), 442–465. PMID: 21321601

Jaligot, E., & Rival, A. (2015). Applying epigenetics in plant breeding: Balancing genome stability and phenotypic plasticity. *Advances in Plant Breeding Strategies: Breeding, Biotechnology and Molecular Tools*, 159-192

Kakoulidou, I., Avramidou, E. V., Baránek, M., Brunel-Muguet, S., Farrona, S., Johannes, F., & Maury, S. (2021). Epigenetics for crop improvement in times of global change. *Biology (Basel)*, *10*(8), 766. PMID: 34439998

Kumar, R. (2014). Role of microRNAs in biotic and abiotic stress responses in crop plants. *Applied Biochemistry and Biotechnology*, *174*, 93–115. PMID: 24869742

Kumar, S. (2019). Epigenetics and epigenomics for crop improvement: Current opinion. *Advances in Biotechnology & Microbiology*, *14*, 555879.

Kumar, S., & Mohapatra, T. (2021). Dynamics of DNA methylation and its functions in plant growth and development. *Frontiers in Plant Science*, *12*, 596236. PMID: 34093600

Kumar, V., Thakur, J. K., & Prasad, M. (2021). Histone acetylation dynamics regulating plant development and stress responses. *Cellular and Molecular Life Sciences : CMLS, 78*, 4467–4486. PMID: 33638653

Liang, Z., Riaz, A., Chachar, S., Ding, Y., Du, H., & Gu, X. (2020). Epigenetic modifications of mRNA and DNA in plants. *Molecular Plant, 13*(1), 14–30. PMID: 31863849

Lindermayr, C., Rudolf, E. E., Durner, J., & Groth, M. (2020). Interactions between metabolism and chromatin in plant models. *Molecular Metabolism, 38*, 100951. PMID: 32199818

Liu, J., Xu, L., Shang, J., Hu, X., Yu, H., Wu, H., & Zhao, Y. (2021). Genome-wide analysis of the maize superoxide dismutase (SOD) gene family reveals important roles in drought and salt responses. *Genetics and Molecular Biology, 44*(3), e20210035. PMID: 34606562

Lucibelli, F., Valoroso, M. C., & Aceto, S. (2022). Plant DNA methylation: An epigenetic mark in development, environmental interactions, and evolution. *International Journal of Molecular Sciences, 23*(15), 8299. PMID: 35955429

Mercé, C., Bayer, P. E., Tay Fernandez, C., Batley, J., & Edwards, D. (2020). Induced methylation in plants as a crop improvement tool: Progress and perspectives. *Agronomy (Basel), 10*(10), 1484.

Miryeganeh, M. (2021). Plants' epigenetic mechanisms and abiotic stress. *Genes, 12*(8), 1106. PMID: 34440280

Mishra, G. P., Dikshit, H. K., Devi, J., Aski, M. S., & Durgesh, K. (2023). Epigenetics of abiotic stress tolerance in legumes. In *Legumes: Physiology and Molecular Biology of Abiotic Stress Tolerance* (pp. 89-114). Singapore: Springer Nature Singapore
Montgomery, S. A., Tanizawa, Y., Galik, B., Wang, N., Ito, T., Mochizuki, T., & Berger, F. (2020). Chromatin organization in early land plants reveals an ancestral association between H3K27me3, transposons, and constitutive heterochromatin. *Current Biology, 30*(4), 573–588.

Mujtaba, M., Wang, D., Carvalho, L. B., Oliveira, J. L., Espirito Santo Pereira, A. D., Sharif, R., & Fraceto, L. F. (2021). Nanocarrier-mediated delivery of miRNA, RNAi, and CRISPR-Cas for plant protection: Current trends and future directions. *ACS Agricultural Science & Technology, 1*(5), 417–435.

Noman, A., Fahad, S., Aqeel, M., Ali, U., Amanullah, Anwar, S., & Zainab, M. (2017). miRNAs: major modulators for crop growth and development under abiotic stresses. *Biotechnology letters, 39*, 685-700

Pikaard, C. S., & Scheid, O. M. (2014). Epigenetic regulation in plants. *Cold Spring Harbor Perspectives in Biology*, *6*(12), a019315. PMID: 25452385

Provart, N. J., Alonso, J., Assmann, S. M., Bergmann, D., Brady, S. M., Brkljacic, J., & McCourt, P. (2016). 50 years of Arabidopsis research: Highlights and future directions. *The New Phytologist*, *209*(3), 921–944. PMID: 26465351

Raza, A., Charagh, S., Karikari, B., Sharif, R., Yadav, V., Mubarik, M. S., & Zhuang, W. (2023). miRNAs for crop improvement. *Plant Physiology and Biochemistry*, •••, 107857. PMID: 37437345

Rehman, M., & Tanti, B. (2020). Understanding epigenetic modifications in response to abiotic stresses in plants. *Biocatalysis and Agricultural Biotechnology*, *27*, 101673.

Sagor, G. H. M., Zhang, S., Kojima, S., Simm, S., Berberich, T., & Kusano, T. (2016). Reducing cytoplasmic polyamine oxidase activity in Arabidopsis increases salt and drought tolerance by reducing reactive oxygen species production and increasing defense gene expression. *Frontiers in Plant Science*, *7*, 214. PMID: 26973665

Saiyed, A. N., Vasavada, A. R., & Johar, S. K. (2022). Recent trends in miRNA therapeutics and the application of plant miRNA for prevention and treatment of human diseases. *Future Journal of Pharmaceutical Sciences*, *8*(1), 24. PMID: 35382490

Sanders, D. M. (2018). *Dissecting the Mechanisms of Epigenetic Regulation in Plant Development and Stress*. The University of Wisconsin-Madison.

Saraswat, S., Yadav, A. K., Sirohi, P., & Singh, N. K. (2017). Role of epigenetics in crop improvement: Water and heat stress. *Journal of Plant Biology*, *60*, 231–240.

Scippa, G. S., Di Michele, M., Onelli, E., Patrignani, G., Chiatante, D., & Bray, E. A. (2004). The histone-like protein H1-S and the response of tomato leaves to water deficit. *Journal of Experimental Botany*, *55*(394), 99–109. PMID: 14645393

Shen, Q., Lin, Y., Li, Y., & Wang, G. (2021). Dynamics of H3K27me3 modification on plant adaptation to environmental cues. *Plants*, *10*(6), 1165. PMID: 34201297

Shi, Y., Zhang, X., Chang, X., Yan, M., Zhao, H., Qin, Y., & Wang, H. (2021). Integrated analysis of DNA methylome and transcriptome reveals epigenetic regulation of CAM photosynthesis in pineapple. *BMC Plant Biology*, *21*, 1–14. PMID: 33407144

Singroha, G., & Sharma, P. (2019). Epigenetic modifications in plants under abiotic stress. *Epigenetics*, •••, 10.

Sun, C., Ali, K., Yan, K., Fiaz, S., Dormatey, R., Bi, Z., & Bai, J. (2021). Exploration of epigenetics for improvement of drought and other stress resistance in crops: A review. *Plants*, *10*(6), 1226. PMID: 34208642

Sun, M., Yang, Z., Liu, L., & Duan, L. (2022). DNA methylation in plant responses and adaption to abiotic stresses. *International Journal of Molecular Sciences*, *23*(13), 6910. PMID: 35805917

Tiwari, R., & Rajam, M. V. (2022). RNA-and miRNA-interference to enhance abiotic stress tolerance in plants. *Journal of Plant Biochemistry and Biotechnology*, *31*(4), 689–704.

Varotto, S., Tani, E., Abraham, E., Krugman, T., Kapazoglou, A., Melzer, R., & Miladinović, D. (2020). Epigenetics: Possible applications in climate-smart crop breeding. *Journal of Experimental Botany*, *71*(17), 5223–5236. PMID: 32279074

Vuković, M., Kutnjak, M., Vitko, S., Tkalec, M., & Vidaković-Cifrek, Ž. (2024). Heat Priming Modifies Heat Stress Response in BPM1-Overexpressing Arabidopsis thaliana (L.) Heynh. *Journal of Plant Growth Regulation*, •••, 1–18.

Wang, P., Zhao, L., Hou, H., Zhang, H., Huang, Y., Wang, Y., & Li, L. (2015). Epigenetic changes are associated with programmed cell death induced by heat stress in seedling leaves of Zea mays. *Plant & Cell Physiology*, *56*(5), 965–976. PMID: 25670712

Wani, S. H., Kumar, V., Khare, T., Tripathi, P., Shah, T., Ramakrishna, C., Aglawe, S., & Mangrauthia, S. K. (2020). miRNA applications for engineering abiotic stress tolerance in plants. *Biologia*, *75*, 1063–1081.

Xu, Z., Huang, J., Qu, C., Chang, R., Chen, J., Wang, Q., & Liu, G. (2020). Functional characterization and expression patterns of PnATX genes under different abiotic stress treatments in Populus. *Tree Physiology*, *40*(4), 520–537. PMID: 32031640

Yan, F., Zhou, W., Zhao, H., Yuan, Z., Wang, Y., Jiang, K., & Zhang, Y. (2013). Geological events play a larger role than P leistocene climatic fluctuations in driving the genetic structure of Q uasipaa boulengeri (A nura: D icroglossidae). *Molecular Ecology*, *22*(4), 1120–1133. PMID: 23216961

Yang, S., Xu, K., Chen, S., Li, T., Xia, H., Chen, L., & Luo, L. (2019). A stress-responsive bZIP transcription factor OsbZIP62 improves drought and oxidative tolerance in rice. *BMC Plant Biology*, *19*, 1–15. PMID: 31208338

Yu, Y., Wang, M., Zhou, X., Du, H., Liu, B., Jiang, L., & Wang, Y. (2022). Unexpected Low DNA Methylation in Transposable Elements at the 5′-CCGG Sites in Three Fruit Tree Cultivars. *Agronomy (Basel)*, *12*(6), 1347.

Zhang, B. (2015). MicroRNA: A new target for improving plant tolerance to abiotic stress. *Journal of Experimental Botany*, *66*(7), 1749–1761. PMID: 25697792

Zhang, Z., Zeng, Y., Hou, J., & Li, L. (2024). Advances in understanding the roles of plant HAT and HDAC in non-histone protein acetylation and deacetylation. *Planta*, *260*(4), 93. PMID: 39264431

Zhao, L., Wang, P., Yan, S., Gao, F., Li, H., Hou, H., & Li, L. (2014). Promoter-associated histone acetylation is involved in the osmotic stress-induced transcriptional regulation of the maize ZmDREB2A gene. *Physiologia Plantarum*, *151*(4), 459–467. PMID: 24299295

Zhao, L., Zhou, Q., He, L., Deng, L., Lozano-Duran, R., Li, G., & Zhu, J. K. (2022). DNA methylation underpins the epigenomic landscape regulating genome transcription in Arabidopsis. *Genome Biology*, *23*(1), 197. PMID: 36127735

Zheng, Y., Ding, Y., Sun, X., Xie, S., Wang, D., Liu, X., & Zhou, D. X. (2016). Histone deacetylase HDA9 negatively regulates salt and drought stress responsiveness in Arabidopsis. *Journal of Experimental Botany*, *67*(6), 1703–1713. PMID: 26733691

Zheng, Y., Joyce, B. T., Liu, L., Zhang, Z., Kibbe, W. A., Zhang, W., & Hou, L. (2017). Prediction of genome-wide DNA methylation in repetitive elements. *Nucleic Acids Research*, *45*(15), 8697–8711. PMID: 28911103

Chapter 13

Epigenetics for Developing Abiotic Stress-Tolerant Plants: Updated Methods and Current Achievements

Muhammad Waseem
https://orcid.org/0000-0001-7947-7722
Hainan University, China

Sana Basharat
Hainan University, China

Iffat Shaheen
Bahauddin Zakariya University, Multan, Pakistan

Pingwu Liu
Hainan University, China

ABSTRACT

The refined use of these methodologies allows for precise epigenetic manipulation, presenting a promising avenue for tailoring plants to withstand abiotic stressors, including drought, salinity, and extreme temperatures. Cutting-edge methods, such as advanced next-generation sequencing, ChIP-seq, and CRISPR-based epigenome editing, take center stage in unraveling the intricate landscape of epigenetic modifications. This chapter comprehensively examines how these tools empower researchers to decode the epigenetic language governing DNA methylation, histone modifications, and small RNA-mediated pathways. Within the context of this chapter, notable

DOI: 10.4018/979-8-3693-8648-4.ch013

achievements in harnessing epigenetic insights to bolster abiotic stress tolerance are highlighted. These accomplishments range from the development of crops with enhanced drought resistance and salinity tolerance to those exhibiting resilience in the face of temperature extremes.

INTRODUCTION

Plant being sessile organism experience a wide range of drastic environmental conditions (temperature extremes, drought, salinity, inadequate light, radiations, flood, nutrient deficiency or toxicity, heavy metal toxicity) compromising final crop productivity and threat to global food security (Abdulraheem et al., 2024). Therefore, plant evolved the mechanism to memorize the stressful experiences for a certain period of time to increase their survival when they are exposed to stress again (Ramakrishnan et al., 2022). The concept of epigenetics was first introduced by Waddington by integrating genetics and epigenesis for explaining plant phenology (Akhter et al., 2021). Unlike animals, plants do not possess nervous system, the stressful memories are regulated through epigenetic modifications which develop appropriate molecular and physiological alterations by sensing and responding to abiotic stress (Ramakrishnan et al., 2022). These epigenetic stress modifications are intergenerational and transgenerational, adding an extra layer of gene expression regulation (Lu & Hyun, 2021).

The plant ability to tolerate stress vary among different species depending in genetics, duration, and intensity of stress (Song et al., 2021). The epigenetic regulation of gene expression could be advantageous in plant breeding to develop crops with desirable traits without comprising the ecosystem (Akhter et al., 2021). These changes are due to modifications in chromatin state such as, DNA methylation (DM), histone modifications (HM), and non-coding RNA modifications (ncRM) causing alterations in DNA structure and associated protein hereby influencing gene activity (Jogam et al., 2022). In *planta*, DM is an essential and critical epigenetic modification and could occurs predominantly at CG context (designated as CG methylation), but it can also occur in CNG and CNN contexts (where N is A, C, or T). Both DMs and HMs modify gene expression by hindering transcription factor binding to regulatory proteins (Kawakatsu et al., 2017; Suelves et al., 2016).

Epigenetic regulations are critical for various plant processes including plant vegetative and reproductive development (Kumar et al., 2021), disease susceptibility, and adoptability to abiotic stresses like drought, temperature, and nutrients deficiency (Li et al., 2021; Qin et al., 2022). For sustainable agriculture, there is a need of efforts to manipulate the epigenetic processes of critical agronomic traits by DM, HMs, and ncRNAs to improve crop productivity and quality (Ashapkin et al., 2020). The epigenetic variations (either short term or long term) play significant role in evolution, survival, and stable phenotypic plasticity.

Epigenetics being potential plant breeding tools could be used for crop improvement under multiple stresses. However, there is need to understand the relationship between plant epigenetics and stress responses. This could be challenging due to lack direct evidence linking epigenetic modification with phenotypic plasticity under multiple stresses (Tu et al., 2022). Recently, DM mediated gene expression regulation have been recognized as a critical component of plant stress adaptation under challenging environmental conditions (Duarte-Aké et al., 2023). Few researches have demonstrated that various enzymes can be used in stress-induced epigenetic tolerance. These include, DNA methyltransferase, histone acetyltransferases and histone deacetylases (Figure 1.) (Sun et al., 2022). To develop stress-tolerance plant it is pivotal to understand the mechanism involved in detail. This chapter comprehensively summarizes the recent advances of plant epigenetic mechanisms in plant abiotic stress responses and insights how plant survive under different environmental stresses and their generational heritability.

Figure 1. Mechanism of epigenetic regulation in plants. These include histone modification, DNA methylation, and small RNA interference (generated with Biorender).

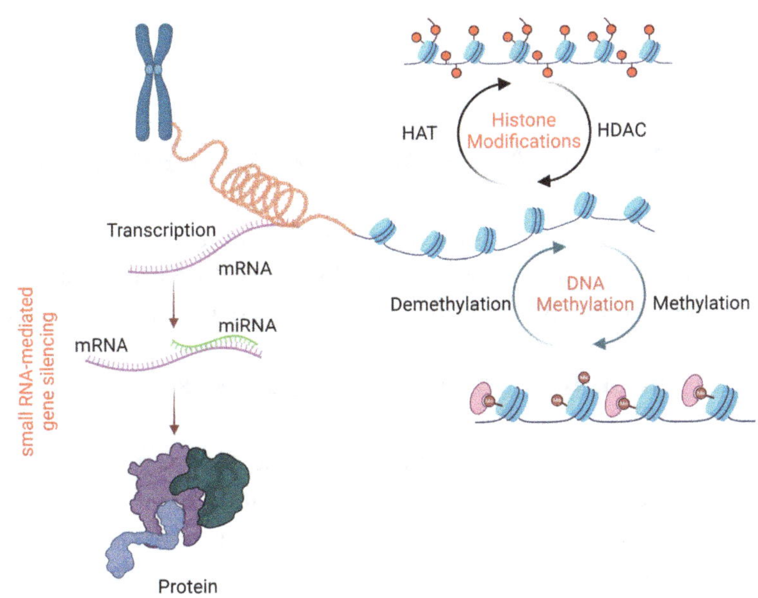

TECHNIQUES TO DETECT PLANT EPIGENETIC ALTERATIONS

Epigenetic regulations play essential role in the conservation of genome integrity and expression regulation. The detection of epigenetic alterations marks is pivotal for developing strategies and understanding the underlying mechanism to enhance productivity and crop tolerance. There are various methods for detection of epigenetic modification like chip-seq, chip-PCR, Chromatin immunoprecipitation (ChiP) techniques, Methylation sensitive Restriction enzymes (MSREs), Methyl-sensitive amplification polymorphism (MSAP), Chop-PCR, and Bisulfite sequencing (Figure 2).

Figure 2. Summary of techniques used to detect epigenetic modifications in plants

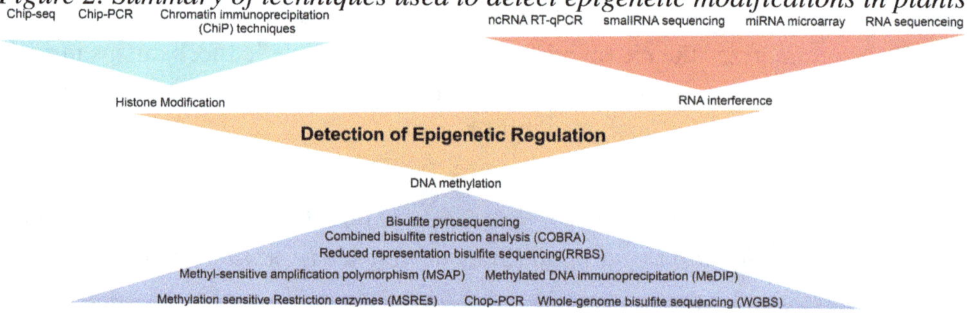

EPIGENETIC REGULATION IN PLANTS STRESS RESPONSES

Plants developed variety of epigenetic mechanisms to cope with abiotic stress by fine tuning gene expression including DM, HMs, ncRMs, and chromatin remodeling. These strategies play essential role in regulating gene expression and genome stability.

DNA Methylation In Plant Stress Responses

In plant DNA methylation dynamic epigenetic modification involves addition of methyl group to cytosine residues and significantly impact stress-induced gene expression regulation (Kovalchuk, 2021). DM certainly prevent gene transcriptional regulation and chromatic remodeling proteins (Sadakierska-Chudy et al., 2015), determining patterns of histone modifications and recruiting repressor complex HD7Acs, DNMTs, and MBD proteins. Unlike, DNA sequence manipulation, DM has potential to develop novel phenotypes with desirable traits in crossbreeding population (Huang et al., 2022), with alterations in methylation mediated genes expression and subsequent phenotypes (Tonosaki et al., 2022). Plant responds to stress by alterations in DM levels, for instance, Arabidopsis AtDM1 gene methylates drought stress tolerance specific genes (Lee et al., 2023). Likewise, pathogen infections can also alter DM levels, suppressing and activating defense related genes. DM can either active or passive(Asensi-Fabado et al., 2017) and can cause RNAi interference, histone modifications, alterations in chromatin structure and repressing gene expression (Suelves et al., 2016).

DM is a complex multienzyme process that alter chromatin structure by adding methyl group to cytosine residue of DNA creates 5-methylcytosine (5mC) (Lee et al., 2023). 5mC act as repressive marker through methylation and demethylation and play pivotal role in genome stability, suppress gene expression, and regulate key plant traits such as leaf morphology, disease susceptibility, and stress tolerance (Lämke & Bäurle, 2017). For instance, Ferreira et al. (2015) revealed that global DM levels play significant role in salinity stress by regulating gene expression in DNA methylome in rice genotypes. The DM levels corelated with salinity levels, for instance, high salinity reduces DM levels by up regulating DNA demethylase, vice versa (Wang et al., 2015). Another study on an apple demonstrated significance of epigenetic modification to seed dormancy by cold stress, with reduced total methylation and resumption of growth (Kumar et al., 2016). Liang et al. (2014) demonstrated that DM methylation alter the expression of various drought stress responsive genes tough and unknow mechanism. Table 1. Demonstrate the summary of epigenetics regulation in response to abiotic stresses.

Table 1. DNA methylation modification responses to plant abiotic stress

Stress	*Plants*	**Processes**	**Mechanisms/Responses**
Heat	*Zea mays*	DM analysis	Enhancing tolerance to heat and increasing methylation (Eichten & Springer, 2015)
	Brassica napus	Msap	Improving DM in heat-tolerant and heat-sensitive genotypes (Shen et al., 2017)
Drought	*Arabidopsis thaliana*	DM	Hypomethylation *(Luo & He, 2020)*
	Linum usitatissimum	DM	Alternative splicing and transcription (Wang et al., 2023)
	Zea mays	Transcriptome, miRNA, DM analysis	Promoting water retention (Min et al., 2016)
	Hordeum vulgare	DM at CKX2.1	Enhance DM (Surdonja et al., 2017)
	Populus trichocarpa	Single-base-resolution methylomes	Enhancing methylated cytosines (Liang et al., 2014)

continued on following page

Table 1. Continued

Stress	Plants	Processes	Mechanisms/Responses
Heavy metals	*Oryza sativa*	DM	Revealed DNA methylation-modified genes differentially expressed (Feng et al., 2016)
	Oryza sativa	DM	hypermethylation of transposable elements adjacent to stress-inducing genes (Secco et al., 2015)
	Trifolium repens	DM	Hypomethylation in tolerant plants upon prolonged exposure (González et al., 2016)
	Scenedesmus acutus	DM	Cytosine methylation (Ferrari et al., 2020)
	Oryza sativa	DM	Increasing DM (Feng et al., 2016)
Cold	*Malus domestica*	DM	Break dormancy and activate growth (Kumar et al., 2016)
Salt	*Glycine max*	Transcription factors expression	Demethylation and hypomethylation tolerant and susceptibility (Chen et al., 2019)
	Oryza sativa	ELISA-based assay	Hypomethylation in intolerant cultivar (Shankar et al., 2016)
	Oryza sativa	Comparative DM	Display differential methylome flexibility (Ferreira et al., 2015; Wang et al., 2015)
	Oryza sativa	DM	hypermethylation and hypomethylation (Pathak et al., 2021)
	Medicago spp	DM	(Al-Lawati et al., 2016)
Defense	*Arabidopsis thaliana*	DDM1	DM, activation of defense related genes (Lee et al., 2023)
	Arabidopsis thaliana	Demethylation	Defense genes are regulated in trans by DNA methylation (López Sánchez et al., 2016)

Histone Modification in Plant Stress Response

Histone modifications (HMs) dynamically regulated gene expression and chromatin modification which trigger gene transcriptional regulation based on genomic context (Nunez-Vazquez et al., 2022). In plants, HMs has demonstrated to play significant role in plant responses to multiple abiotic stresses. For instance, histone deacetylase inhibitor offer less susceptibility to *Botrytis cinerea* infection in *A. thaliana* (Hu et al., 2019). Research shows that drought stress activate histone H3 acetylation levels, leading to enhanced expression of stress-responsive genes in *A. thaliana* (Wollmann et al., 2017). Likewise, alterations in HM patters is crucial for salinity stress and genes associated with osmotic and ion homeostasis regulation (Ueda & Seki, 2020). Bivalent histone modifications (H3K4me3/ H3K27me3) by SHORT LIFE, a plant specific methylation reader protein, is crucial in cold stress responses in potato (*Solanum tuberosum*) (Zeng et al., 2019). This reflects the dynamic nature of HM to temperature extremes. Histone deacetylation (HDA) is another dynamic aspects of HM to active or repress drought responsive genes (Nunez-Vazquez et al., 2022). Histone methylation is regarded as complex modifications modulate gene expression depends on the level of methylation and specific residues modulation (Hu et al., 2019). Conversely, protein kinase mediated histone phosphorylation crucial for plant multifaced stress responses. For instance, MAPK (mitogen-activated protein kinase) cascade controls stress response through histone phosphorylation and stress induced expression regulation (Mozgova et al., 2019). HMs like methylation, acetylation, ADP-ribosylation, sumoylation, and ubiquitination alter stress-signaling pathways gene expression and modify chromatin structure, however, the extend of function in plant stress response needs further understanding (Wang & Qiao, 2020).

Although the precise mechanism of HM still being elucidate, however, few studies tried to illustrate the potential mechanism of HM in different plant species. For global food security, it is critical to develop tolerant crops with ability to flourished challenging environment, and to do so, it is crucial to delve more deeper into how plant respond to stress and how HM alter gene expression (Ramakrishnan et al., 2022). For instance, repression of Arabidopsis WRKY TF by H3K27me3 activates the expression of stress-responsive genes (Zheng et al., 2020). Wu et al. (2019) HM (H3K27me3) plays an important role in regulation of *AcSVP2* expression associated with downstream regulation of drought/osmotic stress regulation. HM influenced by multiple stresses such as temperature extremes, drought, salinity in Arabidopsis (Zhu, 2016). The histone variants play critical role is stress responses in specific region of genome – stress responsive genes. In planta, HM are categorized into various genetics families. These HM variants include histone lysine/arginine pairs of methylation sites such as H3R2/H3K4; H3R8/H3K9; H3R26/H3K27, where both residues are methylated (Di Lorenzo & Bedford, 2011). Furthermore, drought stress in Arabidopsis results in H3K4me3, H3K9ac 3K4me1, H3K4me2, and H3K4me3 modifications in the promoter regions of stress responsive genes like RD29B, RAP2.4, RD29A, RD20 (Begum, 2022; Hu et al., 2015). However, when rice seedlings experience drought stress, several genes exhibited differential H3K4me3 modified (Zhang et al., 2020). The HMs is associated with upregulation of stress-responsive genes, and offers insights into underlying regulatory mechanism linked with plant ability to adopt to multiple stresses.

Small RNA-Mediated Gene Silencing

RNA interference and chromatin modification is another pivotal aspect of epigenetic mechanism involving microRNA (miRNAs) and small interfering RNAs (siRNAs) regulating various downstream stress responsive genes (Begum, 2022). siRNAs are derived from exogenous/endogenous ddRNAs, while miRNAs are generated by hairpin-shaped precursors. The small RNAs cause translational inhibition or degradation through based pairing with target mRNAs via RNA-induced silencing complex (RISC) (Waseem et al., 2020). Recent evidences suggested that ncRNAs play critical role in plant adaptation to abiotic stresses such as cold (Waseem et al., 2020; Waseem et al., 2022). Plants develop strategies to combat the damaging effect of reactive oxygen species in response to abiotic stress such as drought and salinity by producing antioxidant enzymes. In few plants, miRNAs such as miRN398 and miR528 are known to regulate osmotic and oxidative stress network in Arabidopsis (Liu et al., 2018).

In additional to role in both translational and posttranslational gene expression regulation, small RNAs, also active in chromatin modification resulting in activation and suppression of transcription activity of various stress responsive genes (Lee et al., 2023). Moreover, mRNAs also interact with variety of histone modifiers at specific genomic regions and modulate gene expression (Jagadhesan et al., 2022). Altogether, the plant response to stresses allows plant to adapt to varying environmental factors reply on sRNA-mediated gene silencing and chromatin alterations.

ADVANCES IN EPIGENETIC REGULATION OF PLANT ABIOTIC STRESS RESPONSE

Owning to potential for crop yield and vigor improvement plant epigenetic regulation garnered significant attentions in last decade. HM and DM are the long-term alteration in DNA structure, which can either suppress or activate gene expression. On the other hand, DNA methyltransferases target cytosines with context to CpNpG and CpG cause temporary alteration in DNA structure (Rehman & Tanti, 2020). Recently studies have shown that drought stress induced intensive DNA methylation leading to suppresses DNA activity. For instance, in Arabidopsis drought stress cause DM in various regions of genome including transposable elements and in CDS of drought responsive genes (Zhang et al., 2018). Conversely, plants grown under optimized conditions, have lower methylation levels with normal gene expression regulation. Likewise, epigenetic landscape in rice revealed that DM and HM play distinct role in differential methylation and HM associated with different stress responsive genes of drought, salinity, and high-temperature (Akhter et al., 2021).

According to Li et al. (2021) and Hu et al. (2019) epigenetic modification enhanced drought tolerance in maize. They found that histone acetylation at specific genomic regions alter the expression of drought responsive genes which in turn assist in water utilization and stress tolerance. In addition to these studies, studies related to effects of non-coding RNAs in epigenetic regulation make significant impacts on stress-induced epigenetic regulations (Ferdous et al., 2015; Statello et al., 2021).

These observations suggest that epigenetic regulatory network offer novel strategies for improving crop adaptations to wide range of environment and stress tolerance.

APPLICATIONS OF PLANT EPIGENETIC REGULATION

Epigenetic regulation of plant stress response has great importance for environmental adoptability and agriculture. Few key potential implications are:

1. Traditional breeding methods for introducing desirable agronomic traits have limitation in terms of time and precisions. Researcher have identified plenty of plant stress responsive makers for crop improvements. By integrating stress associated epigenetic genes, the crop ability to cope with a stress could lead to development of stress-resilient crops with comprising productivity and ensuing global food security.
2. By uncovering the specific regulatory network or biological molecular pathways at epigenetic levels provide a potential avenue for agriculture sustainability. For instance, modulating the epigenetic markers associated with salinity stress, crops can be cultivated in saline soil without impending crops productivity (Yeo, 1998). Such epigenetic modulation provides invaluable ways to design strategies for crop improvement under saline environment.
3. Epigenetic modifications can assist in decisions about crop management practices and precision agriculture techniques. Such precisions include optimize irrigation, application of targeted treatments, site specific strategies implementations to enhance crop productivity with minimum resource utilizations.
4. Global warming significantly hinders agriculture crop productivity. Epigenetic mechanism influences specific genetic traits and molecular mechanism that mitigate and adapt plant to fluctuating environmental changes.
5. The knowledge of epigenetic modification can be applied to plant conservation and biodiversity to ecosystem restoration. Such insights can aim to biodiversity preservation and promoting ecological resilience.

REFERENCES

Abdulraheem, M. I., Xiong, Y., Moshood, A. Y., Cadenas-Pliego, G., Zhang, H., & Hu, J. (2024). Mechanisms of plant epigenetic regulation in response to plant stress: Recent discoveries and implications. *Plants*, *13*(2), 163. PMID: 38256717

Akhter, Z., Bi, Z., Ali, K., Sun, C., Fiaz, S., Haider, F. U., & Bai, J. (2021). In response to abiotic stress, DNA methylation confers epigenetic changes in plants. *Plants*, *10*(6), 1096. PMID: 34070712

Al-Lawati, A., Al-Bahry, S., Victor, R., Al-Lawati, A. H., & Yaish, M. W. (2016). Salt stress alters DNA methylation levels in alfalfa (Medicago spp). *Genetics and Molecular Research*, *15*(1), 15018299. PMID: 26985924

Asensi-Fabado, M.-A., Amtmann, A., & Perrella, G. (2017). Plant responses to abiotic stress: The chromatin context of transcriptional regulation. *Biochimica et Biophysica Acta (BBA)-. Gene Regulatory Mechanisms*, *1860*(1), 106–122. PMID: 27487458

Ashapkin, V. V., Kutueva, L. I., Aleksandrushkina, N. I., & Vanyushin, B. F. (2020). Epigenetic mechanisms of plant adaptation to biotic and abiotic stresses. *International Journal of Molecular Sciences*, *21*(20), 7457. PMID: 33050358

Begum, Y. (2022). Regulatory role of microRNAs (miRNAs) in the recent development of abiotic stress tolerance of plants. *Gene*, *821*, 146283. PMID: 35143944

Chen, R., Li, M., Zhang, H., Duan, L., Sun, X., Jiang, Q., & Hu, Z. (2019). Continuous salt stress-induced long non-coding RNAs and DNA methylation patterns in soybean roots. *BMC Genomics*, *20*, 1–12. PMID: 31606033

Di Lorenzo, A., & Bedford, M. T. (2011). Histone arginine methylation. *FEBS Letters*, *585*(13), 2024–2031. PMID: 21074527

Duarte-Aké, F., Us-Camas, R., & De-la-Peña, C. (2023). Epigenetic Regulation in Heterosis and Environmental Stress: The Challenge of Producing Hybrid Epigenomes to Face Climate Change. *Epigenomes*, *7*(3), 14. PMID: 37489402

Eichten, S. R., & Springer, N. M. (2015). Minimal evidence for consistent changes in maize DNA methylation patterns following environmental stress. *Frontiers in Plant Science*, *6*, 308. PMID: 25999972

Feng, S. J., Liu, X. S., Tao, H., Tan, S. K., Chu, S. S., Oono, Y., & Yang, Z. M. (2016). Variation of DNA methylation patterns associated with gene expression in rice (Oryza sativa) exposed to cadmium. *Plant, Cell & Environment*, *39*(12), 2629–2649. PMID: 27412910

Ferdous, J., Hussain, S. S., & Shi, B. J. (2015). Role of micro RNA s in plant drought tolerance. *Plant Biotechnology Journal, 13*(3), 293–305. PMID: 25583362

Ferrari, M., Torelli, A., Marieschi, M., & Cozza, R. (2020). Role of DNA methylation in the chromium tolerance of Scenedesmus acutus (Chlorophyceae) and its impact on the sulfate pathway regulation. *Plant Science, 301*, 110680. PMID: 33218643

Ferreira, L. J., Azevedo, V., Maroco, J., Oliveira, M. M., & Santos, A. P. (2015). Salt tolerant and sensitive rice varieties display differential methylome flexibility under salt stress. *PLoS One, 10*(5), e0124060. PMID: 25932633

González, A. P. R., Chrtek, J., Dobrev, P. I., Dumalasová, V., Fehrer, J., Mráz, P., & Latzel, V. (2016). Stress-induced memory alters growth of clonal offspring of white clover (Trifolium repens). *American Journal of Botany, 103*(9), 1567–1574. PMID: 27620180

Hu, Y., Lu, Y., Zhao, Y., & Zhou, D.-X. (2019). Histone acetylation dynamics integrates metabolic activity to regulate plant response to stress. *Frontiers in Plant Science, 10*, 1236. PMID: 31636650

Hu, Z., Song, N., Zheng, M., Liu, X., Liu, Z., Xing, J., & Peng, H. (2015). Histone acetyltransferase GCN 5 is essential for heat stress-responsive gene activation and thermotolerance in Arabidopsis. *The Plant Journal, 84*(6), 1178–1191. PMID: 26576681

Huang, H., Wu, N., Liang, Y., Peng, X., & Shu, J. (2022). SLNL: A novel method for gene selection and phenotype classification. *International Journal of Intelligent Systems, 37*(9), 6283–6304.

Jagadhesan, B., Das, S., Singh, D., Jha, S. K., Durgesh, K., & Sathee, L. (2022). Micro RNA mediated regulation of nutrient response in plants: The case of nitrogen. *Plant Physiology Reports, 27*(3), 345–357.

Jogam, P., Sandhya, D., Alok, A., Peddaboina, V., Allini, V. R., & Zhang, B. (2022). A review on CRISPR/Cas-based epigenetic regulation in plants. *International Journal of Biological Macromolecules, 219*, 1261–1271. PMID: 36057300

Kawakatsu, T., Nery, J. R., Castanon, R., & Ecker, J. R. (2017). Dynamic DNA methylation reconfiguration during seed development and germination. *Genome Biology, 18*, 1–12. PMID: 28911331

Kovalchuk, I. (2021). Role of DNA methylation in genome stability. In *Genome Stability* (pp. 435–452). Elsevier.

Kumar, G., Rattan, U. K., & Singh, A. K. (2016). Chilling-mediated DNA methylation changes during dormancy and its release reveal the importance of epigenetic regulation during winter dormancy in apple (Malus x domestica Borkh.). *PLoS One*, *11*(2), e0149934. PMID: 26901339

Kumar, V., Thakur, J. K., & Prasad, M. (2021). Histone acetylation dynamics regulating plant development and stress responses. *Cellular and Molecular Life Sciences : CMLS*, *78*, 4467–4486. PMID: 33638653

Lämke, J., & Bäurle, I. (2017). Epigenetic and chromatin-based mechanisms in environmental stress adaptation and stress memory in plants. *Genome Biology*, *18*, 1–11. PMID: 28655328

Lee, S., Choi, J., Park, J., Hong, C. P., Choi, D., Han, S., & Hwang, I. (2023). DDM1-mediated gene body DNA methylation is associated with inducible activation of defense-related genes in Arabidopsis. *Genome Biology*, *24*(1), 106. PMID: 37147734

Li, S., He, X., Gao, Y., Zhou, C., Chiang, V. L., & Li, W. (2021). Histone acetylation changes in plant response to drought stress. *Genes*, *12*(9), 1409. PMID: 34573391

Liang, D., Zhang, Z., Wu, H., Huang, C., Shuai, P., Ye, C.-Y.,...Wang, J. (2014, 2014). Single-base-resolution methylomes of Populus trichocarpa reveal the association between DNA methylation and drought stress.

Liu, Y., Zhang, A., Yin, H., Meng, Q., Yu, X., Huang, S., & Xu, Z. Y. (2018). Trithorax-group proteins ARABIDOPSIS TRITHORAX4 (ATX4) and ATX 5 function in abscisic acid and dehydration stress responses. *The New Phytologist*, *217*(4), 1582–1597. PMID: 29250818

López Sánchez, A., Stassen, J. H. M., Furci, L., Smith, L. M., & Ton, J. (2016). The role of DNA (de) methylation in immune responsiveness of Arabidopsis. *The Plant Journal*, *88*(3), 361–374. PMID: 27341062

Lu, X., & Hyun, T. K. (2021). The role of epigenetic modifications in plant responses to stress. *Botanica Serbica*, *45*(1), 3–12.

Luo, X., & He, Y. (2020). Experiencing winter for spring flowering: A molecular epigenetic perspective on vernalization. *Journal of Integrative Plant Biology*, *62*(1), 104–117. PMID: 31829495

Min, H., Chen, C., Wei, S., Shang, X., Sun, M., Xia, R., & Xie, Q. (2016). Identification of drought tolerant mechanisms in maize seedlings based on transcriptome analysis of recombination inbred lines. *Frontiers in Plant Science*, *7*, 1080. PMID: 27507977

Mozgova, I., Mikulski, P., Pecinka, A., & Farrona, S. (2019). Epigenetic mechanisms of abiotic stress response and memory in plants. *Epigenetics in plants of agronomic importance: Fundamentals and applications: Transcriptional regulation and chromatin remodelling in plants*, 1-64.

Nunez-Vazquez, R., Desvoyes, B., & Gutierrez, C. (2022). Histone variants and modifications during abiotic stress response. *Frontiers in Plant Science, 13*, 984702. PMID: 36589114

Pathak, H., Kumar, M., Molla, K. A., & Chakraborty, K. (2021). Abiotic stresses in rice production: impacts and management.

Qin, X., Zhang, K., Fan, Y., Fang, H., Nie, Y., & Wu, X.-L. (2022). The bacterial MtrAB two-component system regulates the cell wall homeostasis responding to environmental alkaline stress. *Microbiology Spectrum, 10*(5), e02311–e02322. PMID: 36073914

Ramakrishnan, M., Zhang, Z., Mullasseri, S., Kalendar, R., Ahmad, Z., Sharma, A., & Wei, Q. (2022). Epigenetic stress memory: A new approach to study cold and heat stress responses in plants. *Frontiers in Plant Science, 13*, 1075279. PMID: 36570899

Rehman, M., & Tanti, B. (2020). Understanding epigenetic modifications in response to abiotic stresses in plants. *Biocatalysis and Agricultural Biotechnology, 27*, 101673.

Sadakierska-Chudy, A., Kostrzewa, R. M., & Filip, M. (2015). A comprehensive view of the epigenetic landscape part I: DNA methylation, passive and active DNA demethylation pathways and histone variants. *Neurotoxicity Research, 27*, 84–97. PMID: 25362550

Secco, D., Wang, C., Shou, H., Schultz, M. D., Chiarenza, S., Nussaume, L.,… Lister, R. (2015). Stress induced gene expression drives transient DNA methylation changes at adjacent repetitive elements. *elife, 4*, e09343.

Shankar, R., Bhattacharjee, A., & Jain, M. (2016). Transcriptome analysis in different rice cultivars provides novel insights into desiccation and salinity stress responses. *Scientific Reports, 6*(1), 23719. PMID: 27029818

Shen, Y., Sun, S., Hua, S., Shen, E., Ye, C. Y., Cai, D., & Fan, L. (2017). Analysis of transcriptional and epigenetic changes in hybrid vigor of allopolyploid Brassica napus uncovers key roles for small RNA s. *The Plant Journal, 91*(5), 874–893. PMID: 28544196

Song, Z. T., Zhang, L. L., Han, J. J., Zhou, M., & Liu, J. X. (2021). Histone H3K4 methyltransferases SDG25 and ATX1 maintain heat-stress gene expression during recovery in Arabidopsis. *The Plant Journal, 105*(5), 1326–1338. PMID: 33278042

Statello, L., Guo, C.-J., Chen, L.-L., & Huarte, M. (2021). Gene regulation by long non-coding RNAs and its biological functions. *Nature Reviews. Molecular Cell Biology*, *22*(2), 96–118. PMID: 33353982

Suelves, M., Carrió, E., Núñez-Álvarez, Y., & Peinado, M. A. (2016). DNA methylation dynamics in cellular commitment and differentiation. *Briefings in Functional Genomics*, *15*(6), 443–453. PMID: 27416614

Sun, M., Yang, Z., Liu, L., & Duan, L. (2022). DNA methylation in plant responses and adaption to abiotic stresses. *International Journal of Molecular Sciences*, *23*(13), 6910. PMID: 35805917

Surdonja, K., Eggert, K., Hajirezaei, M.-R., Harshavardhan, V. T., Seiler, C., Von Wirén, N., & Kuhlmann, M. (2017). Increase of DNA methylation at the HvCKX2. 1 promoter by terminal drought stress in barley. *Epigenomes*, *1*(2), 9.

Tonosaki, K., Fujimoto, R., Dennis, E. S., Raboy, V., & Osabe, K. (2022). Will epigenetics be a key player in crop breeding? *Frontiers in Plant Science*, *13*, 958350. PMID: 36247549

Tu, Y.-T., Chen, C.-Y., Huang, Y.-S., Chang, C.-H., Yen, M.-R., Hsieh, J.-W. A., & Wu, K. (2022). HISTONE DEACETYLASE 15 and MOS4-associated complex subunits 3A/3B coregulate intron retention of ABA-responsive genes. *Plant Physiology*, *190*(1), 882–897. PMID: 35670741

Ueda, M., & Seki, M. (2020). Histone modifications form epigenetic regulatory networks to regulate abiotic stress response. *Plant Physiology*, *182*(1), 15–26. PMID: 31685643

Wang, L., & Qiao, H. (2020). Chromatin regulation in plant hormone and plant stress responses. *Current Opinion in Plant Biology*, *57*, 164–170. PMID: 33142261

Wang, L., Wang, L., Tan, M., Wang, L., Zhao, W., You, J., & Wang, W. (2023). The pattern of alternative splicing and DNA methylation alteration and their interaction in linseed (Linum usitatissimum L.) response to repeated drought stresses. *Biological Research*, *56*(1), 12. PMID: 36922868

Wang, W., Huang, F., Qin, Q., Zhao, X., Li, Z., & Fu, B. (2015). Comparative analysis of DNA methylation changes in two rice genotypes under salt stress and subsequent recovery. *Biochemical and Biophysical Research Communications*, *465*(4), 790–796. PMID: 26319557

Waseem, M., Liu, Y., & Xia, R. (2020). Long non-coding RNAs, the dark matter: An emerging regulatory component in plants. *International Journal of Molecular Sciences*, *22*(1), 86. PMID: 33374835

Waseem, M., Yang, X., Aslam, M. M., Li, M., Zhu, L., Chen, S., & Liu, P. (2022). Genome-wide identification of long non-coding RNAs in two contrasting rapeseed (Brassica napus L.) genotypes subjected to cold stress. *Environmental and Experimental Botany*, *201*, 104969.

Wollmann, H., Stroud, H., Yelagandula, R., Tarutani, Y., Jiang, D., Jing, L., & Nie, X. (2017). The histone H3 variant H3. 3 regulates gene body DNA methylation in Arabidopsis thaliana. *Genome Biology*, *18*, 1–10. PMID: 28521766

Wu, R., Wang, T., Richardson, A. C., Allan, A. C., Macknight, R. C., & Varkonyi-Gasic, E. (2019). Histone modification and activation by SOC1-like and drought stress-related transcription factors may regulate AcSVP2 expression during kiwifruit winter dormancy. *Plant Science*, *281*, 242–250. PMID: 30824057

Yeo, A. (1998). Molecular biology of salt tolerance in the context of whole-plant physiology. *Journal of Experimental Botany*, *49*(323), 915–929.

Zeng, Z., Zhang, W., Marand, A. P., Zhu, B., Buell, C. R., & Jiang, J. (2019). Cold stress induces enhanced chromatin accessibility and bivalent histone modifications H3K4me3 and H3K27me3 of active genes in potato. *Genome Biology*, *20*, 1–17. PMID: 31208436

Zhang, H., Guo, F., Qi, P., Huang, Y., Xie, Y., Xu, L., & Bian, H. (2020). OsHDA710-mediated histone deacetylation regulates callus formation of rice mature embryo. *Plant & Cell Physiology*, *61*(9), 1646–1660. PMID: 32592489

Zhang, H., Lang, Z., & Zhu, J.-K. (2018). Dynamics and function of DNA methylation in plants. *Nature Reviews. Molecular Cell Biology*, *19*(8), 489–506. PMID: 29784956

Zheng, Y., Ge, J., Bao, C., Chang, W., Liu, J., Shao, J., & Zhou, D.-X. (2020). Histone deacetylase HDA9 and WRKY53 transcription factor are mutual antagonists in regulation of plant stress response. *Molecular Plant*, *13*(4), 598–611. PMID: 31891777

Zhu, J.-K. (2016). Abiotic stress signaling and responses in plants. *Cell*, *167*(2), 313–324. PMID: 27716505

Chapter 14
Epigenomics:
A High Throughput Approach to Study Plant Stress Adaptations

Amber Jabeen
https://orcid.org/0009-0001-4501-3345
Quaid-i-Azam University, Islamabad, Pakistan

Muhammad Kaleem
https://orcid.org/0009-0003-6548-3920
Quaid-i-Azam University, Islamabad, Pakistan

Lubna Anjum Minhas
Quaid-i-Azam University, Islamabad, Pakistan

Abdul Samad Mumtaz
https://orcid.org/0000-0002-7034-9203
Quaid-i-Azam University, Islamabad, Pakistan

ABSTRACT

Epigenomic processes i.e. DNA methylation, histone modification and RNA-mediated regulatory processes are pivotal for the adaptive gene expression responses of plants to environmental stressors. Bisulfite sequencing, chromatin immunoprecipitation sequencing, and epiGWAS, facilitate the comprehensive mapping of dynamic modifications across the genome. These advanced technologies are elucidating the processes through which plants develop "stress memories" that bolster resilience across generations, thereby illuminating the intricate regulatory networks that enable plant survival in challenging environments. Epigenomic knowledge with breeding initiatives enable scientists to develop crop varieties exhibiting enhanced stress

DOI: 10.4018/979-8-3693-8648-4.ch014

tolerance, thereby contributing to sustainable agricultural practices. This chapter outlines current methods, key findings, and future directions in the field, highlighting the potential of epigenomic research to revolutionize plant resilience strategies.

INTRODUCTION

In recent times, epigenomics has emerged as a pivotal domain of investigation aimed at elucidating the mechanisms by which plants navigate environmental challenges, thereby providing insights that are especially vital in relation to climate change. Environmental stressors such as drought, salinity, and temperature variability pose significant challenges to plant systems on a global scale, and epigenomic research endeavors to decipher the modalities through which plants perceive, react to, and retain memory of these stress events (Abdul Raheem et al., 2024). Essential to this adaptable mechanism are epigenetic changes, including the methylation of DNA, histone alterations, and chromatin reconfiguration—all of which are vital in regulating gene activity in response to stress. These modifications, in contrast to alterations in DNA sequences, are reversible and can be dynamically modified as plants encounter and adapt to environmental challenges (Das & Adhikari, 2023, Ahmad et al., 2022a). Thus, epigenomic investigations have established a foundational framework for devising strategies aimed at enhancing plant resilience, thereby addressing the urgent need for sustainable agricultural practices amidst global environmental instability.

Recent advancements in technology have significantly accelerated the progress of epigenomics, thereby enabling researchers to investigate these intricate modifications on a genome-wide basis. High-throughput sequencing, chromatin immunoprecipitation (ChIP), and transcriptomic methodologies have emerged as fundamental technologies that facilitate the comprehensive exploration of epigenetic responses across entire plant genomes (Wang KC et al., 2018). These tools support the identification of specific genes and epigenetic markers that are associated with stress adaptation, thereby illuminating the multifaceted regulatory networks that orchestrate plant responses to environmental stressors. For example, transcriptomic analyses permit the quantification of alterations in gene expression under stress conditions, whereas ChIP sequencing can be employed to discern histone modifications and DNA-binding proteins that are integral to stress responses (Schmitz & Zhang, 2011, Ahmad et al., 2022b). Such meticulous molecular mapping elucidates the mechanisms by which plants regulate and prioritize specific genes during episodes of stress, thereby providing a conceptual framework for comprehending the intricate interactions between genetic and epigenetic factors that empower plants to adapt to unfavorable conditions.

One of the most compelling dimensions of epigenetic regulation is the concept of epigenetic memory, which refers to the mechanism by which plants "remember" prior encounters with stress and can demonstrate enhanced resilience during subsequent exposures (Sun C et al., 2021). This memory is facilitated by stable epigenetic modifications that can endure across cellular divisions, thereby enabling plants to retain adaptive traits. Methodologies such as bisulfite sequencing and chromatin profiling allow for the precise mapping of DNA methylation patterns, thereby revealing how these modifications are either stably preserved or dynamically altered in response to environmental fluctuations (Rehman & Tanti, 2020). The notion of epigenetic memory is particularly noteworthy as it implies that adaptive responses to stress may be transmittable to subsequent generations, thereby offering a potential mechanism for long-term resilience in plant populations. This phenomenon, termed transgenerational epigenetic inheritance, opens avenues for breeding programs that could leverage these adaptive traits, ultimately contributing to the advancement of stress-resistant crop varieties (Varroto S. et al., 2021, Ahmad et al., 2024a).

Within the domain of agricultural innovation, high-throughput epigenomic methodologies present considerable potential for augmenting crop resilience. By elucidating the specific epigenetic modifications correlated with stress tolerance, researchers can pinpoint prospective molecular targets for genetic engineering and selective breeding initiatives (Varroto S. et al., 2021). This approach is not only critical for enhancing our understanding of stress physiology in plants but also represents a proactive strategy for alleviating the repercussions of climate change on food security. Crop species that possess engineered or selected epigenetic characteristics hold significant promise for demonstrating enhanced growth and yield consistency under challenging environmental conditions, thereby facilitating global agricultural sustainability (Kakoulidou, I. et al., 2021, Ahmad et al., 2024b).

The incorporation of epigenomic information into plant breeding initiatives presents a revolutionary methodology for crop enhancement, which is predicated not solely on genetic variability but also on the dynamic and adaptive capabilities afforded by epigenetic modulation (Tonosaki. et al., 2022, Ali et al., 2022). Prospective investigations in plant epigenomics are expected to deepen our comprehension of the molecular foundations underpinning stress adaptation, ultimately aiding in the cultivation of climate-resilient crops capable of enduring a diverse range of environmental challenges (Kakoulidou, I. et al., 2021). By elucidating the function of epigenetics in the responses of plants to stressors, researchers are unveiling novel pathways toward agricultural resilience and adaptation amidst an era characterized by unparalleled environmental transformations (Petereit et al., 2022).

EPIGENETIC MECHANISM IN PLANT STRESS RESPONSE

Epigenetic regulation is fundamentally crucial in coordinating plant reactions to environmental stressors. Fundamental mechanisms such as DNA methylation, the mediation of small RNAs, alterations in histones, and the reorganization of chromatin serve to effectively regulate gene expression without the need to modify the DNA sequence (Singh et al., 2021, Ijaz et al., 2023), thereby equipping plants with the essential adaptability for a range of environmental challenges (As illustrated in **Figure 1**).

Figure 1. Schematic presentation of epigenetic mechanisms regulated by either biotic or abiotic stress

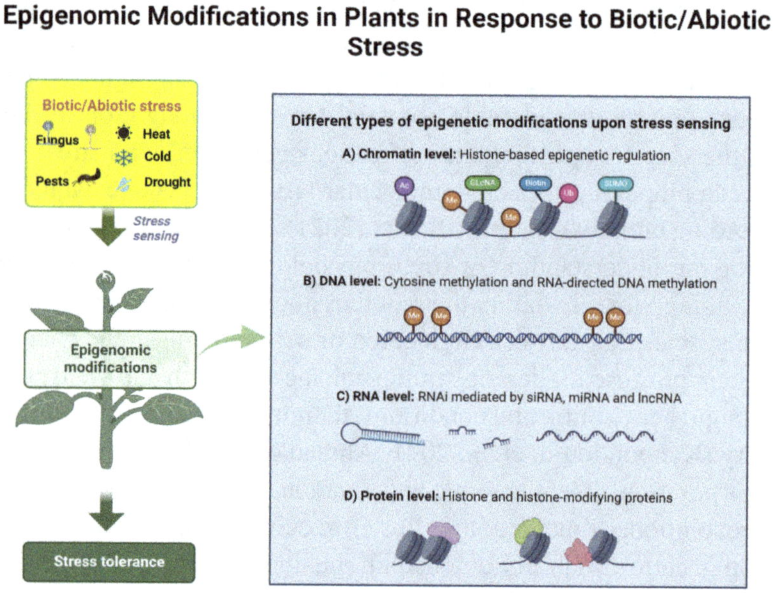

DNA methylation constitutes a pivotal epigenetic modification characterized by the addition of methyl groups to cytosine residues within the DNA framework. This modification possesses the capacity to silence genes, particularly those associated with stress responses, by limiting transcriptional access to designated regions within the genome. The significance of DNA methylation is crucial in managing the expression of genes that respond to stress and mobile genetic elements when faced with abiotic challenges like drought, high salinity, and extreme temperature scenarios (Kyung J. et al., 2022). In the presence of stressful stimuli, specific genomic loci may experience demethylation, which serves to activate gene expression in support of stress tolerance (Chinnusamy & Zhu, 2009, Ijaz et al., 2024a). Histone modifications, which include acetylation, phosphorylation, and methylation significantly impact chromatin accessibility, thereby governing gene expression (Tsuji et al., 2006). For instance, the acetylation of histones H3 and H4 typically results in chromatin relaxation, subsequently fostering gene expression during stress responses (Boycheva et al., 2014, Ijaz et al., 2024b). In contrast, histone deacetylation and methylation may culminate in gene silencing (Sawicka & Seiser, 2012). These modifications are indispensable in shaping plant responses to environmental signals, as they facilitate rapid adjustments in transcriptional activity in reaction to stresses such as heat or drought (Luo et al., 2020).

Small RNAs i.e. small interfering RNAs (siRNAs) and microRNAs (miRNAs) are responsible for gene silencing through the mechanism of RNA interference (RNAi). These molecules are essential to both transcriptional and translational regulation of gene expression in the context of stress conditions (Nunez-Vazquez et al., 2022). For example, miRNAs are capable of targeting stress-responsive genes for degradation or repression, while siRNAs contribute to the processes of DNA methylation and histone modifications, thereby reinforcing gene silencing mechanisms during adaptive stress responses (Chinnusamy & Zhu, 2009, Ullah et al., 2023a). Chromatin remodeling pertains to the dynamic repositioning of nucleosomes, which allows for the modulation of accessibility of specific DNA regions to transcriptional machinery. This intricate process is meticulously regulated by chromatin remodelers, which can either compact or relax the chromatin structure in accordance with environmental stimuli (Avramova, 2015). Under conditions of stress, chromatin remodeling may facilitate the swift activation or repression of genes pertinent to adaptive responses, thereby providing a nuanced flexibility in transcriptional regulation (Bhadouriya et al., 2021). Therefore, epigenetic mechanisms, including DNA methylation, histone modifications, small RNAs, and chromatin remodeling, are fundamentally integral to the methodologies through which plants navigate stress responses, thereby providing a versatile and heritable framework for adaptation to environmental fluctuations.

HIGH-THROUGHPUT TECHNOLOGIES IN PLANT EPIGENOMICS

High-throughput technologies have emerged as essential tools in the investigation of the epigenetic regulation pertaining to plant stress adaptation. These methodologies facilitate genome-wide mapping and analysis of DNA modifications, histone alterations, and chromatin accessibility, thereby providing comprehensive insights into the mechanisms by which plants react to environmental stimuli (Rehman &

Tanti, 2020, Schmitz & Zhang, 2011, Ullah et al., 2023b). Below, we examine some of the most prevalently utilized high-throughput technologies within the domain of plant epigenomics and functionalities and challenges associated with these technologies are presented in **table 1.**

Bisulfite Sequencing (BS-Seq)

Bisulfite sequencing (BS-Seq) represents a state-of-the-art technique in the domain of epigenomic investigation, particularly esteemed for its proficiency in discerning DNA methylation configurations at single-base resolution. DNA methylation, a pivotal epigenetic modification, plays a fundamental role in modulating gene expression in response to various environmental stressors, including drought, salinity, and extreme temperature fluctuations. The BS-Seq methodology is integral for elucidating the alterations in methylation patterns that occur under such stress conditions, thereby offering significant insights into the molecular adaptive mechanisms employed by plants (Adusumalli et al., 2015). The foundational principle of BS-Seq resides in its capacity to differentiate between methylated and unmethylated cytosines. Following bisulfite treatment, unmethylated cytosines are converted to uracil, while methylated cytosines remain intact (**Figure 2**). This selective conversion facilitates the precise mapping of methylation sites throughout the genome by contrasting treated and untreated DNA sequences (Al Harrasi et al., 2018). This degree of specificity is exceptionally advantageous for epigenomic research, as it empowers investigators to ascertain methylation patterns down to the level of individual cytosines, thereby generating high-resolution data that is critical for comprehending the intricacies of epigenetic regulation (Urich et al., 2015, Haider et al., 2021a).

Whole-genome bisulfite sequencing (WGBS) has emerged as a preferred methodology in plant epigenomic studies due to its exhaustive perspective on the methylome, defined as the entirety of DNA methylation modifications present within a genome (As illustrated in Figure 2). WGBS is capable of profiling methylation states across all cytosines, thereby providing a holistic view of how methylation patterns adapt to environmental stressors (Gupta & Garg, 2020). Such information is imperative for identifying genes that respond to stress through methylation alterations, thereby laying the groundwork for understanding how plants recalibrate their gene expression in response to varying environmental conditions. For instance, investigations employing WGBS have charted genome-wide methylation shifts in plants subjected to drought stress, elucidating specific genomic regions where methylation changes correspond with the upregulation or downregulation of stress-responsive genes. These insights contribute to delineating the molecular pathways that plants activate to endure abiotic stress, thereby enhancing our understanding of the mechanisms underlying stress adaptation (Das & Adhikari, 2023).

Figure 2. Graphical illustration of principle of bisulphite sequencing

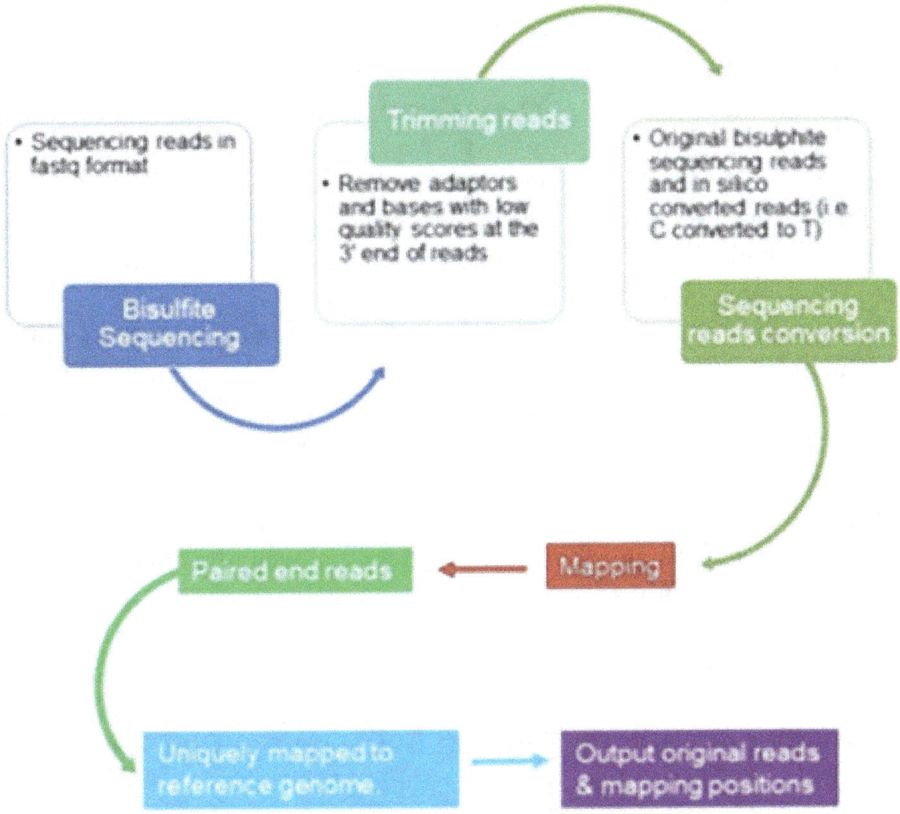

With the emergence of increasingly complex and voluminous sequencing datasets, bioinformatics tools have become essential for the analysis and interpretation of BS-Seq data. A notable tool, ViewBS, facilitates the visualization and examination of extensive bisulfite sequencing datasets, enabling researchers to more efficiently interpret methylation patterns across entire genomes (Huang et al., 2018). ViewBS allows users to visualize methylation states, monitor changes over temporal scales or across varying conditions, and compare methylomes amongst distinct samples (Huang et al., 2018). This functionality is particularly pertinent in studies of stress adaptation, where comparative analyses of stressed and unstressed samples can uncover differential methylation patterns associated with stress tolerance (Yaish et al., 2018).

The insights derived from BS-Seq datasets possess extensive implications for plant breeding and the enhancement of crop resilience. By identifying critical methylation sites correlated with stress tolerance, researchers can target potential candidates for genetic engineering or selective breeding endeavors (Cong et al., 2019). For example, methylation sites that exhibit consistent alterations in response to drought or elevated salinity levels may serve as valuable markers for breeding initiatives aimed at improving stress resilience (Rajkumar et al., 2020). Moreover, given that DNA methylation possesses the capacity to be inherited across successive generations via epigenetic mechanisms, the insights derived from bisulfite sequencing (BS-Seq) may substantiate breeding methodologies designed to cultivate crops exhibiting inheritable traits of stress tolerance (Kou et al., 2022; Al Harrasi et al., 2018, Haider et al., 2021b).

In short, bisulfite sequencing has emerged as an essential technique within the discipline of plant epigenomics, providing exceptional resolution in the delineation of DNA methylation patterns and elucidating their contributions to plant stress responses (Lu et al., 2017). The capability to scrutinize methylation states across the entire genome empowers researchers to unravel the molecular frameworks that underpin plant adaptation to environmental stressors, thereby enhancing both theoretical insights and practical applications in the agricultural sector (Grehl et al., 2020). As advancements in technology and bioinformatics tools persist, it is anticipated that BS-Seq will assume an increasingly pivotal role in the development of resilient crop varieties capable of flourishing amid the challenges imposed by climate change.

Chromatin Immunoprecipitation Sequencing (ChIP-Seq)

Chromatin Immunoprecipitation Sequencing (ChIP-Seq) has solidified its status as a fundamental methodology in the realm of epigenomic research, proving particularly instrumental in deciphering the molecular responses of plants to environmental stress (Patten et al., 2018). This technique furnishes vital insights into the interactions between proteins and DNA, as well as modifications to histones throughout the genome, thereby unveiling the regulatory landscapes that dictate gene expression (Kaufmann et al., 2010). By utilizing specific antibodies that target DNA-binding proteins or histone modifications, ChIP-Seq facilitates the precise localization of sites where these regulatory proteins or epigenetic marks interact with DNA. Such interactions frequently underpin the processes of gene activation or repression, rendering ChIP-Seq essential for comprehending the epigenetic modulation of stress-response genes (Cortijo et al., 2018, Haider et al., 2022a).

Methodology and Process of ChIP-Seq

ChIP-Seq integrates chromatin immunoprecipitation (ChIP) with high-throughput sequencing, thereby enabling the identification of DNA regions associated with particular proteins, including transcription factors or histone modifications. The procedure initiates with the cross-linking of proteins to DNA within chromatin, succeeded by chromatin fragmentation. Subsequently, an antibody specific to the protein or histone modification of interest is employed to selectively precipitate the DNA segments linked to the targeted protein or modification (Patten et al., 2018). The resultant DNA-protein complexes undergo isolation and purification, after which the DNA is sequenced to yield a comprehensive mapping of binding sites dispersed across the genome. This technique is particularly efficacious for detecting histone modifications, such as methylation or acetylation, which are critical to plant stress responses (Cortijo et al., 2018). For instance, histone acetylation is frequently correlated with gene activation and has been documented at genes associated with drought tolerance in plants (Song et al., 2016). Through the mapping of these modifications, ChIP-Seq enables researchers to identify genes that are actively regulated in response to environmental stressors. This process not only elucidates which genes are activated or repressed during stress but also highlights the specific histone modifications or transcription factors involved, thereby providing a high-resolution perspective on the operational gene regulatory mechanisms.

Applications in Plant Stress Response Research

In case of plant stress adaptation, Chromatin Immunoprecipitation followed by Sequencing (ChIP-Seq) serves a critical role in elucidating the regulatory networks that govern responses to abiotic stressors, including drought, elevated salinity, and extreme thermal conditions. Under stress conditions, specific modifications to histones and binding events of transcription factors are initiated, leading to alterations in gene expression patterns that facilitate plant adaptation to hostile environmental conditions (Wu et al., 2022, Haider et al., 2022b). For example, investigations employing ChIP-Seq have elucidated how distinct histone modifications can either promote or obstruct the interaction of transcription factors with genes responsive to stress. These modifications are frequently context-dependent, manifesting solely in reaction to particular stress stimuli, thereby enabling a flexible and adaptive response to fluctuating environmental conditions (Nakato & Sakata, 2021). ChIP-Seq analyses have uncovered that genes responsive to stress are frequently characterized by specific histone modifications, such as H3K4me3 (indicative of active transcription) or H3K27me3 (associated with gene repression). By delineating these modifications across the genomic landscape, ChIP-Seq investigations can elucidate the genes that are either activated or repressed in response to stressors and identify the regulatory proteins implicated in these processes (Zhu et al., 2012). This knowledge is crucial for comprehending how plants intricately modulate their stress responses at a molecular level, thereby illuminating the adaptive strategies that may be exploited in crop breeding initiatives aimed at enhancing resilience.

Data Analysis and Reproducibility with CoBRA

The vast datasets produced by ChIP-Seq experiments necessitate the application of robust bioinformatics tools to ensure effective analysis and precise interpretation. The Containerized Bioinformatics Workflow for Reproducible ChIP/ATAC-seq Analysis (CoBRA) has emerged as an indispensable resource in assuring reproducibility and uniformity in the analysis of ChIP-Seq data (Qiu et al., 2021). CoBRA provides a comprehensive pipeline for the analysis of ChIP-Seq data, incorporating tools for quality assessment, alignment, peak identification, and visualization (Qiu et al., 2020). This pipeline empowers researchers to produce standardized results, thereby facilitating more coherent interpretations of protein-DNA interactions and histone modifications across diverse studies (Qiu et al., 2021). The methodology of CoBRA guarantees that ChIP-Seq analyses are not only reproducible but also efficient and accessible, a feature that is particularly advantageous in extensive studies where consistency is paramount. Such a degree of standardization is vital for comparative analyses of plant stress responses across various taxa or environmental circumstances, promoting a more profound understanding of both universal and species-specific adaptive mechanisms (Das et al., 2022).

The ChIP-Seq technology, especially when integrated with sophisticated bioinformatics workflows such as CoBRA, possesses significant potential for agricultural advancements. By illuminating the epigenetic regulatory frameworks that underlie stress responses, ChIP-Seq data can provide insights that inform breeding programs designed to enhance crop resilience. The identification of pivotal transcription factors and histone modifications correlated with stress tolerance paves the way for targeted genetic modifications or selective breeding aimed at incorporating these resilient traits into crop cultivars (Bailey et al., 2013). In conclusion, ChIP-Seq functions as a fundamental instrument in plant epigenomic research, offering extraordinary insights into the regulatory mechanisms that govern gene expression alterations in response to environmental stressors. With the evolution of data analysis tools such as CoBRA, the influence of ChIP-Seq on the understanding and enhancement of plant resilience is poised for expansion, presenting avenues for sustainable agricultural practices in the context of climate change.

Assay for Transposase-Accessible Chromatin with High-Throughput Sequencing (ATAC-Seq)

The ATAC-Seq technique serves as an analytical approach for evaluating chromatin accessibility, thereby enabling scholars to delineate open chromatin domains that are potentially involved in the regulation of gene expression. This methodology utilizes the hyperactive Tn5 transposase enzyme (**Figure 3**) to cleave and integrate sequencing adapters into the accessible genomic regions, yielding critical information regarding transcription factor interaction sites and nucleosome arrangement (Buenrostro et al., 2015). This technique proves particularly advantageous in the domain of population epigenomics by facilitating the examination of chromatin accessibility alterations in response to environmental stressors (Augusto et al., 2020).

Figure 3. Mechanism of ATAC-sequencing

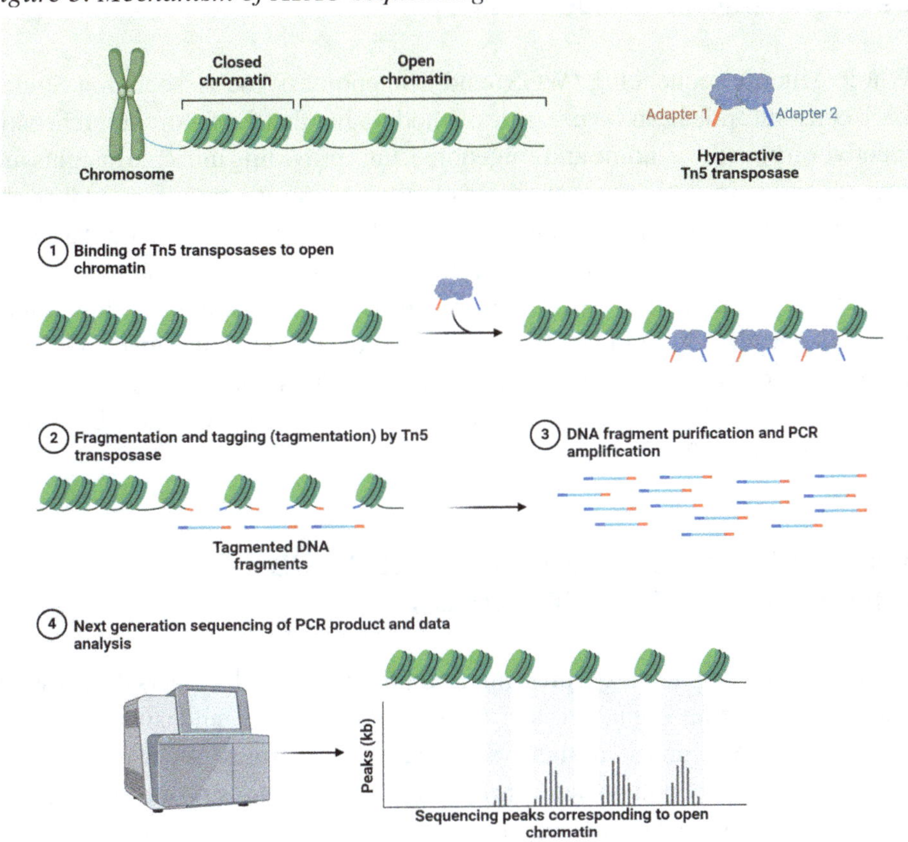

RNA Sequencing (RNA-Seq) Coupled with Small RNA Sequencing

RNA-Seq functions as a profiling tool for gene expression through the sequencing of the total RNA extracted from a biological sample, thereby imparting significant insights into the mechanisms by which plants modulate gene expression in reaction to stress. Complementarily, small RNA sequencing specifically targets small regulatory RNA molecules, including microRNAs (miRNAs) and small interfering RNAs (siRNAs), which are fundamental to the processes of post-transcriptional regulation. The integration of RNA-Seq with small RNA sequencing facilitates a comprehensive investigation of the dynamic interactions between gene expression and regulatory small RNAs under conditions of stress (Hendrickson et al., 2018).

Whole-Genome Sequencing and Epigenome-Wide Association Studies (EWAS)

Whole-genome sequencing (WGS) and Epigenome-Wide Association Studies (EWAS) constitute potent and synergistic methodologies that furnish a comprehensive perspective on both the genome and epigenome, thus providing intricate insights into the genetic and epigenetic governance of plant responses to environmental stress. Collectively, WGS and EWAS empower researchers to discern genetic variations and correlate epigenetic modifications with specific phenotypic characteristics, particularly those pertinent to stress adaptation. These high-throughput methodologies are revolutionary in advancing our understanding of the molecular mechanisms by which plants acclimatize to challenges such as drought, salinity, and extreme temperature events, all of which are increasingly prevalent and exacerbated by climate change (Zhang et al., 2021).

WHOLE-GENOME SEQUENCING (WGS) AND ITS ROLE IN PLANT EPIGENOMICS

Whole-genome sequencing represents a methodology that deciphers the complete genetic sequence of an organism, thereby providing a comprehensive map of all genetic elements. Within the framework of plant epigenomics, WGS is indispensable for identifying not only genetic variations but also regions prone to epigenetic modifications, including DNA methylation and histone alterations. These modifications are frequently pivotal in determining how plants modulate gene expression in the face of environmental stressors (de Kort et al., 2022). By furnishing a detailed genetic blueprint, WGS enables scholars to differentiate between genetic and epigenetic influences on phenotypic traits, thereby elucidating the intricate interplay between an organism's genome and its environmental context. For example, WGS can uncover single nucleotide polymorphisms (SNPs) or structural variants that may engage with epigenetic mechanisms to influence gene expression (de Kort et al., 2022). Furthermore, WGS provides insights into the sequence context of methylation sites and histone modifications, which may aid in predicting the regulatory dynamics of these epigenetic markers under stress conditions (Zhang et al., 2021).

Epigenome-Wide Association Studies (EWAS) for Linking Epigenetic Marks to Plant Stress Responses

Epigenome-wide association studies (EWAS) rigorously investigate the correlation between distinct epigenetic markers such as DNA methylation patterns and histone modification states and observable phenotypic traits or specific environmental parameters (Rakyan et al., 2011). EWAS is of particular significance within the domain of plant epigenomics as it facilitates the establishment of connections between discrete epigenetic alterations and stress responses. Through the genomic analysis of these correlations, EWAS plays a crucial role in pinpointing "epigenetic hotspots" where variations in methylation or histone modifications are markedly related to traits associated with stress resilience, thereby providing insights into the fundamental adaptive mechanisms (Akhter et al., 2021). In the field of plant stress research, EWAS can elucidate the manner in which epigenetic modifications instigated by environmental stressors, such as drought or elevated salinity, affect gene expression and ultimately govern phenotypic manifestations. For instance, EWAS has been employed to identify genomic regions exhibiting consistent alterations in DNA methylation levels in plants exposed to water scarcity, thereby establishing a linkage between these modifications and the genes implicated in drought tolerance (Akhter et al., 2021). By delineating these associations, EWAS offers profound insights into the strategies employed by plants to utilize epigenetic mechanisms for adaptation to environmental shifts, thereby empowering researchers to identify candidate genes and pathways amenable to manipulation in crop enhancement initiatives (Can et al., 2021).

Integrative Power of WGS and EWAS in Plant Adaptation Research

The amalgamation of Whole Genome Sequencing (WGS) and EWAS presents a holistic framework for investigating plant adaptation, as it integrates data on genetic variability alongside epigenetic modification profiles. This synthesis is particularly beneficial as it enables researchers to disentangle the relative influences of genetic and epigenetic components on plant responses to stress (Ramakrishnan et al., 2022), (Can et al., 2021). For example, while WGS can elucidate genetic variants that predispose certain plant species to effectively endure drought conditions, EWAS can underscore particular methylation landscapes or histone modification patterns that augment the expression of drought-resilient characteristics, even among genetically akin plants. This integrative methodology also facilitates the exploration of epigenetic inheritance, a phenomenon whereby stress-induced epigenetic modifications are transmitted to succeeding generations (Liu et al., 2019). By monitoring both genetic and epigenetic alterations across generations through WGS and EWAS, scholars can ascertain whether adaptive traits exhibit heritability, thereby rendering them potentially advantageous for selective breeding purposes (Ramakrishnan et al., 2021).

Applications in Crop Resilience and Future Directions

WGS and EWAS are revolutionizing the field of plant breeding by elucidating genetic and epigenetic markers associated with stress resilience, thereby equipping breeders with tools necessary for the development of climate-resilient crops. For instance, by concentrating on epigenetic modifications that bolster tolerance to elevated temperatures or drought conditions, crop breeders can harness selective breeding or gene-editing methodologies to incorporate these traits into commercially viable crop varieties (Miladinovic et al., 2021). This capacity is indispensable for the creation of crops capable of thriving amidst increasingly erratic climatic conditions. Furthermore, the high-resolution data afforded by WGS and EWAS engender opportunities for predictive modeling within agricultural contexts, enabling breeders to anticipate crop performance under distinct environmental scenarios (Dong et al., 2021). Future inquiries will likely concentrate on the refinement of these technological frameworks and the development of bioinformatics instruments to systematically analyze the intricate datasets produced by WGS and EWAS.

Table 1. The functionalities and challenges of different high-throughput methods for plant epigenomic analysis

Technology	Applications	Strengths	Limitations	References
DNA Methylation Sequencing (WGBS, RRBS)	Mapping cytosine methylation across genome or regions	High-resolution genome-wide methylation data	Expensive, complex bioinformatics, high DNA input requirement	Schmitz & Zhang, 2011; Yadav et al., 2018
Chromatin Immunoprecipitation Sequencing (ChIP-Seq)	Mapping histone modifications and transcription factor binding	Captures protein-DNA interactions, important for histone modification studies	Requires high cell input, prone to noise, requires validated antibodies	Lee, 2023
RNA Sequencing (RNA-Seq)	Transcriptome profiling, differential expression analysis	High sensitivity, detects rare transcripts, low cost per sample	Bioinformatic complexity, limited for studying low-abundance transcripts	Wang et al., 2018; Wang et al., 2010; Hao et al., 2015
ATAC-Seq (Assay for Transposase-Accessible Chromatin)	Identifying open chromatin regions, regulatory element mapping	Low input requirement, high sensitivity for active chromatin regions	Limited to fresh/frozen tissues, lower resolution than other methods	Stunnenberg & Vermeulen, 2011; Hsieh J. et al., 2024
Bisulfite-Converted Single-Cell Sequencing	Single-cell methylation profiling for cellular heterogeneity	High resolution at single-cell level, minimal DNA input	High cost, complex data analysis, limited throughput for population studies	Ji, 2014
3D Carbene Chip-Based SPRi Technology	Profiling binding kinetics of epigenetic interactions	Low nonspecific binding, reusable, functional immobilization of biomolecules	Limited to surface-binding molecules, requires specialized equipment	Zhao et al., 2017

EPIGENETIC MEMORY AND TRANSGENERATIONAL STRESS ADAPTATION

Epigenetic memory denotes the capacity of plants to encode and recall previous environmental stressors, thereby enhancing their efficiency in responding to subsequent encounters with similar stressors. This memory is sustained through heritable epigenetic alterations, including DNA methylation, histone modifications, and small RNA-mediated transformations (Chinnusamy & Zhu, 2009; Liu et al., 2020). In certain instances, these modifications are transmittable to progeny, thus facilitating the transgenerational propagation of stress resilience (Tricker, 2015; Ou et al., 2012; Kinoshita & Seki, 2014).

Concept of Epigenetic Memory

Epigenetic memory encapsulates the biological phenomenon wherein plants possess the capability to retain recollections of previous environmental conditions and adapt their responses accordingly, even across successive generations. This concept is paramount for plant survival as it fosters more effective reactions to recurrent environmental stressors (Brukhin & Albertini, 2021). The mechanisms underpinning epigenetic memory predominantly include DNA methylation and histone modifications, which preserve altered gene expression profiles through cellular divisions and, at times, across generational lines (Kinoshita & Seki, 2014). This adaptive response empowers plants to dynamically modulate their reactions to environmental stimuli, as the accumulation of distinctive epigenetic marks facilitates the retention of a "memory" concerning prior adversities. A subset of these modifications is reversible, whereas others exhibit stability and establish the foundation of stress memory (Lämke & Bäurle, 2017). Certain epigenetic marks possess sufficient stability to transcend generational boundaries without undergoing resetting, thereby contributing to the phenomenon termed "transgenerational epigenetic inheritance." This occurrence is frequently observed in plants, wherein modifications induced by stress are occasionally conveyed from parents to offspring, conferring evolutionary advantages in persistently challenging environments (Tricker, 2015). For instance, particular chromatin modifications linked to cold stress memory can be inherited, thus facilitating subsequent generations in enduring analogous low-temperature scenarios (Brukhin & Albertini, 2021). Although some epigenetic modifications exhibit stability, plants are equipped with mechanisms to reset these marks to prevent the accumulation of potentially detrimental "memories" associated with prior stressors. For example, throughout the reproductive phase of plants, numerous stress-induced epigenetic alterations are eradicated, a process referred to as reprogramming, which guarantees that progeny commence with a "clean slate" unless advantageous traits are deemed essential (Lin & Hsieh, 2017).

The concept of epigenetic memory is particularly vital for plants owing to their sessile characteristics; they lack mobility to evade stressors and must consequently cultivate robust internal mechanisms to manage such adversities. Through processes such as DNA methylation and histone modification, plants forge a flexible yet enduring response system that facilitates the prompt activation of stress-responsive genes (Tricker, 2015). Furthermore, epigenetic memory equips plants with the ability to precondition their defenses for recurrent stress episodes, thereby augmenting their survival prospects in fluctuating environments (Iwasaki & Paszkowski,

2014). Thus, epigenetic memory within plants constitutes a dynamic and adaptive mechanism that enables responses to environmental stressors. By modulating DNA methylation, histone alterations, and utilizing small RNAs, plants establish a durable yet malleable response to stress that may persist across generations (Lin & Hsieh, 2017). This memory mechanism not only underpins immediate stress responses but also holds considerable implications for long-term adaptation and evolutionary resilience in plants.

Transgenerational Inheritance of Stress Tolerance

The phenomenon of transgenerational inheritance of stress tolerance in plant species pertains to the capacity of offspring to acquire adaptive characteristics from parental plants that have undergone exposure to environmental stressors. This concept holds particular significance for plant organisms due to their inherent immobility in adapting to fluctuating environmental conditions (Iwasaki & Paszkowski, 2014). Through the transmission of stress-related adaptations via mechanisms such as DNA methylation, histone modifications, and small RNAs, plants can equip future generations to confront analogous environmental adversities. Comprehending these mechanisms not only bears implications for the domains of plant evolution and ecology but also opens potential pathways for enhancing agricultural productivity through epigenetic interventions (Bilichak & Kovalchuk, 2016).

Mechanisms of Transgenerational Stress Inheritance

DNA methylation constitutes a pivotal epigenetic mechanism implicated in the transgenerational inheritance of stress tolerance. Alterations in methylation can endure in the progeny, thereby influencing gene expression in a manner that augments resilience to stressors such as drought, salinity, and pathogenic invasion. For instance, investigations involving *Oryza sativa* (rice) have elucidated that DNA methylation patterns instigated by heavy metal exposure were reliably passed across generations, thereby conferring increased tolerance to similar stressors in offspring (Ou et al., 2012). Beyond DNA methylation, histone modifications serve as a mechanism for stress memory by modifying chromatin architecture, thereby either facilitating or constraining access to specific genes associated with stress response (Molinier et al., 2006). Such modifications are occasionally stable enough to endure through both mitotic and meiotic divisions, thereby reaching progeny. Research conducted on *Arabidopsis thaliana* has demonstrated that stress-induced histone modifications can elevate rates of homologous recombination, a process that promotes genomic adaptability in response to environmental stressors (Molinier et al., 2006).

Small RNAs, particularly small interfering RNAs (siRNAs), play a crucial role in the maintenance of DNA methylation and histone modifications, particularly under stressful conditions. This mechanism assists in the sustained expression of genes conferring stress tolerance across generations. For instance, *Nicotiana benthamiana* plants afflicted by viral infections exhibited enhanced resilience to salinity and osmotic pressures in their offspring, attributable to the action of small RNAs, indicating that stress tolerance induced by viral exposure can transcend a singular generation (Hernández-Walias et al., 2022).

Examples of Transgenerational Stress Tolerance

The exposure of *Arabidopsis thaliana* to saline and thermal stress has been empirically demonstrated to enhance the tolerance of its progeny; however, this phenomenon does not consistently endure over time. In a controlled experimental framework, *A. thaliana* specimens subjected to thermal stress displayed expedited flowering in their offspring, representing an adaptive mechanism for swift reproductive success under adverse conditions. Nevertheless, after two generations devoid of stress exposure, these transgenerational manifestations were predominantly reversible, thereby underscoring the malleable and context-sensitive characteristics of stress memory (Suter & Widmer, 2013). Additionally, viral infection can act as a catalyst for transgenerational stress tolerance. Research has indicated that *Nicotiana* species afflicted with specific viral pathogens acquired a heritable resilience to subsequent abiotic stressors, including salinity and drought. This virus-induced epigenetic adaptation illuminates the capacity of pathogens to prepare plants for heightened resilience in arduous environments, albeit it also introduces fitness trade-offs, exemplified by diminished seed yield in ensuing generations (Hernández-Walias et al., 2022). In *Oryza sativa*, exposure to heavy metal stress has been shown to alter DNA methylation patterns across generations, with these epigenetic modifications bestowing augmented resilience to analogous stressors. Offspring of stressed plants exhibited superior growth and survival metrics upon re-exposure to heavy metals, thereby exemplifying the manner in which specific environmental stimuli can foster inheritable epigenetic adaptations that enhance survival rates (Ou et al., 2012). In short, the investigation of transgenerational inheritance of stress tolerance is imperative for elucidating plant adaptation and resilience in response to environmental adversities. Through mechanisms such as DNA methylation, histone modifications, and small RNAs, plants possess the capability to transmit adaptive traits to their progeny, thereby facilitating ongoing survival and improved functionality under recurring stress conditions. By leveraging these epigenetic insights, agricultural biotechnology can further investigate methodologies to breed crops resilient to stress, thereby contributing to sustainable food production (Iwasaki & Paszkowski, 2014).

Applications and Future Perspectives

Epigenomics presents considerable potential for enhancing crop resilience and productivity in the face of environmental adversities. The subsequent sections delve into pioneering applications and prospective avenues in epigenetic research, encompassing epigenetic engineering, multi-omics integration, and epigenomic breeding.

EPIGENENTIC ENGINEERING

Epigenetic engineering constitutes a groundbreaking domain that focuses on the modification of epigenetic marks to regulate gene expression without necessitating alterations to the DNA sequence itself. In contrast to conventional genetic engineering, which entails direct modifications to the DNA, epigenetic engineering emphasizes reversible alterations such as DNA methylation and histone modifications that influence gene expression dynamics (Sun et al., 2021). These modifications can be meticulously calibrated to either activate or repress specific genes, providing an advanced methodology for augmenting plant resilience to environmental stresses, including drought, salinity, and temperature variability. As environmental challenges escalate due to climate change, epigenetic engineering has emerged as a promising strategy for developing climate-resilient crops endowed with enhanced stress tolerance (Varotto et al., 2020).

Mechanisms of Epigenetic Engineering

The fundamental objectives of epigenetic engineering encompass two predominant categories of modifications: DNA methylation and histone modifications. DNA methylation predominantly acts to inhibit gene expression through the addition of methyl groups to cytosine bases, thereby obstructing the accessibility of transcription factors to the DNA. In contrast, histone modifications—such as the acetylation, methylation, or phosphorylation of histone proteins—have the capacity to either enhance or suppress gene expression by modifying chromatin architecture (Vaissiere, Sawan, & Herceg, 2008). Acetylation, for instance, typically facilitates gene expression by relaxing chromatin structure, while methylation may either activate or repress gene activity contingent upon the specific histone and modification site involved (Cheung & Lau, 2005). Through the manipulation of these epigenetic markers, researchers can dynamically and reversibly influence gene expression, rendering epigenetic engineering a versatile strategy for the management of plant stress responses (Agarwal et al., 2020).

CRISPR-Based Epigenome Editing Tools

Recent innovations in CRISPR-based technologies have revolutionized the field of epigenetic engineering, permitting precise governance over particular gene loci. CRISPR-based epigenome editing employs a modified iteration of the CRISPR-Cas9 system, referred to as dCas9 (deactivated Cas9), which is devoid of DNA-cleaving capabilities (Fadul, Arshad, & Mehmood, 2023). When conjoined with diverse epigenetic modifiers, dCas9 can be strategically directed to specific genomic regions to either introduce or eliminate epigenetic marks, thereby modulating gene expression without instigating permanent alterations to the DNA sequence (Brocken & Dame, 2018). For example, dCas9 can be associated with DNA methyl-transferases to augment DNA methylation and suppress target gene expression, or with histone acetyltransferases to facilitate gene activation (Vojta et al., 2016). The specificity and adaptability of CRISPR-based epigenome editing instruments empower researchers to meticulously calibrate gene regulation in manners that were previously elusive. By selectively targeting genes responsive to stress, CRISPR-dCas9 methodologies can be employed to enhance the expression of genes implicated in drought or salt tolerance, or to silence genes detrimental to stress resilience (Moradpour & Abdullah, 2019). This targeted strategy provides a compelling mechanism for the development of crops that are more proficient in coping with environmental stresses, potentially fostering more resilient agricultural systems.

Applications in Plant Stress Resilience

A particularly promising domain of epigenetic engineering resides in augmenting plant resilience to abiotic stresses. For instance, the enhancement of drought tolerance in plants has been associated with the activation of genes that govern water retention and stress signaling pathways. By employing CRISPR-based epigenome editing to upregulate these genes, researchers can enhance a plant's capacity to conserve water and endure conditions of drought. Likewise, genes that facilitate salt tolerance can be activated to enable plants to flourish in saline soils, which are increasingly prevalent in arid and semi-arid regions (Nazir et al., 2022). Beyond the direct enhancement of tolerance to abiotic stresses, epigenetic engineering can also be utilized to stabilize crop yields under fluctuating environmental conditions. For example, by targeting genes that modulate flowering time or root architecture, researchers can develop crop varieties that are optimally adapted to specific climatic or soil conditions (Chennakesavulu et al., 2021). This application harbors substantial potential for acclimating crops to localized environmental contexts, diminishing dependence on chemical inputs, and promoting sustainable agricultural practices (Bhat et al., 2021).

Epigenetic Memory and Heritability of Engineered Traits

A captivating dimension of epigenetic modifications pertains to their capacity for "epigenetic memory," wherein environmental or engineered epigenetic alterations can endure through cellular divisions and, in certain instances, be transmitted to subsequent generations (Iwasaki & Paszkowski, 2014). Although not all epigenetic modifications exhibit heritability, this phenomenon implies that traits imbued with stress resilience via epigenetic engineering possess the potential to be inherited, thereby fostering multi-generational resilience without necessitating continual modifications (Wibowo et al., 2016). While the stability and heritability of epigenetically engineered traits in flora necessitate further investigation, the prospect of transmitting advantageous traits through epigenetic memory signifies a notable benefit for sustainable agricultural practices (Bond & Baulcombe, 2014).

Challenges and Future Directions

In spite of its considerable potential, epigenetic engineering in plants encounters numerous obstacles. The intricate interplay among various epigenetic marks and the environmental specificity associated with epigenetic responses complicate the prediction of outcomes resulting from engineered modifications (Ahmad et al., 2021). Furthermore, although CRISPR-based methodologies exhibit a high degree of precision, off-target effects remain a significant concern, particularly when epigenetic modifications may inadvertently influence non-target genes (Guo et al., 2023). The pursuit of more sophisticated delivery systems and the comprehension of the long-term stability of these modifications are active domains of research, essential for the safe execution of epigenetic engineering in the enhancement of crops (Lee et al., 2020). Ethical considerations and regulatory challenges also represent substantial factors in the acceptance of epigenetically modified crops. Since epigenetic engineering does not modify the foundational DNA, it may encounter fewer regulatory obstacles in comparison to traditional genetic modification; nevertheless, public perception and acceptance will exert a considerable influence on its assimilation into agricultural practices (Gao, 2018).

INTEGRATION OF MULTI-OMICS

The amalgamation of multi-omics technologies—genomics, transcriptomics, proteomics, metabolomics, and epigenomics—constitutes a revolutionary methodology for elucidating the intricate molecular networks that underpin plant responses to environmental stressors (Yang et al., 2021). By encompassing diverse molecular layers, multi-omics furnishes an extensive overview of how plants perceive, react to, and acclimatize to stresses such as drought, salinity, heat, and nutrient deficiencies (Roychowdhury et al., 2023). This comprehensive perspective not only augments our understanding of the fundamental mechanisms governing plant resilience but also facilitates the identification of critical biomarkers, gene regulatory networks, and metabolic pathways associated with stress tolerance (Jamil et al., 2020). By integrating multi-omics data with advanced computational methodologies, including machine learning and network analysis, researchers are enabled to predict complex traits and expedite breeding strategies, thereby propelling the advancement of climate-resilient and high-yield crops (Mahmood et al., 2022).

Key Components of Multi-Omics and Their Roles in Stress Response

Genomics

Genomic investigations establish the critical basis for multi-omics by delineating the comprehensive DNA sequence of a plant, elucidating genetic variants, and ascertaining the architecture of genes associated with stress responses. Genomics is indispensable for the identification of single nucleotide polymorphisms (SNPs) and gene families that are integral to stress tolerance, thereby providing the essential groundwork for the selection of advantageous traits in breeding initiatives (Cushman & Bohnert, 2000).

Transcriptomics

Transcriptomic assessments scrutinize the complete ensemble of RNA transcripts expressed under particular conditions, thereby illuminating alterations in gene expression in response to environmental stressors. Utilizing methodologies such as RNA sequencing, transcriptomics can discern genes that are either upregulated or downregulated and unveil regulatory elements that play a role in stress responses, thereby offering insights into the dynamic shifts in gene expression that occur within challenging environments (Wang et al., 2020).

Proteomics

Proteomics concentrates on the protein composition within cells, examining protein abundance, post-translational modifications, and interactions. Given that proteins represent the functional entities that instigate cellular responses, proteomic analyses yield direct insights into the molecular mechanisms underpinning stress responses, accentuating the roles of enzymes, transcription factors, and structural proteins that facilitate adaptation (Hakeem et al., 2012).

Metabolomics

Metabolomics encompasses the profiling of small-molecule metabolites that act as intermediates and terminal products of cellular processes (Goufo et al., 2017). Metabolites can serve as indicators of the physiological condition of plants and unveil metabolic adjustments undertaken in response to stress, such as the accumulation of osmoprotectants during drought conditions. Consequently, metabolomics provides a snapshot of the biochemical pathways that are activated in response to stress stimuli (Llanes et al., 2018).

Epigenomics

Epigenomic investigations delve into modifications of the genome, encompassing DNA methylation and histone modifications, which regulate gene expression without necessitating alterations to the DNA sequence itself (Forestan et al., 2019). Epigenomics is vital for comprehending how plants "remember" prior stress exposures and adapt over time through mechanisms such as epigenetic inheritance, thereby contributing an additional regulatory layer that complements genetic information (Nguyen et al., 2022).

Integration and Analysis of Multi-Omics Data

The integration of data across diverse omics layers provides a more nuanced perspective of plant stress responses, as each omics layer presents distinctive insights into various molecular mechanisms. By consolidating these layers, researchers can acquire a systems-level comprehension of plant adaptation, identify traits indicative of stress resilience, and pinpoint candidate genes or pathways for the enhancement of crops. For instance, the integration of genomics and transcriptomic data can correlate genetic variations with specific gene expression profiles under stress conditions, thereby facilitating the identification of candidate genes that contribute to stress tolerance (Jamil et al., 2020). The incorporation of proteomics and metabolomics data further empowers researchers to trace these gene expression changes back to the proteins and metabolites they yield, thereby elucidating the biochemical pathways activated in response to stress (Pinu et al., 2019). Concurrently, epigenomic data can elucidate the regulatory mechanisms governing gene expression through modifications to DNA and chromatin structure, demonstrating how plants dynamically modulate their responses at the epigenetic level. Advanced computational methodologies, including machine learning, network analysis, and statistical modeling, play an instrumental role in the analysis of multi-omics data. Machine learning algorithms are capable of detecting patterns and associations within extensive datasets, facilitating the prediction of intricate traits based on multi-omics markers (Picard et al., 2021). Conversely, network analysis can elucidate interactions among genes, proteins, and metabolites, identifying critical regulatory nodes and pathways implicated in stress tolerance (Rajeev et al., 2024). These analytical tools empower the extraction of significant insights from vast datasets, assisting researchers in constructing predictive models of plant responses to environmental stressors.

Applications in Crop Improvement Programs

The incorporation of multi-omics methodologies into crop enhancement initiatives provides substantial benefits, particularly in the generation of cultivars exhibiting improved resilience and agricultural productivity. Through the identification of biomarkers linked to stress tolerance, multi-omics facilitates the selection and propagation of resilient phenotypes by breeders with increased efficacy (Zenda et al., 2021). This data-centric methodology expedites the breeding timeline, thereby diminishing the duration necessary to cultivate crops capable of flourishing under fluctuating climatic conditions (Mahmood et al., 2022). A prominent application encompasses marker-assisted selection (MAS), wherein molecular markers obtained from multi-omics investigations steer the selection of advantageous traits (Yang et al., 2021). Multi-omics datasets are adept at identifying specific gene variants, proteins, or metabolites that are associated with desirable traits, thereby affording breeders the opportunity to select for these attributes with heightened accuracy (Zhang et al., 2022). For example, the identification of a metabolite associated with osmotic stress tolerance in a drought-resistant plant variety can provide critical insights for breeding initiatives aimed at enhancing water-use efficiency in agricultural crops (Jain et al., 2019). Additionally, the integration of multi-omics supports the creation of transgenic or gene-edited plants featuring targeted modifications. By elucidating essential regulatory genes and pathways that convey stress tolerance, researchers are empowered to employ techniques such as CRISPR-Cas9 to introduce or modify genes pertinent to these traits, thereby augmenting resilience (Pathak et al., 2018). Furthermore, epigenetic markers discerned through epigenomic analyses can also direct gene editing efforts, offering potential targets for the activation of stress-responsive genes without necessitating alterations to the fundamental DNA sequence (Roychowdhury et al., 2023).

Challenges and Future Directions

Despites its potential, the consolidation of multi-omics data entails numerous challenges. The sheer volume and intricacy of multi-omics data render effective management, processing, and analysis challenging, necessitating a robust bioinformatics infrastructure and specialized expertise. Moreover, issues related to data compatibility and standardization present formidable obstacles, as each omics technology produces data in distinct formats and scales, thereby complicating integration endeavors. Furthermore, multi-omics investigations demand considerable computational resources and sophisticated algorithms that are capable of accommodating the high dimensionality of the data. The formulation of predictive models that accurately encapsulate plant responses within authentic field conditions continues to be a persisting challenge, given that laboratory settings may not entirely replicate natural environments. Nevertheless, with ongoing progress in data analytics, artificial intelligence, and machine learning, these challenges are progressively being mitigated. As multi-omics platforms advance, future investigative efforts are anticipated to concentrate on refining data integration methodologies and crafting more precise predictive models for crop enhancement. The emergence of cloud computing and accessible bioinformatics platforms is projected to democratize multi-omics research, rendering it more attainable for research teams globally to exploit these technologies within plant breeding programs.

CONCLUSION

The amalgamation of multi-omics technologies delivers an extensive and robust framework for elucidating plant stress responses and informing crop enhancement strategies. By assimilating genetic, transcriptomic, proteomic, metabolomic, and epigenomic datasets, multi-omics facilitates a systemic comprehension of the mechanisms by which plants acclimatize to environmental adversities, thereby aiding the discernment of biomarkers, regulatory networks, and pathways crucial for resilience. This integrative methodology, in conjunction with sophisticated data analytics, is revolutionizing the domain of plant breeding, presenting a data-centric trajectory for the development of cultivars capable of enduring the challenges posed by a fluctuating climate.

REFERENCES

Abdulraheem, M. I., Xiong, Y., Moshood, A. Y., Cadenas-Pliego, G., Zhang, H., & Hu, J. (2024). Mechanisms of Plant Epigenetic Regulation in Response to Plant Stress: Recent Discoveries and Implications. *Plants (Basel, Switzerland)*, *13*(2), 163. DOI: 10.3390/plants13020163 PMID: 38256717

Adhikari, D., & Adhikari, D. (2023). Role of epigenetics and the high-throughput sensing techniques to detect stress adaptation mechanisms in crop plants: A mini review. *Innovations in Agriculture*, *6*, e32855. DOI: 10.25081/ia.2023-1986

Adusumalli, S., Mohd Omar, M. F., Soong, R., & Benoukraf, T. (2015). Methodological aspects of whole-genome bisulfite sequencing analysis. *Briefings in Bioinformatics*, *16*(3), 369–379. PMID: 24867940

Agarwal, G., Kudapa, H., Ramalingam, A., Choudhary, D., Sinha, P., Garg, V., Singh, V., Patil, G., Pandey, M., Nguyen, H., Guo, B., Sunkar, R., Niederhuth, C., & Varshney, R. (2020). Epigenetics and epigenomics: Underlying mechanisms, relevance, and implications in crop improvement. *Functional & Integrative Genomics*, *20*, 739–761. DOI: 10.1007/s10142-020-00756-7 PMID: 33089419

Ahmad, A., Ashraf, S., Munawar, N., Jamil, A., Ghaffar, A., & Shahbaz, M. (2021). CRISPR/Cas-Mediated Abiotic Stress Tolerance in Crops. 177-211. https://doi.org/ .DOI: 10.1007/978-981-15-7142-8_6

Ahmad, Y., Haakim, Z., Iqbal, J., Abbasi, B. A., Mahmood, T., & Kazi, M. (2024b). Technological Innovations for Abiotic Stress Resistance in Horticultural Crops. OMICs-based Techniques for Global Food Security, 233-244.

Ahmad, Y., Haider, S., Iqbal, J., Abbasi, B. A., Yaseen, T., & Mahmood, T. (2022a). The mechanisms of genome editing technologies in crop plants. In *Principles and Practices of OMICS and Genome Editing for Crop Improvement* (pp. 295–313). Springer International Publishing.

Ahmad, Y., Haider, S., Iqbal, J., Abbasi, B. A., Yaseen, T., & Mahmood, T. (2022b). The mechanisms of genome editing technologies in crop plants. In *Principles and Practices of OMICS and Genome Editing for Crop Improvement* (pp. 295–313). Springer International Publishing.

Ahmad, Y., Haider, S., Iqbal, J., Naseer, S., Attia, K. A., Mohammed, A. A., & Mahmood, T. (2024a). In-silico analysis and transformation of OsMYB48 transcription factor driven by CaMV35S promoter in model plant–Nicotiana tabacum L. conferring abiotic stress tolerance. *GM Crops and Food: Biotechnology in Agriculture and the Food Chain*, *15*(1), 130–149. PMID: 38551174

Akhter, Z., Bi, Z., Ali, K., Sun, C., Fiaz, S., Haider, F. U., & Bai, J. (2021). In Response to Abiotic Stress, DNA Methylation Confers EpiGenetic Changes in Plants. *Plant, 10*, 1096. https://doi.org/DOI: 10.3390/plants10061096

Al-Harrasi, I., Al-Yahyai, R., & Yaish, M. W. (2018). Differential DNA methylation and transcription profiles in date palm roots exposed to salinity. *PLoS One*, *13*(1), e0191492. PMID: 29352281

Ali, I., Salah, K. B. H., Sher, H., Ali, H., Ullah, Z., Ali, A., & Mahmood, T. (2022). Drought stress enhances the efficiency of floral dip method of Agrobacterium-mediated transformation in Arabidopsis thaliana. *Brazilian Journal of Biology*, *84*, e259326. PMID: 35703626

Avramova, Z. (2015). Transcriptional 'memory'of a stress: Transient chromatin and memory (epigenetic) marks at stress-response genes. *The Plant Journal*, *83*(1), 149–159. PMID: 25788029

Bailey, T., Krajewski, P., Ladunga, I., Lefebvre, C., Li, Q., Liu, T., & Zhang, J. (2013). Practical guidelines for the comprehensive analysis of ChIP-seq data. *PLoS Computational Biology*, *9*(11), e1003326. PMID: 24244136

Bhadouriya, S. L., Mehrotra, S., Basantani, M. K., Loake, G. J., & Mehrotra, R. (2021). Role of Chromatin Architecture in Plant Stress Responses: An Update. *Frontiers in Plant Science*, *11*, 603380. DOI: 10.3389/fpls.2020.603380 PMID: 33510748

Bhat, M., Mir, R., Kumar, V., Shah, A., Zargar, S., Rahman, S., & Jan, A. (2021). Mechanistic insights of CRISPR/Cas mediated genome editing towards enhancing abiotic stress tolerance in plants. *Physiologia Plantarum*. Advance online publication. DOI: 10.1111/ppl.13359 PMID: 33576013

Bilichak, A., & Kovalchuk, I. (2016). Transgenerational response to stress in plants and its application for breeding. *Journal of Experimental Botany*, *67*(7), 2081–2092. PMID: 26944635

Bond, D., & Baulcombe, D. (2014). Small RNAs and heritable epigenetic variation in plants. *Trends in Cell Biology*, *24*(2), 100–107. DOI: 10.1016/j.tcb.2013.08.001 PMID: 24012194

Boycheva, I., Vassileva, V., & Iantcheva, A. (2014). Histone Acetyltransferases in Plant Development and Plasticity. *Current Genomics*, *15*, 28–37. PMID: 24653661

Brocken, D., Tark-Dame, M., & Dame, R. (2018). dCas9: A Versatile Tool for Epigenome Editing. *Current Issues in Molecular Biology*, *26*, 15–32. DOI: 10.21775/cimb.026.015 PMID: 28879853

Brukhin, V., & Albertini, E. (2021). Epigenetic modifications in plant development and reproduction. *Epigenomes*, *5*(4), 25. PMID: 34968249

Buenrostro, J. D., Wu, B., Chang, H. Y., & Greenleaf, W. J. (2015). ATAC-seq: A method for assaying chromatin accessibility genome-wide. *Current Protocols in Molecular Biology*, *109*(1), 21–29. PMID: 25559105

Can, S. N., Nunn, A., Galanti, D., Langenberger, D., Becker, C., Volmer, K., & Rensing, S. A. (2021). The EpiDiverse plant epigenome-wide association studies (EWAS) pipeline. *Epigenomes*, *5*(2), 12. PMID: 34968299

Chennakesavulu, K., Singh, H., Trivedi, P., Jain, M., & Yadav, S. (2021). State-of-the-Art in CRISPR Technology and Engineering Drought, Salinity, and Thermo-tolerant crop plants. *Plant Cell Reports*, *41*, 815–831. DOI: 10.1007/s00299-021-02681-w PMID: 33742256

Cheung, P., & Lau, P. (2005). Epigenetic regulation by histone methylation and histone variants. *Molecular Endocrinology (Baltimore, Md.)*, *19*(3), 563–573. DOI: 10.1210/ME.2004-0496 PMID: 15677708

Chinnusamy, V., & Zhu, J. K. (2009). Epigenetic regulation of stress responses in plants. *Current Opinion in Plant Biology*, *12*(2), 133–139. PMID: 19179104

Cong, W., Miao, Y., Xu, L., Zhang, Y., Yuan, C., Wang, J., & Ou, X. (2019). Transgenerational memory of gene expression changes induced by heavy metal stress in rice (Oryza sativa L.). *BMC Plant Biology*, *19*, 1–14. PMID: 31248374

Cortijo, S., Charoensawan, V., Roudier, F., & Wigge, P. A. (2018). Chromatin immunoprecipitation sequencing (ChIP-Seq) for transcription factors and chromatin factors in Arabidopsis thaliana roots: from material collection to data analysis. *Root Development: Methods and Protocols*, 231-248.

Cushman, J., & Bohnert, H. (2000). Genomic approaches to plant stress tolerance. *Current Opinion in Plant Biology*, *3*(2), 117–124. DOI: 10.1016/S1369-5266(99)00052-7 PMID: 10712956

Das, C., Adhikari, S., Bhattacharya, A., Chakraborty, S., Mondal, P., Yadav, S. S., & Pandita, T. K. (2023). Epigenetic–Metabolic Interplay in the DNA Damage Response and Therapeutic Resistance of Breast Cancer. *Cancer Research*, *83*(5), 657–666. PMID: 36661847

de Carvalho Augusto, R., Rey, O., Cosseau, C., Chaparro, C., Vidal-Dupiol, J., Allienne, J. F., & Grunau, C. (2020). A simple ATAC-seq protocol for population epigenetics. *Wellcome open research, 5*.

De Kort, H., Toivainen, T., Van Nieuwerburgh, F., Andrés, J., Hytönen, T. P., & Honnay, O. (2022). Signatures of polygenic adaptation align with genome-wide methylation patterns in wild strawberry plants. *The New Phytologist*, *235*(4), 1501–1514. PMID: 35575945

Dong, H., Huang, Y., & Wang, K. (2021). The development of herbicide resistance crop plants using CRISPR/Cas9-mediated gene editing. *Genes*, *12*(6), 912. PMID: 34204760

Fadul, S., Arshad, A., & Mehmood, R. (2023). CRISPR-based epigenome editing: Mechanisms and applications. *Epigenomics*. Advance online publication. DOI: 10.2217/epi-2023-0281 PMID: 37990877

Forestan, C., Farinati, S., Zambelli, F., Pavesi, G., Rossi, V., & Varotto, S. (2019). Epigenetic signatures of stress adaptation and flowering regulation in response to extended drought and recovery in Zea mays. *Plant, Cell & Environment*. Advance online publication. DOI: 10.1111/pce.13660 PMID: 31677283

Gao, C. (2018). The future of CRISPR technologies in agriculture. *Nature Reviews. Molecular Cell Biology*, *19*, 275–276. DOI: 10.1038/nrm.2018.2 PMID: 29382940

Goufo, P., Moutinho-Pereira, J., Jorge, T., Correia, C., Oliveira, M., Rosa, E., António, C., & Trindade, H. (2017). Cowpea (Vigna unguiculata L. Walp.) Metabolomics: Osmoprotection as a Physiological Strategy for Drought Stress Resistance and Improved Yield. *Frontiers in Plant Science*, *8*. Advance online publication. DOI: 10.3389/fpls.2017.00586 PMID: 28473840

Grehl, C., Wagner, M., Lemnian, I., Glaser, B., & Grosse, I. (2020). Performance of mapping approaches for whole-genome bisulfite sequencing data in crop plants. *Frontiers in Plant Science*, *11*, 176. PMID: 32256504

Guo, C., Ma, X., Gao, F., & Guo, Y. (2023). Off-target effects in CRISPR/Cas9 gene editing. *Frontiers in Bioengineering and Biotechnology*, *11*. Advance online publication. DOI: 10.3389/fbioe.2023.1143157 PMID: 36970624

Gupta, K., & Garg, R. (2020). Method for bisulfite sequencing data analysis for whole-genome level DNA methylation detection in legumes. *Legume Genomics: Methods and Protocols*, 127-145.

Haider, S., Iqbal, J., Naseer, S., Shaukat, M., Abbasi, B. A., Yaseen, T., & Mahmood, T. (2022b). Unfolding molecular switches in plant heat stress resistance: A comprehensive review. *Plant Cell Reports*, *41*(3), 775–798. PMID: 34401950

Haider, S., Iqbal, J., Naseer, S., Yaseen, T., Shaukat, M., Bibi, H., & Mahmood, T. (2021a). Molecular mechanisms of plant tolerance to heat stress: Current landscape and future perspectives. *Plant Cell Reports*, ***•••***, 1–25. PMID: 33890138

Haider, S., Iqbal, J., Shaukat, M., Naseer, S., & Mahmood, T. (2021b). The epigenetic chromatin-based regulation of somatic heat stress memory in plants. *Plant Gene*, *27*, 100318.

Haider, S., Raza, A., Iqbal, J., Shaukat, M., & Mahmood, T. (2022a). Analyzing the regulatory role of heat shock transcription factors in plant heat stress tolerance: A brief appraisal. *Molecular Biology Reports*, *49*(6), 5771–5785. PMID: 35182323

Hakeem, K., Chandna, R., Ahmad, P., Iqbal, M., & Ozturk, M. (2012). Relevance of proteomic investigations in plant abiotic stress physiology. *OMICS: A Journal of Integrative Biology*, *16*(11), 621–635. DOI: 10.1089/omi.2012.0041 PMID: 23046473

Hao, Z., Fan, C., Cheng, T., Su, Y., Wei, Q., & Li, G. (2015). Genome-wide identification, characterization and evolutionary analysis of long intergenic noncoding RNAs in cucumber. *PLoS One*, *10*(3), e0121800. PMID: 25799544

Hendrickson, D. G., Soifer, I., Wranik, B. J., Botstein, D., & Scott McIsaac, R. (2018). Simultaneous profiling of DNA accessibility and gene expression dynamics with ATAC-seq and RNA-seq. *Computational Cell Biology: Methods and Protocols*, 317-333.

Hernández-Walias, F. J., García, M., Moreno, M., Giannoukos, I., González, N., Sanz-García, E., & Tenllado, F. (2022). Transgenerational tolerance to salt and osmotic stresses induced by plant virus infection. *International Journal of Molecular Sciences*, *23*(20), 12497. PMID: 36293354

Hsieh, J. W. A., Lin, P. Y., Wang, C. T., Lee, Y. J., Chang, P., Lu, R. J. H., & Wang, C. J. R. (2024). Establishing an optimized ATAC-seq protocol for the maize. *Frontiers in Plant Science*, *15*, 1370618. PMID: 38863553

Huang, K. Y. Y., Huang, Y. J., & Chen, P. Y. (2018). BS-Seeker3: Ultrafast pipeline for bisulfite sequencing. *BMC Bioinformatics*, *19*, 1–4. PMID: 29614954

Ijaz, S., Iqbal, J., Abbasi, B. A., Tufail, A., Ullah, Z., Sharifi-Rad, J., & Uddin, S. (2023). Biofortification: Lessons from the Past and Strategies for Future Food Security. In *Legumes Biofortification* (pp. 521–545). Springer International Publishing.

Ijaz, S., Iqbal, J., Abbasi, B. A., Yaseen, T., Rehman, S., Kazi, M., & Mahmood, T. (2024a). Role of OMICS-Based Technologies in Plant Sciences. OMICs-based Techniques for Global Food Security, 45-66.

Ijaz, S., Iqbal, J., Abbasi, B. A., Yaseen, T., Rehman, S., Kazi, M. & Mahmood, T. (2024b). Role of OMICS-Based Technologies in Plant Sciences. OMICs-based Techniques for Global Food Security, 45-66.

Iwasaki, M., & Paszkowski, J. (2014). Epigenetic memory in plants. *The EMBO Journal, 33*(18), 1987–1998. PMID: 25104823

Jain, D., Ashraf, N., Khurana, J., & Kameshwari, M. (2019). *The 'Omics' Approach for Crop Improvement Against Drought Stress*. Sustainable Development and Biodiversity., DOI: 10.1007/978-3-319-91956-0_8

Jamil, I., Remali, J., Azizan, K., Muhammad, N., Arita, M., Goh, H., & Aizat, W. (2020). Systematic Multi-Omics Integration (MOI) Approach in Plant Systems Biology. *Frontiers in Plant Science, 11*. Advance online publication. DOI: 10.3389/fpls.2020.00944 PMID: 32754171

Ji, L., Sasaki, T., Sun, X., Ma, P., Lewis, Z. A., & Schmitz, R. J. (2014). Methylated DNA is over-represented in whole-genome bisulfite sequencing data. *Frontiers in Genetics, 5*, 341. PMID: 25374580

Kakoulidou, I., Avramidou, E. V., Baránek, M., Brunel-Muguet, S., Farrona, S., Johannes, F., Kaiserli, E., Lieberman-Lazarovich, M., Martinelli, F., Mladenov, V., Testillano, P. S., Vassileva, V., & Maury, S. (2021). Epigenetics for Crop Improvement in Times of Global Change. *Biology (Basel), 10*(8), 766. DOI: 10.3390/biology10080766 PMID: 34439998

Kaufmann, K., Muino, J. M., Østerås, M., Farinelli, L., Krajewski, P., & Angenent, G. C. (2010). Chromatin immunoprecipitation (ChIP) of plant transcription factors followed by sequencing (ChIP-SEQ) or hybridization to whole genome arrays (ChIP-CHIP). *Nature Protocols, 5*(3), 457–472. PMID: 20203663

Kinoshita, T., & Seki, M. (2014). Epigenetic memory for stress response and adaptation in plants. *Plant & Cell Physiology, 55*(11), 1859–1863. PMID: 25298421

Kou, S., Gu, Q., Duan, L., Liu, G., Yuan, P., Li, H., & Liu, L. (2022). Genome-wide bisulphite sequencing uncovered the contribution of DNA methylation to rice short-term drought memory formation. *Journal of Plant Growth Regulation, 41*(7), 2903–2917.

Kyung, J., Jeon, M., & Lee, I. (2022). Recent advances in the chromatin-based mechanism of FLOWERING LOCUS C repression through autonomous pathway genes. *Frontiers in Plant Science, 13*, 964931. PMID: 36035698

Lämke, J., & Bäurle, I. (2017). Epigenetic and chromatin-based mechanisms in environmental stress adaptation and stress memory in plants. *Genome Biology*, *18*, 1–11. PMID: 28655328

Lee, J., Mazarei, M., Pfotenhauer, A., Dorrough, A., Poindexter, M., Hewezi, T., Lenaghan, S., Graham, D., & Stewart, C. (2020). Epigenetic Footprints of CRISPR/Cas9-Mediated Genome Editing in Plants. *Frontiers in Plant Science*, *10*. Advance online publication. DOI: 10.3389/fpls.2019.01720 PMID: 32117329

Lee, S. C., Adams, D. W., Ipsaro, J. J., Cahn, J., Lynn, J., Kim, H. S., & Martienssen, R. A. (2023). Chromatin remodeling of histone H3 variants underlies epigenetic inheritance of DNA methylation. *bioRxiv*.

Lin, J. Y., & Hsieh, T. F. (2017). Epigenetic reprogramming during plant reproduction. *Plant Epigenetics*, 405-425.

Liu, D., Zhao, L., Wang, Z., Zhou, X., Fan, X., Li, Y., & Jiang, Y. (2019). EWASdb: Epigenome-wide association study database. *Nucleic Acids Research*, *47*(D1), D989–D993. PMID: 30321400

Liu, J., & He, Z. (2020). Small DNA methylation, big player in plant abiotic stress responses and memory. *Frontiers in Plant Science*, *11*, 595603. PMID: 33362826

Llanes, A., Andrade, A., Alemano, S., & Luna, V. (2018). Metabolomic Approach to Understand Plant Adaptations to Water and Salt Stress., 133-144. https://doi.org/.DOI: 10.1016/B978-0-12-812689-9.00006-6

Lu, J., Johnston, A., Berichon, P., Ru, K. L., Korbie, D., & Trau, M. (2017). PrimerSuite: A high-throughput web-based primer design program for multiplex bisulfite PCR. *Scientific Reports*, *7*(1), 41328. PMID: 28117430

Luo, C., Fernie, A. R., & Yan, J. (2020). Single-cell genomics and epigenomics: Technologies and applications in plants. *Trends in Plant Science*, *25*(10), 1030–1040. PMID: 32532595

Mahmood, U., Li, X., Fan, Y., Chang, W., Niu, Y., Li, J., & Lu, K. (2022). Multi-omics revolution to promote plant breeding efficiency. *Frontiers in Plant Science*, *13*, 1062952. PMID: 36570904

Miladinovic, D., Antunes, D., Yildirim, K., Bakhsh, A., Cvejić, S., Kondić-Špika, A., & Hilioti, Z. (2021). Targeted plant improvement through genome editing: From laboratory to field. *Plant Cell Reports*, *40*(6), 935–951. PMID: 33475781

Molinier, J., Ries, G., Zipfel, C., & Hohn, B. (2006). Transgeneration memory of stress in plants. *Nature*, *442*(7106), 1046–1049. PMID: 16892047

Moradpour, M., & Abdulah, S. (2019). CRISPR/dCas9 platforms in plants: Strategies and applications beyond genome editing. *Plant Biotechnology Journal, 18*, 32–44. DOI: 10.1111/pbi.13232 PMID: 31392820

Nakato, R., & Sakata, T. (2021). Methods for ChIP-seq analysis: A practical workflow and advanced applications. *Methods (San Diego, Calif.), 187*, 44–53. PMID: 32240773

Nazir, R., Mandal, S., Mitra, S., Ghorai, M., Das, N., Jha, N., Majumder, M., Pandey, D., & Dey, A. (2022). CRISPR/Cas genome-editing toolkit to enhance salt stress tolerance in rice and wheat. *Physiologia Plantarum, 13642*. Advance online publication. DOI: 10.1111/ppl.13642 PMID: 35099818

Nguyen, N., Vu, N., & Cheong, J. (2022). Transcriptional Stress Memory and Transgenerational Inheritance of Drought Tolerance in Plants. *International Journal of Molecular Sciences, 23*. Advance online publication. DOI: 10.3390/ijms232112918 PMID: 36361708

Nunez-Vazquez, R., Desvoyes, B., & Gutierrez, C. (2022). Histone variants and modifications during abiotic stress response. *Frontiers in Plant Science, 13*, 984702. PMID: 36589114

Ou, X., Zhang, Y., Xu, C., Lin, X., Zang, Q., Zhuang, T., ... & Liu, B. (2012). Transgenerational inheritance of modified DNA methylation patterns and enhanced tolerance induced by heavy metal stress in rice (Oryza sativa L.).

Pathak, R., Baunthiyal, M., Pandey, D., & Kumar, A. (2018). Augmentation of crop productivity through interventions of omics technologies in India: challenges and opportunities. *3 Biotech*, 8, 1-28. https://doi.org/.DOI: 10.1007/s13205-018-1473-y

Patten, D. K., Corleone, G., & Magnani, L. (2018). Chromatin immunoprecipitation and high-throughput sequencing (ChIP-Seq): tips and tricks regarding the laboratory protocol and initial downstream data analysis. *Epigenome Editing: Methods and Protocols*, 271-288.

Petereit, J., Bayer, P. E., Thomas, W. J., Tay Fernandez, C. G., Amas, J. C., Zhang, Y., Batley, J., & Edwards, D. (2022). Pangenomics and Crop Genome Adaptation in a Changing Climate. *Plants*, •••, 11. PMID: 35956427

Picard, M., Scott-Boyer, M., Bodein, A., Périn, O., & Droit, A. (2021). Integration strategies of multi-omics data for machine learning analysis. *Computational and Structural Biotechnology Journal, 19*, 3735–3746. DOI: 10.1016/j.csbj.2021.06.030 PMID: 34285775

Pinu, F., Beale, D., Paten, A., Kouremenos, K., Swarup, S., Schirra, H., & Wishart, D. (2019). Systems Biology and Multi-Omics Integration: Viewpoints from the Metabolomics Research Community. *Metabolites*, *9*. Advance online publication. DOI: 10.3390/metabo9040076 PMID: 31003499

Qiu, X., Feit, A. S., Feiglin, A., Xie, Y., Kesten, N., Taing, L., & Long, H. W. (2021). CoBRA: Containerized bioinformatics workflow for reproducible ChIP/ATAC-seq analysis. *Genomics, Proteomics & Bioinformatics*, *19*(4), 652–661. PMID: 34284136

Qiu, Y., & Huang, S. (2020). CTCF-mediated genome organization and leukemogenesis. *Leukemia*, *34*(9), 2295–2304. PMID: 32518417

Rajeev, A.C., G, P. & Pan, A. (2024). Integrated gene expression and network analysis identify drought-response genes and pathways in *Solanum*: A computational study. *Discover Applied Sciences*, *6*, 468. DOI: 10.1007/s42452-024-06025-7

Rajkumar, M. S., Shankar, R., Garg, R., & Jain, M. (2020). Bisulphite sequencing reveals dynamic DNA methylation under desiccation and salinity stresses in rice cultivars. *Genomics*, *112*(5), 3537–3548. PMID: 32278023

Rakyan, V. K., Down, T. A., Balding, D. J., & Beck, S. (2011). Epigenome-wide association studies for common human diseases. *Nature Reviews. Genetics*, *12*(8), 529–541. PMID: 21747404

Ramakrishnan, M., Satish, L., Kalendar, R., Narayanan, M., Kandasamy, S., Sharma, A., & Zhou, M. (2021). The dynamism of transposon methylation for plant development and stress adaptation. *International Journal of Molecular Sciences*, *22*(21), 11387. PMID: 34768817

Ramakrishnan, M., Zhang, Z., Mullasseri, S., Kalendar, R., Ahmad, Z., Sharma, A., & Wei, Q. (2022). Epigenetic stress memory: A new approach to study cold and heat stress responses in plants. *Frontiers in Plant Science*, *13*, 1075279. PMID: 36570899

Rehman, M., & Tanti, B. (2020). Understanding epigenetic modifications in response to abiotic stresses in plants. *Biocatalysis and Agricultural Biotechnology*, *27*, 101673.

Roychowdhury, R., Das, S., Gupta, A., Parihar, P., Chandrasekhar, K., Sarker, U., Kumar, A., Ramrao, D., & Sudhakar, C. (2023). Multi-Omics Pipeline and Omics-Integration Approach to Decipher Plant's Abiotic Stress Tolerance Responses. *Genes*, *14*. Advance online publication. DOI: 10.3390/genes14061281 PMID: 37372461

Sawicka, A., & Seiser, C. (2012). Histone H3 phosphorylation – A versatile chromatin modification for different occasions. *Biochimie*, *94*, 2193–2201. PMID: 22564826

Schmitz, R. J., & Zhang, X. (2011). High-throughput approaches for plant epigenomic studies. *Current Opinion in Plant Biology, 14*(2), 130–136. PMID: 21470901

Singh, V., Chaudhary, P. R., Taunk, J., Singh, C. K., Sharma, S., Singh, V., Singh, D., Chinnusamy, V., Yadav, R., & Pal, M. (2021). Plant epigenomics for extenuation of abiotic stresses: Challenges and future perspectives. *Journal of Experimental Botany*. PMID: 34302734

Song, L., Koga, Y., & Ecker, J. R. (2016). Profiling of transcription factor binding events by chromatin immunoprecipitation sequencing (ChIP-seq). *Current Protocols in Plant Biology, 1*(2), 293–306. DOI: 10.1002/cppb.20014 PMID: 28782043

Stunnenberg, H. G., & Vermeulen, M. (2011). Towards cracking the epigenetic code using a combination of high-throughput epigenomics and quantitative mass spectrometry-based proteomics. *BioEssays, 33*(7), 547–551. PMID: 21607996

Sun, C., Ali, K., Yan, K., Fiaz, S., Dormatey, R., Bi, Z., & Bai, J. (2021). Exploration of Epigenetics for Improvement of Drought and Other Stress Resistance in Crops: A Review. *Plants (Basel, Switzerland), 10*(6), 1226. DOI: 10.3390/plants10061226 PMID: 34208642

Suter, L., & Widmer, A. (2013). Environmental heat and salt stress induce transgenerational phenotypic changes in Arabidopsis thaliana. *PLoS One, 8*(4), e60364. PMID: 23585834

Tonosaki, K., Fujimoto, R., Dennis, E. S., Raboy, V., & Osabe, K. (2022). Will epigenetics be a key player in crop breeding? *Frontiers in Plant Science, 13*, 958350. PMID: 36247549

Tricker, P. J. (2015). Transgenerational inheritance or resetting of stress-induced epigenetic modifications: Two sides of the same coin. *Frontiers in Plant Science, 6*, 699. PMID: 26442015

Tsuji, H., Saika, H., Tsutsumi, N., Hirai, A., & Nakazono, M. (2006). Dynamic and reversible changes in histone H3-Lys4 methylation and H3 acetylation occurring at submergence-inducible genes in rice. *Plant & Cell Physiology, 47*(7), 995–1003. PMID: 16774928

Ullah, Z., Iqbal, J., Abbasi, B. A., Akhtar, W., Kanwal, S., Ali, I., & Mahmood, T. (2023b). Assessment of Gus Expression Induced by Anti-Sense Os PPO Gene Promoter and Antioxidant Enzymatic Assays in Response to Drought and Heavy Metal Stress in Transgenic Arabidopsis thaliana. *Sustainability, 15*(17), 12783.

Ullah, Z., Iqbal, J., Abbasi, B. A., Ijaz, S., Sharifi-Rad, J., Yaseen, T., & Mahmood, T. (2023a). How Genome Editing Can Be Helpful in the Biofortification of Legumes. In *Legumes Biofortification* (pp. 207–232). Springer International Publishing.

Urich, M. A., Nery, J. R., Lister, R., Schmitz, R. J., & Ecker, J. R. (2015). MethylC-seq library preparation for base-resolution whole-genome bisulfite sequencing. *Nature Protocols*, *10*(3), 475–483. PMID: 25692984

Vaissiere, T., Sawan, C., & Herceg, Z. (2008). Epigenetic interplay between histone modifications and DNA methylation in gene silencing. *Mutation Research*, *659*, 1–2, 40–48. PMID: 18407786

Varotto, S., Tani, E., Abraham, E., Krugman, T., Kapazoglou, A., Melzer, R., Radanović, A., & Miladinović, D. (2020). Epigenetics: Possible applications in climate-smart crop breeding. *Journal of Experimental Botany*, *71*(17), 5223–5236. DOI: 10.1093/jxb/eraa188 PMID: 32279074

Vojta, A., Dobrinić, P., Tadić, V., Bočkor, L., Korać, P., Julg, B., Klasić, M., & Zoldoš, V. (2016). Repurposing the CRISPR-Cas9 system for targeted DNA methylation. *Nucleic Acids Research*, *44*, 5615–5628. DOI: 10.1093/nar/gkw159 PMID: 26969735

Wang, K. C., & Chang, H. Y. (2018). Epigenomics: Technologies and Applications. *Circulation Research*, *122*(9), 1191–1199. DOI: 10.1161/CIRCRESAHA.118.310998 PMID: 29700067

Wang, L., Li, P., & Brutnell, T. P. (2010). Exploring plant transcriptomes using ultra high-throughput sequencing. *Briefings in Functional Genomics*, *9*(2), 118–128. PMID: 20130067

Wang, L., Xi, Y., Sung, S., & Qiao, H. (2018). RNA-seq assistant: Machine learning based methods to identify more transcriptional regulated genes. *BMC Genomics*, *19*, 1–13. PMID: 30029596

Wang, X., Li, N., Li, W., Gao, X., Cha, M., Qin, L., & Liu, L. (2020). Advances in Transcriptomics in the Response to Stress in Plants. *Global Medical Genetics*, *7*, 30–34. DOI: 10.1055/s-0040-1714414 PMID: 32939512

Wibowo, A., Becker, C., Marconi, G., Durr, J., Price, J., Hagmann, J., Papareddy, R., Putra, H., Kageyama, J., Becker, J., Weigel, D., & Gutierrez-Marcos, J. (2016). Hyperosmotic stress memory in Arabidopsis is mediated by distinct epigenetically labile sites in the genome and is restricted in the male germline by DNA glycosylase activity. *eLife*, *5*. Advance online publication. DOI: 10.7554/eLife.13546 PMID: 27242129

Wu, J., Gao, T., Hu, J., Zhao, L., Yu, C., & Ma, F. (2022). Research advances in function and regulation mechanisms of plant small heat shock proteins (sHSPs) under environmental stresses. *The Science of the Total Environment, 825*, 154054. PMID: 35202686

Yadav, C. B., Pandey, G., Muthamilarasan, M., & Prasad, M. (2018). Epigenetics and epigenomics of plants. *Plant Genetics and Molecular Biology*, 237-261.

Yaish, M. W., Al-Lawati, A., Al-Harrasi, I., & Patankar, H. V. (2018). Genome-wide DNA Methylation analysis in response to salinity in the model plant caliph medic (Medicago truncatula). *BMC Genomics, 19*, 1–17. PMID: 29361906

Yang, Y., Saand, M., Huang, L., Abdelaal, W., Zhang, J., Wu, Y., Li, J., Sirohi, M., & Wang, F. (2021). Applications of Multi-Omics Technologies for Crop Improvement. *Frontiers in Plant Science, 12*. Advance online publication. DOI: 10.3389/fpls.2021.563953 PMID: 34539683

Zenda, T., Liu, S., Dong, A., Li, J., Wang, Y., Liu, X., Wang, N., & Duan, H. (2021). Omics-Facilitated Crop Improvement for Climate Resilience and Superior Nutritive Value. *Frontiers in Plant Science, 12*. Advance online publication. DOI: 10.3389/fpls.2021.774994 PMID: 34925418

Zhang, Y., Tan, D., Zhang, X., Yan, M., Zhang, Y., & Chen, Q. (2021). PANDORA-seq expands the repertoire of regulatory small RNAs by overcoming RNA modifications. *Nature Cell Biology, 23*(4), 424–436. PMID: 33820973

Zhao, S., Yang, M., Zhou, W., Zhang, B., Cheng, Z., Huang, J., & Li, H. (2017). Kinetic and high-throughput profiling of epigenetic interactions by 3D-carbene chip-based surface plasmon resonance imaging technology. *Proceedings of the National Academy of Sciences of the United States of America, 114*(35), E7245–E7254. PMID: 28808021

Zhu, J. Y., Sun, Y., & Wang, Z. Y. (2012). Genome-wide identification of transcription factor-binding sites in plants using chromatin immunoprecipitation followed by microarray (ChIP-chip) or sequencing (ChIP-seq). *Plant Signalling Networks: Methods and Protocols*, 173-188.

Chapter 15
Case Studies on Epigenetic Regulation in Model Plant Species

Muhammad Awais Ibrahim
https://orcid.org/0000-0002-1642-3636
Quaid-i-Azam University, Islamabad, Pakistan

Muhammad Muzamil Sattar
Quaid-i-Azam University, Islamabad, Pakistan

Rifat Bashir
Quaid-i-Azam University, Islamabad, Pakistan

Muhammad Kaleem
https://orcid.org/0009-0003-6548-3920
Quaid-i-Azam University, Islamabad, Pakistan

ABSTRACT

Epigenetics, the examination of inheritable modifications in gene function absent variations in DNA sequence, has become a crucial domain in plant biology. This chapter examines the processes of epigenetic control in model plants, including Arabidopsis thaliana, rice (Oryza sativa), and maize (Zea mays), emphasizing their distinct roles in elucidating DNA methylation, histone modifications, and RNA-directed DNA methylation (RdDM). Epigenetic systems are essential for plant development, stress responses, and adaptation to environmental changes. The investigation of epigenetic regulation is being transformed by recent developments in single-cell epigenomics, CRISPR-based epigenome editing technologies, and multi-omics integration. The chapter discusses the challenges of fully grasping transgenerational

DOI: 10.4018/979-8-3693-8648-4.ch015

inheritance and the stability of epigenetic markers. Connecting advanced research with agricultural applications highlights the possibility of utilizing epigenetics for crop enhancement and sustainable agriculture in response to climate change.

INTRODUCTION TO EPIGENETICS IN PLANTS

Epigenetics denotes inheritable modifications in gene function that transpire without changing the fundamental DNA sequence. These alterations govern gene expression by processes including DNA methylation, histone modification, and small RNA interference. In plants, these epigenetic alterations play crucial roles in development, environmental adaptability, and stress response. For example, DNA methylation, a significant epigenetic mark, silences transposable elements, ensuring genomic integrity and allowing plants to respond dynamically to their environment (Iwasaki & Paszkowski, 2014; Ahmad et al., 2010). Histone changes further affect chromatin structure, impacting accessibility for transcription and helping in cellular memory of external cues (Pikaard & Mittelsten Scheid, 2014, Ahmad et al., 2022a). This epigenetic regulation underpins many essential processes such as flowering, stress responses, and even somaclonal variation in clonally grown plants (Rival et al., 2010, Ahmad et al., 2022b).

UNIQUE EPIGENETIC FEATURES IN PLANTS

Plants show unique epigenetic features that distinguish them from mammals, including a high degree of cellular totipotency and flexible epigenetic reprogramming. Unlike animals, where the germline is determined early in development, plant cells retain the capacity to recreate complete organisms throughout their life cycle. This adaptability is assisted by plant-specific processes such as RNA-directed DNA methylation (RdDM), which targets specific genomic areas for silencing via short RNAs (Brukhin & Albertini, 2021). Additionally, histone modifications such as H3K27me3 in plants display different patterns that connect with environmental stress adaption and developmental signals (Feng et al., 2010). Another plant-specific epigenetic trait is the high heritability of epigenetic marks across generations. This permits plants to pass on adaptive features without permanent genetic modifications, boosting their phenotypic flexibility (Anastasiadi et al., 2021). Furthermore, plants can "erase" epigenetic markers during gametogenesis or embryogenesis, permitting fresh epigenetic programming in offsprings (Feng et al., 2010).

OVERVIEW OF PLANT-SPECIFIC MECHANISMS

Plants have developed specific epigenetic processes (**Figure 1**) to regulate gene expression, many of which are distinct from those in animals. One of the most notable is RNA-directed DNA methylation (RdDM), a mechanism specific to plants. RdDM exploits small interfering RNAs (siRNAs) to guide the addition of methyl groups to specific DNA sequences, primarily targeting transposable elements and repetitive sections. This method serves to suppress potentially damaging genetic elements and modulate gene expression in response to environmental stimuli (Brukhin & Albertini, 2021) (Feng & Jacobsen, 2011).

Another essential element of plant epigenetics is the distinct pattern of histone modifications. While both plants and animals share basic histone modification types, such as methylation and acetylation, plants exhibit specific modifications such H3K27me3, which plays a critical role in the regulation of development and stress responses. This mark is actively controlled throughout important developmental stages and in response to environmental stimuli, leading to plant plasticity (Feng et al., 2010; Pikaard & Scheid, 2014). Plants can reset epigenetic markers during gametogenesis and embryogenesis to permit new patterns of gene expression in progeny. Unlike animals, this reprogramming frequently retains plasticity, enabling for quick adaptability to environmental changes throughout generations (Brukhin & Albertini, 2021, Ahmad et al., 2024a).

Figure 1. Overview of plant-specific mechanisms

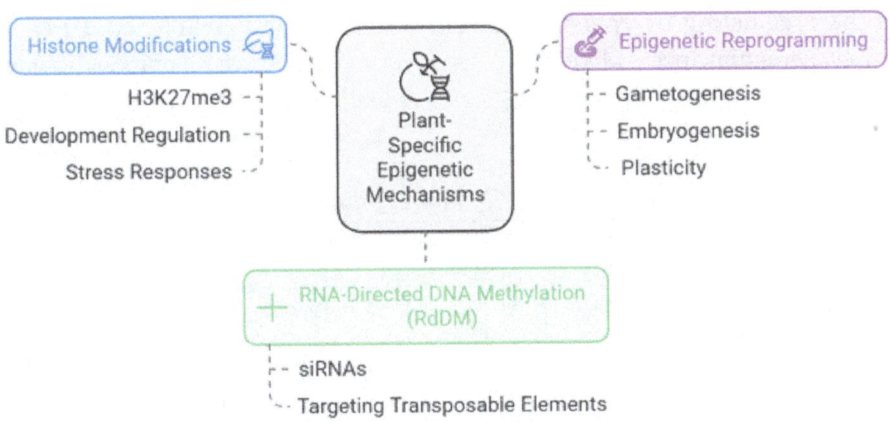

MODEL SPECIES IN PLANT EPIGENETIC STUDIES

Importance of Model Species in Understanding Epigenetic Regulation

Model species serve a significant role in expanding our understanding of epigenetic mechanisms in plants. These species, with their well-annotated genomes and genetic tools, provide a foundation for analyzing complicated epigenetic processes including as DNA methylation, histone modification, and RNA-mediated gene silencing. *Arabidopsis thaliana* is the prototypical model plant, affording insights into the epigenetic control of development and stress responses due to its small genome, simplicity of genetic manipulation, and extensive mutant libraries (Schmitz & Ecker, 2012). Similarly, species like rice (*Oryza sativa*) and maize (*Zea mays*) serve as examples for monocots, (Shi et al., 2015). These species provide insights into conserved and distinct epigenetic traits throughout plant lineages (Henderson & Jacobsen, 2007; Rapp & Wendel, 2005).

Brief Introduction to Key Model Species

Among the model species, *Arabidopsis thaliana* stands out due to its pioneering role in epigenetic research, contributing to discoveries in DNA methylation patterns and histone modifications (Initiative, 2000). Rice (*Oryza sativa*), as a monocot model, provides essential insights into crop-specific epigenetic control, particularly in features like blooming time and stress (Shi et al., 2015). Maize (*Zea mays*), with its huge and complex genome, contributes in understanding transposable elements and their regulation by epigenetic mechanisms (Pikaard & Scheid, 2014). Additionally, perennial species like grapevine offer a perspective on epigenetic memory and flexibility in long-lived plants (Fortes & Gallusci, 2017, Ahmad et al., 2024b).

Table 1. Comparison of epigenetics in model plants

Plant Species	Epigenetic Features	Key Genes Studied	Research Applications	Reference
Arabidopsis	DNA methylation, histone modification, RdDM	FLC, AtHAC1, PRC2	Flowering time regulation, stress responses	Schmitz & Ecker, 2012; Feng et al., 2010
Rice	DNA methylation, histone modification	GS3, GW2, Sub1A	Grain size regulation, submergence tolerance	Shi et al., 2015; Wang et al., 2015
Maize	DNA methylation, histone modifications	ZmCHB101, FLC	Developmental timing, transposon regulation	Pikaard & Scheid, 2014; Yu et al., 2016

OBJECTIVES

The primary purpose of this chapter is to describe how investigations in model plants increase our knowledge of epigenetic regulation. These findings include understanding how epigenetic marks are established, maintained, and modified in response to developmental cues and environmental stimuli. Research in *Arabidopsis* and other models has discovered important regulatory networks that govern gene expression without affecting the underlying DNA sequence (Schmitz & Ecker, 2012). This work offers the basis for utilizing epigenetic understanding to improve crop resilience and (Gallusci et al., 2017).

RELEVANCE TO AGRICULTURE, CROP IMPROVEMENT, AND CLIMATE RESILIENCE

Understanding epigenetic regulation in plants is vital for addressing difficulties in agriculture. Epigenetic mechanisms offer methods for creating crops with higher tolerance to abiotic stresses such as drought and salinity, as well as resistance to pests and diseases (Kakoulidou et al., 2021, Ali et al., 2022). Moreover, they present options for enhancing productivity and quality by modifying gene expression in response to environmental changes (Rapp & Wendel, 2005b). The ability to utilize epigenetic variety can lead to more sustainable agricultural methods, leading to food security in a changing environment (Richards et al., 2017).

MECHANISMS OF EPIGENETIC REGULATION IN PLANTS

DNA Methylation

DNA methylation is a critical epigenetic alteration in plants (**Figure 2**), requiring the addition of a methyl group to the 5th carbon of cytosine bases. This process is mediated by a family of DNA methyltransferases (MTases), which are categorized into three primary types: MET1, responsible for maintenance methylation in the CG context; CMTs, which target CHG sites; and DRM2, which performs *de novo* methylation in CHH contexts via the RNA-directed DNA methylation (RdDM) pathway (Finnegan & Kovac, 2000; Du, 2016). This alteration serves a vital function in silencing transposable elements (TEs) and preserving genome stability, thereby safeguarding the plant genome from potentially harmful insertions (Zhong et al., 2014).

DNA methylation in plants occurs in three sequence contexts: CG, CHG, and CHH (where H = A, T, or C), each playing distinct regulatory roles. CG methylation is generally connected with gene body methylation, which may enhance transcriptional elongation, while CHG and CHH methylation are primarily linked to heterochromatin and silencing of TEs (Zhang et al., 2018). Functionally, DNA methylation modulates gene expression by suppressing or silencing genes and TEs, critical throughout development and environmental stress responses. For instance, methylation of promoter regions can inhibit transcription factor binding, effectively suppressing gene expression. This epigenetic regulation is crucial in developmental processes such as seed germination, blooming, and stress adaption (Meyer, 2010; Xie & Yu, 2015, Ijaz et al., 2023).

Histone Modifications

Histone modifications, including methylation, acetylation, phosphorylation, ubiquitination, and sumoylation, play key roles in controlling chromatin structure and gene expression (**Figure 2**). These post-translational modifications are managed by particular enzymes: histone methyltransferases (HMTs) add methyl groups, histone acetyltransferases (HATs) add acetyl groups, and histone deacetylases (HDACs) remove them. For instance, acetylation generally promotes transcriptional activation by loosening chromatin structure, but methylation can either stimulate or repress gene expression depending on the specific histone residues changed (Zhao et al., 2019; Zhang et al., 2007). In Arabidopsis, H3K27me3, a repressive mark, is regulated by the Polycomb repressive complex 2 (PRC2), affecting genes involved in blooming and stress response. Similarly, in rice, H3K4me3 and H3K36me3 marks enhance transcriptional activity, regulating developmental changes like flowering (Berr et al., 2011; Pfluger & Wagner, 2007). These histone alterations are critical in modifying plant responses to environmental changes, acting as an interface between the genome and environmental signals.

Figure 2. Epigenetics regulation in plants

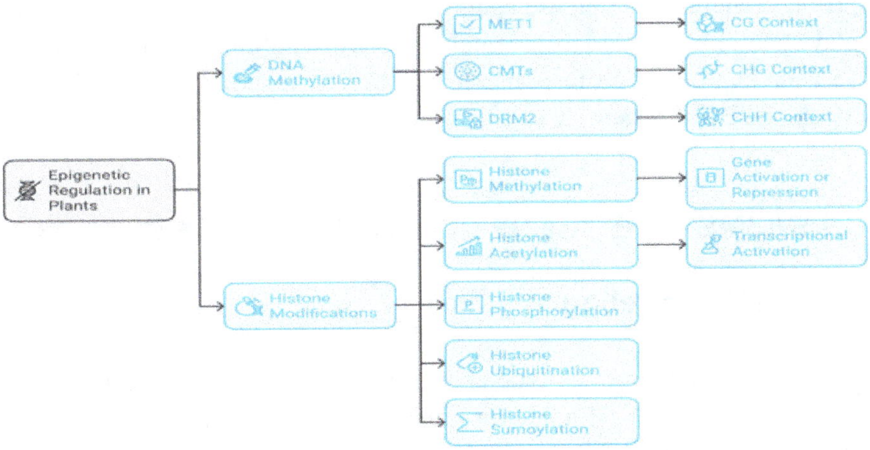

CASE STUDIES IN ARABIDOPSIS AND RICE

In *Arabidopsis*, histone acetylation by AtHAC1 controls the blooming time by altering blooming LOCUS C (FLC) expression. Loss of function in AtHAC1 slows flowering under long-day conditions, demonstrating the significance of histone acetylation in developmental timing (Deng et al., 2007). In rice, the histone methyltransferase SDG724 promotes blooming by regulating H3K36 methylation at specific flowering loci, demonstrating a conserved role of histone methylation in monocots and dicots (Sun et al., 2012).

NON-CODING RNAS AND RNA-DIRECTED DNA METHYLATION (RDDM)

The RNA-directed DNA methylation (RdDM) pathway is a plant-specific process involving short RNAs (sRNAs) that guide DNA methylation to specific genomic locations. This process is predominantly regulated by small interfering RNAs (siRNAs) and long non-coding RNAs (lncRNAs), which attract the DOMAINS REARRANGED METHYLTRANSFERASE (DRM) to target locations, generating or sustaining methylation marks (Brukhin & Albertini, 2021). siRNAs come from double-stranded RNA processed by DICER-LIKE enzymes and are loaded onto ARGONAUTE proteins to guide the RdDM machinery. This route is critical for transposon silencing and defense against viral elements (Xie & Yu, 2015). RdDM also plays a vital role in environmental responses, enabling plants to dynamically change gene expression in response to abiotic challenges including drought and salinity (Matzke & Mosher, 2014).

MECHANISMS OF EPIGENETIC REGULATION IN PLANTS

Chromatin Remodeling

Chromatin remodeling is a crucial step in epigenetic regulation that limits the accessibility of the DNA template for transcription, replication, and repair. This is performed through chromatin remodeling complexes, which use ATP hydrolysis to reposition, expel, or restructure nucleosomes. These complexes include SWI/SNF, ISWI, CHD, and INO80 families, each performing specific tasks. For instance, SWI/SNF complexes modify nucleosome location to permit transcriptional activation, while ISWI complexes facilitate nucleosome assembly, contributing to chromatin compaction (Ojolo et al., 2018; Han et al., 2015). These complexes work in combination with histone changes to fine-tune gene expression, generating a dynamic chromatin state that balances gene activation and repression. Their activity ensures that specific genes are accessible during important embryonic phases or in response to environmental cues.

Mechanisms of Chromatin Accessibility

Chromatin remodeling complexes work by sliding nucleosomes along the DNA, evicting histones, or integrating histone variations, thus regulating DNA accessibility. For instance, the INO80 and SWR1 complexes replace canonical histones with the H2A.Z variation, which affects chromatin stability and enhances transcriptional control. Similarly, the CHD family remodelers reposition nucleosomes to expose regulatory regions important for transcription factor binding (Shang & He, 2021; Wang et al., 2019). These remodeling processes are crucial during developmental changes, such as flowering, and in stress adaption, where rapid transcriptional reprogramming is required.

Specific Complexes in Model Plants

In Arabidopsis, the SWI/SNF complex has been intensively researched for its role in maintaining meristem function and regulating flowering timing. The BRAHMA (BRM) subunit of SWI/SNF enhances the expression of floral repressors, delaying flowering under non-inductive conditions (Jarillo et al., 2009). In maize, the core component ZmCHB101 regulates vegetative and reproductive development by altering chromatin accessibility at developmental genes (Yu et al., 2016). Moreover, the INO80 complex plays a vital function in stress responses, particularly in integrating environmental cues into chromatin structural modifications, boosting plant survival under harsh conditions (Ma et al., 2011, Ijaz et al., 2024a).

CASE STUDIES IN MODEL PLANT SPECIES

Arabidopsis Thaliana

Flowering time in *Arabidopsis thaliana* is closely regulated by a complex interaction of genetic and epigenetic factors, with the FLOWERING LOCUS C (FLC) gene being a prominent role (**Figure 3**). FLC encodes a MADS-box transcription factor that serves as a floral repressor by suppressing the expression of FLOWERING LOCUS T (FT) and SUPPRESSOR OF OVEREXPRESSION OF CO 1 (SOC1), important genes driving flowering. The inhibition of FLC during vernalization (prolonged exposure to cold) is a prime example of epigenetic control. This process is assisted by histone changes, such as trimethylation of histone H3 at lysine 27 (H3K27me3), mediated by the Polycomb Repressive Complex 2 (PRC2). PRC2 components like VERNALIZATION 2 (VRN2) and CURLY LEAF (CLF) progressively deposit repressive marks at FLC, suppressing its expression to facilitate flowering (Finnegan et al., 2005; He, 2009).

DNA methylation also helps to FLC regulation. The FWA gene, for instance, is silenced by CG and non-CG methylation in wild-type plants, but its demethylation leads to late blooming phenotypes. This indicates how DNA methylation might alter flowering processes by changing gene activity (Schmitz et al., 2008). Environmental variables, notably photoperiod and temperature, strongly impact flowering through epigenetic modifications. Long photoperiods generate CONSTANS (CO) expression, which promotes FT activation, while cold exposure prompts FLC silencing. These dynamic alterations highlight how plants integrate environmental stimuli into their developmental processes through epigenetic pathways (Pien et al., 2008; Berr et al., 2011).

Stress Responses

Plants continually undergo abiotic stressors such as drought, salt, and temperature extremes, which need quick and reversible gene expression changes. Epigenetic processes, including histone modifications, DNA methylation, and RNA-directed DNA methylation (RdDM), are necessary for this plasticity. Under drought and salinity stress, histone deacetylases (HDACs) like HDA6 and HDA9 regulate the chromatin state of stress-responsive genes. Loss of HDA9 activity resulted in elevated acetylation levels at promoter regions of stress-related genes, enhancing their expression. This enhances resistance to osmotic stress by enhancing water consumption efficiency (Zheng et al., 2016). In contrast, histone methylation markers like as H3K4me3 and H3K27me3 are regulated to either activate or repress stress-responsive regions.

The RdDM system is critical in suppressing transposable elements (TEs) under stress, limiting genomic instability. In *Arabidopsis*, siRNAs lead DNA methylation to TEs and other repetitive regions, ensuring their repression. This pathway is crucial under salinity stress, because TE activation could affect gene function (Pan, 2013). Stress-induced epigenetic alterations are generally heritable, giving kids with a "memory" of past environmental stressors. For instance, plants subjected to drought have altered DNA methylation patterns that persist into later generations. This transgenerational epigenetic memory promotes stress tolerance in offspring by proactively activating stress-response pathways (Ganguly et al., 2017, Ijaz et al., 2024b).

Case Study: Vernalization and FLC Repression

During vernalization, FLC repression is initiated by cold-induced production of VERNALIZATION INSENSITIVE 3 (VIN3), which interacts with PRC2. This interaction establishes H3K27me3 marks across the FLC locus, keeping its silent status long after the cold period ends. The consistent suppression of FLC ensures that blooming is initiated only when favorable conditions return. This epigenetic method illustrates how environmental stimuli are entrenched in the plant's developmental program through histone changes (He, 2009).

Case Study: Stress-Induced Epigenetic Memory

Arabidopsis plants subjected to drought stress exhibit alterations in DNA methylation and histone modifications at stress-responsive genes. This epigenetic reprogramming promotes the expression of genes like *RD29A*, which encodes a desiccation-responsive protein. Remarkably, these alterations can persist over generations, preparing the progeny for comparable stress situations. Such memory effects are mediated by short RNAs and the RdDM pathway, which promote stress-responsive gene expression (Ganguly et al., 2017, Ullah et al., 2023a).

Figure 3. Epigenetics regulations in Arabidopsis thaliana

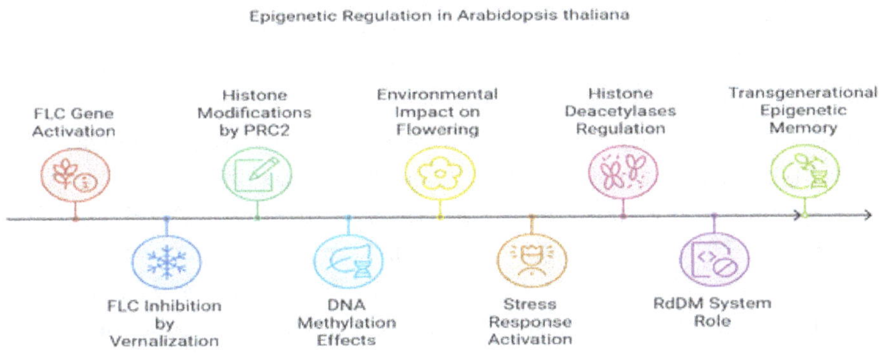

CASE STUDIES IN MODEL PLANT SPECIES

Oryza sativa (Rice)

Grain size is a crucial driver of rice yield, controlled by many epigenetic variables (**Figure 4**). DNA methylation at key regulatory loci has a critical function in altering grain size. For instance, the quantitative trait loci (QTLs) *GS3*, *GW2*, and *GS5* are strongly regulated through methylation patterns, changing their expression levels and eventually influencing grain morphology and yield. Variations in the methylation of the *GS3* gene lead to varied grain sizes, with hypomethylation linked with larger grains (Li et al., 2011). Similarly, the *GW2* locus is controlled by histone acetylation, which affects chromatin accessibility, promoting cell proliferation in the developing grain (Shi et al., 2020). Histone alterations like as H3K4me3 and H3K27me3 have also been involved in nutrient signaling pathways that impact grain development. These alterations control genes implicated in the gibberellin pathway, necessary for cell division and elongation during grain filling.

The histone demethylase JMJ706 eliminates repressive marks at crucial developmental sites, enabling grain expansion (Qi et al., 2012).

ABIOTIC STRESS TOLERANCE

Rice undergoes several abiotic challenges such as submergence, salt, and temperature fluctuations, which greatly effect yield. Epigenetic processes, particularly RNA-directed DNA methylation (RdDM) and histone acetylation, play key roles in improving stress resilience. The Sub1A locus, a critical factor in submergence tolerance, is epigenetically controlled to facilitate survival during prolonged flooding. DNA methylation patterns at Sub1A influence its expression, which in turn regulates quiescence and energy conservation processes, critical for submergence survival (Sarkar & Ray, 2015). Under high salinity, histone acetylation at stress-responsive genes boosts their expression, enabling better ion homeostasis and osmotic balance. Studies on rice genotypes suggest that tolerant cultivars have higher histone acetylation at the *NHX1* gene, which encodes a sodium-proton antiporter, promoting ion compartmentalization and salt tolerance (Karan et al., 2012). Furthermore, transgenerational epigenetic memory has been shown in rice, where stress-induced methylation alterations are transmitted, priming later generations for improved stress tolerance (Wang et al., 2015, Ullah et al., 2023b).

Figure 4. Epigenetics regulation in Oryza sativa

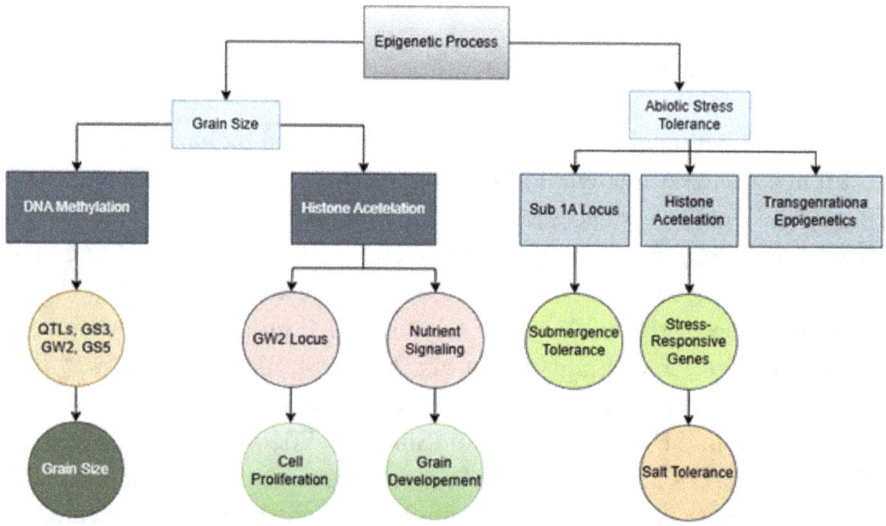

APPLICATIONS AND IMPLICATIONS OF EPIGENETIC STUDIES IN MODEL PLANTS

Epigenetic research in model plants such as Arabidopsis thaliana, rice (Oryza sativa), and maize (Zea mays) has increased our understanding of heritable gene control and its promise for agricultural innovation. By understanding mechanisms like as DNA methylation, histone modifications, and RNA-directed DNA methylation, this research have deep applicability in solving difficulties including climate change, food security, and ethical considerations in crop production.

Epigenetic Modification for Crop Improvement

Epigenetic alterations give a toolset for boosting agricultural attributes such as yield, disease resistance, and stress tolerance without affecting the DNA sequence. For example, DNA methylation and histone changes can regulate stress-responsive genes, helping plants to tolerate harsh environments like drought and salinity (Malhotra et al., 2020). Studies on rice have established the significance of chromatin modification in enhancing drought resistance, indicating possible implications in developing climate-resilient crops (Sun et al., 2021).

High-throughput technologies, such as genome-wide methylation profiling, have proven essential in uncovering epigenetic markers that predict phenotypic features, making them valuable for precision breeding. For instance, single-base resolution methylome mapping in Arabidopsis and rice has revealed insights into the functional significance of epigenetic markers (Jain et al., 2021). These developments suggest that epigenome-based breeding tactics, or "epibreeding," could supplement standard genetic procedures (Samantara et al., 2021, Haider et al., 2021a).

Table 2. Applications of epigenetics in crop

Target Trait	Epigenetic Mechanism Utilized	Implementation Method	Agricultural Impact	Future Research Areas
Stress Tolerance	DNA Methylation, Histone Modification	Selective breeding, epigenome editing	Improved drought and salinity tolerance	Identifying optimal epigenetic markers for crop resilience
Crop Yield Improvement	Histone Acetylation, DNA Methylation	Epigenetic breeding strategies	Increased grain size, improved yield potential	Utilizing multi-omics to improve yield prediction
Disease Resistance	RNA-Directed DNA Methylation (RdDM)	CRISPR-based epigenome editing	Enhanced resistance to pathogens	Engineering disease-resistant crops through epigenetics

Epigenetic Responses to Climate Change

Epigenetics plays a vital role in plants' ability to adapt to climate change by changing gene expression in response to environmental challenges (**Figure 5**). For example, histone changes and DNA methylation allow plants to fine-tune their transcriptional responses to temperature fluctuations and other environmental stresses (Varotto et al., 2020). These alterations can contribute to phenotypic plasticity, enabling plants to adapt fast to their surroundings (Fortes & Gallusci, 2017). In perennial plants like grapevines, epigenetic regulation has been found to mediate stress responses, allowing these plants to live in various terroirs. Such findings underline the possibility for adopting epigenetics to minimize the impacts of climate change on agriculture (Fortes & Gallusci, 2017, Haider et al., 2021b).

Transgenerational Epigenetic Inheritance

One of the most exciting features of plant epigenetics is transgenerational inheritance, where epigenetic alterations caused by environmental stimuli are transferred to following generations (**Figure 5**). Studies have revealed how "stress memory" is encoded through heritable epigenetic marks, enabling plants to prepare for repeating stress situations (Thiebaut et al., 2019). For instance, drought-stressed plants have been seen to pass on features of improved stress tolerance to their progeny, mediated by stable DNA methylation patterns (Sun et al., 2021). While promising, such inheritance is not always consistent, and further research is needed to understand the stability and mechanisms behind epigenetic memory (Gallusci et al., 2017).

ETHICAL AND REGULATORY CONSIDERATIONS

The application of epigenetics in agricultural enhancement creates ethical and regulatory issues (Figure 5). Unlike genetic alteration, epigenetic treatments do not alter DNA sequences, but the potential for undesired off-target effects and ecological repercussions must be carefully addressed. Regulatory frameworks for epigenetic applications are still growing, and there is a need for clear standards to ensure the proper use of this technology (Tomkova & Berr, 2019). Public perception of epigenetic breeding is another key issue. Transparent communication regarding the science and safety of epigenetic changes is vital to acquire societal acceptance and enable policy formation (Agarwal et al., 2020).

Figure 5. Applications of epigenetics studies in model plants

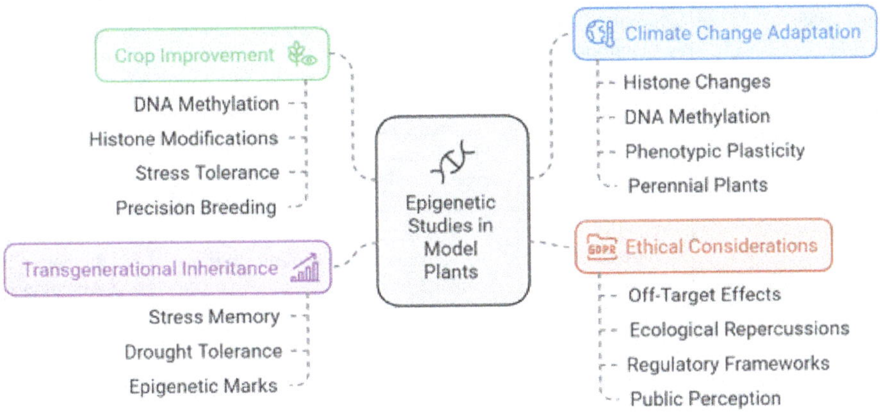

FUTURE DIRECTIONS IN PLANT EPIGENETIC RESEARCH

Emerging Techniques and Technologies

Recent years have witnessed revolutionary breakthroughs in tools for epigenetic study and manipulation in plant species. Single-cell epigenomics has emerged as a significant tool, allowing researchers to examine cell-specific chromatin states, histone modifications, and DNA methylation patterns in unprecedented depth. The merging of technologies like ATAC-seq and ChIPmentation provides greater insights into chromatin accessibility and histone landscapes (Carlberg & Molnár, 2018). CRISPR-based technologies have moved beyond genome editing into precision epigenome editing, using catalytically inactive dCas9 coupled to epigenetic modifiers. These methods enable targeted alterations such as DNA methylation and histone acetylation, enabling potent strategies for functional research and gene regulation in plants (Vojta et al., 2016). Chromatin conformation studies, utilizing technologies like Hi-C and Capture-C, have clarified the three-dimensional architecture of the genome, revealing insights into long-range regulatory connections and their epigenetic control (Nakamura et al., 2021).

Table 3. Technological advances in plant epigenetics studies

Technology/Method	Purpose	Advantages	Limitations	Case Studies
Single-cell Epigenomics	Investigating chromatin states in individual cells	High resolution, cell-specific data	Complex data analysis, high cost	Arabidopsis stress response analysis
CRISPR-based Epigenome Editing	Precision modification of epigenetic marks	Targeted gene regulation, reversible edits	Off-target effects, incomplete understanding of long-term effects	Off-target effects, incomplete understanding of long-term effects
Hi-C/ Capture-C	Mapping three-dimensional chromatin structure	Reveals long-range gene regulation networks	Limited to specific genomic regions	Epigenetic regulation in Arabidopsis flowering

Integrating Multi-Omics Approaches

The convergence of epigenomics with transcriptomics, proteomics, and metabolomics constitutes a frontier in plant science. Multi-omics techniques enable the building of complete models relating epigenetic regulation to gene expression, protein function, and metabolic pathways. For instance, combining these datasets has revealed how plants dynamically change their epigenomes in response to environmental stress (Carlberg & Molnár, 2018). Case studies in model plants, such as Arabidopsis thaliana, highlight the power of multi-omics in unraveling complex features. For example, research correlating DNA methylation profiles with metabolomic alterations under drought stress have discovered epigenetic markers connected with stress tolerance, indicating their potential for agricultural development (Song & Koo, 2021, Haider et al., 2022a).

CHALLENGES

Despite the development, the area of plant epigenetics confronts major hurdles. One key shortcoming is the inadequate understanding of transgenerational inheritance of epigenetic markers. While some epigenetic alterations are heritable, their stability over generations remains variable and context-dependent (Lo & Qi, 2017, Haider et al., 2022b). Open questions also concern the reversibility of epigenetic changes. While certain marks, like histone acetylation, are dynamically regulated, others, such as DNA methylation, may persist longer than anticipated, challenging efforts to predict their biological effects. Developing models that effectively predict the functional implications of epigenetic alterations will require further exploration (Kungulovski & Jeltsch, 2015b).

CONCLUSION

Plant epigenetics constitutes a cornerstone of modern plant biology, bringing insights into the regulation of gene expression through processes such as DNA methylation, histone modifications, and RNA-mediated pathways. Studies in model plants have shown how these epigenetic pathways influence important developmental processes, stress responses, and heritable adaptations, providing a platform for agricultural advancements. The merging of sophisticated technologies, such as single-cell sequencing and CRISPR-based epigenome editing, with multi-omics methods has permitted unprecedented resolution in understanding plant epigenomes. These methods are showing the intricate interplay between the genome and environment, as well as the transgenerational influence of epigenetic alterations.

Despite substantial gains, difficulties exist. The poor understanding of epigenetic inheritance, the reversibility of alterations, and limits in the specificity of tools for targeted epigenetic editing remain areas of current inquiry. Addressing these problems will require continuing improvement of methodology and extended study in varied plant species. Looking forward, the application of epigenetics in agriculture holds transformative promise. By exploiting epigenetic control, crops can be engineered for better resilience to climate stress, improved nutrient usage efficiency, and sustainable productivity. The aim for epigenetics in agriculture extends beyond enhancing crop performance to developing practices that accord with ecological and societal goals, fostering a future of sustainable food security.

REFERENCES

Agarwal, G., Kudapa, H., Ramalingam, A., Choudhary, D., Sinha, P., Garg, V., Singh, V. K., Patil, G. B., Pandey, M. K., Nguyen, H. T., Guo, B., Sunkar, R., Niederhuth, C. E., & Varshney, R. K. (2020). Epigenetics and epigenomics: Underlying mechanisms, relevance, and implications in crop improvement. *Functional & Integrative Genomics, 20*(6), 739–761. DOI: 10.1007/s10142-020-00756-7 PMID: 33089419

Ahmad, A., Zhang, Y., & Cao, X. (2010). Decoding the epigenetic language of plant development. *Molecular Plant, 3*(4), 719–728. DOI: 10.1093/mp/ssq026 PMID: 20663898

Ahmad, Y., Haakim, Z., Iqbal, J., Abbasi, B. A., Mahmood, T., & Kazi, M. (2024b). Technological Innovations for Abiotic Stress Resistance in Horticultural Crops. OMICs-based Techniques for Global Food Security, 233-244.

Ahmad, Y., Haider, S., Iqbal, J., Abbasi, B. A., Yaseen, T., & Mahmood, T. (2022a). The mechanisms of genome editing technologies in crop plants. In *Principles and Practices of OMICS and Genome Editing for Crop Improvement* (pp. 295–313). Springer International Publishing.

Ahmad, Y., Haider, S., Iqbal, J., Abbasi, B. A., Yaseen, T., & Mahmood, T. (2022b). The mechanisms of genome editing technologies in crop plants. In *Principles and Practices of OMICS and Genome Editing for Crop Improvement* (pp. 295–313). Springer International Publishing.

Ahmad, Y., Haider, S., Iqbal, J., Naseer, S., Attia, K. A., Mohammed, A. A., & Mahmood, T. (2024a). In-silico analysis and transformation of OsMYB48 transcription factor driven by CaMV35S promoter in model plant–Nicotiana tabacum L. conferring abiotic stress tolerance. *GM Crops and Food: Biotechnology in Agriculture and the Food Chain, 15*(1), 130–149. PMID: 38551174

Ali, I., Salah, K. B. H., Sher, H., Ali, H., Ullah, Z., Ali, A., & Mahmood, T. (2022). Drought stress enhances the efficiency of floral dip method of Agrobacterium-mediated transformation in Arabidopsis thaliana. *Brazilian Journal of Biology, 84*, e259326. PMID: 35703626

Anastasiadi, D., Venney, C. J., Bernatchez, L., & Wellenreuther, M. (2021). Epigenetic inheritance and reproductive mode in plants and animals. *Trends in Ecology & Evolution, 36*(12), 1124–1140. DOI: 10.1016/j.tree.2021.08.006 PMID: 34489118

Berr, A., Shafiq, S., & Shen, W. (2011). Histone modifications in transcriptional activation during plant development. *Biochimica Et Biophysica Acta (BBA) -. Gene Regulatory Mechanisms, 1809*(10), 567–576. DOI: 10.1016/j.bbagrm.2011.07.001 PMID: 21777708

Brukhin, V., & Albertini, E. (2021). Epigenetic modifications in plant development and reproduction. *Epigenomes, 5*(4), 25. DOI: 10.3390/epigenomes5040025 PMID: 34968249

Carlberg, C., & Molnár, F. (2018). Methods and Applications of Epigenomics. In *Springer eBooks* (pp. 19–38). https://doi.org/DOI: 10.1007/978-981-10-7614-5_2

Deng, W., Liu, C., Pei, Y., Deng, X., Niu, L., & Cao, X. (2007). Involvement of the histone acetyltransferase ATHAC1 in the regulation of flowering time via repression ofFLOWERING LOCUS cin arabidopsis. *Plant Physiology, 143*(4), 1660–1668. DOI: 10.1104/pp.107.095521 PMID: 17416640

Du, J. (2016). Structure and mechanism of plant DNA methyltransferases. *Advances in Experimental Medicine and Biology*, •••, 173–192. DOI: 10.1007/978-3-319-43624-1_8 PMID: 27826839

Feng, S., & Jacobsen, S. E. (2011). Epigenetic modifications in plants: An evolutionary perspective. *Current Opinion in Plant Biology, 14*(2), 179–186. DOI: 10.1016/j.pbi.2010.12.002 PMID: 21233005

Feng, S., Jacobsen, S. E., & Reik, W. (2010). Epigenetic reprogramming in plant and animal development. *Science, 330*(6004), 622–627. DOI: 10.1126/science.1190614 PMID: 21030646

Finnegan, E., & Kovac, K. (2000). Plant DNA methyltransferases. *Plant Molecular Biology, 43*(2/3), 189–201. DOI: 10.1023/a:1006427226972 PMID: 10999404

Finnegan, E. J., Kovac, K. A., Jaligot, E., Sheldon, C. C., Peacock, W. J., & Dennis, E. S. (2005). The downregulation of FLOWERING LOCUS C (FLC) expression in plants with low levels of DNA methylation and by vernalization occurs by distinct mechanisms. *The Plant Journal, 44*(3), 420–432. DOI: 10.1111/j.1365-313x.2005.02541.x PMID: 16236152

Fortes, A. M., & Gallusci, P. (2017). Plant stress Responses and Phenotypic Plasticity in the Epigenomics Era: Perspectives on the Grapevine Scenario, a model for perennial crop plants. *Frontiers in Plant Science, 08*. Advance online publication. DOI: 10.3389/fpls.2017.00082 PMID: 28220131

Fortes, A. M., & Gallusci, P. (2017). Plant Stress Responses and Phenotypic Plasticity in the Epigenomics Era: Perspectives on the Grapevine Scenario, a Model for Perennial Crop Plants. *Frontiers in Plant Science*, *08*. Advance online publication. DOI: 10.3389/fpls.2017.00082 PMID: 28220131

Gallusci, P., Dai, Z., Génard, M., Gauffretau, A., Leblanc-Fournier, N., Richard-Molard, C., Vile, D., & Brunel-Muguet, S. (2017). Epigenetics for Plant Improvement: Current knowledge and modeling Avenues. *Trends in Plant Science*, *22*(7), 610–623. DOI: 10.1016/j.tplants.2017.04.009 PMID: 28587758

Ganguly, D. R., Crisp, P. A., Eichten, S. R., & Pogson, B. J. (2017). The Arabidopsis DNA Methylome Is Stable under Transgenerational Drought Stress. *Plant Physiology*, *175*(4), 1893–1912. DOI: 10.1104/pp.17.00744 PMID: 28986422

Haider, S., Iqbal, J., Naseer, S., Shaukat, M., Abbasi, B. A., Yaseen, T., & Mahmood, T. (2022b). Unfolding molecular switches in plant heat stress resistance: A comprehensive review. *Plant Cell Reports*, *41*(3), 775–798. PMID: 34401950

Haider, S., Iqbal, J., Naseer, S., Yaseen, T., Shaukat, M., Bibi, H., & Mahmood, T. (2021a). Molecular mechanisms of plant tolerance to heat stress: Current landscape and future perspectives. *Plant Cell Reports*, **•••**, 1–25. PMID: 33890138

Haider, S., Iqbal, J., Shaukat, M., Naseer, S., & Mahmood, T. (2021b). The epigenetic chromatin-based regulation of somatic heat stress memory in plants. *Plant Gene*, *27*, 100318.

Haider, S., Raza, A., Iqbal, J., Shaukat, M., & Mahmood, T. (2022a). Analyzing the regulatory role of heat shock transcription factors in plant heat stress tolerance: A brief appraisal. *Molecular Biology Reports*, *49*(6), 5771–5785. PMID: 35182323

Han, S., Wu, M., Cui, S., & Wagner, D. (2015). Roles and activities of chromatin remodeling ATPases in plants. *The Plant Journal*, *83*(1), 62–77. DOI: 10.1111/tpj.12877 PMID: 25977075

He, Y. (2009). Control of the transition to flowering by chromatin modifications. *Molecular Plant*, *2*(4), 554–564. DOI: 10.1093/mp/ssp005 PMID: 19825638

Henderson, I. R., & Jacobsen, S. E. (2007). Epigenetic inheritance in plants. *Nature*, *447*(7143), 418–424. DOI: 10.1038/nature05917 PMID: 17522675

Ijaz, S., Iqbal, J., Abbasi, B. A., Tufail, A., Ullah, Z., Sharifi-Rad, J., & Uddin, S. (2023). Biofortification: Lessons from the Past and Strategies for Future Food Security. In *Legumes Biofortification* (pp. 521–545). Springer International Publishing.

Ijaz, S., Iqbal, J., Abbasi, B. A., Yaseen, T., Rehman, S., Kazi, M., & Mahmood, T. (2024a). Role of OMICS-Based Technologies in Plant Sciences. OMICs-based Techniques for Global Food Security, 45-66.

Ijaz, S., Iqbal, J., Abbasi, B. A., Yaseen, T., Rehman, S., Kazi, M. & Mahmood, T. (2024b). Role of OMICS-Based Technologies in Plant Sciences. OMICs-based Techniques for Global Food Security, 45-66.

Initiative, A. G. (2000). Analysis of the genome sequence of the flowering plant Arabidopsis thaliana. *Nature*, *408*(6814), 796–815. DOI: 10.1038/35048692 PMID: 11130711

Iwasaki, M., & Paszkowski, J. (2014). Epigenetic memory in plants. *The EMBO Journal*, *33*(18), 1987–1998. DOI: 10.15252/embj.201488883 PMID: 25104823

Jain, N., Taak, Y., Choudhary, R., Yadav, S., Saini, N., Vasudev, S., & Yadava, D. (2021). Advances and prospects of epigenetics in plants. In *Elsevier eBooks* (pp. 421–444). https://doi.org/DOI: 10.1016/b978-0-323-85652-2.00013-0

Jarillo, J. A., Pineiro, M., Cubas, P., Martinez-Zapater, J. M., Jarillo, J. A., Pineiro, M., Cubas, P., & Martinez-Zapater, J. M. (2009). Chromatin remodeling in plant development. *The International Journal of Developmental Biology*, *53*(8-9–10), 1581–1596. https://doi.org/DOI: 10.1387/ijdb.072460jj

Kakoulidou, I., Avramidou, E. V., Baránek, M., Brunel-Muguet, S., Farrona, S., Johannes, F., Kaiserli, E., Lieberman-Lazarovich, M., Martinelli, F., Mladenov, V., Testillano, P. S., Vassileva, V., & Maury, S. (2021). Epigenetics for crop improvement in times of global change. *Biology (Basel)*, *10*(8), 766. DOI: 10.3390/biology10080766 PMID: 34439998

Karan, R., DeLeon, T., Biradar, H., & Subudhi, P. K. (2012). Salt stress induced variation in DNA methylation pattern and its influence on gene expression in contrasting rice genotypes. *PLoS One*, *7*(6), e40203. DOI: 10.1371/journal.pone.0040203 PMID: 22761959

Kungulovski, G., & Jeltsch, A. (2015b). Epigenome Editing: State of the Art, Concepts, and Perspectives. *Trends in Genetics*, *32*(2), 101–113. DOI: 10.1016/j.tig.2015.12.001 PMID: 26732754

Li, Y., Fan, C., Xing, Y., Jiang, Y., Luo, L., Sun, L., Shao, D., Xu, C., Li, X., Xiao, J., He, Y., & Zhang, Q. (2011). Natural variation in GS5 plays an important role in regulating grain size and yield in rice. *Nature Genetics*, *43*(12), 1266–1269. DOI: 10.1038/ng.977 PMID: 22019783

Lo, A., & Qi, L. (2017). Genetic and epigenetic control of gene expression by CRISPR–Cas systems. *F1000 Research, 6*, 747. DOI: 10.12688/f1000research.11113.1 PMID: 28649363

Ma, K., Flores, C., & Ma, W. (2011). Chromatin configuration as a battlefield in Plant-Bacteria interactions. *Plant Physiology, 157*(2), 535–543. DOI: 10.1104/pp.111.182295 PMID: 21825106

Malhotra, P., Verma, G., Sidhu, G., & Duhan, N. (2020). EPIGENOMICS: ROLE, APPROACHES AND APPLICATIONS IN PLANTS. *The Journal of Animal and Plant Sciences, 30*(5). Advance online publication. DOI: 10.36899/japs.2020.5.0122

Meyer, P. (2010). DNA methylation systems and targets in plants. *FEBS Letters, 585*(13), 2008–2015. DOI: 10.1016/j.febslet.2010.08.017 PMID: 20727353

Nakamura, M., Gao, Y., Dominguez, A. A., & Qi, L. S. (2021). CRISPR technologies for precise epigenome editing. *Nature Cell Biology, 23*(1), 11–22. DOI: 10.1038/s41556-020-00620-7 PMID: 33420494

Ojolo, S. P., Cao, S., Priyadarshani, S. V. G. N., Li, W., Yan, M., Aslam, M., Zhao, H., & Qin, Y. (2018). Regulation of Plant Growth and Development: A Review from a chromatin Remodeling perspective. *Frontiers in Plant Science, 9*. Advance online publication. DOI: 10.3389/fpls.2018.01232 PMID: 30186301

Pan, L. (2013). Epigenetic regulation of abiotic stress response in plants to improve the stress tolerance. *Hereditas (Beijing), 35*(6), 745–751. DOI: 10.3724/sp.j.1005.2013.00745 PMID: 23774019

Pfluger, J., & Wagner, D. (2007). Histone modifications and dynamic regulation of genome accessibility in plants. *Current Opinion in Plant Biology, 10*(6), 645–652. DOI: 10.1016/j.pbi.2007.07.013 PMID: 17884714

Pikaard, C. S., & Scheid, O. M. (2014). Epigenetic regulation in plants. *Cold Spring Harbor Perspectives in Biology, 6*(12), a019315. DOI: 10.1101/cshperspect.a019315 PMID: 25452385

Qi, P., Lin, Y., Song, X., Shen, J., Huang, W., Shan, J., Zhu, M., Jiang, L., Gao, J., & Lin, H. (2012). The novel quantitative trait locus GL3.1 controls rice grain size and yield by regulating Cyclin-T1;3. *Cell Research, 22*(12), 1666–1680. DOI: 10.1038/cr.2012.151 PMID: 23147796

Rapp, R. A., & Wendel, J. F. (2005). Epigenetics and plant evolution. *The New Phytologist, 168*(1), 81–91. DOI: 10.1111/j.1469-8137.2005.01491.x PMID: 16159323

Richards, C. L., Alonso, C., Becker, C., Bossdorf, O., Bucher, E., Colomé-Tatché, M., Durka, W., Engelhardt, J., Gaspar, B., Gogol-Döring, A., Grosse, I., van Gurp, T. P., Heer, K., Kronholm, I., Lampei, C., Latzel, V., Mirouze, M., Opgenoorth, L., Paun, O., & Verhoeven, K. J. F. (2017). Ecological plant epigenetics: Evidence from model and non-model species, and the way forward. *Ecology Letters*, *20*(12), 1576–1590. DOI: 10.1111/ele.12858 PMID: 29027325

Samantara, K., Shiv, A., De Sousa, L. L., Sandhu, K. S., Priyadarshini, P., & Mohapatra, S. R. (2021). A comprehensive review on epigenetic mechanisms and application of epigenetic modifications for crop improvement. *Environmental and Experimental Botany*, *188*, 104479. DOI: 10.1016/j.envexpbot.2021.104479

Sarkar, R. K., & Ray, A. (2015). Submergence-tolerant rice withstands complete submergence even in saline water: Probing through chlorophyll a fluorescence induction O-J-I-P transients. *Photosynthetica*, *54*(2), 275–287. DOI: 10.1007/s11099-016-0082-4

Schmitz, R. J., & Ecker, J. R. (2012). Epigenetic and epigenomic variation in Arabidopsis thaliana. *Trends in Plant Science*, *17*(3), 149–154. DOI: 10.1016/j.tplants.2012.01.001 PMID: 22342533

Schmitz, R. J., Tamada, Y., Doyle, M. R., Zhang, X., & Amasino, R. M. (2008). Histone H2B deubiquitination is required for transcriptional activation of FLOWERING LOCUS C and for proper control of flowering in arabidopsis. *Plant Physiology*, *149*(2), 1196–1204. DOI: 10.1104/pp.108.131508 PMID: 19091875

Shang, J., & He, X. (2021). Chromatin-remodeling complexes: Conserved and plant-specific subunits in Arabidopsis. *Journal of Integrative Plant Biology*, *64*(2), 499–515. DOI: 10.1111/jipb.13208 PMID: 34964264

Shi, C., Dong, N., Guo, T., Ye, W., Shan, J., & Lin, H. (2020). A quantitative trait locus GW6 controls rice grain size and yield through the gibberellin pathway. *The Plant Journal*, *103*(3), 1174–1188. DOI: 10.1111/tpj.14793 PMID: 32365409

Shi, J., Dong, A., & Shen, W. (2015). Epigenetic regulation of rice flowering and reproduction. *Frontiers in Plant Science*, *5*. Advance online publication. DOI: 10.3389/fpls.2014.00803 PMID: 25674094

Song, M., & Koo, T. (2021). Recent advances in CRISPR technologies for genome editing. *Archives of Pharmacal Research*, *44*(6), 537–552. DOI: 10.1007/s12272-021-01336-4 PMID: 34164771

Sun, C., Ali, K., Yan, K., Fiaz, S., Dormatey, R., Bi, Z., & Bai, J. (2021). Exploration of Epigenetics for Improvement of Drought and Other Stress Resistance in Crops: A Review. *Plants*, *10*(6), 1226. DOI: 10.3390/plants10061226 PMID: 34208642

Sun, C., Fang, J., Zhao, T., Xu, B., Zhang, F., Liu, L., Tang, J., Zhang, G., Deng, X., Chen, F., Qian, Q., Cao, X., & Chu, C. (2012). The Histone methyltransferase SDG724 mediates H3K36ME2/3 deposition at MADS50 and RFT1 and promotes flowering in rice. *The Plant Cell*, *24*(8), 3235–3247. DOI: 10.1105/tpc.112.101436 PMID: 22892321

Thiebaut, F., Hemerly, A. S., & Ferreira, P. C. G. (2019). A Role for Epigenetic Regulation in the Adaptation and Stress Responses of Non-model Plants. *Frontiers in Plant Science*, *10*. Advance online publication. DOI: 10.3389/fpls.2019.00246 PMID: 30881369

Tomkova, D., & Berr, A. (2019). Book Review: Epigenetics in Plants of Agronomic Importance: Fundamentals and Applications. *Frontiers in Plant Science*, *10*. Advance online publication. DOI: 10.3389/fpls.2019.00882

Ullah, Z., Iqbal, J., Abbasi, B. A., Akhtar, W., Kanwal, S., Ali, I., & Mahmood, T. (2023b). Assessment of Gus Expression Induced by Anti-Sense Os PPO Gene Promoter and Antioxidant Enzymatic Assays in Response to Drought and Heavy Metal Stress in Transgenic Arabidopsis thaliana. Sustainability, 15(17), 12783.ang, J., Gao, S., Peng, X., Wu, K., & Yang, S. (2019). Roles of the INO80 and SWR1 Chromatin remodeling complexes in plants. *International Journal of Molecular Sciences*, *20*(18), 4591. DOI: 10.3390/ijms20184591

Ullah, Z., Iqbal, J., Abbasi, B. A., Ijaz, S., Sharifi-Rad, J., Yaseen, T., & Mahmood, T. (2023a). How Genome Editing Can Be Helpful in the Biofortification of Legumes. In *Legumes Biofortification* (pp. 207–232). Springer International Publishing.

Varotto, S., Tani, E., Abraham, E., Krugman, T., Kapazoglou, A., Melzer, R., Radanović, A., & Miladinović, D. (2020). Epigenetics: Possible applications in climate-smart crop breeding. *Journal of Experimental Botany*, *71*(17), 5223–5236. DOI: 10.1093/jxb/eraa188 PMID: 32279074

Vojta, A., Dobrinić, P., Tadić, V., Bočkor, L., Korać, P., Julg, B., Klasić, M., & Zoldoš, V. (2016). Repurposing the CRISPR-Cas9 system for targeted DNA methylation. *Nucleic Acids Research*, *44*(12), 5615–5628. DOI: 10.1093/nar/gkw159 PMID: 26969735

Wang, W., Huang, F., Qin, Q., Zhao, X., Li, Z., & Fu, B. (2015). Comparative analysis of DNA methylation changes in two rice genotypes under salt stress and subsequent recovery. *Biochemical and Biophysical Research Communications, 465*(4), 790–796. DOI: 10.1016/j.bbrc.2015.08.089 PMID: 26319557

Xie, M., & Yu, B. (2015). SiRNA-directed DNA methylation in plants. *Current Genomics, 16*(1), 23–31. DOI: 10.2174/1389202915666141128002211 PMID: 25937811

Yu, X., Jiang, L., Wu, R., Meng, X., Zhang, A., Li, N., Xia, Q., Qi, X., Pang, J., Xu, Z., & Liu, B. (2016). The core subunit of a Chromatin-Remodeling complex, ZMCHB101, plays essential roles in maize growth and development. *Scientific Reports, 6*(1). Advance online publication. DOI: 10.1038/srep38504 PMID: 27917953

Zhang, H., Lang, Z., & Zhu, J. (2018). Dynamics and function of DNA methylation in plants. *Nature Reviews. Molecular Cell Biology, 19*(8), 489–506. DOI: 10.1038/s41580-018-0016-z PMID: 29784956

Zhang, K., Sridhar, V. V., Zhu, J., Kapoor, A., & Zhu, J. (2007). Distinctive Core Histone Post-Translational Modification Patterns in Arabidopsis thaliana. *PLoS One, 2*(11), e1210. DOI: 10.1371/journal.pone.0001210 PMID: 18030344

Zhao, T., Zhan, Z., & Jiang, D. (2019). Histone modifications and their regulatory roles in plant development and environmental memory. *Journal of Genetics and Genomics. Journal of Genetics and Genomics, 46*(10), 467–476. DOI: 10.1016/j.jgg.2019.09.005 PMID: 31813758

Zheng, Y., Ding, Y., Sun, X., Xie, S., Wang, D., Liu, X., Su, L., Wei, W., Pan, L., & Zhou, D. (2016). Histone deacetylase HDA9 negatively regulates salt and drought stress responsiveness in Arabidopsis. *Journal of Experimental Botany, 67*(6), 1703–1713. DOI: 10.1093/jxb/erv562 PMID: 26733691

Zhong, X., Du, J., Hale, C. J., Gallego-Bartolome, J., Feng, S., Vashisht, A. A., Chory, J., Wohlschlegel, J. A., Patel, D. J., & Jacobsen, S. E. (2014). Molecular Mechanism of action of plant DRM de novo DNA methyltransferases. *Cell, 157*(5), 1050–1060. DOI: 10.1016/j.cell.2014.03.056 PMID: 24855943

Chapter 16
Plant Epigenetics for Crop Improvement

Shabir H. Wani
https://orcid.org/0000-0002-7456-4090
Sher-e-Kashmir University of Agricultural Sciences and Technology of Kashmir, India

Gopal Wasudeo Narkhede
School of Agriculture, SR University, Warangal, India

G. Harish Kumar
School of Agriculture, SR University, Warangal, India

M. Arun Kumar
School of Agriculture, SR University, Warangal, India

Bijendra Kumar
Madhav University, Pindwara, India

T. Soujanya
School of Agriculture, SR University, Warangal, India

ABSTRACT

Plant epigenetics research is fast progressing, revealing advanced processes that govern gene expression beyond DNA sequence modifications. Future directions include a better knowledge of epigenetic alterations such as DNA methylation, histone modifications, and small RNA pathways in plant growth, stress response, and adaptability. Advances in high-throughput sequencing and CRISPR/Cas-based epigenome editing provide new tools for investigating these mechanisms. However, hurdles remain, including the need to combine epigenetic data with genomic and phenotypic information, as well as to understand the context-specific impacts of epigenetic modifications. Furthermore, transform epigenetic discoveries into useful applications for agricultural development and sustainability remains a considerable challenge. Addressing these issues will be critical for using plant epigenetics to increase agricultural output and resilience in the face of environmental change.

DOI: 10.4018/979-8-3693-8648-4.ch016

INTRODUCTION

Epigenetics refers to changes in gene expression or cellular characteristics that do not entail modifications to the underlying DNA sequence. In plants, these changes are usually mediated by chemical modifications to DNA or histones, which can be altered by environmental factors, developmental signals, and stress. Plants' epigenetic systems include DNA methylation, histone modification, and small RNA pathways.

Key Mechanisms of Epigenetic Regulation in Plants

- **DNA Methylation:** The addition of methyl groups to DNA, usually at cytosine residues, can inhibit gene expression. DNA methylation patterns can be constant and heritable, impacting gene expression over generations (Agius et al., 2023).
- **Histone Modifications:** Histones are proteins that wrap around the DNA. Chemical alterations to histones, such as acetylation, methylation, and phosphorylation, can change chromatin structure, making it more open (active gene expression) or more compact (gene silence).
- **Small RNAs (e.g., siRNAs, miRNAs):** Small RNA molecules can drive DNA methylation or histone alterations to specific genomic regions, thereby silencing genes. They regulate gene expression at the post-transcriptional stage (Cecere, 2021).

Significance of Plant Epigenetics

- **Development and Growth:** Epigenetic mechanisms control plant growth, such as flowering time, seed development, and organ development. Epigenetic alterations can help plants respond to developmental cues and environmental changes.
- **Stress Response:** Drought, severe temperatures, and disease are all examples of environmental challenges that plants face. Plants can quickly respond and adapt to these challenges because of epigenetic changes that modify gene expression patterns. Some of these changes can be trans-generational, assisting future generations in dealing with comparable challenges (Alonso et al., 2019).
- **Adaptation and Evolution:** Epigenetic modifications allow for fast adaptability without changing the genetic code. This enables plants to rapidly evolve novel characteristics, which is particularly significant in response to environmental changes and climate change (Jin et al. 2024).

- **Crop Improvement:** Crop attributes including yield, disease resistance, and stress tolerance can all be influenced by epigenetic changes. Understanding plant epigenetics gives up new possibilities for increasing agricultural output and resilience *via* non-genetic strategies like epigenetic breeding (Alfalahi et al., 2022).
- **Gene Regulation and Plasticity:** Epigenetics enhances the plasticity of plant genomes by allowing for fine-tuning of gene expression in response to both internal and external cues. This improves the plant's ability to perform optimally in a variety of settings (Fallet et al., 2023).

Evolution of Plant Epigenetics Research: From DNA Sequence to Epigenetic Modifications

The evolution of plant epigenetics research indicates a significant transition in understanding of plant biology, from a concentration on the genetic code (DNA sequence) to taking into account the broader regulatory environment that controls gene expression. This transformation has resulted in significant findings about how plants adapt to their surroundings, evolve, and display complex characteristics (Berr and Alvarez venegas, 2023). Below is an overview of how this field has evolved over time:

The Early Years: Genetic Focus and Classical Genetics

Much of the early history of genetics focused on the sequence of DNA and how genes were inherited. Classical genetics was mainly based on the core dogma: DNA is transcribed into RNA, which is then translated into proteins that execute biological functions. Researchers linked genes to specific characteristics and began to comprehend Mendelian inheritance, which was supposed to explain the majority of plant heredity (Alfalahi et al., 2022).

Introduction of Epigenetics: Beyond DNA Sequence

In the 1940s and 1950s, Conrad Waddington invented the term "epigenetics" to describe how external factors can alter gene expression while the underlying DNA sequence remained intact. However, the importance of epigenetics in plants was not well recognized at this time. In the 1980s and 1990s, research in other creatures began to demonstrate that genetic information could be altered or regulated without affecting the DNA sequence itself. These modifications, known as epigenetic changes, comprised DNA methylation, histone modifications, and RNA interference (Alonso et al., 2019).

Key Developments

- **DNA Methylation:** DNA methylation in plants was the first widely known epigenetic modification, and it was proven to regulate gene expression without changing the gene's sequence. DNA methylation is a frequent mechanism of gene silencing that can be carried down across generations (Agius et al., 2023).
- **Histone Modifications:** Histone proteins, which wind around DNA, can also be modified by acetylation and methylation. These changes can affect how tightly DNA is packed, limiting gene accessibility for transcription.
- **RNA-based Silencing:** The discovery of short RNAs and their role in gene silencing (miRNAs and siRNAs) revealed an additional layer of epigenetic regulation (da Costa et al., 2024).

Technological Advancements: Mapping Epigenetic Marks

The early 2000s saw significant technology developments that enabled researchers to better map and understand epigenetic alterations in plants.

- **High-throughput Sequencing:** The advent of next-generation sequencing (NGS) enabled extensive analysis of the plant genome and epigenome. Scientists used whole-genome bisulfite sequencing, chromatin immuneprecipitation sequencing (ChIP-seq), and other techniques to map global DNA methylation, histone changes, and chromatin structure.
- **Epigenetic Landscape of Development:** Using these methods, researchers began to understand how epigenetic alterations influenced plant development and cell differentiation. These alterations were found to be important in flowering time, seed development, and stress responses (Eriksson et al., 2020).

Epigenetics and Stress Responses: Adaptation and Plasticity

As the field advanced, epigenetics studies in plants became more focused on how environmental elements such as light, temperature, drought, and nutrient availability affect epigenetic changes. One of the most intriguing areas of research understands how plants use epigenetics to adapt to their surroundings.

Key Insights:

- **Epigenetic Memory:** Plants can "remember" previous environmental circumstances through epigenetic modifications, even when the stressors are no longer present. This mechanism, known as epigenetic memory, has implications for how plants adapt to changing climates and grow in unsuitable environments.

- **Transgenerational Epigenetic Inheritance:** Certain epigenetic alterations can be handed down to kids without changing the DNA sequence. This has resulted in a novel concept of heredity, with epigenetic indicators acting as an additional layer of genetic transmission.
- **Environmental Epigenetics:** Plants can quickly respond to environmental stimuli by altering their epigenome, giving them increased adaptability. For example, drought can cause DNA methylation modifications that assist plants withstand future water constraint (Gaude et al., 2024).

Integrating Epigenetics with Genomics and Systems Biology

With the increased availability of plant genome sequences and epigenomic data, researchers can now investigate plant biology in a more integrated manner, combining genomics, transcriptomics, proteomics, and epigenomics to understand how all of these layers of information interact.

- **Epi-transcriptomics:** The study of RNA modifications (such as methylation) that influence gene expression is a developing field. These alterations can have an impact on RNA processing, stability, and translation, impacting plant development and stress responses (Garcia-Garcia et al., 2022).
- **Systems Biology:** The integration of multidimensional data (genomics, epigenomics, transcriptomics, etc.) provides a comprehensive understanding of the biology of plants. Artificial intelligence and machine learning are being used to predict how epigenetic changes affect gene expression and complex characteristics in plants (Gogolev et al., 2021).

The evolution of plant epigenetics research has shifted the emphasis from a solely genetic understanding of plant biology to a much more sophisticated approach that incorporates epigenetic alterations as essential regulators of gene expression and plant behavior. The role of epigenetics in plant growth, environmental adaptability, and evolution is a fascinating and fast expanding field of study, providing new insights into plant breeding, agriculture, and climate resilience (Comai, 2023).

ADVANCEMENTS IN EPIGENETIC TECHNOLOGIES

Plant epigenetic technologies have advanced rapidly, opening up new opportunities for improving crop attributes and tackling agricultural concerns. Epigenetics refers to changes in gene expression that do not entail modifications to the underlying DNA sequence yet can be passed down through generations. These changes are frequently impacted by environmental conditions, and understanding them has important consequences for crop breeding, disease resistance, and climate adaptation (Gogolev et al., 2021).

Epigenome Editing

Epigenome editing tools enable the exact alteration of epigenetic marks (such as DNA methylation, histone modifications, and non-coding RNAs) without changing the DNA sequence itself. Some notable technologies include:

- **CRISPR/Cas-based Epigenetic Tools:** CRISPR/Cas9 has been developed to change epigenetic indicators. For example, utilizing catalytically inactive Cas9 (dCas9) coupled to various effector proteins, genes can be silenced or activated by targeting certain genomic regions.
- **dCas9-DNMT3A:** For adding DNA methylation.
- **dCas9-TET1:** For de-methylation of DNA regions.
- **dCas9-p300:** For histone acetylation, activating genes.

These technologies have been used to improve stress tolerance, increase agricultural yields, and introduce desirable characteristics in crops such as rice, maize, and wheat (Comai, 2023).

DNA Methylation and Histone Modification

Epigenetic alterations, notably DNA methylation and histone modifications, play an important role in regulating plant growth, development, and stress responses. Researchers have used techniques like as bisulfite sequencing and ChIP-seq (chromatin immuno-precipitation sequencing) to map these alterations in depth (Agius et al., 2023). Advancements include:

- **Targeted DNA Methylation:** Researchers have successfully targeted certain genes for methylation to improve characteristics such as seed size, disease resistance, and stress tolerance.

- **Histone Modification Mapping:** Understanding how histones are changed has revealed information about how gene expression is regulated at different developmental stages or in response to environmental stress (Hewezi, 2018).

Small RNAs in Epigenetic Regulation

Small RNAs (miRNAs) and siRNAs (small interfering RNAs) play an important role in epigenetic control in plants. These molecules can silence genes by interacting with certain mRNA molecules or altering chromatin. Researchers are investigating:

- **Engineering Small RNA Pathways:** Modifying short RNA pathways to silence or activate genes involved in disease resistance, drought tolerance, and flowering time has demonstrated potential.
- **Small RNA-based Epigenetic Memory:** Some plants can "remember" past stress circumstances *via* small RNA-mediated epigenetic alterations, which can be passed down to future generations (Abdulraheem et al., 2024).

Epi-transcriptomics

Epi-transcriptomics is the study of chemical alterations to RNA molecules that influence gene expression and plant responses to environmental stress. Recent advances include:

- **RNA Methylation:** Methylation of RNA molecules, particularly messenger RNA (mRNA), can affect stability, translation, and splicing. These changes play an essential role in plant stress responses.
- **m6A Methylation:** The methylation of the nitrogen base adenosine (m6A) in RNA molecules has been found to influence plants' responses to abiotic stresses such as drought and cold (Agius et al. 2023; Fig 1).

Epigenetic Memory and Stress Response

Plants may "remember" environmental challenges such as drought or disease, which leads to improved stress tolerance in future generations. This phenomenon is known as epigenetic memory. According to new research, epigenetic changes in response to stress (such as DNA methylation) can be inherited, allowing crops to adapt to changing conditions without the need for genetic alterations (Fig 1).

- **Transgenerational Epigenetic Inheritance:** Certain stress-induced epigenetic alterations in plants can be transferred to offspring, increasing their ability to deal with similar environmental stressors (Kakoulidou et al., 2021; Jin et al., 2024).

Epigenetic Regulation of Flowering and Fruit Development

Epigenetic alterations can influence important agricultural characteristics such as flowering time and fruit growth. Targeting epigenetic markers that affect genes involved in these processes allows researchers to:

- Speed up or delay flowering times to suit particular growing conditions.
- Increase fruit yield and quality by modifying the expression of key developmental genes (Liang et al. 2020).

High-Throughput Epigenetic Mapping

Advances in sequencing technologies such as ChIP-seq, RNA-seq, and Hi-C (which maps 3D chromatin structure) have made it easier to map and interpret large-scale epigenetic alterations. This has resulted in the

- Identification of epigenetic networks that regulate complex characteristics in crops
- Identifying epigenetic changes linked to improved agronomic features like drought tolerance, disease resistance, and yield (Liu et al., 2023).

Gene Expression Regulation through Non-Coding RNAs

Non-coding RNAs (ncRNAs), including long non-coding RNAs (lncRNAs), have been shown to have important roles in epigenetic regulation in plants. These compounds can interact with chromatin to regulate gene expression, and understanding their role could aid in improving agricultural characteristics (Liu et al. 2024).

Synthetic Epigenetics and Synthetic Biology

Synthetic biology approaches, which integrate genetic engineering and epigenetic technologies, have resulted in advances such as synthetic epigenetic switches. These switches allow for the modulation of gene expression in response to environmental inputs, making them useful for managing crop performance under different situations (Abdulraheem et al., 2024).

Applications in Crop Improvement

- **Disease Resistance:** Epigenetic modification of plant genes related to immune responses can enhance disease resistance without altering the genetic code.
- **Abiotic Stress Tolerance:** Modifying epigenetic factors can make plants more resilient to drought, heat, and salinity by reprogramming the expression of stress-related genes.
- **Improved Yield and Quality:** Through epigenetic modification, plants can be made to have higher productivity, better nutritional profiles, and longer shelf life (Alfalahi et al., 2022).

EPIGENETIC MECHANISMS IN PLANT DEVELOPMENT AND STRESS RESPONSES

Epigenetic mechanisms regulate plant development and responses to environmental stressors. Unlike genetic changes, which alter the DNA sequence, epigenetic modifications regulate gene expression without changing the underlying DNA sequence. These alterations are reversible and can be passed down; giving plants a flexible and dynamic response mechanism for responding to changing conditions (Kumar and Mohapatra, 2024).

DNA Methylation

DNA methylation is the insertion of a methyl group to the 5' position of cytosine residues in DNA, which often results in gene silence. In plants, DNA methylation regulates gene expression during development and stress responses (Agius et al., 2023; Rajpal et al., 2022).

- **Developmental Regulation:** DNA methylation patterns vary throughout plant development, influencing processes such as seed germination, blooming, and fruit ripening.
- **Stress Responses:** Drought, salt, and temperature variations can all alter DNA methylation patterns, resulting in the activation or silencing of genes that respond to stress. For example, stress-induced methylation can silence genes that would otherwise be detrimental to the plant's survival in harsh conditions (Shaikh et al., 2022; Samantara et al., 2021).

Histone Modifications

Histones are proteins that wrap DNA, and their changes (such as acetylation, methylation, and phosphorylation) influence chromatin shape and gene expression.

- **Acetylation:** Histone acetylation enhances gene expression by relaxing the chromatin structure, giving transcription factors easy access to the DNA (Lucibelli et al., 2022; Ramkrishnan et al., 2022).
- **Methylation:** Histone methylation can activate or mute genes, depending on which amino acid residues are changed. For example, trimethylation of histone H3 at lysine 4 (H3K4me3) correlates with active transcription, whereas trimethylation at lysine 27 (H3K27me3) correlates with gene silence (Agius et al., 2023).
- **Stress Responses:** Histone alterations play a major role in the plant's response to environmental challenges such as heat shock and pathogen attack. Stress-induced histone changes can alter the plant's gene expression profile, increasing stress tolerance.

Non-Coding RNAs (ncRNAs)

ncRNAs, which include miRNAs and long non-coding RNAs, regulate gene expression at both the transcriptional and post-transcriptional levels.

- **MicroRNAs:** These short RNAs control gene expression by binding to complementary mRNA regions, causing mRNA degradation or translational repression. MiRNAs, for example, play a role in regulating genes associated with water retention and stress tolerance during droughts (Chang et al., 2020; Brukhin and Albertini, 2021).
- **Long Non-Coding RNAs:** lncRNAs regulate chromatin remodelling and gene silencing, impacting developmental processes and stress responses. Some lncRNAs are involved in the development of stress-induced chromatin structures, which help plants cope with environmental challenges (Hemenway & Gehring, 2023).

EPIGENETIC REGULATION IN PLANT DEVELOPMENT

Seed Germination and Dormancy

Epigenetic control is necessary for seed germination and dormancy. DNA methylation and histone changes help to keep the seed dormant or promote germination once the environmental conditions are right (Liu et al., 2024; Chang et al., 2020).

Flowering and Flower Development

Epigenetic markings are crucial for controlling blooming time, which is controlled by both inherent genetic programming and environmental influences. Vernalisation (the process of cold exposure to stimulate flowering) is regulated by epigenetic alterations at specific loci involved in flowering time control, such as histone modifications and DNA methylation (Abdulraheem et al. 2024).

Organogenesis and Leaf Development

Epigenetic mechanisms affect leaf patterning, stem cell maintenance, and organ development. Changes in histone modifications, DNA methylation, and non-coding RNA regulation all help to regulate these activities precisely in time and space (Abdulraheem et al., 2024; Sun et al., 2021).

EPIGENETIC REGULATION IN STRESS RESPONSES

Abiotic Stress (Drought, Salinity, Temperature)

Plants are continually exposed to environmental stresses such as drought, salt, and temperature fluctuations. Plants can adapt to these challenges by epigenetic changes, which reprogramme gene expression (Abdulraheem et al. 2024).

- **Drought Stress:** DNA methylation and histone changes regulate genes involved in water retention and osmotic stress tolerance.
- **Salinity Stress:** Epigenetic regulation of ion transporters and other salt-responsive genes helps plants manage ionic balance and reduce damage from high salinity.
- **Heat Stress:** Stress-induced histone modifications enable the expression of heat shock proteins (HSPs), which help maintain cellular integrity under heat stress (Talarico et al., 2024; Tang et al., 2020).

Biotic Stress (Pathogen Attack, Herbivory)

Plants respond to pathogen attacks by activating immunological pathways, and epigenetic modifications play an important role in regulating this response. DNA methylation and histone changes can either suppress or activate defense-related genes. For example, resistance (R) genes that give disease resistance can be epigenetically suppressed in the absence of pathogens but activated when infected (Tonosaki et al., 2022; Varotto et al., 2022).

- **Pathogen Resistance:** Specific histone modifications and short RNAs control plant immunological responses, including the systemic acquired resistance (SAR) mechanism, which gives long-term protection following an initial infection (Abdulraheem et al. 2024).

MEMORY OF STRESS

One notable aspect of epigenetic control is "epigenetic memory." This refers to plants' ability to "remember" previous stress exposures and respond more effectively to potential stress events. Plants that have been subjected to drought, for example, may have increased tolerance to successive droughts as a result of epigenetic modifications over several generations (Miryeganeh, 2021; Roman – Reyna & Crandall, 2024).

Plant epigenetic systems allow for the sophisticated regulation of gene expression in response to developmental cues and environmental challenges. Plants can fine-tune gene activity by modulating DNA methylation, histone proteins, and non-coding RNAs, allowing them to evolve and adapt to changing conditions. Understanding these pathways may lead to the production of crops with increased stress resistance and growth, providing potential answers to climate change and other environmental concerns (Singroha & Sharma, 2019; Subramanian et al., 2024).

CHALLENGES IN PLANT EPIGENETICS RESEARCH

Plant epigenetics is a rapidly expanding area that offers fascinating opportunities for learning about how plants respond to environmental challenges, develop, and change throughout time. Despite substantial progress, plant epigenetics research faces numerous significant challenges that limit its full potential. These issues can be divided into three categories: technical, biological, and conceptual, each of which offers a barrier to acquiring a better knowledge of how epigenetic mechanisms work in plants (Zhang et al., 2024; Zhang et al., 2022).

Complexity of Epigenetic Regulation

Plant genomes are extremely complicated, with a large number of genes regulated by epigenetic mechanisms such DNA methylation, histone modifications, and non-coding RNAs. The interplay of these processes is not always well understood, particularly in terms of environmental stress responses and development (Yamaguchi, 2022; Yajnik et al., 2024; Zhang et al., 2022).

- **Epigenetic Networks:** Understanding how diverse epigenetic mechanisms (such as DNA methylation and histone modifications) interact to build complex networks is a big challenge. These interactions are context-dependent, varying according to tissue type, developmental stage, and environmental factors (Varotto et al., 2022; Tirnaz & Batley, 2019; Vaschetto, 2024; Vaschetto, 2024).
- **Plasticity and Memory:** One of the most exciting elements of epigenetics is the flexibility of gene expression and the possibility of "epigenetic memory" (heritable changes in gene expression caused by past environmental exposures). Disentangling the molecular basis of this memory, particularly how it can be passed on through generations or reset, remains a significant challenge (Thiebaut et al., 2019; Liu et al., 2023).

Technical Limitations

Plant epigenetics research need advanced techniques for studying DNA modifications, histone changes, and non-coding RNAs, as well as mapping how these events affect gene expression.

- **High-Throughput Sequencing Technologies:** While next-generation sequencing (NGS) methods like as ChIP-Seq (Chromatin Immunoprecipitation sequencing) and bisulfite sequencing have made significant advances in epigenetic research, they still have limits in terms of resolution, cost, and the complexity of interpreting vast volumes of data. For example, distinguishing between adjustments that are directly linked to changes in gene expression and those that are purely correlationally can be challenging (Gogolev et al., 2021; Brunel – Muguet et al., 2022; Shaikh et al., 2022).
- **Resolution and Sensitivity:** Techniques such as ChIP-Seq and ATAC-Seq (Assay for Transposase-Accessible Chromatin utilizing Sequencing) are strong tools, but they show low sensitivity in specific tissues or under certain conditions, particularly in plant species with complicated genomes. The capacity to map fine-scale epigenetic alterations is limited by the imperfect res-

olution of chromatin dynamics (for example, identifying individual histone modifications at a single base-pair level) (Xie & duan, 2023; Xie et al., 2023).

- **Tissue-Specific Epigenomics:** Plants have tissue-specific epigenetic control, which means that epigenetic marks can differ drastically between organs, stages of development, and in response to environmental stimuli. Isolating and analyzing specific tissues (e.g., leaves, roots, seeds) might be problematic because many plants are multi-cellular and have heterogeneous tissues that require careful dissection for accurate epigenetic research (Tang et al., 2020; Chao et al., 2023).
- **Epigenetic Editing:** While CRISPR-Cas9-based methods have transformed genetic research, similar approaches to epigenetic modification (such as targeting DNA methylation or histone marks) are still in their early phases. Developing effective methods for targeted epigenetic change while preserving the underlying DNA sequence is an immense task (Zhang et al., 2024; Zhang et al., 2022).

Environmental Context and Stress Effects

Environmental conditions, such as abiotic stresses like drought, temperature extremes, and salt, as well as biotic stresses like pathogen infection, can have a substantial impact on epigenetic regulation. Understanding how these factors interact with epigenetic processes is quite challenging.

- **Environmental Heterogeneity:** Plants respond to environmental stressors in a highly context-dependent manner, with the same stress producing diverse epigenetic responses depending on the timing, intensity, and duration of exposure. Heat stress, for example, may activate heat shock proteins through histone changes, although the exact response may vary between species or even different kinds of the same species (Talarico et al., 2024; Singroha & Sharma, 2019).
- **Epigenetic Plasticity:** Plants' ability to respond to environmental changes by altering epigenetic marks frequently involves stochastic processes, which can make it difficult to identify deterministic mechanisms in stress responses. This brings heterogeneity into experimental data, making it difficult to identify particular epigenetic pathways that influence stress tolerance or development (Liu & Zhong, 2024; Jin et al., 2024).

Inheritance and Epigenetic Transmission

Another key problem in plant epigenetics research is understanding the inheritance of epigenetic marks across generations.

- **Maternal *vs*. Paternal Inheritance:** Many plants acquire epigenetic modifications from their maternal or paternal, but the mechanisms governing this inheritance remain unknown. While certain epigenetic alterations can be inherited permanently, others may reset during reproduction or fertilization, and it is not always obvious which epigenetic marks are heritable (Eriksson et al., 2020; Chang et al., 2020).
- **Transgenerational Epigenetic Inheritance:** While some research indicates that plants can "remember" stress exposure via epigenetic markers, the molecular basis for this trans-generational inheritance is not fully known. The durability and reversibility of epigenetic alterations across generations is a contentious issue, particularly in non-model plant species where these changes are difficult to trace over numerous generations (Yajnik et al., 2024).

Species-Specific and Comparative Challenges

Plant species differ significantly in their epigenetic regulation, challenging the development of generalizable models and techniques.

- **Species-Specific Epigenetic Mechanisms:** While much of the research on plant epigenetics has been done in model species like Arabidopsis thaliana, applying findings to crop species or wild plants is often difficult. Different plant species have distinct epigenetic landscapes, and specific mechanisms (for example, how DNA methylation influences gene expression) may not be immediately transferable between species (Tonosaki et al., 2022; Tang et al., 2020).
- **Polyploidy and Epigenetics:** Many key crop species, including wheat, cotton, and tobacco, are polyploid (have multiple sets of chromosomes), which complicates epigenetic regulation. Polyploids' epigenetic mechanisms differ from those of diploids, and knowing how these epigenetic alterations are preserved or altered over many sets of chromosomes is a major problem (Hemenway & Gehring, 2023).

Data Interpretation and Computational Challenges

High-throughput sequencing and other approaches generate huge volumes of data, which are challenging to analyze and integrate into a cohesive understanding of plant epigenetics.

- **Data Integration:** Epigenetic data is frequently derived from many sources, such as DNA methylation profiling, ChIP-Seq for histone modifications, RNA-Seq for gene expression, and other chromatin accessibility experiments. Integrating these diverse data sources to gain a complete understanding of gene regulation presents a significant computational challenge (Kakoulidou et al., 2021).
- **Predicting Epigenetic Effects:** While scientists can assess changes in epigenetic markers, predicting their functional implications (i.e., whether an alteration will result in gene activation or repression) is difficult. Epigenetic modifications, unlike genetic mutations, can be context-dependent, meaning that the same modification may have different effects in various tissues or under different environmental situations (Hemenway and Gehring, 2023).

Lack of Comprehensive Plant Epigenome Databases

Although certain plant species have been thoroughly investigated in terms of epigenetic control (for example, Arabidopsis and rice), there is a scarcity of complete and readily accessible databases for the epigenomes of many important crops and non-model species. This makes it difficult to compare epigenetic variations between species or apply lessons learnt from model systems to agricultural environments.

INTEGRATING EPIGENETICS WITH OTHER "OMICS" TECHNOLOGIES

Integrating epigenetics with other "omics" technologies, such as genomics, transcriptomics, proteomics, and metabolomics, represents a promising new area in plant research, allowing for a more comprehensive understanding of plant biology. This multi-layered method allows for the investigation of how epigenetic modifications affect gene expression, protein synthesis, and metabolic pathways in response to environmental stimuli, developmental signals, and stress conditions. However, successfully integrating requires overcoming a number of technological and analytical hurdles (Lucibelli et al., 2022; Fallet et al., 2023).

Epigenetics and Genomics

Genomics focuses on DNA sequences to provide information about an organism's genetic makeup. Epigenetics, on the other hand, is concerned with genome alterations that affect gene expression and function without changing the DNA sequence itself. Combining genomic data with epigenetic data allows researchers to:

- Identify epigenetic markers that are associated with specific genetic characteristics or disease resistance.
- Investigate gene-environment interactions in more depth, including how epigenetic modifications can influence genetic expression in response to environmental stresses (such as drought, salinity, or temperature variations).
- Understand non-genetic inheritance, in which epigenetic changes passed down through generations, influencing the evolution of plant populations (Berr & Alvarez venegas, 2023)

Epigenetics and Transcriptomics

Transcriptomics is the study of RNA transcripts to determine gene expression patterns. Epigenetic alterations, such as DNA methylation and histone modifications, directly control gene expression, hence combining the two technologies is critical. Key aspects of integration are:

- Identifying how DNA methylation and histone changes influence transcription in response to distinct developmental stages or stressors.
- Understanding the function of long non-coding RNAs (lncRNAs) and their epigenetic regulation in modulating gene expression
- Investigating alternative splicing and transcriptional silencing, this could be affected by epigenetic variables (Kumar and Mohapatra, 2024; Liang et al., 2020; Rajpal et al., 2022).

Epigenetics and Proteomics

Proteomics is the study of all the proteins expressed in a cell or organism. Epigenetic alterations influence protein expression, function, and relationships, hence proteomics and epigenetics are extremely complimentary. Integrating the two provides insights into:

- How epigenetic modifications influence the proteome, leading to changes in protein abundance, folding, or post-translational modifications.

- Proteomic alterations in response to environmental stress, and how epigenetic pathways influence these responses to promote plant resilience.
- Understanding the relationships between histone proteins and other regulatory proteins involved in chromatin remodeling, which influences gene expression patterns (Tonosaki et al., 2022).

Epigenetics and Metabolomics

Metabolomics is the study of tiny molecules (metabolites) that represent the biological activity of an organism. Genetics and epigenetics can both have an impact on metabolic pathways. Combining epigenetic and metabolomic data improves:

- Identify how epigenetic changes affect metabolic pathways, which may be important for stress adaption and crop productivity.
- Investigate how environmental influences might cause epigenetic alterations in plants, impacting processes like photosynthesis, secondary metabolite biosynthesis, and nutrient intake.
- Identify biochemical feedback loops in which metabolic changes might lead to additional epigenetic modifications, hence contributing to plant adaptation (Yajnik et al., 2024).

Epigenetics and Phenomics

Phenomics is the large-scale measurement of phenotypes, which are frequently linked to growth, yield, stress resistance, and other agronomically relevant aspects. Epigenetics can play an important role in influencing phenotypic expression, and by combining phenomic data with epigenetic profiles, researchers can:

- Discover epigenetic changes connected with important agronomic properties as drought tolerance, pest resistance, and nutrient efficiency.
- Use high-throughput phenotyping techniques to evaluate phenotypic responses to epigenetic changes, potentially revealing novel paths for increasing crop performance (Hewezi, 2018).

Epigenetics and Metagenomics

Metagenomics is the study of microbial communities and genetic material in a specific environment. Plants are continually interacting with a diverse group of microorganisms, including *rhizobacteria* and *mycorrhizal* fungus, which influence plant growth and stress resistance. Integrating epigenetic and metagenomic data allows for:

- Understanding how epigenetic changes in plants affect the composition of the microbiome, potentially leading to greater plant health and resilience.
- Investigating how microbial communities affect the plant epigenome, which suggests a feedback loop between epigenetic alterations and microbiome dynamics (Eriksson et al., 2020).

TOOLS AND TECHNOLOGIES FOR INTEGRATION

Several technical tools are used to efficiently combine epigenetics with other omics data, including:

- High-throughput sequencing technologies (such as RNA-seq, ChIP-seq, and bisulfite sequencing) allow for the creation of enormous databases of genomic, transcriptomic, and epigenetic information.
- Bioinformatics tools, such as multi-omics integration software (MetaboAnalyst and Cytoscape), enable the analysis of complicated data and the detection of correlations between epigenetic and other molecular profiles.
- Machine learning and artificial intelligence (AI) are rapidly being used to anticipate plant responses based on multi-omics data, as well as to find essential regulators and biomarkers for plant breeding and biotechnology (Alfalahi et al., 2022; Ramkrishnan et al., 2022).

APPLICATIONS IN PLANT BREEDING AND BIOTECHNOLOGY

By integrating epigenetics with other omics data, plant scientists can:

- Develop crop varieties with enhanced characteristics (e.g., stress tolerance, higher yield) based on an understanding of how epigenetic changes affect phenotypic outcomes.
- Use epigenetic markers to aid in marker-assisted selection in breeding programs, increasing the accuracy with which desirable traits are selected.
- Enhance biotechnology uses such as transgenic plants and gene-editing techniques (e.g., CRISPR-Cas9), in which epigenetic changes can be utilized to control gene expression without changing the DNA sequence (Alfalahi et al., 2022).

THE FUTURE OF PLANT EPIGENETICS RESEARCH

The future of plant epigenetics research is bright, as it continues to investigate the complex processes that regulate gene expression and inheritance in plants without altering the DNA sequence. This field has the potential to fundamentally alter agricultural operations, conservation efforts, and our understanding of plant biology overall.

Improved Crop Yield and Resilience

- **Stress Response and Adaptation:** Epigenetic alterations such as DNA methylation, histone modification, and short RNAs influence how plants respond to environmental challenges like drought, salinity, and severe temperatures. Understanding these pathways better may lead to the development of crops that are more adaptable to climate change, hence improving food security (Agius et al., 2023).
- **Epigenetic Breeding:** Researchers can improve desired characteristics such as disease resistance, photosynthetic efficiency, and nutrient content by modifying epigenetic markers rather than changing the underlying genetic coding. This approach could supplement traditional breeding approaches, resulting in the generation of crops with improved attributes and performance (Alfalahi et al., 2022).

Epigenetic Memory and Transgenerational Inheritance

- **Epigenetic Inheritance:** One of the most exciting aspects of plant epigenetics is how environmental influences can cause epigenetic modifications that are passed down to future generations. This "epigenetic memory" could help plants better adapt to their surroundings, perhaps speeding up evolutionary processes.
- **Environmental Influence:** Studies have shown that plants can "remember" stressful events (such as herbivore assaults or drought) and adapt more efficiently in the future. Researchers hope to learn how such epigenetic markers are passed down, which could lead to techniques for enhancing plant resilience across generations (Vaschetto, 2024).

Epigenomic Tools and Technology

- **Advanced Epigenomic Techniques:** The advent of methods such as CRISPR-based epigenetic editing, high-throughput sequencing technologies, and single-cell epigenomics is changing the way we study plant epigenomics. These technologies will enable precise epigenetic mapping and the identification of key regulators of plant growth and development.

- **Synthetic Epigenetics:** Researchers are also looking into synthetic biology ways to build plant epigenomes for specific properties. This could lead to more sustainable agricultural practices by lowering reliance on traditional genetic changes, potentially providing a non-GMO solution for crop improvement (Yajnik et al., 2024).

Environmental and Ecosystem Impact

- **Conservation Genetics:** Epigenetic modifications can affect plant adaptation, making epigenetic research essential for conservation. Understanding the role of epigenetics in plant adaptability may allow conservationists to protect endangered species by encouraging genetic resilience through epigenetic interventions.
- **Plant-Microbe Interactions:** There is a growing interest in how epigenetic mechanisms influence plant interactions with microorganisms including beneficial bacteria and fungi. Epigenetic alterations may influence how plants maintain their microbiomes, affecting nutrient uptake, disease resistance, and general health (Thiebaut et al., 2019; Tonosaki et al., 2022).

Ethical and Regulatory Considerations

- **Ethical Implications of Epigenetic Editing:** As plant epigenetic manipulation advances, ethical concerns may develop, particularly about transgenerational impacts and unforeseen consequences. Regulators will need to strike a balance between innovation and safety to guarantee that epigenetic alterations do not harm ecosystems or human health.
- **Regulation of Epigenetic-Based Crops:** Unlike genetically modified organisms (GMOs), epigenetically modified crops may be subject to different regulatory systems. This raises concerns about how these new technologies will be regulated in various nations and what standards will be used to evaluate their safety (Roman – Reyna & Crandall, 2024; Sun et al., 2021).

INTEGRATION WITH OTHER OMICS APPROACHES

- **Multi-Omic Integration:** Plant epigenetics will continue to be integrated with genomes, transcriptomics, proteomics, and metabolomics, resulting in a more comprehensive understanding of plant biology. By merging these disciplines, researchers can learn how epigenetic modifications affect gene expression, protein production, and metabolism, resulting in more precise control over plant properties.

- **Systems Biology Models:** As data and computer capacity increase, systems biology models that combine epigenetic data with other biological information will most likely be developed in the future to anticipate and manipulate plant behavior in a more systematic manner (da Costa et al., 2024).

PRECISION PLANT BREEDING

Epigenetic insights can aid in the development of more individualized crop breeding plans that are adapted to specific environments or farmer requirements. Breeders can design crops that are well-suited to specific geographies or farming practices by studying how epigenetic alterations react to various environmental variables (Alfalahi et al., 2022).

CONCLUSION

The study of plant epigenetics has opened up new avenues for understanding how plants respond to their surroundings and how these responses might be handed on to future generations. As we move forward, epigenetic mechanisms have enormous potential for driving agricultural innovation and improving plant tolerance to climate change, pests, and diseases. Plant epigenetics research continues to focus on developing stress-tolerant crops that are optimized for sustainability and yield. Advances in precision epigenetic editing technologies, like as CRISPR-based systems, will determine future field directions by enabling researchers to change epigenetic markers with remarkable accuracy.

This could result in the production of crops with specific, desirable characteristics without changing their underlying genetic structure. The combination of epigenetics and other omics methods, like as genomics, transcriptomics, and metabolomics, will provide more detailed insights into how gene expression is regulated and how plants adapt to their environments at the molecular level. However, these tremendous opportunities come with enormous challenges. The complexity of epigenetic regulation, the durability of transgenerational epigenetic modifications, and the difficulty of forecasting long-term results all provide challenges that must be solved. Furthermore, ethical and regulatory issues will be essential in establishing how epigenetic changes are used in agricultural techniques, as well as how these innovations are regulated to ensure their safety and sustainability.

As plant epigenetics advances, interdisciplinary collaboration in genomics, bioinformatics, and agricultural sciences will be critical. Researchers must also widen their research beyond model species to include a variety of crops and wild plants, each with distinct epigenetic properties. Only by addressing these obstacles will plant epigenetics reach its full potential, resulting in practical applications that provide food security, environmental sustainability, and biodiversity preservation. While the road ahead is full with complications, the future of plant epigenetics looks promising. By improving our understanding and use of epigenetic mechanisms, we can usher in a new era of plant science that has the potential to alter agriculture and contribute to a more resilient and sustainable future.

REFERENCES

Abdulraheem, M. I., Xiong, Y., Moshood, A. Y., Cadenas-Pliego, G., Zhang, H., & Hu, J. (2024). Mechanisms of plant epigenetic regulation in response to plant stress: Recent discoveries and implications. *Plants, 13*(2), 163. PMID: 38256717

Agius, D. R., Kapazoglou, A., Avramidou, E., Baranek, M., Carneros, E., Caro, E., & Vassileva, V. (2023). Exploring the crop epigenome: A comparison of DNA methylation profiling techniques. *Frontiers in Plant Science, 14*, 1181039. PMID: 37389288

Alfalahi, A. O., Hussein, Z. T., Khalofah, A., Sadder, M. T., Qasem, J. R., Al-Khayri, J. M., Jain, S. M., & Almehemdi, A. F. (2022). Epigenetic variation as a new plant breeding tool: A review. *Journal of King Saud University. Science, 34*(8), 102302.

Alonso, C., Ramos-Cruz, D., & Becker, C. (2019). The role of plant epigenetics in biotic interactions. *The New Phytologist, 221*(2), 731–737. PMID: 30156271

Berr, A., & Alvarez Venegas, R. (2023). Recent advances of epigenetics in crop biotechnology, volume II. *Frontiers in Plant Science, 14*, 1175539. PMID: 37035043

Brukhin, V., & Albertini, E. (2021). Epigenetic modifications in plant development and reproduction. *Epigenomes, 5*(4), 25. PMID: 34968249

Brunel-Muguet, S., Vetukuri, R. R., & Testillano, P. S. (2022). Epigenetics for crop adaptation to climate change. *Physiologia Plantarum*, 174. PMID: 36572663

Cecere, G. (2021). Small RNAs in epigenetic inheritance: From mechanisms to trait transmission. *FEBS Letters, 595*(24), 2953–2977. PMID: 34671979

Chang, Y. N., Zhu, C., Jiang, J., Zhang, H., Zhu, J. K., & Duan, C. G. (2020). Epigenetic regulation in plant abiotic stress responses. *Journal of Integrative Plant Biology, 62*(5), 563–580. PMID: 31872527

Chao, H., Zhang, S., Hu, Y., Ni, Q., Xin, S., Zhao, L., Ivanisenko, V. A., Orlov, Y. L., & Chen, M. (2023). Integrating omics databases for enhanced crop breeding. *Journal of Integrative Bioinformatics, 20*(4), 20230012. PMID: 37486120

Comai, L. (2023). Triggers and mediators of epigenetic remodeling in plants. *Frontiers in Epigenetics and Epigenomics, 1*, 1188733. DOI: 10.3389/freae.2023.1188733

da Costa, G. S., Cerqueira, A. F., de Brito, C. R., Mielke, M. S., & Gaiotto, F. A. (2024). Epigenetics Regulation in Responses to Abiotic Factors in Plant Species: A Systematic Review. *Plants, 13*(15), 2082. PMID: 39124200

Eriksson, M. C., Szukala, A., Tian, B., & Paun, O. (2020). Current research frontiers in plant epigenetics: An introduction to a virtual issue. *The New Phytologist*, *226*(2), 285. PMID: 32180259

Fallet, M., Blanc, M., Di Criscio, M., Antczak, P., Engwall, M., Bosagna, C. G., Ruegg, J., & Keiter, S. H. (2023). Present and future challenges for the investigation of transgenerational epigenetic inheritance. *Environment International*, *172*, 107776. PMID: 36731188

García-García, I., Méndez-Cea, B., Martín-Gálvez, D., Seco, J. I., Gallego, F. J., & Linares, J. C. (2022). Challenges and perspectives in the epigenetics of climate change-induced forests decline. *Frontiers in Plant Science*, *12*, 797958. PMID: 35058957

Gaude, A. A., Siqueira, R. H., Botelho, S. B., & Jalmi, S. K. (2024). Epigenetic arsenal for stress mitigation in plants. *Biochimica et Biophysica Acta* (BBA)-. *General Subjects*, *1868*(7), 130620. PMID: 38636616

Gogolev, Y. V., Ahmar, S., Akpinar, B. A., Budak, H., Kiryushkin, A. S., Gorshkov, V. Y., Henzel, G., Demchenko, K. N., Kovalchuk, I., Poblete, F. M., Muslu, T., Tsers, I. D., Yadav, N. S., & Korzun, V. (2021). OMICs, epigenetics, and genome editing techniques for food and nutritional security. *Plants*, *10*(7), 1423. PMID: 34371624

Hemenway, E. A., & Gehring, M. (2023). Epigenetic regulation during plant development and the capacity for epigenetic memory. *Annual Review of Plant Biology*, *74*(1), 87–109. PMID: 36854474

Hewezi, T. (2018). Epigenetic regulation of plant development and stress responses. *Plant Cell Reports*, *37*, 1–2. PMID: 29159566

Jin, Q., Chachar, M., Ali, A., Chachar, Z., Zhang, P., Riaz, A., Ahmed, N., & Chachar, S. (2024). Epigenetic Regulation for Heat Stress Adaptation in Plants: New Horizons for Crop Improvement under Climate Change. *Agronomy (Basel)*, *14*(9), 2105.

Kakoulidou, I., Avramidou, E. V., Baránek, M., Brunel-Muguet, S., Farrona, S., Johannes, F., Kaiserli, E., Lazarovich, M. L., Martinelli, F., Mladenov, V., Tetillano, P. S., Vassileva, V., & Maury, S. (2021). Epigenetics for crop improvement in times of global change. *Biology (Basel)*, *10*(8), 766. PMID: 34439998

Kumar, S., & Mohapatra, T. (2024). Epigenetic Modifications in Genome Help Remembering the Stress Tolerance Strategy Adopted by the Plant. *Frontiers in Bioscience-Landmark*, *29*(3), 126. PMID: 38538276

Liang, Z., Riaz, A., Chachar, S., Ding, Y., Du, H., & Gu, X. (2020). Epigenetic modifications of mRNA and DNA in plants. *Molecular Plant, 13*(1), 14–30. PMID: 31863849

Liu, J., & Zhong, X. (2024). Population epigenetics: DNA methylation in the plant omics era. *Plant Physiology, 194*(4), 2039–2048. PMID: 38366882

Liu, S., He, M., Lin, X., & Kong, F. (2023). Epigenetic regulation of photoperiodic flowering in plants. *The Plant Genome, 16*(4), e20320. PMID: 37013370

Liu, X., Zhu, K., & Xiao, J. (2023). Recent advances in understanding of the epigenetic regulation of plant regeneration. *aBIOTECH, 4*(1), 31–46. PMID: 37220541

Lucibelli, F., Valoroso, M. C., & Aceto, S. (2022). Plant DNA Methylation: An Epigenetic Mark in Development, Environmental Interactions, and Evolution. *International Journal of Molecular Sciences, 23*, 8299. DOI: 10.3390/ijms23158299 PMID: 35955429

Miryeganeh, M. (2021). Plants epigenetic mechanisms and abiotic stress. *Genes, 12*(8), 1106. PMID: 34440280

Rajpal, V. R., Rathore, P., Mehta, S., Wadhwa, N., Yadav, P., Berry, E., Goel, S., Bhat, V., & Raina, S. N. (2022). Epigenetic variation: A major player in facilitating plant fitness under changing environmental conditions. *Frontiers in Cell and Developmental Biology, 10*, 1020958. PMID: 36340045

Roman-Reyna, V., & Crandall, S. G. (2024). Seeing in the dark: A metagenomic approach can illuminate the drivers of plant disease. *Frontiers in Plant Science, 15*, 1405042. PMID: 39055364

Samantara, K., Shiv, A., de Sousa, L. L., Sandhu, K. S., Priyadarshini, P., & Mohapatra, S. R. (2021). A comprehensive review on epigenetic mechanisms and application of epigenetic modifications for crop improvement. *Environmental and Experimental Botany, 188*, 104479.

Shaikh, A. A., Chachar, S., Chachar, M., Ahmed, N., Guan, C., & Zhang, P. (2022). Recent Advances in DNA Methylation and Their Potential Breeding Applications in Plants. *Horticulturae, 8*, 562. DOI: 10.3390/horticulturae8070562

Singroha, G., & Sharma, P. (2019). Epigenetic modifications in plants under abiotic stress. *Epigenetics*, •••, 10.

Subramanian, A. T., Roy, P., Aravind, B., Kumar, A. P., & Mohannath, G. (2024). Epigenome editing strategies for plants: A mini review. *Nucleus (Austin, Tex.)*, •••, 1–13.

Sun, C., Ali, K., Yan, K., Fiaz, S., Dormatey, R., Bi, Z., & Bai, J. (2021). Exploration of epigenetics for improvement of drought and other stress resistance in crops: A review. *Plants*, *10*(6), 1226. PMID: 34208642

Talarico, E., Zambelli, A., Araniti, F., Greco, E., Chiappetta, A., & Bruno, L. (2024). Unravelling the Epigenetic Code: DNA Methylation in Plants and Its Role in Stress Response. *Epigenomes*, *8*(3), 30. PMID: 39189256

Tang, D., Gallusci, P., & Lang, Z. (2020). Fruit development and epigenetic modifications. *The New Phytologist*, *228*(3), 839–844. PMID: 32506476

Thiebaut, F., Hemerly, A. S., & Ferreira, P. C. G. (2019). A role for epigenetic regulation in the adaptation and stress responses of non-model plants. *Frontiers in Plant Science*, *10*, 246. PMID: 30881369

Tirnaz, S., & Batley, J. (2019). Epigenetics: Potentials and challenges in crop breeding. *Molecular Plant*, *12*(10), 1309–1311. PMID: 31541738

Tonosaki, K., Fujimoto, R., Dennis, E. S., Raboy, V., & Osabe, K. (2022). Will epigenetics be a key player in crop breeding? *Frontiers in Plant Science*, *13*, 958350. PMID: 36247549

Varotto, S., Krugman, T., Aiese Cigliano, R., Kashkush, K., Kondić-Špika, A., Aravanopoulos, F. A., Pradillo, M., Consiglio, F., Aversano, R., Pecinka, A., & Miladinović, D. (2022). Exploitation of epigenetic variation of crop wild relatives for crop improvement and agro-biodiversity preservation. *Theoretical and Applied Genetics*, *135*(11), 3987–4003. PMID: 35678824

Vaschetto, L. M. (2024). Epigenetics and its Role in Plant Evolution. In *Epigenetics in Crop Improvement: Safeguarding Food Security in an Ever-Changing Climate* (pp. 3–12). Springer Nature Switzerland.

Vaschetto, L. M. (2024). The Role of Epigenetic Variation in Plant Adaptation. In *Epigenetics in Crop Improvement: Safeguarding Food Security in an Ever-Changing Climate* (pp. 29–47). Springer Nature Switzerland.

Xie, S. S., & Duan, C. G. (2023). Epigenetic regulation of plant immunity: From chromatin codes to plant disease resistance. *aBIOTECH*, *4*(2), 124–139. PMID: 37581024

Xie, Y., Chan, L. Y., Cheung, M. Y., Li, M. W., & Lam, H. M. (2023). Current technical advancements in plant epi-transcriptomic studies. *The Plant Genome*, *16*(4), e20316. PMID: 36890704

Yajnik, K. N., Singh, I. K., & Singh, A. (2024). lncRNAs and epigenetics regulate plant's resilience against biotic stresses. *Plant Physiology and Biochemistry*, •••, 108892. PMID: 38964086

Yamaguchi, N. (2022). Epigenetics in Plant Development. *Frontiers in Plant Science*, *13*, 864945. PMID: 35295634

Zhang, D., Zhang, D., Zhang, Y., Li, G., Sun, D., Zhou, B., & Li, J. (2024). Insights into the Epigenetic Basis of Plant Salt Tolerance. *International Journal of Molecular Sciences*, *25*(21), 11698. PMID: 39519250

Zhang, R., Zhang, C., Yu, C., Dong, J., & Hu, J. (2022). Integration of multi-omics technologies for crop improvement: Status and prospects. *Frontiers in Bioinformatics*, *2*, 1027457. PMID: 36438626

Zhang, Y., Andrews, H., Eglitis-Sexton, J., Godwin, I., Tanurdžić, M., & Crisp, P. A. (2022). Epigenome guided crop improvement: Current progress and future opportunities. *Emerging Topics in Life Sciences*, *6*(2), 141–151. PMID: 35072210

Chapter 17
Future Directions and Challenges in Plant Epigenetics Research

Shreya Anand
Mizoram University, India

Rahul Kumar
https://orcid.org/0009-0005-4262-5521
Mahatma Gandhi Central University, India

Prakash Kumar Sarangi
Central Agricultural University, India

Garima Khaspuria
https://orcid.org/0009-0007-8927-580X
Vivekananda Global University, India

Akhilesh Kumar Singh
Mahatma Gandhi Central University, India

Saurabh Singh Rathore
https://orcid.org/0000-0002-6997-487X
Mahatma Gandhi Central University, India

ABSTRACT

The field of plant epigenetics is expanding, revealing how plants regulate gene expression and adapt to their environments. This chapter discusses the future directions and challenges in plant epigenetics, highlighting advanced technologies like single-cell epigenomics and CRISPR-based editing. Overcoming challenges such as lack of reference epigenomes for diverse species and ethical concerns in agricultural applications is key. Interdisciplinary collaboration and innovation are crucial to advancing research and ensuring sustainable crop production.

DOI: 10.4018/979-8-3693-8648-4.ch017

INTRODUCTION

The field of plant epigenetics is expanding at a rapid pace, revealing diverse mechanisms by which plants regulate gene expression and adapt to their environments. This chapter explores future directions and challenges in plant epigenetics research, highlighting the integration of advanced technologies and the necessity for interdisciplinary approaches. Understanding epigenetic modifications, such as DNA methylation, histone modifications, and non-coding RNAs, is essential for advancing agricultural biotechnology and ensuring sustainable crop production. Epigenetic modifications include processes like DNA methylation, histone modifications, and the involvement of non-coding RNAs. DNA methylation involves the addition of a methyl group to the DNA molecule, typically acting to suppress gene expression. This process can be influenced by environmental factors, enabling plants to respond to stress conditions like drought or pathogen attacks. Histone modifications, on the other hand, involve chemical changes to the proteins around which DNA is wrapped, affecting how tightly or loosely the DNA is packaged and, consequently, its accessibility for transcription. Non-coding RNAs are RNA molecules that do not encode proteins but play crucial roles in regulating gene expression at various levels, including chromatin structure, transcription, and post-transcriptional processing. The study of plant epigenetics has significantly advanced our understanding of how plants regulate gene expression and adapt to a wide range of environmental conditions (Baulcombe et al., 2014). As a vital aspect of gene regulation, epigenetics plays a key role in influencing plant growth, development, stress responses, and long-term adaptability (Grativol et al., 2012). Through epigenetic mechanisms, plants can precisely modulate gene expression without altering the underlying DNA sequence (Chang et al., 2020). offering a dynamic regulatory layer capable of responding to internal signals and environmental stimuli (Walther, 2020), particularly amid changing environmental conditions, has become increasingly evident.

Epigenetics in plants encompasses several key mechanisms, including DNA methylation, histone modifications, and non-coding RNAs (Ali et al., 2024) (Figure 1). Each of these mechanisms plays a distinct role in shaping gene expression patterns and chromatin organization. Notably, they do not alter the DNA sequence itself but function as molecular marks that can be inherited through cell divisions or reset during specific developmental phases (Tikhodeyev, 2018). This heritable yet reversible nature of epigenetic modifications provides an impressive layer of flexibility and control over the plant genome (Galindo et al., 2018).

This chapter delves into the future directions and challenges in plant epigenetic research, emphasizing the integration of state-of-the-art technologies and the necessity for interdisciplinary approaches. Understanding epigenetic modifications is crucial for advancing agricultural biotechnology and ensuring sustainable crop production. By manipulating epigenetic marks, scientists can develop crops that are more resilient to environmental stresses, diseases, and pests, ultimately contributing to global food security. One key future direction in plant epigenetics involves the advancement and application of high-resolution and high-throughput technologies. Techniques such as single-cell epigenomics, CRISPR-based epigenome editing, and advanced imaging methods are positioned to revolutionize our ability to map and manipulate epigenetic landscapes with unprecedented precision. Single-cell epigenomics allows researchers to study epigenetic modifications at the level of individual cells, providing insights into cellular heterogeneity and how different cells within a plant may respond differently to environmental cues. This level of detail is essential for understanding complex traits and developing targeted interventions. CRISPR-based epigenome editing is a powerful tool that enables precise modifications to the epigenome, without altering the underlying DNA sequence. By targeting specific regions of the genome, researchers can activate or silence genes in a controlled manner, providing a deeper understanding of gene function and regulation. Advanced imaging methods, such as super-resolution microscopy, allow scientists to visualize epigenetic modifications in real-time, offering a dynamic view of how these changes occur and propagate through cells and tissues.

Despite notable advancements, there are still significant challenges to be addressed in the field of plant epigenetics. A major hurdle is the absence of comprehensive reference epigenomes for a wider range of plant species, particularly those of agricultural importance. Most of the current research has focused on model plants like Arabidopsis thaliana, which, while valuable, does not fully represent the diversity of plant species used in agriculture. Developing reference epigenomes for crops such as wheat, rice, maize, and other staple foods is essential for translating epigenetic research into practical applications. Overcoming this challenge will necessitate not only technological innovation but also extensive collaboration and data sharing within the scientific community. Establishing international consortia and data-sharing platforms can facilitate the collection and dissemination of epigenetic data across different plant species and research groups. This collaborative approach can accelerate the identification of key epigenetic markers associated with desirable traits and enhance the breeding of resilient and high-yielding crops.

Furthermore, the functional characterization of epigenetic modifications and their dynamic nature under various environmental conditions poses substantial experimental and analytical challenges. Epigenetic marks are often context-dependent, meaning their effects can vary based on factors such as tissue type, developmental stage, and environmental conditions. This complexity requires sophisticated experimental designs and robust analytical tools to accurately interpret the data. Advances in bioinformatics and computational biology are crucial for integrating large-scale epigenetic datasets and extracting meaningful insights. In addition to technical challenges, there are regulatory, ethical, and societal difficulties that impede the practical application of basic epigenetic research in agriculture. The safe and efficient use of epigenome editing technologies to enhance crops necessitates robust risk assessment structures and active public involvement. Regulatory frameworks must be established to ensure that the use of these technologies is safe for both the environment and human health. This includes evaluating potential off-target effects, unintended consequences, and long-term impacts on ecosystems. Active public involvement is crucial to tackle potential apprehensions and ensure transparency in applying epigenetic research in agriculture. Public engagement initiatives can help demystify the science behind epigenetics and foster informed discussions about the benefits and risks associated with epigenome editing. Building trust and addressing ethical concerns are essential for gaining public acceptance and support for the adoption of these technologies.

In conclusion, the future of plant epigenetic research holds enormous potential for deepening our understanding of plant biology and enhancing crop resilience and productivity. Overcoming the significant technical, logistical, and societal challenges is essential to realizing this potential. By fostering interdisciplinary collaboration and leveraging innovative technologies, the scientific community can undoubtedly drive transformative advancements in plant epigenetics and its application to sustainable agriculture. The integration of high-resolution and high-throughput technologies, coupled with extensive data sharing and collaboration, will empower researchers to unravel the complex interplay between genetic and epigenetic factors. This will provide valuable insights into plant development, stress responses, and adaptation, ultimately leading to the development of crops that are more resilient to environmental challenges. As we continue to explore the epigenetic landscape of plants, it is essential to address regulatory, ethical, and societal concerns to ensure the safe and responsible application of these technologies. The future of plant epigenetics is bright, with the potential to revolutionize agricultural practices and contribute to global food security. Recent advances in plant epigenetics and what the future may hold for this growing field have also been discussed here. As the world's agricultural needs increase due to a rising population and climate change, the demand for strong, high-yield crops is more important than ever. Research in epigenetics shows great potential for tackling these issues. By using epigenetic regulation, scientists can create new ways to improve crop performance, increase resistance to stress, and promote sustainable food production.

Figure 1. Plant epigenetics, future and its challenges

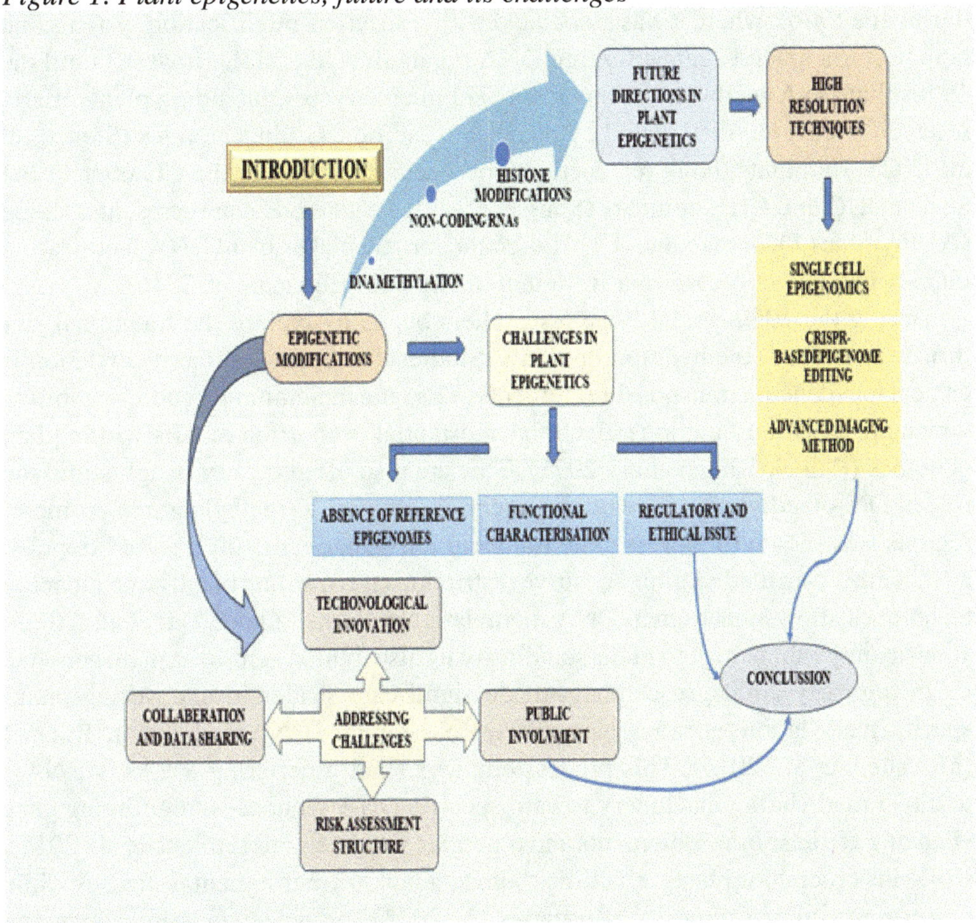

A BRIEF HISTORY OF KEY EPIGENETIC MECHANISMS AND CRISPR TECHNOLOGY

DNA methylation, histone modifications, non-coding RNAs, and CRISPR technology have changed how we understand gene regulation, epigenetics, and genome editing. Below is a brief history of each area, focusing on key discoveries and milestones that have shaped these fields (Table 1). The study of DNA methylation in plants has a rich and evolving history. Initially recognized as a mechanism for gene regulation and genome stability, it has now become a central focus of plant epigenetics (Zhang et al., 2018). DNA methylation refers to the addition of a methyl group to cytosine bases in DNA (Kumar et al., 2018). This process was first examined in bacteria during the 1940s, where it was associated with restriction-modification systems that help bacteria defend against foreign DNA (Ganguli et al., 2021). It wasn't until the 1970s that DNA methylation was identified in eukaryotes, including plants. Early research uncovered methylated cytosines at specific CG dinucleotides (Stevens et al., 2022). In plants, however, methylation occurs not only in the CG context but also in CHG and CHH contexts (Song et al., 2017), where H can represent adenine (A), thymine (T), or cytosine (C). This characteristic makes plant DNA methylation unique and distinctly different from that of animals (Elhamamsy, 2016).

During the 1980s and 1990s, researchers began to explore the functional significance of DNA methylation in plants. Studies demonstrated that methylation is critical for silencing transposable elements (TEs) and maintaining genomic stability, which are essential functions given the substantial proportion of TEs within plant genomes (Ramakrishnan et al., 2021). This era also offered early insights into the role of DNA methylation in regulating gene expression, as methylation in promoter regions was shown to repress gene transcription (Dhar et al., 2021). Additionally, advancements in methylation-sensitive restriction enzymes and bisulfite sequencing techniques allowed us to map DNA methylation patterns (Khodadadi et al., 2021), significantly enhancing our understanding of its distribution across the plant genome.

In the early 2000s, researchers made significant strides in uncovering plant-specific methylation pathways, particularly RNA-directed DNA methylation (RdDM) (Movahedi et al., 2015). This process employs small interfering RNAs (siRNAs) to direct methylation machinery toward specific DNA sequences, reinforcing gene silencing at transposable elements and repetitive DNA regions (Holoch et al., 2015). RdDM is crucial for plants, enabling them to adapt to environmental stresses while preserving genome integrity (Kumar et al., 2023). Advancements in sequencing technology during the 2010s facilitated high-resolution, genome-wide mapping of DNA methylation in plants (Hong et al., 2023). These studies revealed that DNA methylation undergoes dynamic changes in response to environmental stimuli, developmental stages, and stress conditions (Kumar et al., 2021). DNA methylation is an important process in plants. It affects key functions like seed development, flowering time, stress tolerance, and memory passed to future generations (Liu et al., 2020). With new CRISPR-based tools, scientists can now change DNA methylation precisely. This advanced technique can help improve crops and support sustainable agriculture. By editing DNA methylation in specific genes or regions, scientists can create crops that are better able to withstand climate change, diseases, and environmental stress (Kakoulidou et al., 2021). The history of DNA methylation research in plants shows a path from early discoveries to significant advancements in plant biotechnology and agriculture.

Table 1. Discoveries and their impact on plant biology

Year	Scientist(s)	Discovery/Contribution	Impact on Plant Biotechnology	Reference
1884	Albrecht Kossel	Identified histones as basic nuclear proteins.	Laid groundwork for understanding chromatin structure.	(Farsetti et al., 2023)
1948	Rollin Hotchkiss	Discovered DNA methylation as a chemical modification.	Initiated the study of DNA methylation in gene regulation.	(Wassenegger et al., 2021)
1964	Vincent Allfrey	Linked histone acetylation and methylation to RNA synthesis.	Showed that histone modifications affect gene expression.	(Chen et al., 2024)
1975	Arthur Riggs, Robin and John Pugh	Proposed that DNA methylation could be inherited, shaping epigenetic regulation theories.	Established methylation as a mechanism for heritable gene expression.	(Tompkins, 2022).
1980s-1990s	Kazuo Kinoshita	Discovered the role of DNA methylation in regulating gene expression in plant reproductive organs.	Demonstrated methylation's importance in plant development.	(Kim et al., 2024)
1990s	David Baulcombe	Showed that small RNAs (siRNAs) in plants could silence genes, key in RNA interference (RNAi).	Opened new avenues for studying gene silencing and stress response.	(Zand & Innes., 2022).
1996	Karolin Luger and Timothy Richmond	Determined the structure of the nucleosome, revealing histone-DNA interactions.	Enabled deeper understanding of chromatin architecture.	(Hondele, 2013).
1999	David Baulcombe	Linked small RNAs with plant defense, introducing RNAi as a regulatory mechanism in plants.	Established RNAi as a tool for genetic and epigenetic regulation.	(Baulcombe, 2023)
2000s	Various (Epigenetic Researchers)	Identified that ncRNAs regulate DNA methylation and gene expression in plants.	Showed how ncRNAs are integral to plant stress responses and development.	(Akhter et al., 2021)
2002	Richard Jorgensen	Demonstrated RNA silencing in plants via transgenes, leading to "co-suppression."	Advanced understanding of gene silencing for crop improvement.	(Rajput et al., 2021)
2012	Jennifer Doudna and Emmanuelle Charpentier	Developed the CRISPR-Cas9 system for programmable gene editing.	Enabled targeted gene editing with unprecedented precision.	(Cao et al., 2016)

continued on following page

Table 1. Continued

Year	Scientist(s)	Discovery/Contribution	Impact on Plant Biotechnology	Reference
2013	Feng Zhang	First demonstrated the application of CRISPR-Cas9 in plant cells.	Established CRISPR as a tool for crop improvement and biotechnology.	(Wang et al 2020)
2014	Daniel Voytas and Others	Refined CRISPR-Cas systems in plants, enhancing gene-editing specificity and efficiency.	Boosted applications in precision breeding and crop resilience.	(Echevarria et al 2021)
Present	Ongoing Research	Advances in single-cell epigenomics, epigenome editing (CRISPR-dCas9), and transgenerational epigenetics in plants.	Enabling next-generation crop improvement and resilience to climate change.	-----------

THE CONCEPT OF EPIGENETICS IN PLANTS

The idea of epigenetics in plants has changed a lot over time. It started with early observations of inherited traits that Mendelian genetics could not explain. Now, it has developed into a deeper understanding of the molecular mechanisms that control gene expression. The term "epigenetics" was first introduced by Conrad Waddington in the 1940s (Kumar et al., 2017). It refers to how the genotype results in the phenotype during development. In plants, scientists saw signs of epigenetic traits even before the term existed. They noticed inherited traits that did not follow Mendelian patterns, such as differences in flower color, growth, and fruit ripening (Jahed et al., 2023). These traits were documented to last across generations. Importantly, these traits could not be linked only to changes in the DNA sequence. This suggests that other regulatory factors influence gene expression as well (Pai et al., 2015).

One of the earliest well-documented instances of epigenetic regulation in plants was Barbara McClintock's discovery of "jumping genes," or transposable elements, in maize during the 1950s (Huang et al., 2017). Her pioneering research demonstrated that these mobile genetic elements could shift their position within the genome, thereby influencing gene expression without changing the DNA sequence itself (Partridge et al., 2018). This foundational work paved the way for subsequent discoveries that uncovered how environmental stimuli and developmental processes could induce heritable changes in gene activity (Boyce, 2016), ultimately establishing the field of modern epigenetics. Over the years, advancements in molecular biology and genomics have allowed researchers to identify key molecular markers involved in epigenetic regulation, including DNA methylation, histone modifications, chromatin remodeling, and small RNA pathways (Clark et al., 2020).

In plants, DNA methylation plays a crucial role in regulating various processes, including transposon silencing, genomic imprinting, stress responses, and developmental timing (Ramakrishnan et al., 2021). This ability to epigenetically modify gene expression in response to environmental factors such as drought, salinity, and temperature has been recognized as a key evolutionary strategy that allows plants to adapt and survive under changing conditions (Jin et al., 2024). The understanding of epigenetics in plants has evolved from being viewed solely as a developmental tool to being recognized as a significant adaptive mechanism with important implications for agriculture and biotechnology. Today, modern plant epigenetics encompasses not only the study of individual epigenetic marks but also the dynamic interactions among these marks throughout the epigenome (Alfalahi et al., 2022).

Recent advances in technologies like bisulfite sequencing, ChIP-seq, and RNA-seq have improved our ability to map epigenetic landscapes accurately. These tools show how epigenetic marks control gene expression based on time and location (Kagohara et al., 2018). A key discovery is epigenetic memory, which means that plants can "remember" environmental stresses, making them better able to cope with future challenges (Sharma et al., 2022). Additionally, the field of transgenerational epigenetics has helped us understand plant inheritance. It demonstrates that some epigenetic changes can be passed down to future generations, affecting the traits of offspring without changing the DNA sequence (Jablonka, 2017). Despite the growing knowledge in plant epigenetics, numerous questions remain unanswered. Scientists are actively exploring how epigenetic marks are stably maintained or reset during plant reproduction, the mechanisms by which environmental cues elicit specific epigenetic responses, and how plants integrate genetic and epigenetic information to optimize their development and survival.

The study of epigenetics has changed from a basic description to a useful tool for improving plants and crops. For instance, CRISPR technology allows scientists to edit specific epigenetic markers (Nakamura et al., 2021). This helps enhance traits like drought resistance, disease resistance, and crop yield. These changes in epigenetics are important for agriculture, especially as we face issues like climate change and food security. In short, plant epigenetics has progressed from early studies on inheritance to a well-developed field that combines molecular, developmental, and environmental biology (Hu et al., 2021). Plants can change gene expression through epigenetics, showing their ability to grow and adapt (Ashapkin et al., 2020). As we learn more about plant epigenetics, we can find new ways to boost agricultural productivity, resilience, and sustainability, helping us meet the challenges of the 21st century.

PRACTICAL IMPLICATIONS OF PLANT EPIGENETICS IN AGRICULTURE

The practical implications of plant epigenetics in agriculture have gained increasing attention as researchers explore how heritable changes in gene expression, without alterations in DNA sequence, can be leveraged to enhance crop performance and sustainability (Soni et al., 2024). With growing global challenges such as climate change, declining soil fertility, and increasing food demand, epigenetic tools offer new avenues to improve crop resilience, productivity, and quality (Kakoulidou et al., 2021). Below are the key trends and emerging practical applications of plant epigenetics in agriculture.

Epigenetic Breeding for Stress Tolerance

Plants are consistently exposed to various environmental stresses, including drought, salinity, temperature fluctuations, and pathogens (Bai et al., 2018; Chaudhry & Sidhu, 2022). Epigenetic mechanisms, such as DNA methylation and histone modifications, allow plants to respond dynamically to these challenges and retain a "memory" of the stress, which enhances their resilience during future exposures (Sharma et al., 2022). Breeding programs are now investigating epigenetic markers linked to stress tolerance in order to develop more potent crop varieties (Tirnaz et al., 2019). By selecting plants that maintain beneficial epigenetic states under stress, breeders can produce crops that perform better in harsh conditions without needing to rely on genetic modifications.

Transgenerational Epigenetic Inheritance

There is increasing interest in the heritability of epigenetic trait, changes in gene expression that can be influenced by environmental factors and passed down to subsequent generations (Nilsson et al., 2018). For instance, research has shown that exposure to drought or nutrient limitations can create heritable epigenetic changes that enhance stress tolerance in offspring (Balao et al., 2018). Utilizing these transgenerational effects in crop breeding has the potential to accelerate the development of climate-resilient varieties without requiring extensive genomic modification (Kumar et al., 2019).

Epigenome Editing for Precision Crop Improvement

The emergence of CRISPR-based epigenome editing tools has opened up exciting possibilities for precisely modifying specific epigenetic markers (Enríquez,

2016). Unlike traditional gene editing, which changes the actual DNA sequence, epigenome editing focuses on how genes are expressed by altering DNA methylation or the configuration of chromatin (Fal et al., 2021). This method enables plant breeders to improve key traits such as flowering time, crop yield, and disease resistance, while avoiding the controversies typically associated with genetically modified organisms (GMOs). For example, researchers are using CRISPR-dCas9 fusion systems to either turn on or turn off key genes that are crucial for traits that matter to agriculture (Zhang et al., 2021). This innovative technology is paving the way for a more sustainable and ethical approach to breeding.

Epigenetics in Plant-Microbe Interactions

Recent research has revealed some fascinating insights into how plants defend themselves against pathogens. It turns out that epigenetic modifications play a key role in regulating these immune responses and can even enhance the way plants interact with beneficial microbes in the soil (Zhu et al., 2016). When a plant's immune signalling genes are controlled epigenetically, it can change not only how well the plant responds to threats but also how long it maintains that response (Ding & Wang, 2015). This has important implications for agriculture, as scientists are exploring ways to harness these modifications to boost disease resistance. The goal is to reduce reliance on chemical pesticides and promote more sustainable farming methods. What's particularly interesting is that some beneficial soil microbes can actually trigger these epigenetic changes, helping plants grow stronger and more resilient (Samantara et al., 2021). This opens up exciting possibilities for developing biofertilizers or microbial inoculants that could enhance plant health and productivity. Overall, these findings point to a deeper understanding of plant biology that could benefit farmers and the environment.

Epigenetic Regulation of Yield and Nutritional Quality

Epigenetic factors play a vital role in critical developmental processes like when plants flower, how fruits ripen, and how seeds develop. These processes significantly influence the yield and quality of our crops (Gady et al., 2017). Researchers are now looking into ways to harness these epigenetic changes to boost both the productivity and nutritional value of food. For example, by altering specific methylation patterns in rice or wheat, scientists can potentially increase the size and yield of the grains (Ren, Ding & Qian, 2023). Similarly, by regulating the pathways that affect carotenoids and flavonoids, we can enhance the nutritional benefits of fruits and vegetables (Saini et al., 2022). This approach offers a fresh perspective alongside traditional breeding methods, allowing for quicker and more precise improvements in the crops that we rely on.

Epigenetics and Plant Adaptation to Climate Change

Climate change poses significant threats to global agriculture, making it crucial to understand how plants adapt epigenetically to changing environmental conditions (Jin et al., 2024). When exposed to extreme heat, drought, or salinity, plants undergo specific epigenetic changes that enable them to survive and thrive (Jin et al., 2024). Scientists are working to identify these epigenetic signatures to predict how plants will respond to future climate scenarios (Richards et al., 2017). This knowledge is being used to breed crops that are better equipped to handle environmental stressors. Additionally, scientists are exploring the possibility of inducing stress memory through controlled environmental exposure in nurseries, which may help pre-condition crops for adverse conditions.

Epigenetic Seed Priming

Seed priming is an interesting technique that enhances the germination and growth of seeds before they are even sown (Farooq et al., 2019). By applying mild stress or specific chemical treatments, seeds can undergo changes at the genetic level that help them sprout more reliably and thrive in tough conditions (Long et al., 2015). These changes can stick with the plants throughout their lives, giving them a better chance of establishing themselves, especially in challenging environments. This method is particularly exciting for regions dealing with poor soil or extreme weather, as it could help farmers grow healthier crops in less-than-ideal situations.

Epigenetic Solutions for Crop Domestication and De-domestication

When we domesticate wild plants into crops, we often lose important traits that help them survive in the wild, like resistance to diseases and the ability to endure droughts (Fita et al., 2015). However, epigenetic research is opening up exciting possibilities to address this issue through a process called de-domestication (Hong et al., 2023). This approach allows us to reintroduce the beneficial traits of wild plants into our cultivated crops without changing their genetic makeup. By looking closely at the epigenetic changes in ancestral species, breeders can bring back the traits that make plants more resilient to today's environmental challenges, all while keeping the qualities we love about our crops (Cao et al., 2024). It's a fascinating way to ensure that agriculture can adapt and thrive in a changing world.

Epigenetics and Crop-Specific Studies

Traditionally, a lot of epigenetic research has centred around model plants like "Arabidopsis thaliana". But nowadays, there's a growing recognition of the importance of exploring epigenetic mechanisms in more complex crop plants like wheat, maize, and soybean. Thanks to advancements in sequencing technologies and computational tools, we can now conduct detailed epigenomic mapping in these crops (Omony et al., 2020). This progress is helping us uncover specific epigenetic regulators that are unique to these important species. Shifting our focus to crop-specific epigenetics is vital for turning basic scientific research into real-world agricultural benefits.

Sustainable Agriculture through Epigenetics

Epigenetics is gaining traction as a key player in sustainable agriculture, helping us reduce our dependence on chemical inputs like fertilizers, pesticides, and herbicides (Jindo et al., 2022). Consider crops that are more resistant to diseases and pests due to their epigenetic traits; these plants can lessen our need for chemical treatments, which is a win for both the environment and our health. Additionally, by exploring how epigenetics affects how plants use nutrients, we can work towards developing varieties that need less fertilizer, making farming practices much more eco-friendly (Badiyal et al., 2024). What's exciting is that the practical applications of plant epigenetics are changing the game for crop improvement. We're seeing innovative solutions that address major global challenges, from creating crops that can thrive under stress to precision breeding techniques that enhance disease resistance. As we continue to delve deeper into the world of plant epigenomes, blending these epigenetic methods with traditional breeding and biotechnology could help us develop robust, high-yield crops that can adapt to our ever-changing climate (Yang et al., 2022). That said, there are still hurdles to overcome. We need better tools to identify and modify epigenetic markers in different crop species, as well as a clearer understanding of how these changes are passed down through generations. Collaborative efforts among plant biologists, geneticists, and agricultural scientists will be essential in truly unlocking the potential of plant epigenetics, paving the way for sustainable food production and better care for our environment.

FUTURE DIRECTIONS IN PLANT EPIGENETICS RESEARCH

The future of plant epigenetics is an exciting and rapidly evolving area of study that holds great promise for tackling pressing issues in agriculture, environmental adaptation, and food security, particularly as we face the challenges of climate change. Researchers are studying how epigenetic processes work in plants. These include DNA methylation, histone modifications, and small RNAs. These processes help plants manage stress, grow, and pass traits to their offspring (Shi et al., 2022). One of the key areas of focus is DNA methylation, which plays a significant role in turning genes on and off and managing transposable elements (Deniz et al., 2019). Scientists are not only interested in how it regulates gene expression but also in how it can help plants become more resilient to environmental stressors. By understanding these complex processes, we can develop better crops that thrive in the face of adversity, ensuring a more sustainable future for food production.

Recent advancements in genome mapping technologies, like single-cell and single-molecule sequencing, have opened up new avenues for researchers to explore methylation patterns in plants (Mo & Jiao, 2022). These tools allow for incredibly detailed analysis at the individual cell level, revealing how methylation can differ across various plant tissues and stages of development. On another front, CRISPR-based epigenetic editing tools, such as dCas9 combined with methyltransferase or demethylase enzymes, are revolutionizing the way we can manipulate specific epigenetic marks on genes (Seem et al., 2024). This technique offers a powerful means to improve crop traits without actually altering the DNA sequence, which could ease some of the regulatory concerns typically associated with genetic modifications. Additionally, scientists are paying closer attention to histone modifications, like acetylation, methylation, and phosphorylation which play a crucial role in determining how accessible plant DNA is for transcription (Rehman & Tanti, 2020). . These modifications are becoming increasingly important targets for improving crops, as they hold great potential to enhance agricultural traits.

Scientists are exploring the "histone code" hypothesis, which suggests that specific combinations of histone modifications influence gene expression outcomes (Prakash et al., 2018). The goal is to understand how these combinations affect stress responses and developmental processes. Efforts are being made to develop plants with heritable epigenetic marks that can enhance productivity, stress tolerance, and adaptability to various environments. This strategy, known as "epigenetic breeding," may supplement traditional breeding methods and genomic selection (Kumari et al., 2023). Moreover, advancements in synthetic biology are allowing for the design of synthetic epigenetic circuits that respond to specific environmental cues. This innovation could lead to the creation of crops capable of autonomously regulating gene expression in response to conditions such as drought, pathogens, or nutrient availability.

Non-coding RNAs (ncRNAs), such as microRNAs (miRNAs) and small interfering RNAs (siRNAs), are important in plant research (Zhou et al., 2020). They help control how genes work after a gene is copied and also affect DNA modifications. By changing these ncRNAs, we can adjust gene activity in plants to improve traits like growth rate, flowering time, and resistance to stress. Additionally, long non-coding RNAs (lncRNAs) also play a role in regulating plant genes, which might help us discover new ways to control complex traits (Zhang et al., 2023). One interesting idea is "epigenetic memory." This means plants can "remember" past environmental stresses. Developing crops that are prepared for tough conditions could make them more resilient. Understanding how plants create and keep this memory can help us breed crops that better withstand repeated stresses and show improved responses.

Transgenerational epigenetic inheritance allows plants to pass on traits developed from stress to their offspring (Boskovic & Rando, 2018). This method shows promise for breeding plants that can better withstand challenges. However, we still need to understand how stable and predictable these inherited changes are. More research is necessary to find out which epigenetic markers are consistently passed down and how they affect future generations. As researchers conduct more epigenome-wide association studies (EWAS) in plants, they will be able to identify markers linked to important traits, like higher yield, pest resistance, and drought tolerance. By combining EWAS with genetic and gene expression data, scientists can create models to help in breeding programs (Xu et al., 2017). This makes it easier to choose plants with the best traits. As these tools improve, epigenetic engineering may provide alternatives to chemical inputs, such as pesticides and fertilizers (Lallawmkimi et al., 2024). This could lead to growing crops that naturally resist pests and use nutrients more efficiently, contributing to more sustainable farming practices.

Scientists are working on "epigenetic-based" herbicides that target specific markers in plants (Zhang et al., 2022). These herbicides aim to control invasive species without harming crops. Another area of study is environmental epigenomics, which looks at how factors like temperature, light, and pollutants affect the epigenomes of plants. Understanding how plants respond to climate change can help us create better conservation strategies and design crops that adapt well to changing conditions. Additionally, combining plant epigenetics with machine learning and artificial intelligence (AI) can improve data analysis (Koh et al., 2024). This approach may help predict how plants respond to different environmental factors and enhance crop trait engineering. Since epigenetic changes are complex and interact with hormone and metabolism pathways, researchers are increasingly using integrated methods (Huang et al., 2018). These include epigenomics, transcriptomics, metabolomics, and proteomics to get a complete picture of how plants respond to their environment.

Multi-omics studies can help us understand the complex links between epigenetic regulation and biological processes (Garg et al., 2021). This understanding can lead to better strategies for improving crops. Combining epigenomics with microbiome research is an exciting area of study. Recent results show that microbes associated with plants can change the plant's epigenetic marks, which affects its growth and immune response (Tiwari et al., 2022). By managing the plant microbiome to encourage beneficial epigenetic changes, we can boost crop health and productivity. However, the field of plant epigenetics has several challenges. We need better tools to study and change epigenetic marks. The nature of epigenetic regulation is complex and depends on context. There are also ethical and regulatory issues related to epigenetically modified organisms (EMOs). To address these challenges, we need continuous investment in both basic and applied research, teamwork across disciplines, and clear regulatory guidelines. As we make progress, plant epigenetics has great potential to create strong, sustainable, and high-yielding crops (Jin et al., 2024). These crops can meet the needs of a growing global population and tackle the challenges of a changing environment.

CHALLENGES IN PLANT EPIGENETIC RESEARCH

The application of epigenetic research in plants encounters several technological and conceptual challenges that must be addressed in order to unlock its full potential in agricultural biotechnology (Sun et al., 2021). A key challenge lies in our incomplete understanding of the complex and dynamic nature of epigenetic modifications, such as DNA methylation, histone modifications, and non-coding RNAs, as well as their roles in regulating gene expression (Lv et al., 2020). While significant advancements have been made in model species like Arabidopsis thaliana, comprehensive reference epigenomes are still lacking for many crucial crops, including rice, wheat, maize, and soybean (Bisht et al., 2019). This gap in knowledge restricts our ability to fully understand how epigenetic changes affect plant traits such as yield, stress tolerance, and resistance to pests and diseases.

To develop high-resolution and species-specific reference epigenomes for a broader range of plants, we will need advanced technologies, extensive data collection, and collaborative efforts across multiple research institutions. The dynamic nature of epigenetic marks, which can vary in response to environmental stimuli, complicates their functional characterization (Pal & Tyler, 2016). Unlike static genetic mutations, epigenetic modifications are reversible, which adds complexity to our understanding of how these marks influence long-term phenotypic changes in plants. Additionally, one significant challenge is the limited capacity to manipulate epigenetic marks in a targeted and predictable manner. While CRISPR-based technologies for epigenome editing hold promise, they are still in their infancy, and the precision, efficiency, and long-term stability of these modifications in plants remain uncertain (Walker & Upadhyaya, 2018).

Using epigenetic manipulation to improve crop traits is challenging, especially for traits related to stress responses. These traits involve complex gene networks. The relationship between genetic and epigenetic factors also adds to this complexity

(Bally et al., 2016). Genetic and epigenetic variations interact in non-linear ways, making it hard to identify their roles in trait expression. This makes breeding for desirable traits difficult when using epigenetic tools (Little & Seebacher, 2021). To address these challenges, researchers need to develop advanced computational models and experimental systems to fully understand these interactions. Additionally, there are concerns about how stable and inheritable epigenetic changes are in plants.

Epigenetic marks can change in response to environmental factors or during reproduction (Wang et al., 2017). This leads to questions about how stable these changes are from one generation to the next. This is especially important in agriculture, where farmers need to maintain traits like drought tolerance and disease resistance over different growing seasons and environments. It is also important to ensure that helpful epigenetic changes do not cause unintended problems, such as affecting other key traits, which can harm plant health or reduce productivity.

A major challenge in improving crops is the rules about using epigenetic modifications. Traditional genetic modifications, like those created through transgenic methods, are closely regulated. However, the rules for epigenetic modifications are still being developed. Since epigenetic changes do not change the DNA sequence, it's unclear how these should be regulated compared to genetically modified organisms (GMOs) (Wani et al., 2022). We need clear guidelines to check the safety, effectiveness, and environmental impact of crops with modified epigenomes. This includes concerns about off-target effects and possible unintended consequences for nature. There are also ethical questions about ownership and intellectual property rights related to epigenetic modifications. As technology for editing the epigenome improves, people may debate whether these changes can be patented and how this might make it harder for small farmers and developing countries to access the technology. Public opinion is essential to the success of epigenetically modified crops (Fiaz et al., 2021). People already have concerns about food safety, environmental impact, and corporate control when it comes to genetically modified crops. These concerns may also apply to crops modified through epigenetics, even though this method does not involve adding foreign DNA.

Clear communication and engaging with the public are crucial for gaining acceptance of epigenetically modified crops. We need to help people understand how these modifications work and what benefits they offer. These benefits include making crops stronger against climate change, improving their nutritional value, and reducing the need for chemicals. To effectively develop and use epigenetic technologies in farming, experts from different areas must work together. This includes plant biologists, geneticists, bioinformaticians, agronomists, regulatory experts, and social scientists. We must address challenges related to how epigenetic regulation works, overcome technical issues in precise editing, and consider social and regulatory concerns. Solving these challenges will require a united effort from all these fields.

New technologies, such as high-throughput sequencing, single-cell epigenomics, and computer modelling, are important for understanding the genetic changes in plants and predicting how these changes affect their traits (Luo et al., 2020). It's also crucial to ensure that everyone can access these technologies and to tackle issues related to monopolization and corporate control. This will help more people adopt epigenetically modified crops. In summary, plant epigenetics has great potential to improve agricultural productivity, resilience, and sustainability (Sun et al., 2021). Addressing these challenges is essential to turn basic research into practical solutions that benefit both farmers and consumers.

FUNCTIONAL CHARACTERIZATION OF EPIGENETIC MODIFICATIONS

A major challenge in plant epigenetics is understanding how epigenetic modifications work. We know where these marks are located in the genome, but we are often unclear about their specific roles in regulating gene expression and plant development. This lack of clarity is particularly true for modifications in non-coding regions of the genome or those that occur due to environmental stress. To solve this challenge, we need new experimental methods to capture the changing nature of epigenetic modifications. For example, techniques that allow us to monitor epigenetic changes in real-time in living plants could help us understand how these marks are created and maintained (Sepers et al., 2019). Additionally, we need to develop targeted tools for editing the epigenome, such as CRISPR technologies, to test the roles of specific epigenetic modifications.

REGULATORY AND ETHICAL CONSIDERATIONS

Using epigenetic research in agriculture raises important questions about safety and ethics. Epigenome editing can help improve crops, but it also creates potential risks that we need to manage (Sun et al., 2021). For example, off-target effects can happen, meaning the changes might unintentionally affect the plant's growth or how it interacts with the environment (Koller et al., 2023). Regulatory agencies need to create guidelines to evaluate how safe and effective these edited crops are, especially regarding their long-term impact on ecosystems and food safety. Additionally, we must consider the ethics of commercializing these crops, including issues about who owns the technology, who can access it, and the risk of relying too much on big agriculture companies.

SOCIETAL ENGAGEMENT AND PUBLIC PERCEPTION

Public engagement and perception are key to the success of epigenetic technologies in agriculture (Bogner & Torgersen, 2018). Many people have mixed feelings about genetically modified organisms (GMOs) due to concerns about food safety, environmental effects, and corporate control. Although epigenetic modifications do not change the DNA sequence like traditional genetic modifications do, they might still face public doubt (Wei et al., 2021). To address these concerns, scientists and policymakers need to interact with the public transparently and inclusively. They should educate people about the science behind epigenetic modifications and their potential benefits to build trust and acceptance. Including stakeholders in decision-making, especially regarding regulations, is also important to ensure that epigenetic technologies are developed and used responsibly.

CONCLUSION

In conclusion, plant epigenetics is a growing field that has great potential to improve our understanding of how plants regulate their genes, respond to stress, and adapt to their environments. This knowledge can help us boost crop performance and promote sustainable farming practices. However, there are several important challenges to address. First, researchers need to unravel the complexities of epigenetic changes, such as DNA methylation, histone modification, and small RNA pathways, across different plant species. Understanding these variations is important for creating broader models of how epigenetics works. We also need to study how these changes relate to environmental factors like temperature, drought, and nutrient levels. This understanding is key to knowing how plants adapt and how we can use these adaptations in agriculture. Future research should focus on developing advanced technologies that allow for quick and detailed analysis of plant epigenomes. These tools will help us better understand the epigenetic landscape and identify important regulatory elements. Additionally, we should improve techniques for accurately changing epigenetic states. This includes enhancing gene-editing technologies and exploring new ways to establish beneficial epigenetic traits in crops. As we learn more about plant epigenetics, it is essential to connect basic research with agricultural practices. This connection will help us create innovative strategies to improve crop resilience and productivity, especially in light of climate change and the growing global population. By encouraging collaboration among researchers, agronomists, and policymakers, we can work toward a more sustainable and productive agricultural future.

REFERENCES

Akhter, Z., Bi, Z., Ali, K., Sun, C., Fiaz, S., Haider, F. U., & Bai, J. (2021). In response to abiotic stress, DNA methylation confers epigenetic changes in plants. *Plants*, *10*(6), 1096. PMID: 34070712

Alfalahi, A. O., Hussein, Z. T., Khalofah, A., Sadder, M. T., Qasem, J. R., Al-Khayri, J. M., & Almehemdi, A. F. (2022). Epigenetic variation as a new plant breeding tool: A review. *Journal of King Saud University. Science*, *34*(8), 102302.

Ali, S., & Tang, Y. (2024). Noncoding RNA-Mediated Regulation of DNA Methylation: Insights into Plant Epigenetic Mechanisms. *Journal of Plant Growth Regulation*, •••, 1–16.

Ashapkin, V. V., Kutueva, L. I., Aleksandrushkina, N. I., & Vanyushin, B. F. (2020). Epigenetic mechanisms of plant adaptation to biotic and abiotic stresses. *International Journal of Molecular Sciences*, *21*(20), 7457. PMID: 33050358

Badiyal, A., Mahajan, R., Rana, R. S., Sood, R., Walia, A., Rana, T., & Jayswal, D. K. (2024). Synergizing biotechnology and natural farming: Pioneering agricultural sustainability through innovative interventions. *Frontiers in Plant Science*, *15*, 1280846. PMID: 38584951

Bai, Y., Kissoudis, C., Yan, Z., Visser, R. G., & van der Linden, G. (2018). Plant behaviour under combined stress: Tomato responses to combined salinity and pathogen stress. *The Plant Journal*, *93*(4), 781–793. PMID: 29237240

Balao, F., Paun, O., & Alonso, C. (2018). Uncovering the contribution of epigenetics to plant phenotypic variation in Mediterranean ecosystems. *Plant Biology*, *20*, 38–49. PMID: 28637098

Bally, A. P., Austin, J. W., & Boss, J. M. (2016). Genetic and epigenetic regulation of PD-1 expression. *The Journal of Immunology : Official Journal of the American Association of Immunologists*, *196*(6), 2431–2437. PMID: 26945088

Baulcombe, D. C. (2023). An RNA world. *Annual Review of Plant Biology*, *74*(1), 1–20. PMID: 36542757

Baulcombe, D. C., & Dean, C. (2014). Epigenetic regulation in plant responses to the environment. *Cold Spring Harbor Perspectives in Biology*, *6*(9), a019471. PMID: 25183832

Bisht, D. S., Bhatia, V., & Bhattacharya, R. (2019, December). Improving plant-resistance to insect-pests and pathogens: The new opportunities through targeted genome editing. [). Academic Press.]. *Seminars in Cell & Developmental Biology*, *96*, 65–76. PMID: 31039395

Bogner, A., & Torgersen, H. (2018). Precaution, responsible innovation and beyond–in search of a sustainable agricultural biotechnology policy. *Frontiers in Plant Science*, *9*, 1884. PMID: 30619432

Boyce, W. T. (2016). Differential susceptibility of the developing brain to contextual adversity and stress. *Neuropsychopharmacology : Official Publication of the American College ofNeuropsychopharmacology*, *41*(1), 142–162. PMID: 26391599

Cao, H. X., Wang, W., Le, H. T., & Vu, G. T. (2016). The power of CRISPR-Cas9-induced genome editing to speed up plant breeding. *International Journal of Genomics*, *2016*(1), 5078796. PMID: 28097123

Cao, S., & Chen, Z. J. (2024). Transgenerational epigenetic inheritance during plant evolution and breeding. *Trends in Plant Science*. PMID: 38806375

Chang, Y. N., Zhu, C., Jiang, J., Zhang, H., Zhu, J. K., & Duan, C. G. (2020). Epigenetic regulation in plant abiotic stress responses. *Journal of Integrative Plant Biology*, *62*(5), 563–580. PMID: 31872527

Chaudhry, S., & Sidhu, G. P. S. (2022). Climate change regulated abiotic stress mechanisms in plants: A comprehensive review. *Plant Cell Reports*, *41*(1), 1–31. PMID: 34351488

Chen, Y., Guo, P., & Dong, Z. (2024). The role of histone acetylation in transcriptional regulation and seed development. *Plant Physiology*, *194*(4), 1962–1979. PMID: 37979164

Clark, J., & Rager, J. E. (2020). Epigenetics: An overview of CpG methylation, chromatin remodeling, and regulatory/noncoding RNAs. *Environmental epigenetics in toxicology and public health*, 3-32.

Deniz, Ö., Frost, J. M., & Branco, M. R. (2019). Regulation of transposable elements by DNA modifications. *Nature Reviews. Genetics*, *20*(7), 417–431. PMID: 30867571

Dhar, G. A., Saha, S., Mitra, P., & Nag Chaudhuri, R. (2021). DNA methylation and regulation of gene expression: Guardian of our health. *Nucleus (Austin, Tex.)*, *64*(3), 259–270. PMID: 34421129

Ding, B., & Wang, G. L. (2015). Chromatin versus pathogens: The function of epigenetics in plant immunity. *Frontiers in Plant Science*, *6*, 675. PMID: 26388882

Elhamamsy, A. R. (2016). DNA methylation dynamics in plants and mammals: Overview of regulation and dysregulation. *Cell Biochemistry and Function, 34*(5), 289–298. PMID: 27003927

Enríquez, P. (2016). Focus: epigenetics: CRISPR-mediated epigenome editing. *The Yale Journal of Biology and Medicine, 89*(4), 471. PMID: 28018139

Fal, K., Tomkova, D., Vachon, G., Chaboute, M. E., Berr, A., & Carles, C. C. (2021). Chromatin manipulation and editing: Challenges, new technologies and their use in plants. *International Journal of Molecular Sciences, 22*(2), 512. PMID: 33419220

Farooq, M., Usman, M., & Nadeem, F., ur Rehman, H., Wahid, A., Basra, S. M., & Siddique, K. H. (2019). Seed priming in field crops: Potential benefits, adoption and challenges. *Crop & Pasture Science, 70*(9), 731–771.

Farsetti, A., Illi, B., & Gaetano, C. (2023). How epigenetics impacts on human diseases. *European Journal of Internal Medicine, 114*, 15–22. PMID: 37277249

Fiaz, S., Wang, X., Younas, A., Alharthi, B., Riaz, A., & Ali, H. (2021). Apomixis and strategies to induce apomixis to preserve hybrid vigor for multiple generations. *GM Crops and Food: Biotechnology in Agriculture and the Food Chain, 12*(1), 57–70. PMID: 32877304

Fita, A., Rodríguez-Burruezo, A., Boscaiu, M., Prohens, J., & Vicente, O. (2015). Breeding and domesticating crops adapted to drought and salinity: A new paradigm for increasing food production. *Frontiers in Plant Science, 6*, 978. PMID: 26617620

Gady, A. L., Alves, C. S., & Nogueira, F. T. (2017). Epigenetics in plant reproductive development: an overview from flowers to seeds. *Plant Epigenetics*, 329-357.

Galindo-González, L., Sarmiento, F., & Quimbaya, M. A. (2018). Shaping plant adaptability, genome structure and gene expression through transposable element epigenetic control: Focus on methylation. *Agronomy (Basel), 8*(9), 180.

Ganguli, S., & Chakraborty, R. (2021). Bacterial epigenetics opens door to novel frontier in Infection biology. *Nucleus (Austin, Tex.), 64*(3), 383–399.

Grativol, C., Hemerly, A. S., & Ferreira, P. C. G. (2012). Genetic and epigenetic regulation of stress responses in natural plant populations. *Biochimica et Biophysica Acta (BBA)-. Gene Regulatory Mechanisms, 1819*(2), 176–185. PMID: 21914492

Holoch, D., & Moazed, D. (2015). RNA-mediated epigenetic regulation of gene expression. *Nature Reviews. Genetics, 16*(2), 71–84. PMID: 25554358

Hondele, M. (2013). An integrated view of the essential eukaryotic chaperone FACT in complex with histones H2A-H2B.

Hong, K., Radian, Y., Manda, T., Xu, H., & Luo, Y. (2023). The development of plant genome sequencing technology and its conservation and application in endangered gymnosperms. *Plants, 12*(23), 4006. PMID: 38068641

Hu, F., Zhong, H., Wu, C., Wang, S., Guo, Z., Tao, M., & Liu, S. (2021). Development of fisheries in China. *Reproduction and Breeding, 1*(1), 64–79.

Huang, J., Lynn, J. S., Schulte, L., Vendramin, S., & McGinnis, K. (2017). Epigenetic control of gene expression in maize. *International Review of Cell and Molecular Biology, 328*, 25–48. PMID: 28069135

Huang, Q., Ma, C., Chen, L., Luo, D., Chen, R., & Liang, F. (2018). Mechanistic insights into the interaction between transcription factors and epigenetic modifications and the contribution to the development of obesity. *Frontiers in Endocrinology, 9*, 370. PMID: 30034368

Jablonka, E. (2017). The evolutionary implications of epigenetic inheritance. *Interface Focus, 7*(5), 20160135. PMID: 28839916

Jahed, K. R., & Hirst, P. M. (2023). Fruit growth and development in apple: A molecular, genomics and epigenetics perspective. *Frontiers in Plant Science, 14*, 1122397. PMID: 37123845

Jin, Q., Chachar, M., Ali, A., Chachar, Z., Zhang, P., Riaz, A., & Chachar, S. (2024). Epigenetic Regulation for Heat Stress Adaptation in Plants: New Horizons for Crop Improvement under Climate Change. *Agronomy (Basel), 14*(9), 2105.

Jindo, K., Goron, T. L., Pizarro-Tobías, P., Sánchez-Monedero, M. Á., Audette, Y., Deolu-Ajayi, A. O., & Tortosa, G. (2022). Application of biostimulant products and biological control agents in sustainable viticulture: A review. *Frontiers in Plant Science, 13*, 932311. PMID: 36330258

Kagohara, L. T., Stein-O'Brien, G. L., Kelley, D., Flam, E., Wick, H. C., Danilova, L. V., & Fertig, E. J. (2018). Epigenetic regulation of gene expression in cancer: Techniques, resources and analysis. *Briefings in Functional Genomics, 17*(1), 49–63. PMID: 28968850

Kakoulidou, I., Avramidou, E. V., Baránek, M., Brunel-Muguet, S., Farrona, S., Johannes, F., & Maury, S. (2021). Epigenetics for crop improvement in times of global change. *Biology (Basel), 10*(8), 766. PMID: 34439998

Khodadadi, E., Fahmideh, L., Khodadadi, E., Dao, S., Yousefi, M., Taghizadeh, S., & Kafil, H. S. (2021). Current advances in DNA methylation analysis methods. *BioMed Research International, 2021*(1), 8827516. PMID: 33824878

Kim, J. S., Kidokoro, S., Yamaguchi-Shinozaki, K., & Shinozaki, K. (2024). Regulatory networks in plant responses to drought and cold stress. *Plant Physiology, 195*(1), 170–189. PMID: 38514098

Koh, E., Sunil, R. S., Lam, H. Y. I., & Mutwil, M. (2024). Confronting the data deluge: How artificial intelligence can be used in the study of plant stress. *Computational and Structural Biotechnology Journal*. PMID: 39415960

Koller, F., & Cieslak, M. (2023). A perspective from the EU: Unintended genetic changes in plants caused by NGT—their relevance for a comprehensive molecular characterisation and risk assessment. *Frontiers in Bioengineering and Biotechnology, 11*, 1276226. PMID: 37965049

Kumar, J., Choudhary, A. K., Gupta, D. S., & Kumar, S. (2019). Towards exploitation of adaptive traits for climate-resilient smart pulses. *International Journal of Molecular Sciences, 20*(12), 2971. PMID: 31216660

Kumar, M., & Rani, K. (2023). Epigenomics in stress tolerance of plants under the climate change. *Molecular Biology Reports, 50*(7), 6201–6216. PMID: 37294468

Kumar, S., Chinnusamy, V., & Mohapatra, T. (2018). Epigenetics of modified DNA bases: 5-methylcytosine and beyond. *Frontiers in Genetics, 9*, 640. PMID: 30619465

Kumar, S., & Mohapatra, T. (2021). Dynamics of DNA methylation and its functions in plant growth and development. *Frontiers in Plant Science, 12*, 596236. PMID: 34093600

Kumar, S., Singh, A. K., & Mohapatra, T. (2017). Epigenetics: History, present status and future perspective. *Indian Journal of Genetics and Plant Breeding, 77*(04), 445–463.

Kumari, V., Thakur, R., Kumari, J., Kumari, A., & Khajuria, D. (2023). Nutritional improvement in soybean (Glycine max (L.) Merrill) through plant breeding and biotechnological interventions. *Crop & Pasture Science, 75*(1).

Lallawmkimi, M. C., Veda, D. S., Yadav, A., Kumar, M., & Rout, A. (2024). Innovative Approaches in Crop Genetic Engineering for Sustainable Agriculture: A Review. *Journal of Advances in Biology & Biotechnology, 27*(8), 615–631.

Little, A. G., & Seebacher, F. (2021). Physiological performance curves: When are they useful? *Frontiers in Physiology, 12*, 805102. PMID: 34925077

Liu, J., & He, Z. (2020). Small DNA methylation, big player in plant abiotic stress responses and memory. *Frontiers in Plant Science, 11*, 595603. PMID: 33362826

Long, R. L., Gorecki, M. J., Renton, M., Scott, J. K., Colville, L., Goggin, D. E., & Finch-Savage, W. E. (2015). The ecophysiology of seed persistence: A mechanistic view of the journey to germination or demise. *Biological Reviews of the Cambridge Philosophical Society, 90*(1), 31–59. PMID: 24618017

Luo, C., Fernie, A. R., & Yan, J. (2020). Single-cell genomics and epigenomics: Technologies and applications in plants. *Trends in Plant Science, 25*(10), 1030–1040. PMID: 32532595

Lv, W., Ren, Y., Hou, K., Hu, W., Yi, Y., Xiong, M., & Zhang, Q. (2020). Epigenetic modification mechanisms involved in keloid: Current status and prospect. *Clinical Epigenetics, 12*, 1–19. PMID: 33243301

Mo, Y., & Jiao, Y. (2022). Advances and applications of single-cell omics technologies in plant research. *The Plant Journal, 110*(6), 1551–1563. PMID: 35426954

Movahedi, A., Sun, W., Zhang, J., Wu, X., Mousavi, M., Mohammadi, K., & Zhuge, Q. (2015). RNA-directed DNA methylation in plants. *Plant Cell Reports, 34*, 1857–1862. PMID: 26183954

Nakamura, M., Gao, Y., Dominguez, A. A., & Qi, L. S. (2021). CRISPR technologies for precise epigenome editing. *Nature Cell Biology, 23*(1), 11–22. PMID: 33420494

Nilsson, E. E., Sadler-Riggleman, I., & Skinner, M. K. (2018). Environmentally induced epigenetic transgenerational inheritance of disease. *Environmental Epigenetics, 4*(2), dvy016. PMID: 30038800

Omony, J., Nussbaumer, T., & Gutzat, R. (2020). DNA methylation analysis in plants: Review of computational tools and future perspectives. *Briefings in Bioinformatics, 21*(3), 906–918. PMID: 31220217

Pai, A. A., Pritchard, J. K., & Gilad, Y. (2015). The genetic and mechanistic basis for variation in gene regulation. *PLOS Genetics, 11*(1), e1004857. PMID: 25569255

Pal, S., & Tyler, J. K. (2016). Epigenetics and aging. *Science Advances, 2*(7), e1600584. PMID: 27482540

Partridge, S. R., Kwong, S. M., Firth, N., & Jensen, S. O. (2018). Mobile genetic elements associated with antimicrobial resistance. *Clinical Microbiology Reviews, 31*(4), 10–1128. PMID: 30068738

Prakash, K., & Fournier, D. (2018). Evidence for the implication of the histone code in building the genome structure. *Bio Systems, 164*, 49–59. PMID: 29158132

Rajput, M., Choudhary, K., Kumar, M., Vivekanand, V., Chawade, A., Ortiz, R., & Pareek, N. (2021). RNA interference and CRISPR/Cas gene editing for crop improvement: Paradigm shift towards sustainable agriculture. *Plants*, *10*(9), 1914. PMID: 34579446

Ramakrishnan, M., Satish, L., Kalendar, R., Narayanan, M., Kandasamy, S., Sharma, A., & Zhou, M. (2021). The dynamism of transposon methylation for plant development and stress adaptation. *International Journal of Molecular Sciences*, *22*(21), 11387. PMID: 34768817

Rehman, M., & Tanti, B. (2020). Understanding epigenetic modifications in response to abiotic stresses in plants. *Biocatalysis and Agricultural Biotechnology*, *27*, 101673.

Ren, D., Ding, C., & Qian, Q. (2023). Molecular bases of rice grain size and quality for optimized productivity. *Science Bulletin*, *68*(3), 314–350. PMID: 36710151

Richards, C. L., Alonso, C., Becker, C., Bossdorf, O., Bucher, E., Colomé-Tatché, M., & Verhoeven, K. J. (2017). Ecological plant epigenetics: Evidence from model and non-model species, and the way forward. *Ecology Letters*, *20*(12), 1576–1590. PMID: 29027325

Saini, R. K., Ranjit, A., Sharma, K., Prasad, P., Shang, X., Gowda, K. G. M., & Keum, Y. S. (2022). Bioactive compounds of citrus fruits: A review of composition and health benefits of carotenoids, flavonoids, limonoids, and terpenes. *Antioxidants*, *11*(2), 239. PMID: 35204122

Samantara, K., Shiv, A., de Sousa, L. L., Sandhu, K. S., Priyadarshini, P., & Mohapatra, S. R. (2021). A comprehensive review on epigenetic mechanisms and application of epigenetic modifications for crop improvement. *Environmental and Experimental Botany*, *188*, 104479.

Seem, K., Kaur, S., Kumar, S., & Mohapatra, T. (2024). Epigenome editing for targeted DNA (de) methylation: A new perspective in modulating gene expression. *Critical Reviews in Biochemistry and Molecular Biology*, •••, 1–30. PMID: 38440883

Sepers, B., van den Heuvel, K., Lindner, M., Viitaniemi, H., Husby, A., & van Oers, K. (2019). Avian ecological epigenetics: Pitfalls and promises. *Journal of Ornithology*, *160*, 1183–1203.

Sharma, M., Kumar, P., Verma, V., Sharma, R., Bhargava, B., & Irfan, M. (2022). Understanding plant stress memory response for abiotic stress resilience: Molecular insights and prospects. *Plant Physiology and Biochemistry*, *179*, 10–24. PMID: 35305363

Shi, M., Wang, C., Wang, P., Zhang, M., & Liao, W. (2022). Methylation in DNA, histone, and RNA during flowering under stress condition: A review. *Plant Science*, *324*, 111431. PMID: 36028071

Song, X., & Cao, X. (2017). Context and complexity: Analyzing methylation in trinucleotide sequences. *Trends in Plant Science*, *22*(5), 351–353. PMID: 28392157

Soni, R., Ashaq, M., Manjunath, P., Ashoka, P., Pandey, S. K., Hosur, S. R., & Raj, R. (2024). Harnessing Genomic Resources and Molecular Breeding Techniques for Advancing Crop Resilience and Productivity. *Plant Cell Biotechnology and Molecular Biology*, *25*(11-12), 124–136.

Stevens, A. J., de Jong, L., & Kennedy, M. A. (2022). The dynamic regulation of G-quadruplex DNA structures by cytosine methylation. *International Journal of Molecular Sciences*, *23*(5), 2407. PMID: 35269551

Sun, C., Ali, K., Yan, K., Fiaz, S., Dormatey, R., Bi, Z., & Bai, J. (2021). Exploration of epigenetics for improvement of drought and other stress resistance in crops: A review. *Plants*, *10*(6), 1226. PMID: 34208642

Tikhodeyev, O. N. (2018). The mechanisms of epigenetic inheritance: How diverse are they? *Biological Reviews of the Cambridge Philosophical Society*, *93*(4), 1987–2005. PMID: 29790249

Tirnaz, S., & Batley, J. (2019). Epigenetics: Potentials and challenges in crop breeding. *Molecular Plant*, *12*(10), 1309–1311. PMID: 31541738

Tiwari, M., Pati, D., Mohapatra, R., Sahu, B. B., & Singh, P. (2022). The impact of microbes in plant immunity and priming induced inheritance: A sustainable approach for crop protection. *Plant Stress*, *4*, 100072.

Tompkins, J. D. (2022). Discovering DNA methylation, the history and future of the writing on DNA. *Journal of the History of Biology*, *55*(4), 865–887. PMID: 36239862

Valle-Echevarria, A. D., Fumia, N., Gore, M. A., & Kantar, M. (2021). Accelerating crop domestication in the era of gene editing. *Plant Breeding Reviews*, *45*, 185–211.

Walker, J. A., & Upadhyaya, M. (2018). Emerging therapeutic targets for neurofibromatosis type 1. *Expert Opinion on Therapeutic Targets*, *22*(5), 419–437. PMID: 29667529

Walther, A. (2020). From responsive to adaptive and interactive materials and materials systems: A roadmap. *Advanced Materials*, *32*(20), 1905111. PMID: 31762134

Wang, D., Zhang, F., & Gao, G. (2020). CRISPR-based therapeutic genome editing: Strategies and in vivo delivery by AAV vectors. *Cell*, *181*(1), 136–150. PMID: 32243786

Wang, Y., Liu, H., & Sun, Z. (2017). Lamarck rises from his grave: Parental environment-induced epigenetic inheritance in model organisms and humans. *Biological Reviews of the Cambridge Philosophical Society*, *92*(4), 2084–2111. PMID: 28220606

Wani, A. K., Akhtar, N., & Shukla, S. (2022). CRISPR/Cas9: Regulations and challenges for law enforcement to combat its dual-use. *Forensic Science International*, *334*, 111274. PMID: 35316773

Wassenegger, M., & Dalakouras, A. (2021). Viroids as a tool to study RNA-directed DNA methylation in plants. *Cells*, *10*(5), 1187. PMID: 34067940

Wei, S., Tao, J., Xu, J., Chen, X., Wang, Z., Zhang, N., & Jiang, Y. (2021). Ten years of EWAS. *Advancement of Science*, *8*(20), 2100727. PMID: 34382344

Xu, Y., Li, P., Zou, C., Lu, Y., Xie, C., Zhang, X., & Olsen, M. S. (2017). Enhancing genetic gain in the era of molecular breeding. *Journal of Experimental Botany*, *68*(11), 2641–2666. PMID: 28830098

Yang, F., Zhang, J., Liu, Q., Liu, H., Zhou, Y., Yang, W., & Ma, W. (2022). Improvement and re-evolution of tetraploid wheat for global environmental challenge and diversity consumption demand. *International Journal of Molecular Sciences*, *23*(4), 2206. PMID: 35216323

Zand Karimi, H., & Innes, R. W. (2022). Molecular mechanisms underlying host-induced gene silencing. *The Plant Cell*, *34*(9), 3183–3199. PMID: 35666177

Zhang, D., Zhang, J., Wang, Y., Wang, G., Tang, P., Liu, Y., & Ouyang, L. (2023). Targeting epigenetic modifications in Parkinson's disease therapy. *Medicinal Research Reviews*, *43*(5), 1748–1777. PMID: 37119043

Zhang, D., Zhang, Z., Unver, T., & Zhang, B. (2021). CRISPR/Cas: A powerful tool for gene function study and crop improvement. *Journal of Advanced Research*, *29*, 207–221. PMID: 33842017

Zhang, H., Lang, Z., & Zhu, J. K. (2018). Dynamics and function of DNA methylation in plants. *Nature Reviews. Molecular Cell Biology*, *19*(8), 489–506. PMID: 29784956

Zhang, L., Lin, T., Zhu, G., Wu, B., Zhang, C., & Zhu, H. (2023). LncRNAs exert indispensable roles in orchestrating the interaction among diverse noncoding RNAs and enrich the regulatory network of plant growth and its adaptive environmental stress response. *Horticulture Research*, *10*(12), uhad234. PMID: 38156284

Zhou, X., Cui, J., Meng, J., & Luan, Y. (2020). Interactions and links among the noncoding RNAs in plants under stresses. *Theoretical and Applied Genetics*, *133*, 3235–3248. PMID: 33025081

Zhu, Q. H., Shan, W. X., Ayliffe, M. A., & Wang, M. B. (2016). Epigenetic mechanisms: An emerging player in plant-microbe interactions. *Molecular Plant-Microbe Interactions*, *29*(3), 187–196. PMID: 26524162

Chapter 18
Epigenetic Mechanisms Underlying Drought Tolerance in Plants

Sobia Kanwal
University of the Punjab, Pakistan, Pakistan

Gul Nawaz
http://orcid.org/0009-0000-5328-0454
University of the Punjab, Pakistan, Pakistan

Muhammad Zeeshan Arshed
University of the Punjab, Pakistan, Pakistan

Ammara Saleem
University of the Punjab, Pakistan, Pakistan

Banzeer Ahsan Abbasi
http://orcid.org/0000-0003-3036-4481
Rawalpindi Women University, Pakistan, Pakistan

ABSTRACT

Epigenetics is a key feature in plants responding to drought stress by regulating plant tolerance or adaptation capacity with physical water deficit. Transgenerational inheritance can occur because epigenetic changes regulate gene expression without changing the genetic code (for example, DNA methylation, histone modification and non-coding RNAs). Such adaptations, allow plants to induce stress-responsive genes, modify metabolic pathways and maintain cellular homeostasis under drought conditions. Epigenetic regulation helps in the rapid adaptation corresponding to

DOI: 10.4018/979-8-3693-8648-4.ch018

developing drought tolerance and memory that has the potential of trans-generational expression. Knowledge of these mechanisms may be useful for breeding and biotechnological strategies to improve crop resistance in water-limited environments.

1. INTRODUCTION

Drought stress is one of the most important constraints in plant development and productivity, severely affecting biomass production at a sustainable level for crop production (Ahmad et al., 2024). They affect plant forms and functions like seed germinations, roots proliferation, biomass elevation and final yield (Ahmad et al., 2022a, Ahmad et al., 2022b, Saleem, Zulfiqar, Zulfiqar, et al., 2024). Previous research work has identified that drought impacts the chlorophyll biosynthesis, the rate of photosynthesis decreases, and a decrease in CO2 assimilation rates are due to the stomatal conductance (Akpinar & Cansev, 2022; Zulfiqar et al., 2023). Additionally, it affects WUE partly and at the same time affects the degree of rachis elongation and leaf area. Several strategies such as tolerance, avoidance and escape are utilized by the plants to cope with drought stress (Lukić et al., 2023). Among the first adaptive responses in plants exposed to drought stress are accumulation of osmolytes and up regulation of antioxidative systems (Mushtaq et al., 2021). Knowledge of such biochemical and molecular changes is essential for the framing of related development strategies aimed at enhancing plant resilience to water stress conditions (Ahmad et al., 2024a, Ahmad et al., 2024b).

Significance of Drought Tolerance for Agricultural Sustainability

The ability to resist drought is key for sustainable agriculture under climate change. Thus, microbes improve plants resistance to drought stress or decrease harsh effects of drought on plants via bioremediation and biofertilization mechanisms (Hussain et al., 2023). Plant growth-promoting rhizobacteria (PGPR) is an environmentally feasible alternative to alleviate crop drought stress (Zulfiqar et al., 2022). Molecular and biochemical bases of drought tolerance in major crops such as maize are very important for developing stress-resistant varieties (Ashapkin et al., 2020, Ali et al., 2022). There are various strategies for engineering drought tolerance which include classical breeding, biotechnology and plant-microbe interactions (Bagheri et al., 2023). Furthermore, it has been suggested that indirect approaches like reducing or preventing soil salinity would impart added resilience to crops and sustainable agriculture (Bachani et al., 2022). Effects of drought stress are displayed in Table 1, below table

Table 1. The effects of drought stress on various plants and their observed responses.

No.	Plant Species	Effect of Drought Stress	Key Findings and Observations	References
1	Wheat (Triticum aestivum)	Reduced grain yield and spikelet sterility	Significant reduction in growth and grain quality	(Ahmad et al., 2022)
2	Maize (Zea mays)	Decreased photosynthesis and kernel weight	Lower leaf area index and reduced chlorophyll content	(Sheoran et al., 2022)
3	Rice (Oryza sativa)	Decreased tillering and panicle initiation	Yield loss due to impaired water and nutrient uptake	(Kim et al., 2020)
4	Soybean (Glycine max)	Reduced seed size and protein content	Delayed flowering and shortened pod-filling duration	(Sintaha et al., 2022)
5	Tomato (Solanum lycopersicum)	Increased leaf abscission and fruit drop	Reduced fruit set and total soluble solids	(Liang et al., 2020)
6	Canola (Brassica napus)	Lowered biomass production and seed oil content	Altered amino acid and carbohydrate metabolism	(Khodabin et al., 2020)
7	Sunflower (Helianthus annuus)	Wilted leaves and reduced seed formation	High oxidative stress leading to lipid peroxidation	(Keipp et al., 2020)
8	Barley (Hordeum vulgare)	Decline in spike development and grain yield	Enhanced proline accumulation as a stress response	(Ahmad et al., 2024)
9	Lentil (Lens culinaris)	Reduced branch number and pod formation	Significant decrease in seed yield under drought	(Choukri et al., 2020)
10	Pea (Pisum sativum)	Reduced biomass and photosynthetic rate	Less efficient water-use and early leaf senescence	(Bagheri et al., 2023)
11	Cotton (Gossypium hirsutum)	Reduced boll size and fiber quality	Decreased reproductive growth and higher flower abortion	(Meshram et al., 2022)
12	Potato (Solanum tuberosum)	Smaller tubers and increased cracking	Reduced carbohydrate storage and altered metabolism	(Nasir & Toth, 2022)
13	Sorghum (Sorghum bicolor)	Stunted growth and delayed flowering	Increased root-to-shoot ratio as adaptation	(Abreha et al., 2022)
14	Chickpea (Cicer arietinum)	Lower pod number and seed set	Decreased photosynthetic efficiency and stomatal closure	(Sachdeva et al., 2022)
15	Sugarcane (Saccharum officinarum)	Reduced sugar content and growth rate	Impaired sucrose metabolism and reduced leaf expansion	(Gomathi et al., 2020)
16	Grapevine (Vitis vinifera)	Smaller berry size and poor fruit quality	Decreased anthocyanin and flavonoid production	(Gambetta et al., 2020)
17	Alfalfa (Medicago sativa)	Decline in forage yield and leaf retention	Reduced protein content in the leaves	(Riasat et al., 2020)
18	Spinach (Spinacia oleracea)	Wilting and reduced leaf area	Lower chlorophyll content and photosynthetic capacity	(Akpinar & Cansev, 2022)

continued on following page

Table 1. Continued

No.	Plant Species	Effect of Drought Stress	Key Findings and Observations	References
19	Lettuce (Lactuca sativa)	Reduced leaf area and water content	Quick bolting and leaf edge browning	(Paim et al., 2020)
20	Broccoli (Brassica oleracea)	Reduced head formation and quality	Stunted growth and lower nutrient content	(Sardar et al., 2024)

Brief Introduction to Epigenetic Regulation and Its Potential Role in Stress Adaptation

Epigenetic regulation is a key feature of plant stress adaptation and response. Such mechanisms include DNA methylation, histone modifications and small RNAs (Figure 1) that mediate effects on gene expression without modifying the underlying DNA sequence (Ashapkin et al., 2020). Such epigenetic alterations enable plants to respond rapidly to environmental stresses and, in some cases, can be passed on for multiple generations probably assisting a long-term adaptation process (Arshed et al., 2024; Saleem, Zulfiqar, Saleem, et al., 2024, Ijaz et al., 2023). Recent developments in sequencing technologies have revealed many details about epigenetic phenomena taking place in plants, with some pathways/gene transcripts that are involved of response to stress conditions were highlighted (Abdulraheem et al., 2024). Although most have focused on model plants such as Arabidopsis, studies performed on non-model species growing in their natural habitats suggest that epigenetic variation may be one way native populations are able to respond and adapt to changing environments associated with climate change (Liu et al., 2024). The accumulated insights and this expanding knowledge base have important agricultural, breeding, and environmental implications (Abdulraheem et al., 2024, Haider et al., 2021a).

2. EPIGENETICS IN PLANTS: AN OVERVIEW

Epigenetics concerns modifications in gene use without alteration in the genetic code of DNA. These involve items such as DNA methylation, histone changes, and non-coding RNA, as identified in Figure 1 below. It is also involved in the differentiation processes of organ development, stress responses, and genome stability in plants (Liu et al., 2024). An example of epigenetic reprogramming is DNA methylation, which occurs through the action of DNA methyltransferases (Akhter et al., 2021, Haider et al., 2021b). Histones differ from other proteins in that they undergo extensive post-synthesis modifications, affecting various cellular activities, including gene expression and DNA repair (Banerjee & Roychoudhury, 2017). Other ncRNAs involved in the regulation of epigenetic modifications include siRNAs and miRNAs, with the latter being involved in RNA-directed DNA methylation. These processes add complexity to how plants can alter their gene expressions. Such epigenetic changes may be inherited by successive generations as epialleles and are potentially significant for helping plants cope with different states and stresses (Ueda & Seki, 2020, Ijaz et al., 2024a, Ijaz et al., 2024b).

Figure 1. Plant's Epigenetic in response to drought stress.

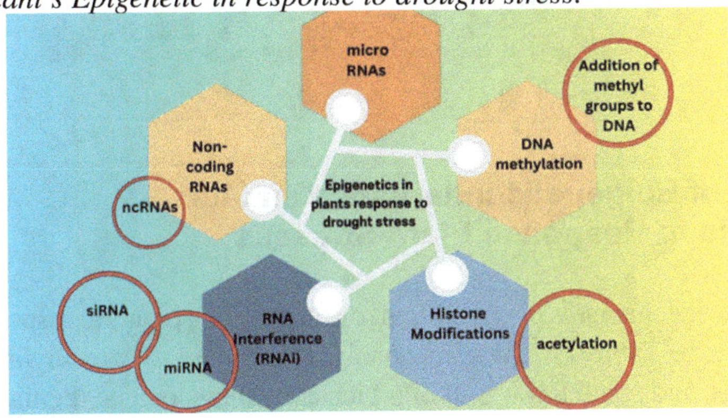

Differences Between Genetic and Epigenetic Mechanisms

Plants employ survival strategies to respond to environmental stresses using both genetic and epigenetic mechanisms (Figure 2). These responses include the activation of heat shock genes that regulate plant behavior under heat stress, influencing adaptive roles in form and function (Andrási et al., 2021). Epigenetic regulation, involving chromatin remodeling, histone modifications, DNA methylation, and non-coding RNAs, controls gene activation during development and in response to stress (Liu et al., 2024, Haider et al., 2022a). These epigenetic changes can lead to trait variations that persist through successive generations, providing opportunities for adaptation in managed environments (Banerjee & Roychoudhury, 2017). While genetic changes are more conserved, epigenetic changes can be transient, helping plants "remember" past stresses. The interplay between genetic and epigenetic strategies is crucial in determining how plants manage various stresses (Alonso et al., 2019, Haider et al., 2021b).

Figure 2. The differences between genetic and epigenetic mechanisms.

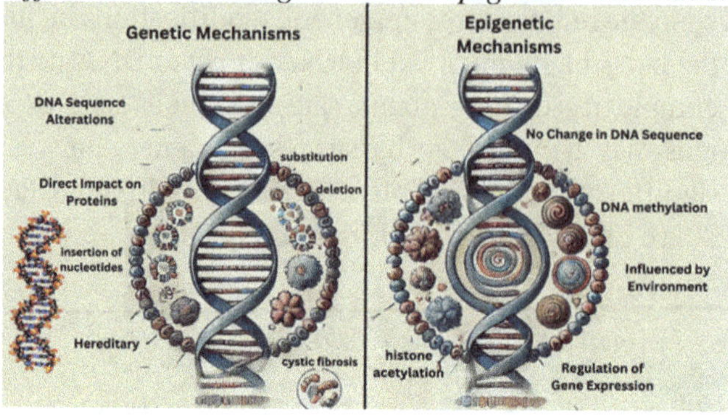

Overview of Epigenetic Inheritance and Its Importance to Response Plant Stresses

Epigenetic mechanisms play a critical role in enabling plants to respond and adapt to stress (refer to Table 2). These mechanisms include DNA methylation, histone modifications, and small RNAs, all of which contribute to gene regulation (Alonso et al., 2019). Remarkably, such modifications can be passed through successive cell divisions and, in some cases, to subsequent generations. These epigenetic modifiers help plants maintain a balance between growth and defense (Banerjee & Roychoudhury, 2017). When a plant experiences stress, it may acquire an epigenetic modification that can be transmitted to the next generation, potentially equipping it to better handle similar stresses. This type of inheritance provides a temporary solution to stress by altering gene expression—affecting phenotypes without altering the DNA sequence (Abdulraheem et al., 2024). Understanding the connection between epigenetics and stress tolerance in plants is essential for advancing crop engineering and developing more resilient plants.

Table 2. Epigenetic inheritance plays a pivotal role in plant adaptation to stress, allowing them to respond, remember, and transmit environmental cues across generations.

Aspect	Description	Importance in Plant Stress Responses	References
Epigenetic Inheritance	Transmission of genetic information through mechanisms that do not involve changes in DNA sequence.	Epigenetic changes are stable and heritable, influencing stress response and adaptation across generations without altering the genetic code.	(Miryeganeh & Saze, 2020)
Key Mechanisms	RNA interference, histone modification, and DNA methylation.	Each of these pathways regulate gene expression, allowing plants to respond to environmental stress by activating or silencing specific stress-related genes.	(Martin & Zhang, 2007)
DNA Methylation	Addition of methyl groups to DNA, typically at cytosine residues, leads to gene silencing.	Methylation patterns can be altered in response to environmental stress, affecting gene expression involved in stress tolerance, e.g., drought or salinity.	(Zhang et al., 2018)
Histone Modifications	Covalent modifications to histones that affect chromatin structure and gene expression.	Histone modifications, such as acetylation or methylation, regulate genes that help plants tolerate stresses like heat or pathogen attacks.	(Zhao et al., 2019)
RNA Interference (RNAi)	Small RNA molecules (e.g., miRNA, siRNA) modify expressions of genes by disabling mRNA/inhibiting translation.	RNAi can fine-tune gene expression during stress responses, helping plants adapt by modulating the expression of stress-related genes.	(Agrawal et al., 2003)
Stress Memory	The plant's ability to "remember" stress through epigenetic modifications across generations.	Epigenetic changes may enable plants to 'inherit' stress tolerance, allowing offspring to respond more effectively to similar environmental challenges.	(Gallusci et al., 2023)
Cross-generational Effects	Epigenetic shifts induced by stress can be transferred to future generations.	Epigenetic inheritance over generations can produce offspring with increased stress tolerance even when is not present.	(Lukić et al., 2023)
Reversible Modifications	Epigenetic changes are reversible, providing flexibility in stress adaptation.	The ability to reverse epigenetic modifications allows plants to adapt quickly to fluctuating environmental conditions, enhancing survival under variable stress.	(Ashapkin et al., 2020)
Implications for Breeding	Understanding epigenetic mechanisms can improve breeding strategies for stress-resilient crops.	Epigenetic breeding could help develop crops with improved tolerance to drought, salinity, or diseases by selecting for favorable epigenetic traits, without changing the underlying genome.	(Tyagi et al., 2024)

3. DNA METHYLATION (DM) AND DROUGHT TOLERANCE

As a result of drought stress, DNA methylation is provoked to modulate gene usage and plant development. These methylation alterations depend on the genotype along the development stage of the tissues and may normalise after the stressful stimuli is eliminated or remain permanently changed (Kumar & Mohapatra, 2021). In maize, drought-tolerant lines with stable proline content show more consistent changes in DNA methylation than drought susceptible lines (Morrison, 2020). Gene body methylation is generally co-ordinately regulated with genes being negatively correlated with gene expression and evidences suggest that it has positive correlation with exon splicing. Small interfering RNAs, linked to DNA methylation also regulate the drought stress (Zi et al., 2024).

In maize, drought changes DNA methylation reducing it in some areas and increasing it in others, for instance in the promoter region of the ZmEXPB2 gene to increase gene expression (D. Zhang et al., 2024). In particular, abscisic acid – a plant hormone responsible for coordinated regulatory reactions to unfavorable environmental conditions – substantially contributes to it: about two thirds of the identified drought-responsive genes of the Arabidopsis thaliana plants are linked with ABA or related compounds such as PBI425 (Huang et al., 2007). The knowledge of these mechanisms offers information that helps INCR understand how crops could be rendered more stress tolerant to drought effect.

Drought brings different changes at the molecular level in plants: there are different proteins activity changes including MAPK and hormones such as jasmonic acid controlling the genes activity and plant stress response. These responses are associated with environmental conditions such as light and recognised as biotic stress factors among other related factors (Salvi et al., 2021).

DNA methylation is involved in plant responses to drought and water stress in an important way, as suggested previously (Kumar & Mohapatra, 2021). Rice, wheat, sesame and other plants have different methylation profile that affect there susceptibility to drought stress (Naderi et al., 2024). In rice, which is a different species to Arabidopsis, there is a putative connection between specific components of DNA methylation and changes in stress-related genes that involve transposition (Naderi et al., 2024). The data presented demonstrate that drought stress treatment increases wheat tissue-specific and developmental stage-dependent levels of demethylation of the wheat varieties with drought-tolerant traits. Sesame increases methylation level when it faces drought stress while water logging decrease methylation level (Kettani et al., 2024). The methylation changes do correlate with gene expression, describing a dynamic system, which might be used for enhancing crops (Singh et al., 2023).

4. HISTONE MODIFICATIONS AND CHROMATIN REMODELING IN DROUGHT RESPONSE

Post-translational modifications of histone proteins and chromatin remodeling are also essential during drought stress adaptation (Sharma et al., 2023). Modifications including acetylation of histones, methylation of histones and phosphorylation affect the DNA and gene transcription (Singh & Prasad, 2022). Among them, acetylation especially allows chromatin to increase gene expression during the drought condition. Many stress-related genes experience fast dynamics in histone modifications, for instance, deposition of H3K9ac during drought stress, which is rapidly replaced upon the relief of stress (Sharma et al., 2023). Fourth, H3K4me3 remains low and possibly as mark of the previous stress. These changes at chromatin level function in conjunction with the changes at the gene regulation level in order to enable the plants to better cope with drought stress and to maintain a memory of stress events (Liu et al., 2014).

A new study brings into focus that how histone modifications make the chromatin structure more accessible or less, that is how it controls gene expression (Gelato & Fischle, 2008). Histones, once synthesized, may also be chemically modified by addition of a variety of residues including acetyl, methyl, phosphate, sumo and ubiquitin groups (Müller & Muir, 2015). These modifications are pivotal for controlling the gene activity; and for maintaining the processes underlying the flow of genetic information. New analysis has pinpointed new varieties for drought stress and sites of these chemical changes, especially at the histone domains governing nucleosome positioning and stability (Huang et al., 2015). These changes pertain to DNA accessibility, nucleosome turnover and gene transcription as well. Histone tails that directly contacted DNA were altered in their state and structure and were linked to chromatin formation and other cellular activities, previously mentioned being the regulation of gene activity, DNA repair, and replication. The combined activity of enzymes that add and remove these marks maintains the dynamic regulation of chromatin compaction and gene expression (Sharma et al., 2023).

To the best of our knowledge, plants do metabolic adjustments and use histones to control the expression of genes more often in response to drought (Chaves et al., 2003). Brachypodium distachyon and Rice show that drought conditions cause histone changes for instance H3K4me3 and H3K9ac signals to be significantly elevated (An, 2021). These modifications are linked with increased accumulation of the transcripts of drought-responsive genes, especially of regions containing proteins. Such changes allow recruitment of the transcription factors and activation of the stress related pathways (Sharma et al., 2023).

Some examples include histone acetylation that leads to chromatin being open and DNA more available for transcription. Like acetylation, methylation will also notify the gene, whether it is to be activated or deactivated depending on the position of the mark. These changes are not permanent, they can quickly respond to any change in the environment such that plants can adapt to drought stress (Ost et al., 2023). Furthermore, certain histone changes may act as epigenetic 'memory' devices, allowing a plant to remember earlier stress and react better in next round of drought. Gradual regulation of chromatin accessibility, histone modifications are involved in plant survival under stress conditions of environment. For instance, in barley plants under drought stress, H3K4me3 levels rise specifically at the HSP17 gene (Ost et al., 2023; Temel et al., 2017). Also, in B. distachyon, higher amounts of H3K9ac are related to increased activity of drought-responsive genes. These changes in how genes are expressed help plants adjust to dry conditions and manage key processes like ABA signaling (Song et al., 2020).

5. NON-CODING RNAS IN DROUGHT TOLERANCE

The small non-coding RNAs are indispensable in enabling the plants to cope with different stress factors including drought stress (Jha et al., 2020). While protein coding RNAs involve in translation of proteins non-coding RNAs control the expression of genes in rather complex ways (Contreras-Cubas et al., 2012). These RNAs are categorized into two main types based on their size and function: small regulatory non-coding RNAs (sRNAs), and long non-coding RNAs (lncRNAs). Each of them performs certain specific functions by which plants are able to cope with the stresses (Zhang et al., 2022).

Micro RNAs or small interfering RNAs of between 20 and 24 nucleotides in size are important in post-transcriptional gene regulation. They are microRNAs (miRNAs) and small interfering RNAs (siRNAs) (He & Hannon, 2004). The difference between miRNA and siRNA is that the former is single stranded and the latter is double stranded. MiRNAs regulate expression of targeted genes to modulate their activity; SiRNAs mainly wrap stress genome fragments and defend plants against viruses and similar threats. They have especially critical functions in managing gene circuits in response to drought spells while ensuring reserve accumulation and stress signaling (Lam et al., 2015).

Long non-coding RNA is any RNA molecule that is more than 200 nucleotides in length and is beginning to attract a lot of attention in plants. These molecules involve themselves in a variety of activities dependent upon their ability to bind genes and proteins at differentiated levels of gene control (Rai et al., 2019). LncRNAs can also be considered as decoy, which complexes with other RNAs, core, which templates the formation of molecular structures, or anchor, which recruits certain factors to focus on certain genes (Segal & Dostie, 2023). The current study also shows that during drought stress, the identified lncRNAs regulated stress-related genes in order to improve the plant tolerance to dryness (Segal & Dostie, 2023).

These studies uphold the harmonization of the miRNAs, siRNAs, and lncRNAs required to modulate the gene expression in response to drought (Tan et al., 2020). This complex web of regulation increases the plant ability to survive unfavourable conditions and thus increases its survival rate. The increasing knowledge about non-coding RNAs suggests that they play crucial roles in plant stress biology and may offer insights towards engineering crop plant stress tolerance.

Role of Micro RNAs in Regulating Drought-Responsive Genes

It is noteworthy that miRNAs play a pivotal role in regulation of drought tolerance since the targets of miRNAs are assigned to genes involved in stress responses such as those of water balance, osmo-resistance and cell preservation. Under drought stress, the expression levels of the identified miRNAs either increased or deactivated their target genes (Amini et al., 2023). This selective regulation is important because it allows the plants to reduce resource use and channel energies towards survival in water-stress environments. For instance, it has been reported that the function of miR398 is regulatory capacity in drought tolerance through targeting genes that code for antioxidant enzymes (Islam et al., 2022). Downregulation of miR398 is due to drought stress which in turn increases the expression of its target genes that intern up the antioxidant activity and reduce oxidative damage in plant cells (Li et al., 2022). Likewise, miR169 was found as downregulated during drought; the latter upregulated the NF-YA transcriptional factor group. It undertakes a significant function in the regulation of downstream stress-responsive genes about root architecture and efficiency of water uptake in drought stress conditions. Furthermore, miR159 and miR160 are bioactive in controlling genes encoding transcription factors associated with abscisic acid (ABA), a key stress-responsive hormone in plants (Singh et al., 2021). ABA is a hormone that is synthesized under drought stress to decrease stomatal aperture and thereby minimize water loss (Muhammad Aslam et al., 2022). By targeting genes in the ABA-dependent regulon, these miRNAs contribute to improving drought tolerance while exerting control over stomatal conductance and water use efficiency (Das & Mondal, 2021).

Case Studies Highlighting the Involvement of Non-Coding RNAs in Plant Drought Tolerance

miR398 and antioxidant regulation in Arabidopsis: One of the best-documented examples of miRNA regulation of drought tolerance is miR398 in Arabidopsis thaliana. Consequently, during drought stress, the expression of miR398 is low and the expression of its target genes: copper/zinc SOD (SOD1 and SOD2) and a copper chaperone for SOD (CCS) are high (Cao et al., 2022). These genes are active in detoxifying reactive oxygen species (ROS) that are usually produced in excess under stress (Sachdev et al., 2021). I have found that the plant increases its ROS-scavenging capacity to alleviate oxidative load and improve drought tolerance.

lncRNA DRIR in rice and Arabidopsis: The previously reported drought-responsive long non-coding RNA DRIR (Drought-Induced lncRNA) has been reported in Oryza sativa and Arabidopsis thaliana. Thus, the current study identifies a novel DRIR function that functions as a positive regulator of drought tolerance through controlling ABA-responsive genes and stress-responsive transcription factors. Some molecular experimental evidence has supported that overexpression of DRIR resulted in improved drought stress tolerance through efficient use of water and minimal water loss in plants (Dong et al., 2022; Feng et al., 2022; Martignago et al., 2020; Wang et al., 2020). This result shows that (lncRNAs) can be used as targets for genetic manipulation aimed at enhancing the drought tolerance of crops.

Mir166 and Root Development in Rice

miR166 also contributes to drought tolerance, particularly regarding the control of root development in rice (Z. Zhang et al., 2024). When drought stress is experienced, the miR166 is downregulated in the process enhancing the HD-ZIP III transcription factor family (Wei et al., 2024). This increases the root to elongate and the plant's capacity to access water down in the soil during dry weather. The observations made in different experiments suggest that the low expression levels of miR166 affect the root system and result in drought resistance, which is why this miRNA can be called an important coordinator of root architecture for drought adaptation.

siRNAs in genome stability and stress adaptation: Furthermore, siRNAs play an important role on drought tolerance based on their function on the stable genome (Sun et al., 2023). Some of them include increased expression of specific siRNAs during drought stress that will silence transposable elements and prevent genomic instability that is occasioned by environmental stresses. In maize, we have seen that through drought-stressed siRNAs, retrotransposons can be strongly down-regulated to prevent genome damage in response to stress and to enable better defense to drought stress (Guo, 2021).

These case studies show how diverse and useful the non-coding RNAs in maintaining plant responses to drought stress are. These non-coding RNAs involved in the regulation of gene expression, root and shoot development and conservation of genome integrity make them important for increasing drought tolerance and a potential focus of breeding and biotechnological modification for increasing crop resistance.

6. EPIGENETIC MEMORY AND DROUGHT TOLERANCE

Concept of Stress Memory in Plants and Its Epigenetic Basis

Unlike animals, plants cannot run away from such stresses forcing the plant to develop sophisticated signal recognition mechanisms that allow the plant to 'recall' past stress such as drought (Kambona, 2023). In this way, stress memory makes plants respond to repeated stress instances using defensive systems more rapidly. This memory can be epigenetic which just means that change in gene expression does not involve the DNA code of the organism (Bernstein, 2022). It even manifests in epigenetic stress memory that is stain begun using modifications like chromatin remodeling, histone modification and DNA methylation that fluctuate how genes will react to any follow-up drought (Akhter et al., 2021). Metabolic and morphological changes attributable to stress memory allow plants to change the epigenetic configuration of their cells so that the plant can respond faster or more intensely were the drought conditions to reoccur. Such priming is particularly useful in ecosystems where drought is cyclical, so, the plants can stand and recuperate from water scarcity conditions with much ease.

Mechanisms of Transgenerational Epigenetic Inheritance of Drought Tolerance Traits

Long-term epigenetic memory can be transmitted to offspring through transgenerational epigenetic memory, where descendants of genotypes that are subjected to drought stress are characterized by improved drought tolerance, although they, miraculously, were not exposed to stresses (Nguyen et al., 2022). Key mechanisms of transgenerational inheritance include:

DNA methylation inheritance: Specific DNA methylation patterns generated by parent plants can be stably transferred across offspring under drought stress (Nguyen et al., 2022). Such 'tags' can either repress or activate debase, a set of genes responsible for drought response in offspring to make a faster response to the stress (Redhu et al., 2023).

Histone modification and chromatin state memory: The putative modifications of histones which are the packaging material of DNA into chromatin can be passed on to descendant cells (Rothi & Greer, 2023). This changes accessibility of genes, retains a memory of stress that is a prerequisite for a quick reaction in case required.

Small RNAs and epigenetic memory transfer: siRNAs can stably transmit epigenetic signals DNA methylation and histone modification across generations to maintain stress-induced epigenetic changes for large number of generations (Akhter et al., 2021). For instance, siRNAs synthesized within the period of drought stress can be inherited by the populace and can direct the correct methylation pattern for the enshrinement of drought-responsive characteristics in the developing offspring.

Examples of Epigenetic Memory Contributing to Enhanced Drought Resilience Across Generations

Arabidopsis and DNA methylation memory of drought stress: DNA methylation changes have also been observed in Arabidopsis thaliana progeny, which are established to memorize genes in connection to drought tolerance in plants exposed to drought stress (Nguyen et al., 2022). It conditions the plant to respond to drought stress more rapidly and effectively with acquisition of drought-responsive genes in successive generations.

Histone modifications in rice for improved drought tolerance: In rice, there study has revealed that H3K27me3 histone modifications elicited by drought stress in parent plants are carried forward to the ensuing generation (Dinkar et al., 2024). These modifications change the chromatin organization around drought regulated genes so that the next generation remains epigenetically primed for quick adaptation to water stress conditions.

Epigenetic memory in tomato plants via siRNAs: Tomato plants have depicted intergenerational epigenetic memory of draught stress via small interfering RNA or siRNAs whereby it associates the DNA methylation of the offspring to stress endurance tracks (Mladenov et al., 2021). This study demonstrates that heir produced by plants that have experienced drought previously are more likely to survive under drought status because epigenetic traits are passed from the parent plant to the offspring.

7. CROSSTALK BETWEEN EPIGENETIC MECHANISMS AND PHYTOHORMONE SIGNALING

Interactions Between Epigenetic Regulation and Phytohormones in Drought Response

Epigenetic regulation in plants and phytohormonal pathways play an important role in plant drought stress adaptation (Kaya et al., 2024). For instance, changes in small RNA, histone modifications and DNA methylation possess the potential to change gene expressions concerning phytohormone biosynthesis, transduction and perception (Li et al., 2020). This "crosstalk" allows plants to synchronize hormonal responses with drought-triggered cellular adjustments very closely.

Abscisic Acid (ABA): ABA is the most prominent hormone involved in drought reactions because, through stomatal control, it helps to minimize water spillage, while osmotic stress is managed (Ali et al., 2020). Expressed changes can modify the ABA biosynthesis and signaling genes to increase the ABA reaction during a water deficit stage (Soma et al., 2021). For instance, DNA methylation on promoters of ABA-responsive genes and certain histone modifications prepare plants to respond rapidly to ABA upon recurring drought.

Auxin: While auxin is said to be responsible for promoting the growth of roots, it is instrumental in establishing drought resistant root systems (Y. Zhang et al., 2024). A role for epigenetic regulation of genes involving auxin signaling particularly in root tissues is to increase root mass and water use efficiency under conditions of drought stress miRNAs similarly, therefore influence the expression of auxin signaling genes regulating root architectural adaptations to water deficit.

Examples of How Epigenetic Modifications Influence Hormone-Related Genes During Drought Stress

ABA biosynthesis and signaling modulation: During drought stress, DNA methylation patterns change near genes involved in ABA biosynthesis and signaling, such as NCED3 (9-cis-epoxycarotenoid dioxygenase) which is key for ABA production (Gu et al., 2021). Methylation reduction in these regions can lead to increased expression, boosting ABA levels and enhancing drought tolerance.

Auxin response modulated by miRNAs: miRNAs like miR160 and miR167 target auxin response factors (ARFs), which are essential for root development and drought adaptation (Jodder, 2020). Under drought stress, changes in miRNA expression reduce the inhibition of ARFs, promoting root elongation and improved water access in soil.

Histone acetylation in jasmonic acid (JA) pathway: Though ABA is dominant in drought responses, jasmonic acid (JA) also plays a supportive role by modulating stress signaling (Hussain et al., 2021). Histone acetylation near JA-responsive genes enhances their expression, indirectly supporting ABA pathways by fortifying cellular defenses under drought stress.

8. APPLICATIONS OF EPIGENETIC RESEARCH IN DEVELOPING DROUGHT-TOLERANT CROPS

Use of epigenetic markers and editing tools for crop improvement: Epigenetic markers and tools for epigenome editing have enabled a new approach to improving crops to improve their drought tolerance (Kakoulidou et al., 2021, Ullah et al., 2023a). Relationship between genes and epigenetic modifications so that certain epigenetic modifications that are connected to drought resilience can be targeted thus producing crops that are well adapted to dry weather.

Epigenetic markers: Histone modification and DNA methylation are two major epigenetic biomarkers that reveal genes or regions that are associated with drought tolerance (Hao et al., 2020). These should be useful in selecting desirable genes for drought tolerance in plant breeding programmes.

CRISPR/dCas9 epigenetic modifiers: While the original CRISPR Cas-9 system cleaves the DNA there is a modified form called dCas9 that can be linked to enzymes that change the way genes are expressed without changing the underlying DNA base pairs (Asmamaw & Zawdie, 2021). For instance, dCas9 fused to DNA methylase or histone acetylase will enhance or suppress expression of drought-responsive genes. This precision tool permits reversible, inheritable alteration in the manifestation of subsequent generations, rendering them a valuable asset for sustainable crop improvement.

Epigenome-Wide Association Studies (EWAS) For Identifying Drought Tolerance Traits

Epigenome-Wide Association Studies (EWAS) are performed to chart the epigenome and relate traits to drought tolerance (Can et al., 2021). Through high-throughput sequencing and bioinformatics enables researchers to discover:

Methylation hotspots: You might think this means that there are some coordinately regulated gene-promoter units across different crop types with regard to drought-induced methylation changes in the different varieties (Woolfson et al., 2022). Such hot spots tend to co-localize with stress-protectant genes and genes related to water conservation and osmotic balance.

Drought-responsive lncRNAs and miRNAs: EWAS also allows for the discovery of non-coding RNAs that have increased or reduced expression under drought stress and which could therefore be used to generate drought-tolerant crops (Tao et al., 2021).

Heritable epigenetic patterns: Moreover, EWAS investigate the DNA methylation level that is sustained across generations and helps in establishing the rationally inherited drought tolerance transgene associated traits, making it a more robust approach in the process of breeding program.

Current Challenges and Future Perspectives in Applying Epigenetics to Crop Breeding

While the application of epigenetic tools in crop breeding is promising, there are several challenges and areas for future development:

Stability of epigenetic modifications: There is challenge of how to reduce variation in epigenetic changes in different generations and under different environmental conditions. Epigenetic marks may be reversible, so the ability to retain desired modifications in crop progeny is still challenging (Kakoulidou et al., 2021).

Precision and Off-target effects: While systems like CRISPR/dCas9 possess required specificity, other genes can get an unwanted epigenetic edit as well. Optimizing these tools to reduce off target effects is again a major focus of active investigation.

Regulatory and public acceptance: Genetic and epigenetic engineering in agriculture tends to attract regulatory constraints and consumer mistrust (Mehlman et al., 2023). Assumption Of epigenetically modified crops by the public is dependent on factual presentation of risks and gains associated with the crops.

Future perspectives: Future developments in machine learning and bioinformatics tools also promise improvement on EWAS using existing gene expression data to improve pinpoint accuracy in the epigenetic features associated with drought tolerance. The combined use of genomics, transcriptomics, and epigenomics data will enable a system-level understanding of drought resilience and call for modulating the system at various levels. Development of epigenetic breeding protocols tailored for specific crops and climates will allow for more adaptable and efficient crop improvement strategies.

9. FUTURE DIRECTIONS AND RESEARCH PROSPECTS

Knowledge Gaps in Understanding Plant Epigenetic Responses to Drought

Despite advancements in understanding plant epigenetics, several knowledge gaps remain, particularly concerning drought tolerance:

Temporal dynamics of epigenetic changes: The developmental schedule of the epigenetic changes during drought stress and the process of its alleviation remain still obscure to a great extent. For example, the authors did not specify how long DNA methylation changes last post-stress or how these changes affect memory during a second round of drought.

Tissue-specific epigenetic responses: While most studies have investigated the overall plant response to drought, at the tissue level, different organs such as the root and leaf respond differently to drought stress (Kim et al., 2020). It is also important to explore more specific tissue epigenetic response in order to obtain more specific rain deficiency tolerance mechanisms description.

Transgenerational epigenetic inheritance mechanisms: It is clear, for instance, that certain forms of epigenetic changes associating with drought stress are transgenerational, but it is yet unclear which epigenetic marks are stably inherited and which factors predefine such outcome (Kakoulidou et al., 2021).

Emerging Technologies for Epigenomic Profiling in Plants Under Stress Conditions

Advances in technology are facilitating more comprehensive and detailed profiling of plant epigenomes under drought conditions:

Single-cell epigenomics: Current platforms in single-cell sequencing enable scientists to study differences in epigenetic regulation in specific cells; this information then reveals how the various cell types may respond individually to drought. This technology might be especially useful for studying reaction to drought in particular tissues, such as cells that take water up from roots.

Nanopore sequencing: The nanopore sequencing approach describes the extended DNA sequencing that can allow mapping of DNA methylation and histone modifications at a single-base resolution in real time. This technology allows for higher resolution for the detection of drought-responsive epigenetic marks on large genomic domains.

CRISPR-based epigenomic tools: Orange also extends the usefulness of CRISPR beyond gene editing and applies them to reading and changing the epigenome. For example, dCas9 can be combined with the proteins that recognize, erased and writes epigenetic signatures, which enable the examination of the particular genes' epigenetic state in response to drought (Kaya et al., 2024, Ullah et al., 2023b).

Prospects for Integrating Epigenetics With Other Omics in Drought Research

The future of drought tolerance research lies in integrating epigenetics with other omics approaches:

Genomics and epigenomics: The integration of VNTR with epigenomic data can indicate genetic regions that are both structurally modified and epigenetically controlled, which are potential sources of stable selection for drought tolerant crops.

Transcriptomics and epigenomics: Combination of transcriptomics with epigenomics allows researchers to see how epigenetic changes affect gene regulation during drought stress (Miryeganeh, 2021). For instance, the outright comparative analysis of up-regulated and down-regulated genes against the changes in methylation profiles provide a much better understanding of how plants modulate stress responses.

Metabolomics and epigenomics: Combining metabolomic data with epigenetics data can provide information on how metabolic changes that takes place due to drought stress is epigenetically regulated so as to provide insights into various physiological adjustments that occurs during the acquisition of drought tolerance.

They believe that such an approach would give a better picture of plant's response to drought stress by reaching a higher dimension and therefore can be applicable in breeding or engineering stress-tolerance traits in plants.

10. CONCLUSION

The field of studying mechanisms has become a tool for improving our understanding of how plants can adapt to and thrive in arid conditions with limited water availability. In this chapter, we looked at how plants deal with not having enough water. They make changes to their DNA and other parts to manage their genes and survive. These changes help plants quickly respond to dry conditions and help them remember these challenges, which makes them stronger over time and across different generations. DNA changes, like adding methyl groups, play a big part in deciding which genes are active when there is not enough water, either turning them off or on as needed. Changes to histones and how the DNA is shaped also help make sure that the genes that react to stress can be accessed easily when needed. Furthermore, non-coding RNAs contribute to post transcriptional regulation with small RNAs and long non coding RNAs serving as regulators of gene networks to enhance drought responses. The idea of memory is crucial in the development of drought crops and offers an exciting opportunity for agricultural advancements, in sustainability by utilizing plants ability to remember stress induced changes at a genetic level. The future development strategies include CRISPR-based epigenetic editing and epigenome-wide association study (EWAS) to possess enormous potential to enhance crop quality. However, there remain certain limitations for instance, the mechanisms of epigenetic alterations under the physiological stress have not been thoroughly studied; the epigenomic strategies for controlled epigenetic alterations are rather limited. In conclusion, the advancements in epigenetic research give a viable platform, for affording with the issues of food insecurity under volatile climatic changes. Thus, the application of epigenetic findings will help speed up the improvements to crops that are capable of sustaining drought as well as the improvement to the agricultural requirements that enhance the sufficient food production for the increasing population. And as we progress in filling those knowledge gaps, and improving some of the technologies, drought-tolerant crops look brighter for a better future and we are one step closer toward improving and ensuring a sustainable and more efficient agricultural environment.

REFERENCES

Abdulraheem, M. I., Xiong, Y., Moshood, A. Y., Cadenas-Pliego, G., Zhang, H., & Hu, J. (2024). Mechanisms of plant epigenetic regulation in response to plant stress: Recent discoveries and implications. *Plants, 13*(2), 163.

Abreha, K. B., Enyew, M., Carlsson, A. S., Vetukuri, R. R., Feyissa, T., Motlhaodi, T., Ng'uni, D., & Geleta, M. (2022). Sorghum in dryland: Morphological, physiological, and molecular responses of sorghum under drought stress. *Planta, 255*, 1–23.

Ahmad, A., Aslam, Z., Javed, T., Hussain, S., Raza, A., Shabbir, R., Mora-Poblete, F., Saeed, T., Zulfiqar, F., & Ali, M. M. (2022). Screening of wheat (Triticum aestivum L.) genotypes for drought tolerance through agronomic and physiological response. *Agronomy (Basel), 12*(2), 287.

Ahmad, M. A., Saleem, A., Tahir, M., Khilji, S. A., Sajid, Z. A., Landry, K. B., El-Sheikh, M. A., & Ahmad, P. (2024). Modulation of the polyamines, osmolytes and antioxidant defense system to ameliorate drought stress tolerance in Hordeum vulgare L. using ascorbic acid. *South African Journal of Botany, 171*, 726–736.

Ahmad, Y., Haakim, Z., Iqbal, J., Abbasi, B. A., Mahmood, T., & Kazi, M. (2024b). Technological Innovations for Abiotic Stress Resistance in Horticultural Crops. OMICs-based Techniques for Global Food Security, 233-244.

Ahmad, Y., Haider, S., Iqbal, J., Abbasi, B. A., Yaseen, T., & Mahmood, T. (2022). The mechanisms of genome editing technologies in crop plants. In *Principles and Practices of OMICS and Genome Editing for Crop Improvement* (pp. 295–313). Springer International Publishing.

Ahmad, Y., Haider, S., Iqbal, J., Naseer, S., Attia, K. A., Mohammed, A. A., & Mahmood, T. (2024a). In-silico analysis and transformation of OsMYB48 transcription factor driven by CaMV35S promoter in model plant–Nicotiana tabacum L. conferring abiotic stress tolerance. *GM Crops and Food: Biotechnology in Agriculture and the Food Chain, 15*(1), 130–149.

Akhter, Z., Bi, Z., Ali, K., Sun, C., Fiaz, S., Haider, F. U., & Bai, J. (2021). In response to abiotic stress, DNA methylation confers epigenetic changes in plants. *Plants, 10*(6), 1096.

Akpinar, A., & Cansev, A. (2022). Physiological and molecular responses of roots differ from those of leaves in spinach plants subjected to short-term drought stress. *South African Journal of Botany, 151*, 9–17.

Ali, I., Salah, K. B. H., Sher, H., Ali, H., Ullah, Z., Ali, A., & Mahmood, T. (2022). Drought stress enhances the efficiency of floral dip method of Agrobacterium-mediated transformation in Arabidopsis thaliana. *Brazilian Journal of Biology, 84*, e259326.

Ali, S., Hayat, K., Iqbal, A., & Xie, L. (2020). Implications of abscisic acid in the drought stress tolerance of plants. *Agronomy (Basel), 10*(9), 1323.

Alonso, C., Ramos-Cruz, D., & Becker, C. (2019). The role of plant epigenetics in biotic interactions. *The New Phytologist, 221*(2), 731–737.

Amini, Z., Salehi, H., Chehrazi, M., Etemadi, M., & Xiang, M. (2023). miRNAs and their target genes play a critical role in response to heat stress in Cynodon dactylon (L.) Pers. *Molecular Biotechnology, 65*(12), 2004–2017.

An, Y. (2021). *Dynamic epigenome changes in response to light in Brachypodium distachyon The Chinese University of Hong Kong*.

Andrási, N., Pettkó-Szandtner, A., & Szabados, L. (2021). Diversity of plant heat shock factors: Regulation, interactions, and functions. *Journal of Experimental Botany, 72*(5), 1558–1575.

Arshed, M. Z., Zulfiqar, A., & Saleem, A. (2024). Alleviation of Chromium Stress in Brassica juncea L. and Soil Remediation by Plant Growth Promoting (PGP) Bacteria. *Pakistan Journal of Scientific and Industrial Research. Series B: Biological Sciences, 67*(1).

Ashapkin, V. V., Kutueva, L. I., Aleksandrushkina, N. I., & Vanyushin, B. F. (2020). Epigenetic mechanisms of plant adaptation to biotic and abiotic stresses. *International Journal of Molecular Sciences, 21*(20), 7457.

Asmamaw, M., & Zawdie, B. (2021). Mechanism and applications of CRISPR/Cas-9-mediated genome editing. *Biologics*, •••, 353–361.

Bachani, J., Mahanty, A., Aftab, T., & Kumar, K. (2022). Insight into calcium signalling in salt stress response. *South African Journal of Botany, 151*, 1–8.

Bagheri, M., Santos, C. S., Rubiales, D., & Vasconcelos, M. W. (2023). Challenges in pea breeding for tolerance to drought: Status and prospects. *Annals of Applied Biology, 183*(2), 108–120.

Banerjee, A., & Roychoudhury, A. (2017). Epigenetic regulation during salinity and drought stress in plants: Histone modifications and DNA methylation. *Plant Gene, 11*, 199–204.

Bernstein, C. (2022). DNA methylation and establishing memory. *Epigenetics Insights*, *15*, 25168657211072499.

Can, S. N., Nunn, A., Galanti, D., Langenberger, D., Becker, C., Volmer, K., Heer, K., Opgenoorth, L., Fernandez-Pozo, N., & Rensing, S. A. (2021). The EpiDiverse plant epigenome-wide association studies (EWAS) pipeline. *Epigenomes*, *5*(2), 12.

Cao, B., Jiang, J., Bai, J., Wang, X., Li, Y., Shao, W., Hu, S., He, Y., & Yu, X. (2022). miR398 Attenuates Heat-Induced Leaf Cell Death via Its Target CSD1 in Chinese Cabbage. *Horticulturae*, *8*(4), 299.

Chaves, M. M., Maroco, J. P., & Pereira, J. S. (2003). Understanding plant responses to drought—From genes to the whole plant. *Functional Plant Biology*, *30*(3), 239–264.

Choukri, H., Hejjaoui, K., El-Baouchi, A., El Haddad, N., Smouni, A., Maalouf, F., Thavarajah, D., & Kumar, S. (2020). Heat and drought stress impact on phenology, grain yield, and nutritional quality of lentil (Lens culinaris Medikus). *Frontiers in Nutrition*, *7*, 596307.

Contreras-Cubas, C., Palomar, M., Arteaga-Vázquez, M., Reyes, J. L., & Covarrubias, A. A. (2012). Non-coding RNAs in the plant response to abiotic stress. *Planta*, *236*, 943–958.

Das, R., & Mondal, S. K. (2021). Plant miRNAs: Biogenesis and its functional validation to combat drought stress with special focus on maize. *Plant Gene*, *27*, 100294.

Dinkar, V., Pandey, S., Kumar, A., Shiv, A., Lal, D., Bharati, A., Joshi, A., Adhikari, S., Singh, A., & Pandey, P. K. (2024). Epigenetic regulations under plant stress: A cereals perspective. *Environmental and Experimental Botany*, ●●●, 105688.

Dong, S.-M., Liang, X., Li, Z.-B., Jie, S., Yan, H.-B., Li, S.-X., Liao, W.-B., & Ming, P. (2022). A novel long non-coding RNA, DIR, increases drought tolerance in cassava by modifying stress-related gene expression. *Journal of Integrative Agriculture*, *21*(9), 2588–2602.

Feng, X., Jia, L., Cai, Y., Guan, H., Zheng, D., Zhang, W., Xiong, H., Zhou, H., Wen, Y., & Hu, Y. (2022). ABA-inducible DEEPER ROOTING 1 improves adaptation of maize to water deficiency. *Plant Biotechnology Journal*, *20*(11), 2077–2088.

Gallusci, P., Agius, D. R., Moschou, P. N., Dobránszki, J., Kaiserli, E., & Martinelli, F. (2023). Deep inside the epigenetic memories of stressed plants. *Trends in Plant Science*, *28*(2), 142–153.

Gambetta, G. A., Herrera, J. C., Dayer, S., Feng, Q., Hochberg, U., & Castellarin, S. D. (2020). The physiology of drought stress in grapevine: Towards an integrative definition of drought tolerance. *Journal of Experimental Botany*, *71*(16), 4658–4676.

Gelato, K. A., & Fischle, W. (2008). Role of histone modifications in defining chromatin structure and function.

Gomathi, R., Krishnapriya, V., Arunkumar, R., Govindaraj, P., & Ram, B. (2020). Physiological traits imparting drought stress tolerance to promising sugarcane (Saccharum spp.) clones. *Plant Physiology Reports*, *25*, 509–515.

Gu, D., Yang, J., Wu, S., Liao, Y., Zeng, L., & Yang, Z. (2021). Epigenetic regulation of the phytohormone abscisic acid accumulation under dehydration stress during postharvest processing of tea (Camellia sinensis). *Journal of Agricultural and Food Chemistry*, *69*(3), 1039–1048.

Guo, W. (2021). RNA-Directed DNA Methylation Prevents Rapid and Heritable Reversal of Transposon Under Heat Stress in Zea Mays [Purdue University].

Haider, S., Iqbal, J., Naseer, S., Shaukat, M., Abbasi, B. A., Yaseen, T., & Mahmood, T. (2022b). Unfolding molecular switches in plant heat stress resistance: A comprehensive review. *Plant Cell Reports*, *41*(3), 775–798.

Haider, S., Iqbal, J., Naseer, S., Yaseen, T., Shaukat, M., Bibi, H., & Mahmood, T. (2021a). Molecular mechanisms of plant tolerance to heat stress: Current landscape and future perspectives. *Plant Cell Reports*, •••, 1–25.

Haider, S., Iqbal, J., Shaukat, M., Naseer, S., & Mahmood, T. (2021b). The epigenetic chromatin-based regulation of somatic heat stress memory in plants. *Plant Gene*, *27*, 100318.

Haider, S., Raza, A., Iqbal, J., Shaukat, M., & Mahmood, T. (2022a). Analyzing the regulatory role of heat shock transcription factors in plant heat stress tolerance: A brief appraisal. *Molecular Biology Reports*, *49*(6), 5771–5785.

Hao, X., Jin, Z., Wang, Z., Qin, W., & Pei, Y. (2020). Hydrogen sulfide mediates DNA methylation to enhance osmotic stress tolerance in Setaria italica L. *Plant and Soil*, *453*, 355–370.

He, L., & Hannon, G. J. (2004). MicroRNAs: Small RNAs with a big role in gene regulation. *Nature Reviews. Genetics*, *5*(7), 522–531.

Huang, D., Jaradat, M. R., Wu, W., Ambrose, S. J., Ross, A. R., Abrams, S. R., & Cutler, A. J. (2007). Structural analogs of ABA reveal novel features of ABA perception and signaling in Arabidopsis. *The Plant Journal*, *50*(3), 414–428.

Huang, H., Lin, S., Garcia, B. A., & Zhao, Y. (2015). Quantitative proteomic analysis of histone modifications. *Chemical Reviews, 115*(6), 2376–2418.

Hussain, , IAshraf, , M. YSaleem, , M. HAshraf, , M. AAli, , BShereen, , AFarid, , GAli, , MShirazi, , M. USaleem, , A. (2023). Alleviating effects of salicylic acid spray on stage-based growth and antioxidative defense system in two drought-stressed rice (Oryza sativa L.) cultivars. *Turkish Journal of Agriculture and Forestry, 47*(1), 79–99.

Hussain, Q., Asim, M., Zhang, R., Khan, R., Farooq, S., & Wu, J. (2021). Transcription factors interact with ABA through gene expression and signaling pathways to mitigate drought and salinity stress. *Biomolecules, 11*(8), 1159.

Ijaz, S., Iqbal, J., Abbasi, B. A., Tufail, A., Ullah, Z., Sharifi-Rad, J., & Uddin, S. (2023). Biofortification: Lessons from the Past and Strategies for Future Food Security. In *Legumes Biofortification* (pp. 521–545). Springer International Publishing.

Ijaz, S., Iqbal, J., Abbasi, B. A., Yaseen, T., Rehman, S., Kazi, M., & Mahmood, T. (2024). Role of OMICS-Based Technologies in Plant Sciences. OMICs-based Techniques for Global Food Security, 45-66.

Islam, W., Idrees, A., Waheed, A., & Zeng, F. (2022). Plant responses to drought stress: microRNAs in action. *Environmental Research, 215*, 114282.

Jha, U. C., Nayyar, H., Jha, R., Khurshid, M., Zhou, M., Mantri, N., & Siddique, K. H. (2020). Long non-coding RNAs: Emerging players regulating plant abiotic stress response and adaptation. *BMC Plant Biology, 20*, 1–20.

Jodder, J. (2020). miRNA-mediated regulation of auxin signaling pathway during plant development and stress responses. *Journal of Biosciences, 45*(1), 91.

Kakoulidou, I., Avramidou, E. V., Baránek, M., Brunel-Muguet, S., Farrona, S., Johannes, F., Kaiserli, E., Lieberman-Lazarovich, M., Martinelli, F., & Mladenov, V. (2021). Epigenetics for crop improvement in times of global change. *Biology (Basel), 10*(8), 766.

Kambona, C. M. (2023). *Memory of past drought stress exposure effects plant responses in subsequent generations in winter wheat (Triticum aestivum L.)*. Universitäts-und Landesbibliothek Bonn.

Kaya, C., Uğurlar, F., & Adamakis, I.-D. S. (2024). Epigenetic Modifications of Hormonal Signaling Pathways in Plant Drought Response and Tolerance for Sustainable Food Security. *International Journal of Molecular Sciences, 25*(15), 8229.

Keipp, K., Hütsch, B. W., Ehlers, K., & Schubert, S. (2020). Drought stress in sunflower causes inhibition of seed filling due to reduced cell-extension growth. *Journal Agronomy & Crop Science, 206*(5), 517–528.

Kettani, R., El Fechtali, M., & Nabloussi, A. (2024). Exploring mechanisms of drought-tolerance and adaptation of selected sesame mutant lines. *Journal of Agriculture and Food Research, 15*, 100911.

Khodabin, G., Tahmasebi-Sarvestani, Z., Rad, A. H. S., & Modarres-Sanavy, S. A. M. (2020). Effect of drought stress on certain morphological and physiological characteristics of a resistant and a sensitive canola cultivar. *Chemistry & Biodiversity, 17*(2), e1900399.

Kim, Y., Chung, Y. S., Lee, E., Tripathi, P., Heo, S., & Kim, K.-H. (2020). Root response to drought stress in rice (Oryza sativa L.). *International Journal of Molecular Sciences, 21*(4), 1513.

Kumar, S., & Mohapatra, T. (2021). Dynamics of DNA methylation and its functions in plant growth and development. *Frontiers in Plant Science, 12*, 596236.

Lam, J. K., Chow, M. Y., Zhang, Y., & Leung, S. W. (2015). siRNA versus miRNA as therapeutics for gene silencing. *Molecular Therapy. Nucleic Acids, •••*, 4.

Li, J., Song, Q., Zuo, Z.-F., & Liu, L. (2022). MicroRNA398: A master regulator of plant development and stress responses. *International Journal of Molecular Sciences, 23*(18), 10803.

Li, T., Gonzalez, N., Inzé, D., & Dubois, M. (2020). Emerging connections between small RNAs and phytohormones. *Trends in Plant Science, 25*(9), 912–929.

Liang, G., Liu, J., Zhang, J., & Guo, J. (2020). Effects of drought stress on photosynthetic and physiological parameters of tomato. *Journal of the American Society for Horticultural Science, 145*(1), 12–17.

Liu, N., Fromm, M., & Avramova, Z. (2014). H3K27me3 and H3K4me3 chromatin environment at super-induced dehydration stress memory genes of Arabidopsis thaliana. *Molecular Plant, 7*(3), 502–513.

Liu, Q., Ton, J., Manavella, P. A., Komiya, R., & Zhai, J. (2024). *Focus on epigenetics* (Vol. 194). Oxford University Press US.

Lukić, N., Schurr, F. M., Trifković, T., Kukavica, B., & Walter, J. (2023). Transgenerational stress memory in plants is mediated by upregulation of the antioxidative system. *Environmental and Experimental Botany, 205*, 105129.

Martignago, D., Rico-Medina, A., Blasco-Escámez, D., Fontanet-Manzaneque, J. B., & Caño-Delgado, A. I. (2020). Drought resistance by engineering plant tissue-specific responses. *Frontiers in Plant Science*, *10*, 1676.

Martin, C., & Zhang, Y. (2007). Mechanisms of epigenetic inheritance. *Current Opinion in Cell Biology*, *19*(3), 266–272.

Mehlman, M. J., Conlon, R. A., & Pearlman, A. (2023). Governing nonconventional genetic experimentation. *Journal of Law and the Biosciences*, *10*(1), lsad003.

Meshram, J. H., Singh, S. B., Raghavendra, K., & Waghmare, V. (2022). Drought stress tolerance in cotton: progress and perspectives. Climate Change and Crop Stress, 135-169.

Miryeganeh, M. (2021). Plants' epigenetic mechanisms and abiotic stress. *Genes*, *12*(8), 1106.

Miryeganeh, M., & Saze, H. (2020). Epigenetic inheritance and plant evolution. *Population Ecology*, *62*(1), 17–27.

Mladenov, V., Fotopoulos, V., Kaiserli, E., Karalija, E., Maury, S., Baranek, M., Segal, N. a., Testillano, P. S., Vassileva, V., & Pinto, G. (2021). Deciphering the epigenetic alphabet involved in transgenerational stress memory in crops. *International Journal of Molecular Sciences*, *22*(13), 7118.

Morrison, R. R. (2020). Genomic mechanisms of adaptation to drought-stress in maize University of Warwick].

Muhammad Aslam, M., Waseem, M., Jakada, B. H., Okal, E. J., Lei, Z., Saqib, H. S. A., Yuan, W., Xu, W., & Zhang, Q. (2022). Mechanisms of abscisic acid-mediated drought stress responses in plants. *International Journal of Molecular Sciences*, *23*(3), 1084.

Müller, M. M., & Muir, T. W. (2015). Histones: At the crossroads of peptide and protein chemistry. *Chemical Reviews*, *115*(6), 2296–2349.

Mushtaq, I., Zulfiqar, A., Ali, M., & Saleem, A. (2021). Effect of sodium fluoride on the growth and productivity of Triticum aestivum L. in the presence of fluoride resistant bacteria. the Presence of Fluoride Resistant Bacteria.

Naderi, S., Maali-Amiri, R., Sadeghi, L., & Hamidi, A. (2024). Physio-biochemical and DNA methylation analysis of the defense response network of wheat to drought stress. *Plant Physiology and Biochemistry*, *209*, 108516.

Nasir, M. W., & Toth, Z. (2022). Effect of drought stress on potato production: A review. *Agronomy (Basel)*, *12*(3), 635.

Nguyen, N. H., Vu, N. T., & Cheong, J.-J. (2022). Transcriptional stress memory and transgenerational inheritance of drought tolerance in plants. *International Journal of Molecular Sciences*, *23*(21), 12918.

Ost, C., Cao, H. X., Nguyen, T. L., Himmelbach, A., Mascher, M., Stein, N., & Humbeck, K. (2023). Drought-Stress-Related Reprogramming of Gene Expression in Barley Involves Differential Histone Modifications at ABA-Related Genes. *International Journal of Molecular Sciences*, *24*(15), 12065.

Paim, B. T., Crizel, R. L., Tatiane, S. J., Rodrigues, V. R., Rombaldi, C. V., & Galli, V. (2020). Mild drought stress has potential to improve lettuce yield and quality. *Scientia Horticulturae*, *272*, 109578.

Rai, M. I., Alam, M., Lightfoot, D. A., Gurha, P., & Afzal, A. J. (2019). Classification and experimental identification of plant long non-coding RNAs. *Genomics*, *111*(5), 997–1005.

Redhu, N. S., Yashveer, S., Taunk, J., Banga, P., Singh, V., Tokas, J., Grewal, S., & Arora, I. (2023). Plant microRNAs: Unexplored biogenesis, prediction tools and applications. *Gene Reports*, *32*, 101799.

Riasat, M., Saed-Mouchehsi, A., & Jafari, A. A. (2020). Effect of drought stress levels on seedling Morpho-physiological traits of alfalfa (Medicago sativa) populations grown in glasshouse. *Journal of Rangeland Science*, *10*(1), 86–97.

Rothi, M. H., & Greer, E. L. (2023). From correlation to causation: The new frontier of transgenerational epigenetic inheritance. *BioEssays*, *45*(1), 2200118.

Sachdev, S., Ansari, S. A., Ansari, M. I., Fujita, M., & Hasanuzzaman, M. (2021). Abiotic stress and reactive oxygen species: Generation, signaling, and defense mechanisms. *Antioxidants*, *10*(2), 277.

Sachdeva, S., Bharadwaj, C., Patil, B. S., Pal, M., Roorkiwal, M., & Varshney, R. K. (2022). Agronomic performance of chickpea affected by drought stress at different growth stages. *Agronomy (Basel)*, *12*(5), 995.

Saleem, A., Zulfiqar, A., Saleem, M. Z., Saleem, M. H., & Fahad, S. (2024). Elevated Expression of YSL Genes and Enhanced Physiological Responses in Rice Subjected to Iron Application. *Journal of Plant Growth Regulation*, •••, 1–19.

Saleem, A., Zulfiqar, A., Zulfiqar, U., & Hussain, S. (2024). Iron Uptake Strategies in Rice: Exploring YSL Genes and Nutritional Enhancement. *Plant Molecular Biology Reporter*, •••, 1–19.

Salvi, P., Manna, M., Kaur, H., Thakur, T., Gandass, N., Bhatt, D., & Muthamilarasan, M. (2021). Phytohormone signaling and crosstalk in regulating drought stress response in plants. *Plant Cell Reports*, *40*, 1305–1329.

Sardar, H., Shafiq, M., Naz, S., Ali, S., Ahmad, R., & Ejaz, S. (2024). Enhancing Drought Tolerance in Broccoli (Brassica oleracea L.) through Melatonin Application: Physiological and Biochemical Insights into Growth, Photosynthesis, and Antioxidant Defense Mechanisms. *Biocatalysis and Agricultural Biotechnology*, •••, 103256.

Segal, D., & Dostie, J. (2023). The talented LncRNAs: Meshing into transcriptional regulatory networks in cancer. *Cancers (Basel)*, *15*(13), 3433.

Sharma, M., Sidhu, A. K., Samota, M. K., Gupta, M., Koli, P., & Choudhary, M. (2023). Post-Translational Modifications in Histones and Their Role in Abiotic Stress Tolerance in Plants. *Proteomes*, *11*(4), 38.

Sheoran, S., Kaur, Y., Kumar, S., Shukla, S., Rakshit, S., & Kumar, R. (2022). Recent advances for drought stress tolerance in maize (Zea mays L.): Present status and future prospects. *Frontiers in Plant Science*, *13*, 872566.

Singh, A., Pandey, H., Pandey, S., Lal, D., Chauhan, D., Aparna, , & Antre, , S. HKumar, , A. (2023). Drought stress in maize: Stress perception to molecular response and strategies for its improvement. *Functional & Integrative Genomics*, *23*(4), 296.

Singh, P., Dutta, P., & Chakrabarty, D. (2021). miRNAs play critical roles in response to abiotic stress by modulating cross-talk of phytohormone signaling. *Plant Cell Reports*, *40*(9), 1617–1630.

Singh, R. K., & Prasad, M. (2022). Delineating the epigenetic regulation of heat and drought response in plants. *Critical Reviews in Biotechnology*, *42*(4), 548–561.

Sintaha, M., Man, C.-K., Yung, W.-S., Duan, S., Li, M.-W., & Lam, H.-M. (2022). Drought stress priming improved the drought tolerance of soybean. *Plants*, *11*(21), 2954.

Soma, F., Takahashi, F., Yamaguchi-Shinozaki, K., & Shinozaki, K. (2021). Cellular phosphorylation signaling and gene expression in drought stress responses: ABA-dependent and ABA-independent regulatory systems. *Plants*, *10*(4), 756.

Song, J., Henry, H., & Tian, L. (2020). Drought-inducible changes in the histone modification H3K9ac are associated with drought-responsive gene expression in Brachypodium distachyon. *Plant Biology*, *22*(3), 433–440.

Sun, X., Xiang, Y., Dou, N., Zhang, H., Pei, S., Franco, A. V., Menon, M., Monier, B., Ferebee, T., & Liu, T. (2023). The role of transposon inverted repeats in balancing drought tolerance and yield-related traits in maize. *Nature Biotechnology*, *41*(1), 120–127.

Tan, X., Li, S., Hu, L., & Zhang, C. (2020). Genome-wide analysis of long non-coding RNAs (lncRNAs) in two contrasting rapeseed (Brassica napus L.) genotypes subjected to drought stress and re-watering. *BMC Plant Biology*, *20*, 1–20.

Tao, X., Li, M., Zhao, T., Feng, S., Zhang, H., Wang, L., Han, J., Gao, M., Lu, K., & Chen, Q. (2021). Neofunctionalization of a polyploidization-activated cotton long intergenic non-coding RNA DAN1 during drought stress regulation. *Plant Physiology*, *186*(4), 2152–2168.

Temel, A., Janack, B., & Humbeck, K. (2017). Drought stress-related physiological changes and histone modifications in barley primary leaves at HSP17 gene. *Agronomy (Basel)*, *7*(2), 43.

Tyagi, A., Mir, Z. A., Almalki, M. A., Deshmukh, R., & Ali, S. (2024). Genomics-Assisted Breeding: A Powerful Breeding Approach for Improving Plant Growth and Stress Resilience. *Agronomy (Basel)*, *14*(6), 1128.

Ueda, M., & Seki, M. (2020). Histone modifications form epigenetic regulatory networks to regulate abiotic stress response. *Plant Physiology*, *182*(1), 15–26.

Ullah, Z., Iqbal, J., Abbasi, B. A., Akhtar, W., Kanwal, S., Ali, I., & Mahmood, T. (2023b). Assessment of Gus Expression Induced by Anti-Sense Os PPO Gene Promoter and Antioxidant Enzymatic Assays in Response to Drought and Heavy Metal Stress in Transgenic Arabidopsis thaliana. *Sustainability*, *15*(17), 12783.

Ullah, Z., Iqbal, J., Abbasi, B. A., Ijaz, S., Sharifi-Rad, J., Yaseen, T., & Mahmood, T. (2023a). How Genome Editing Can Be Helpful in the Biofortification of Legumes. In *Legumes Biofortification* (pp. 207–232). Springer International Publishing.

Wang, L., Wu, Y., Tian, Y., Dai, T., Xie, G., Xu, Y., & Chen, F. (2020). Overexpressing Jatropha curcas CBF2 in Nicotiana benthamiana improved plant tolerance to drought stress. *Gene*, *742*, 144588.

Wei, B., Wang, Y., Ruan, Q., Zhu, X., Wang, X., Wang, T., Zhao, Y., & Wei, X. (2024). Mechanism of action of microRNA166 on nitric oxide in alfalfa (Medicago sativa L.) under drought stress. *BMC Genomics*, *25*(1), 316.

Woolfson, K. N., Esfandiari, M., & Bernards, M. A. (2022). Suberin biosynthesis, assembly, and regulation. *Plants*, *11*(4), 555.

Zhang, D., Gan, Y., Le, L., & Pu, L. (2024). Epigenetic variation in maize agronomical traits for breeding and trait improvement. *Journal of Genetics and Genomics*.

Zhang, H., Lang, Z., & Zhu, J.-K. (2018). Dynamics and function of DNA methylation in plants. *Nature Reviews. Molecular Cell Biology*, *19*(8), 489–506.

Zhang, Y., Wu, X., Wang, X., Dai, M., & Peng, Y. (2024). Crop root system architecture in drought response. *Journal of Genetics and Genomics*.

Zhang, Y., Zhou, Y., Zhu, W., Liu, J., & Cheng, F. (2022). Non-coding RNAs fine-tune the balance between plant growth and abiotic stress tolerance. *Frontiers in Plant Science*, *13*, 965745.

Zhang, Z., Yang, T., Li, N., Tang, G., & Tang, J. (2024). MicroRNA166: Old Players and New Insights into Crop Agronomic Traits Improvement. *Genes*, *15*(7), 944.

Zhao, T., Zhan, Z., & Jiang, D. (2019). Histone modifications and their regulatory roles in plant development and environmental memory. *Journal of Genetics and Genomics*, *46*(10), 467–476.

Zi, N., Ren, W., Guo, H., Yuan, F., Liu, Y., & Fry, E. (2024). DNA methylation participates in drought stress memory and response to drought in Medicago ruthenica. *Genes*, *15*(10), 1286.

Zulfiqar, A., Fatima, R., Ahmed, S., Saleem, A., Sardar, R., Ahmad, M. N., & Yasin, N. A. (2023). Mechanistic insights into the interaction of fluoride resistant bacteria with wheat roots toward enhancing plant productivity by alleviating fluoride stress. *Fluoride*, *56*(3).

Zulfiqar, A., Mushtaq, I., Ahmed, S., Ali, M., Saleem, A., Sadar, R., & Ahmad, M. N. (2022). Fluoride resistant bacteria alleviate fluoride stress in Triticum aestivum l. Through modulating gas exchange characteristics and enhanced plant growth. *Fluoride*, *55*(3), 280–294.

Chapter 19
Epigenetic Mechanisms Underlying Salinity Tolerance in Plants

Muhammad Zeeshan Arshed
University of the Punjab, Pakistan, Pakistan

Asia Maqbool
University of the Punjab, Pakistan, Pakistan

Ammara Saleem
University of the Punjab, Pakistan, Pakistan

Farishta Zarshan
http://orcid.org/0009-0007-8445-2950
Bacha Khan University, Charsadda, Pakistan, Pakistan

Anum Ijaz
University of the Punjab, Pakistan, Pakistan

Madiha Imtiaz
Bacha Khan University, Charsadda, Pakistan, Pakistan

ABSTRACT

Stress by salts is one of the important environmental factors that negatively impacts plant growth, development and yields especially on cereal crops. The use of DNA methylation, histone modification and chromatin remodeling in plant responses to salinity stress cannot be overemphasized. These mechanisms regulate genes that allow plants to cope with osmotic, ionic and oxidative stresses under salt stress that can cause salt sensitivity or tolerance in crops. Additionally, cell signaling under salinity stress leads to formation of stress memory which improves current and future coping with similar stress. Finally, this chapter is focused on discussing the possibilities of epigenetics to improve salinity stress tolerance and crop adaptation as well as evolution of salt tolerant crops. Knowledge of epigenetic regulation of salinity stress may enhance crop productivity in salt affected areas to support sustainable agriculture an food security.

DOI: 10.4018/979-8-3693-8648-4.ch019

1. INTRODUCTION

The numerous environmental stresses that affect plants are threats that compromise the plant growth, yield, and existence in daily basis (Ahmad et al., 2022a, Arshed et al., 2024). Of these, biotic stress factors consist of salinity stresses, drought stress, high and low temperature, and toxic levels of heavy metals, all of which are major forces that hinder agricultural productivity at global level (Figure 1). Among the factors, salinity alone can be indicated that it is one of the most dangerous stresses which influences approximately 20% irrigated land in the world. This stress mainly originate from due to the development of soluble salts in the soil, which may be generated by natural evolutions including the weathering of rocks or man-made events including the frequent use of irrigation and inadequate drainage systems (Ahmad et al., 2024; Akhter et al., 2021).

Salinity involves two elements which hurt plants: osmotic stress and ion toxicity. These are conditions whereby availability of water in the soil to the plant reduces due to high salts therefore the term osmotic stress is used since water has been reduced to plants in the same way it is reduced in a stress (Isayenkov & Maathuis, 2019). At the same time, ionic toxicity occurs when plant tissues are overloaded with ions which include sodium and chloride affecting cellular and metabolic balance. All these stresses accumulate their effects and lead to low photosynthesis rates, slow growth as well poor crop yields (Farouk & Al-Huqail, 2022).

A molecular and cellular level, high salinity caused oxidative stress through production of ROS which can affect proteins, lipids and DNA (Zulfiqar et al., 2022; Zulfiqar, Naseer, et al., 2023). Besides, salinity interferes with the uptake of the nutrients by competing with important ions such as potassium ion (K^+) (Isayenkov & Maathuis, 2019). The seedlings adapt to salinity stress through god's physiologic, chemical and molecular behaviors such as ion distribution, osmotic adjustment, antioxidant mechanisms and stress signaling genes activation. Due to increased evidence of the dangers of salinity to food production, there is considerable interest in the processes by which plants respond to saline environments (Chen et al., 2020; Zulfiqar, Fatima, et al., 2023).

Figure 1. Overview of different abiotic stresses in plants.

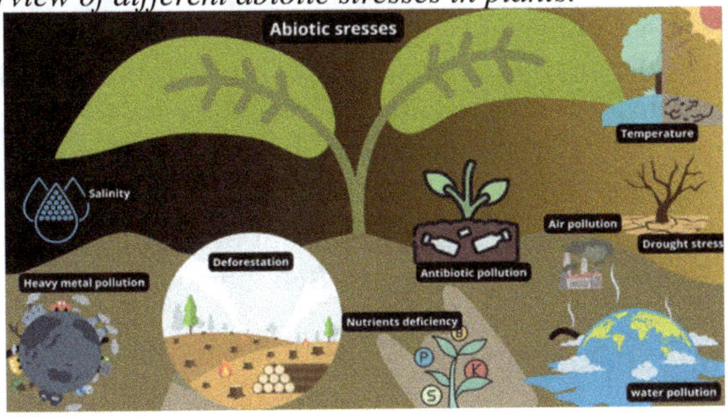

2. PLANT RESPONSE TO SALINITY STRESS

In response to salinity stress, plants have shown diverse features morphologically, physiologically and biochemically to work and cope with unfavorable conditions. These responses are facilitated by elaborate processes that enable the prevention of osmotic, ionic and oxidative impacts of high salinity (Zhang & Dai, 2019).

2.1. Morphological Responses

Stress can affect plant morphology notably, and one of the areas of manifestation includes salinity stress that causes minimum growth, lesser leaf surface area and ill-developed root systems. Osmotic stress produced by saline condition hamper water uptake thus slows down cell expansion mostly the shoot elongation (Soltabayeva et al., 2021). For example, in rice crop, high salinity affects the total tiller number and plant height, thus the biomass yield (Ghosh et al., 2016). Some also adjust the root systems through additional branching or going deep in the ground in search of low saline water supply in some species. Nevertheless, high concentration of salt within the root zone causes necrosis and additional impairment of roots growth and activity (Zhang & Dai, 2019).

2.2. Physiological Responses

He concluded that low level physiological criteria regarding salinity depend on a balance of water in and out of stomata as well as balancing the rates of photosynthesis and ion absorption. The first effect seen is a decrease in water consumption leading to something similar to a water-deficit stress which leads to stomatal closure to reduce water loss (Ghosh et al., 2016). This, however, reduces CO_2 uptake and decreases rates of photosynthesis in crops such as wheat and maize. Ion uptake and compartmentalization also modified in plants for nontoxic levels of sodium (Na^+) and chloride (Cl^-) (Balasubramaniam et al., 2023). For example, halophytes pump excess of salt into the vacuoles with the aim, of avoiding ion toxicity in the cytoplasm. Another is the retention of potassium which is important in enzymatic activities and osmoregulation under stress (Balasubramaniam et al., 2023).

2.3. Biochemical Responses

Molecular mechanism of stress tolerance includes antioxidant defence system and osmolyte in plants in response to ROS under salinity stress. Others of small osmolytes include proline, glycine betaine, trehalose to regulate osmotic stress and stabilize proteins and membranes (Saleem et al., 2022). For instance, tomatoes also enhance accumulation of proline under saline environments to help preserve membranes and enzymes. Antioxidant enzymes like superoxide dismutase (SOD), catalase (CAT), and peroxidase (POD) are activated to neutralize ROS, reducing oxidative damage to proteins and lipids. Plants also alter secondary metabolite production, such as phenolics and flavonoids, to enhance stress tolerance (Alshegaihi et al., 2023; Saleem et al., 2024).

2.4. Adaptive Strategies

Phytohormones such as abscisic acid (ABA), ethylene, and jasmonic acid (JA) play pivotal roles in modulating plant responses to salinity (Fahad et al., 2015). ABA, for instance, is crucial for stomatal regulation and activating stress-responsive genes. Additionally, plants may reallocate energy resources toward reproductive processes, ensuring seed production as a survival strategy (Riemann et al., 2015).

Advances in omics technologies have provided deeper insights into the genetic and molecular mechanisms underlying salt tolerance. Understanding these pathways has become vital for developing salt-tolerant crops through genetic engineering and biotechnological interventions (Ullah et al., 2022).

Overall, plants' responses to salinity are multi-faceted, involving intricate cross-talk between morphological adaptations, physiological adjustments, and biochemical defenses to thrive under challenging environments. Table 1 below table showcasing the effects of salinity stress.

Table 1. The effects of salinity stress on various plants and their observed responses.

No.	Plant Species	Effect of Salinity Stress	Key Findings and Observations	References
1	Wheat (*Triticum aestivum*)	Reduced germination rate and root elongation	Sodium accumulation in leaves, leading to chlorosis	(El Sabagh et al., 2021)
2	Rice (*Oryza sativa*)	Decreased tillering and grain yield	Sensitive at seedling and reproductive stages	(Hoang et al., 2016)
3	Tomato (*Solanum lycopersicum*)	Stunted growth and fruit quality reduction	Reduced potassium uptake and increased osmotic stress	(Ro ca et al., 2023)
4	Soybean (*Glycine max*)	Reduced nodule formation and nitrogen fixation	Decreased seed yield and altered amino acid metabolism	(Hasanuzzaman et al., 2022)
5	Maize (*Zea mays*)	Declined photosynthesis and plant height	Reduced chlorophyll synthesis and increased proline	(Sabagh et al., 2021)
6	Barley (*Hordeum vulgare*)	Enhanced tolerance with reduced yield	Accumulation of osmoprotectants like glycine betaine	(Jadidi et al., 2022)
7	Cotton (*Gossypium hirsutum*)	Decreased boll weight and lint quality	Altered ionic balance affecting growth	(Maryum et al., 2022)
8	Canola (*Brassica napus*)	Reduced oil content and biomass	Increased antioxidant enzyme activity as a defense	(Naveed et al., 2020)
9	Sugarcane (*Saccharum officinarum*)	Reduced cane length and sucrose content	Chloride and sodium toxicity leading to metabolic changes	(Simoes et al., 2020)
10	Potato (*Solanum tuberosum*)	Reduced tuber yield and starch content	Disruption of water uptake and leaf wilting	(Sabbour & Hussein, 2024)
11	Grapevine (*Vitis vinifera*)	Poor berry set and reduced sugar accumulation	Decreased photosynthetic rate and impaired growth	(Al-Taey & abd Al-Ameer, 2023)
12	Alfalfa (*Medicago sativa*)	Decreased forage quality and leaf drop	Reduced leaf area and impaired protein synthesis	(Bhattarai et al., 2020)
13	Sunflower (*Helianthus annuus*)	Stunted growth and lower seed yield	High Na+ accumulation affecting cellular function	(Abd El-Hameid & Sadak, 2020)
14	Lettuce (*Lactuca sativa*)	Reduced leaf area and chlorosis	Reduced water content and quick bolting	(Shin et al., 2020)
15	Broccoli (*Brassica oleracea*)	Reduced head formation and plant vigor	Stressed plants show lower nutrient uptake	(Ali et al., 2022)
16	Lentil (*Lens culinaris*)	Lower seed set and plant height	Ion toxicity and oxidative stress response	(Singh et al., 2020)
17	Chickpea (*Cicer arietinum*)	Stunted growth and fewer pods	Decreased photosynthetic efficiency and metabolic changes	(Yousef et al., 2020)
18	Sorghum (*Sorghum bicolor*)	Reduced plant height and leaf expansion	Salt exclusion mechanism leading to better adaptation	(Mansour et al., 2021)
19	Spinach (*Spinacia oleracea*)	Chlorosis and leaf curling	High salt retention causing cellular damage	(Ramezanifar et al., 2022)
20	Pea (*Pisum sativum*)	Decreased biomass and delayed flowering	Lower water-use efficiency and proline accumulation	(Yousef et al., 2020)

3. EPIGENETICS IN PLANTS: AN OVERVIEW

Epigenetics refers to the study of heritable changes in gene expression that occur without altering the underlying DNA sequence. Imprinting, that is changes that are transmitted through generations but are not directly encoded on the genome are known as epigenetics (Ahmad et al., 2022b, Inbar-Feigenberg et al., 2013). It is worth emphasizing that these changes are regulated by a large number of molecular factors: DNA methylation, histone modifications, chromatin remodeling and regulation of gene expression through non-coding RNA (Haider et al., 2021a, Jablonka & Lamb, 2002). In simple terms, epigenetic modifications facilitate the relationship between stimuli in the haplophytic plant environment and the operation of the plant genome so that plants can survive changing conditions (Inbar-Feigenberg et al., 2013, Ahmad et al., 2024a). In contrast to genetic changes that are permanent, epigenetic modifications are reversible and therefore a dynamic process by which plants respond to abiotic stress including salinity, drought and extremes in temperature.

Abiotic stress, including salinity, imposes a significant burden on plants by disrupting physiological and metabolic processes. In response, plants employ epigenetic modifications to regulate stress-responsive genes, optimize resource allocation, and enhance tolerance mechanisms (Miryeganeh, 2021, Haider et al., 2021b). Understanding the interplay between epigenetics and abiotic stress responses provides valuable insights for improving stress resilience in crops. Beyond salinity, epigenetics underpins plant responses to a wide range of abiotic stresses, including drought, heat, cold, and heavy metal toxicity (Miryeganeh, 2021, Ahmad et al., 2024b). Each stress triggers unique epigenetic reprogramming, as well as shared mechanisms, to optimize plant survival. Epigenetic mechanisms regulate genes involved in water uptake, stomatal regulation, and osmotic adjustment. DNA methylation and histone acetylation modulate salinity-responsive genes, enabling plants to conserve water and tolerate dehydration (Alonso et al., 2019, Ali et al., 2022). Epigenetic changes under heat or cold stress influence genes associated with heat shock proteins and cold acclimation. Histone modifications and miRNA activity facilitate rapid adjustments in gene expression to maintain cellular integrity. Toxicity induced by heavy metals such as cadmium and arsenic is mitigated through epigenetic regulation of genes involved in metal sequestration, antioxidant defense, and detoxification pathways.

4. OVERVIEW OF EPIGENETICS AND SALINITY STRESS

Salinity stress is a major abiotic challenge, particularly in arid and semi-arid regions where soil salinization is prevalent. It exerts osmotic stress, ionic toxicity, and oxidative damage on plants, leading to reduced growth, lower photosynthetic efficiency, and impaired productivity (Banerjee & Roychoudhury, 2017, Ijaz et al., 2023). Epigenetics plays a critical role in modulating plant responses to salinity by fine-tuning the expression of stress-responsive genes and pathways (Figure 2).

Figure 2. Epigenetic mechanism in plants exposed to salt stress.

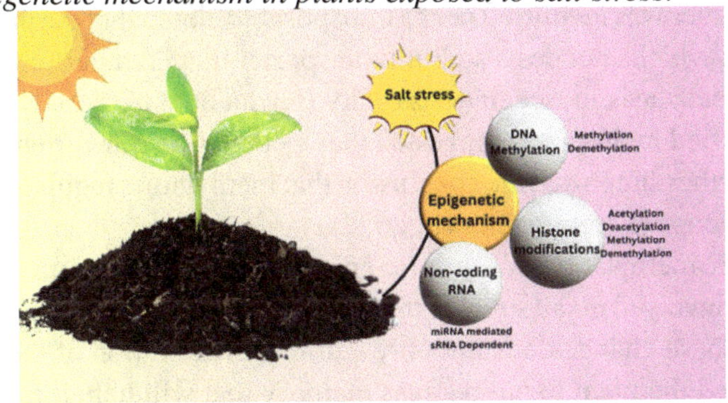

4.1. DNA Methylation and Its Role in Salinity Stress in Plants

DNA methylation is an important epigenetic event owning to its role in controlling the gene expression in plants especially under stress conditions such as salinity. It primarily involves the addition of a methyl group to specific DNA bases, commonly 5-methylcytosine (5mC), across three sequence contexts: Available options included CG, CHG, and CHH where the H can be A, T, or C (Rashid et al., 2022). These methylation patterns are mediated by specific methyltransferase enzymes: As for CG context, MET1 was significantly up-regulated; for CHG context, CMT3 and CMT2 were up-regulated; for CHH context, DRM2 was up-regulated. However, deamination, which counteracts DNA methylation, is catalyzed by DNA demethylases, inclusive of ROS1 and DML2 (Mohamed, 2022; Saradadevi et al., 2021, Haider et al., 2022a).

A physiological phenomenon observed under salinity stress is the fluctuation in DNA methylation signature to regulate gene expression. For instance, methylation of promoter regions is negatively linked with gene activity, while gene body methylation enhances gene expression. This dual regulatory role enables plant to temporal regulate of stress-responsive genes (Mohamed, 2022, Ijaz et al., 2023).

The extent to which salt-induced methylation occurs differs with the type of plant and its tissues. In *Hordeum vulgare* (barley), authors have reported that salt stress induced more DNA methylation in leaves rather than root tissues showing that plants have organ specific memory (Konate et al., 2018). Likewise, salinity stress affected methylation levels in *Glycine max* (soybean) roots.All contexts namely CG, CHG, and CHH were less methylated than corresponding controls, indicating species wise differences in methylome changes (Sun et al., 2022, Haider et al., 2021b).

Thus, it has been demonstrated that the rice *O. sativa* utilizes methylation to modulate primary salt tolerance genes. For example, among CHH and CHG methylation, the former was identified near a transposable element in the promoter region of OsHKT1;5, which encodes a sodium transporter that is important for regulation of Na^+/K^+ homeostasis in rice under salinity (Karan et al., 2012). Higher methylation was identified in this area, suggested that methylation down-regulated the gene expression, further supported the conclusion that methylation regulates ion transport and stress response (Pérez-Pizá et al., 2022).

In all, DNA methylation is a plastic epigenetic process that may up or down-regulate genes involved in salt stress response depending on the adaptive requirements of the plant. These change does not only aid in the restoration of the homeostasis, but also allowed the plant to build stress memory that will help it to better address future issues concerning salinity (Mancia, 2022, Ijaz et al., 2024a).

4.2. Histones Modification and Their Role in Salinity Stress

Histones are proteins that serve as the scaffolding of chromatin materials and are used as a base to which DNA compounds (Karan et al., 2012). They are mostly composed of lysine and arginine amino acids, a characteristic that imparts a positive charge to this protein so that it can associate with negative DNA molecules. C-terminal ends of histone molecules, particularly at N-terminal regions, are preferred targets for chemical restructuring such as acetylation, methylation, ubiquitination and phosphorylation. These modifications dictate how many turns of DNA occurs around the histones effectively controlling genes (Wang et al., 2021, (Mancia, 2022, Ijaz et al., 2024b).

In different plants, Histone modifications are proved to have an essential role in regulating the stress responsive genes under salinity stress. For instance, variations in histone structure have been associated with the variations in the expression of OsBZ8 gene in salt-tolerant and salt-sensitive rice genotypes (Paul et al., 2017). Chromatin remodeling thus allows the plant to prepare for future salinity stress by adding or removing importance gene functions cues to histones.

Histone-modifying enzymes can alter the structure and charged characters of the histones, by means of acetylation, for instance, on H3K9 and H4 the binding affinity of DNA to histones is weakened. This allows the chromatin structure to become less compact, thus DNA becomes more accessible to transcriptional machinery and expression is enhanced. There are the enzymes, histone acetyltransferases (HATs) that put the acetyl group and histone deacetylases (HDACs) that remove it, so establish a regulatory cycle (Gagnidze & Pfaff, 2022, Ullah et al., 2023a).

In *Nicotiana tabacum* (tobacco), *Zea mays* (maize), and *Beta vulgaris* (sugar beet) plants, histone acetylation has been observed to be linked with the inducement of salt-responsive genes. For instance, the acetylation of peroxidase gene promoters in *Beta vulgaris* leads to the formation of antioxidant enzymes in order to minimize accumulation of ROS and improved salt tolerance (Weyen, 2021, Ullah et al., 2023b).

For instance in *Arabidopsis thaliana*, the genes regulated by salt stress acetylation encompass AtSOS1 and AtSOS3 that enables the of the chromatin and activation of pathways that involves ion homeostasis (Ketehouli et al., 2019). Likewise, polyploid wheat, TaHAG bound enzyme regulates histone 3 acetylation genes to enhance ROS scavenging and osmotic protectant as key components of salinity tolerance (Du et al., 2023). Histone modifications indicates there exists significant area for genetic manipulation of histones to enhance stress. They consist primarily of lysine and arginine, which give them a positive charge, allowing interaction with negatively charged DNA. The tails of histones, located at the N-terminal, are hotspots for chemical modifications such as acetylation, methylation, ubiquitination, and phosphorylation. These modifications influence how tightly DNA is wrapped around histones, thereby regulating gene activity (Du et al., 2023; Gagnidze & Pfaff, 2022).

Under salinity stress, histone modifications play a critical role in controlling the expression of stress-responsive genes in various plants. For example, changes in histone structure have been linked to differences in the expression of the OsBZ8 gene in salt-tolerant and salt-sensitive rice varieties. By modifying histones, plants can "prime" themselves to better handle salinity stress by activating or repressing critical genes (Paul et al., 2017).

4.3. Histone Acetylation

Acetylation of histones, particularly on lysine residues like H3K9 and H4, reduces the binding strength between DNA and histones (Müller & Muir, 2015). This loosens the chromatin structure, making DNA more accessible to transcriptional machinery and leading to increased gene expression. Enzymes called histone acetyltransferases (HATs) add acetyl groups, while histone deacetylases (HDACs) remove them, providing a dynamic regulatory system (Hodawadekar & Marmorstein, 2007).

In plants like *Nicotiana tabacum* (tobacco), *Zea mays* (maize), and *Beta vulgaris* (sugar beet), increased histone acetylation has been associated with the activation of salt-responsive genes. For instance, acetylation of peroxidase gene promoters in *Beta vulgaris* triggers the production of antioxidant enzymes, reducing reactive oxygen species (ROS) accumulation and enhancing salt tolerance (Feng et al., 2022; Nicolas-Francès et al., 2018).

In *Arabidopsis thaliana*, salt stress-induced acetylation of genes like AtSOS1 and AtSOS3 relaxes chromatin, facilitating the activation of pathways essential for maintaining ion homeostasis (Singroha et al., 2022). Similarly, in polyploid wheat, the TaHAG enzyme promotes H3 acetylation, upregulating genes involved in ROS detoxification and osmotic regulation, aiding in salinity tolerance. These histone modifications offer potential targets for genetic manipulation to improve stress (Singroha et al., 2022).

4.4. Role of Histone Acetyltransferases (HATs) and Histone Deacetylases (HDACs) in Salinity Stress

In previous study it is established that while histone acetyltransferases (HATs) promote the histone acetylation required for plant response to salinity stress, histone deacetylases (HDACs) remove the acetyl groups thus limiting favorable plant responses (Table 2) (Schneider et al., 2013). HATs catalyse the addition of acetyl groups to histones which weakens the chromatin structure and triggers gene expression. On the other hand, HDACs can take off these groups resulting in a tight coiling of chromatin and suppression of gene (Lee et al., 2021).

Several researchers have attributed a role for HATs in the regulation of salt-responsive genes where they have been shown to increase histone acetylation at specific promoter and coding region of genes such as ZmEXPANSIN B2 and ZmXYLOGLUCAN in maize and GCN5 in Arabidopsis thaliana. Through QRT-PCR analysis Wheat's TaHAG1 moderates genes which are implicated in detoxification of ROS and osmotic balance under salt stress (Thakur & Prasad, 2024; Ueda & Seki, 2020).

However, HDAC often plays a negative role in maintaining salt tolerance gene by decreasing the acetylation of histone, OsSOS1 and LEA1 in rice etc (He et al., 2020; Zafar et al., 2017). This paper has identified the specific roles that individual classes and members of Arabidopsis HDACs play – whereas class I HDACs such as HDA19 is beneficial for salt tolerance, class II HDACs such as HDA5 and HDA15 are detrimental to stress tolerance (Ueda et al., 2017). In rice, HDA710 represses salt stress genes by regulate acetylation of H4K5 and H4K16. Regulatory mechanisms of Histone Acetyltransferases (HATs) and Histone Deacetylases (HDACs) that modulated histone acetylation mobiles form an essential baseline for enhancing salinity tolerance using improved crop improvement techniques (Ullah et al., 2021).

Table 2. Role of HATs and HDACs in Salinity Stress

Category	Examples	Role	Plant Species	Mechanism	References
HATs	ZmHATB, GCN5, TaHAG1	Activate salt-responsive genes by increasing histone acetylation	*Maize, Wheat, Arabidopsis*	Promote chromatin relaxation, upregulate genes related to cell wall biosynthesis, ROS detoxification, and osmotic regulation.	(Sharma et al., 2023)
HDACs	OsHDA1, HDA710	Repress salt tolerance genes by reducing histone acetylation	*Rice, Hibiscus*	Compact chromatin structure, downregulate stress-responsive genes (*SOS1, LEA1*) by altering acetylation at H4K5 and H4K16.	(Yung et al., 2021)
Dual Roles	HDA19, HDA5/15	Class I HDACs promote positive salinity responses, while class II HDACs negatively influence stress responses	*Arabidopsis*	Fine-tune histone acetylation to balance stress signaling and salt stress responses.	(Sanjuan-Badillo et al., 2024)

4.5. Role microRNAs in Salinity Stress

MicroRNAs (miRNAs) and Small Interfering RNAs (siRNAs) plays a central role in genes regulation in plants especially during stress such as salinity (Table 3). Small non-coding RNA molecules, (20–24 nucleotide) are significant for regulating different biological processes and stress response pathways. Participation in the regulation of the plant response to salinity has attracted interest because of their use in enhancing crop tolerance (Khraiwesh et al., 2012).

4.5.1. Biogenesis of miRNAs

MiRNAs originate from the primary miRNA (pri-miRNA) that is produced from the miRNA (MIR) genes by the DNA polymerase II enzyme (Zhang et al., 2015). These pri-miRNAs fold into precursor miRNAs forming a stem loop. The pre-miRNA is then cleaved by DCL1 protein yielding a miRNA–miRNA* duplex (20–22 nt), out of which the miRNA strand remains functional in gene silencing while the miRNA* strand is usually target doomed for degradation (Jodder, 2021). The mature miRNA then becomes methylated by HUA ENHANCER1 (HEN1), and transported from nucleus to cytoplasm by exportin proteins. Here, the miRNA unfolds to single strand and gets incorporate in the RNA-induced silencing that is contained in the complex (RISC), and where the target mRNAs are rendered inapt through complementary base pairing (Sanobar et al., 2021).

4.5.2. Specific Small Interfering RNAs (siRNAs)

Like miRNAs, another class of small RNAs called SiRNAs are 20–25nt in length and also function by interacting with the same pathways (Bhan et al., 2017). They are transcribed from dsRNA precursors, which can be of viral and non-viral in origin including siRNA targeting products of miRNA mediated cleavage (Samarskaya et al., 2023). These dsRNAs are then cleaved by RNA-dependent RNA polymerases (RDRs) and DCL proteins into siRNAs which are used in gene silence. Subcategories of siRNAs are ta-siRNAs, which result from miRNA-guided cleavage of target mRNA; and nat-siRNAs that originate from cis-antisense gene pairs (Chowdhury, 2018).

4.5.3. Role in Stress Response

Thus it turns out that miRNAs and siRNAs are the two classes of non-coding RNA molecules that regulate the responses of plants in any contexts especially under salinity stress. They act as modulators of a number of important stress related genes such as those involved in ion homoeostasis, transcription and signal transducing systems. Thus, these small RNAs can also regulate augmenting or inhibiting specific mRNAs to enable plants to cope up with high saline stress (Yadav et al., 2024).

4.5.4. Plant-Specific Salt Responses

***Arabidopsis thaliana*:** In salt stress, plant expression of some key miRNAs include miR156, miR158, miR159, miR167 and miR396 with up-regulation while miR398 has down-regulation. These miRNAs have roles in modulating stress adaptation genes such as transcription factor and other target genes (Xie et al., 2005).

Radish (*Raphanus sativus*): New salt-responsive miRNAs that control plant development, ion homeostasis and signaling circuits under salt stress have been discovered (Sun et al., 2015).

Flax (*Linum usitatissimum*): In total, the roles of 29 miRNA-target pairs were confirmed in salt-tolerant pathways, and these two molecules had reverse expression patterns under salinity (Paliwal et al., 2023).

Sugarcane (Saccharum sp.): Comparable alteration of miRNA responses to salt stress were observed, such as changes in miR156, miR159, miR166, and miR396, which differ in terms of species, stress and growth conditions (Mazalmazraei et al., 2023).

4.5.5. Role of Small RNAs to Regulation in Glycophytes and Halophytes

Many salt-responsive miRNAs have been identified in glycophytic plants such as *A. thaliana, O. sativa, T. aestivum*, and *Z. mays* as well as in halophytic plants such as *Thellungiella halophila* and *Spartina alterniflora*. These plant miRNAs are involved in processes of ion homeostasis, transcription, and signal transduction pertinent to plant salt tolerance. These data demonstrate species-specific regulation of miRNA in glycophytic and halophytic plants with regard to salt stress (Himabindu et al., 2016; Kumar et al., 2018).

6. TOOLS TO DISCOVER MIRNA AND FOR MODIFYING PLANTS TO BETTER TOLERATE ABIOTIC STRESS

Along with the technology expanding its external forms of miRNA with next-generation sequencing (NGS), there has been better identification and differentiation of conserved and new miRNAs. Several computational resources and databases with respect to miRNA prediction and target validation have been developed. Due to their action on stress-responsive genes, the use of miRNAs in modifying plants targeting creation of resistance to abiotic stresses including salinity is expected (Hu et al., 2017; Tam et al., 2014).

Table 3. Role of Plant miRNAs and lncRNAs in Salt Stress

Category	Molecule	Plant Species	Target Genes/ Pathways	Response to Salt Stress	Reference
Small RNAs	miR164a/b/c/d	*Zea mays*	NAC, MYB, HD-ZIP transcription factors	Regulates transcription factors involved in salt tolerance	(Ding et al., 2010)
Small RNAs	miR156	*A. thaliana, Z. mays*	Transcription factors	Up-regulated in *A. thaliana*; down-regulated in *Z. mays*	(Zheng et al., 2019)
Small RNAs	miR396	*A. thaliana, Z. mays, O. sativa*	Transcription factors	Up-regulated in *A. thaliana* and *Z. mays*; down-regulated in *O. sativa*	(Bari et al., 2014; Yang et al., 2013)
Small RNAs	miR-250, miR-205, miR-330, miR-17	*Z. mays*	Casein kinase II, GPX, P5CS, IF-1	Down-regulated under salt stress, enhancing expression of target genes essential for survival	(Fu et al., 2017)

continued on following page

Table 3. Continued

Category	Molecule	Plant Species	Target Genes/ Pathways	Response to Salt Stress	Reference
24-nt siRNAs	-	*A. thaliana*	MYB74	Regulates MYB74 transcription via RdDM pathway under salt stress	(Xu et al., 2015)
Long Non-coding RNAs	NPC60	*Glycine max*	Salt-responsive	Expression increased 100x under salt stress	(Das et al., 2023)
Long Non-coding RNAs	lncRNA973	*Cotton*	Salt-responsive genes	Overexpression enhances salt tolerance by modulating salt-responsive genes	(Das et al., 2023)
Long Non-coding RNAs	Tissue-specific lncRNAs	*Poplar*	Species-specific salt responses	Differential expression under varying salt conditions	(Ma et al., 2019)

7. CONCLUSION AND FUTURE PROSPECTS

Epigenetic regulation is particularly involved in directing how plants respond to salt stress, and it offers a strong tool with which to improve salt tolerance in plants. Stress responsive genes are regulated by using epigenetic features that involve the DNA methylation, histone modification, and chromatin remodeling. For instance, some of these changes can help plants change their ability to deal with salt stress, by modifying gene accessibility and usage patterns.

The possibility to have the epigenetic marks inherited from one generation to another is an advantage for plants to deal with environmental stress factors. Epigenetic changes are quicker than genetic mutations and most importantly, they cumulative and reversible making it easier for an individual to adapt to stress. But these changes are constrained by the plant's genome, so both genetic and epigenetic changes are at play at the same time.

Further research should investigate the concrete epigenetic processes that modulate gene expression which is affected by salt stress. Further work should be done to find out exactly how these changes to the epigenome are passed on from one generation to the next and on whether these changes enhance stress resilience in plants over the long term.

REFERENCES

Abd El-Hameid, A. R., & Sadak, M. S. (2020). Impact of glutathione on enhancing sunflower growth and biochemical aspects and yield to alleviate salinity stress. *Biocatalysis and Agricultural Biotechnology*, *29*, 101744.

Ahmad, M. A., Saleem, A., Tahir, M., Khilji, S. A., Sajid, Z. A., Landry, K. B., El-Sheikh, M. A., & Ahmad, P. (2024). Modulation of the polyamines, osmolytes and antioxidant defense system to ameliorate drought stress tolerance in Hordeum vulgare L. using ascorbic acid. *South African Journal of Botany*, *171*, 726–736.

Ahmad, Y., Haakim, Z., Iqbal, J., Abbasi, B. A., Mahmood, T., & Kazi, M. (2024b). Technological Innovations for Abiotic Stress Resistance in Horticultural Crops. OMICs-based Techniques for Global Food Security, 233-244.

Ahmad, Y., Haider, S., Iqbal, J., Abbasi, B. A., Yaseen, T., & Mahmood, T. (2022). The mechanisms of genome editing technologies in crop plants. In *Principles and Practices of OMICS and Genome Editing for Crop Improvement* (pp. 295–313). Springer International Publishing.

Ahmad, Y., Haider, S., Iqbal, J., Naseer, S., Attia, K. A., Mohammed, A. A., & Mahmood, T. (2024a). In-silico analysis and transformation of OsMYB48 transcription factor driven by CaMV35S promoter in model plant–Nicotiana tabacum L. conferring abiotic stress tolerance. *GM Crops and Food: Biotechnology in Agriculture and the Food Chain*, *15*(1), 130–149.

Akhter, Z., Bi, Z., Ali, K., Sun, C., Fiaz, S., Haider, F. U., & Bai, J. (2021). In response to abiotic stress, DNA methylation confers epigenetic changes in plants. *Plants*, *10*(6), 1096.

Al-Taey, D. K., & abd Al-Ameer, A. (2023). Effect of Salinity on the Growth and Yield of Grapes: A review. IOP Conference Series: Earth and Environmental Science.

Ali, , LShaheen, , M. RIhsan, , M. ZMasood, , SZubair, , MShehzad, , F. (2022). Growth, photosynthesis and antioxidant enzyme modulations in broccoli (Brassica oleracea L. var. italica) under salinity stress. *South African Journal of Botany*, *148*, 104–111.

Ali, I., Salah, K. B. H., Sher, H., Ali, H., Ullah, Z., Ali, A., & Mahmood, T. (2022). Drought stress enhances the efficiency of floral dip method of Agrobacterium-mediated transformation in Arabidopsis thaliana. *Brazilian Journal of Biology*, *84*, e259326.

Alonso, C., Ramos-Cruz, D., & Becker, C. (2019). The role of plant epigenetics in biotic interactions. *The New Phytologist*, *221*(2), 731–737.

Alshegaihi, R. M., Saleem, M. H., Saleem, A., Ali, B., Aziz, H., Fahad, S., Alataway, A., Dewidar, A. Z., & Elansary, H. O. (2023). Silicon and titanium dioxide mitigate copper stress in wheat (Triticum aestivum L.) through regulating antioxidant defense mechanisms. *Journal of Plant Growth Regulation*, •••, 1–17.

Arshed, M. Z., Zulfiqar, A., & Saleem, A. (2024). Alleviation of Chromium Stress in Brassica juncea L. and Soil Remediation by Plant Growth Promoting (PGP) Bacteria. *Pakistan Journal of Scientific and Industrial Research. Series B: Biological Sciences*, *67*(1).

Balasubramaniam, T., Shen, G., Esmaeili, N., & Zhang, H. (2023). Plants' response mechanisms to salinity stress. *Plants*, *12*(12), 2253.

Banerjee, A., & Roychoudhury, A. (2017). Epigenetic regulation during salinity and drought stress in plants: Histone modifications and DNA methylation. *Plant Gene*, *11*, 199–204.

Bari, A., Sagaidak, A., Pinskii, I., Orazova, S., & Ivashchenko, A. (2014). Binding of miR396 to mRNA of genes encoding growth-regulating transcription factors of plants. *Russian Journal of Plant Physiology: a Comprehensive Russian Journal on Modern Phytophysiology*, *61*(6), 807–810.

Bhan, A., Deb, P., Soleimani, M., & Mandal, S. S. (2017). The Short and Medium Stories of Noncoding RNAs: microRNA and siRNA. *Gene Regulation, Epigenetics and Hormone Signaling*, 137-168.

Bhattarai, S., Biswas, D., Fu, Y.-B., & Biligetu, B. (2020). Morphological, physiological, and genetic responses to salt stress in alfalfa: A review. *Agronomy (Basel)*, *10*(4), 577.

Chen, T.-W., Gomez Pineda, I. M., Brand, A. M., & Stützel, H. (2020). Determining ion toxicity in cucumber under salinity stress. *Agronomy (Basel)*, *10*(5), 677.

Chowdhury, P. (2018). MicroRNA (miRNA) and Small Interfering RNA (siRNA): Biogenesis and Functions in Plants. *Biotic and abiotic stress tolerance in plants*, 323-348.

Das, P., Agarwala, N., Gill, S. S., & Varshney, R. K. (2023). Emerging role of plant long non coding RNAs (lncRNAs) in salinity stress response. *Plant Stress*, •••, 100265.

Ding, M., Hou, P., Shen, X., Wang, M., Deng, S., Sun, J., Xiao, F., Wang, R., Zhou, X., & Lu, C. (2010). Salt-induced expression of genes related to Na+/K+ and ROS homeostasis in leaves of salt-resistant and salt-sensitive poplar species. *Plant Molecular Biology*, *73*, 251–269.

Du, L., Ma, Z., & Mao, H. (2023). Duplicate genes contribute to variability in abiotic stress resistance in allopolyploid wheat. *Plants*, *12*(13), 2465.

El Sabagh, A., Islam, M. S., Skalicky, M., Ali Raza, M., Singh, K., Anwar Hossain, M., Hossain, A., Mahboob, W., Iqbal, M. A., & Ratnasekera, D. (2021). Salinity stress in wheat (Triticum aestivum L.) in the changing climate: Adaptation and management strategies. *Frontiers in Agronomy*, *3*, 661932.

Fahad, S., Hussain, S., Matloob, A., Khan, F. A., Khaliq, A., Saud, S., Hassan, S., Shan, D., Khan, F., & Ullah, N. (2015). Phytohormones and plant responses to salinity stress: A review. *Plant Growth Regulation*, *75*, 391–404.

Farouk, S., & Al-Huqail, A. A. (2022). Sustainable biochar and/or melatonin improve salinity tolerance in borage plants by modulating osmotic adjustment, antioxidants, and ion homeostasis. *Plants*, *11*(6), 765.

Feng, P., Sun, X., Liu, X., Li, Y., Sun, Q., Lu, H., Li, M., Ding, X., & Dong, Y. (2022). Epigenetic regulation of plant tolerance to salt stress by histone acetyltransferase GsMYST1 from wild soybean. *Frontiers in Plant Science*, *13*, 860056.

Fu, R., Zhang, M., Zhao, Y., He, X., Ding, C., Wang, S., Feng, Y., Song, X., Li, P., & Wang, B. (2017). Identification of salt tolerance-related microRNAs and their targets in maize (Zea mays L.) using high-throughput sequencing and degradome analysis. *Frontiers in Plant Science*, *8*, 864.

Gagnidze, K., & Pfaff, D. W. (2022). Epigenetic mechanisms: DNA methylation and histone protein modification. In *Neuroscience in the 21st Century: From Basic to Clinical* (pp. 2677-2716). Springer.

Ghosh, B., Ali, M. N., & Saikat, G. (2016). Response of rice under salinity stress: A review update. *J Res Rice*, *4*(2), 167–174.

Haider, S., Iqbal, J., Naseer, S., Shaukat, M., Abbasi, B. A., Yaseen, T., & Mahmood, T. (2022b). Unfolding molecular switches in plant heat stress resistance: A comprehensive review. *Plant Cell Reports*, *41*(3), 775–798.

Haider, S., Iqbal, J., Naseer, S., Yaseen, T., Shaukat, M., Bibi, H., & Mahmood, T. (2021a). Molecular mechanisms of plant tolerance to heat stress: Current landscape and future perspectives. *Plant Cell Reports*, •••, 1–25.

Haider, S., Iqbal, J., Shaukat, M., Naseer, S., & Mahmood, T. (2021b). The epigenetic chromatin-based regulation of somatic heat stress memory in plants. *Plant Gene*, *27*, 100318.

Haider, S., Raza, A., Iqbal, J., Shaukat, M., & Mahmood, T. (2022a). Analyzing the regulatory role of heat shock transcription factors in plant heat stress tolerance: A brief appraisal. *Molecular Biology Reports*, *49*(6), 5771–5785.

Hasanuzzaman, M., Parvin, K., Anee, T. I., Masud, A. A. C., & Nowroz, F. (2022). *Salt stress responses and tolerance in soybean*. Plant Stress Physiology-Perspectives in Agriculture. IntechOpen.

He, S., Hao, Y., Zhang, Q., Zhang, P., Ji, F., Cheng, H., Lv, D., Sun, Y., Hao, F., & Miao, C. (2020). Histone deacetylase inhibitor SAHA improves high salinity tolerance associated with hyperacetylation-enhancing expression of ion homeostasis-related genes in cotton. *International Journal of Molecular Sciences*, *21*(19), 7105.

Himabindu, Y., Chakradhar, T., Reddy, M. C., Kanygin, A., Redding, K. E., & Chandrasekhar, T. (2016). Salt-tolerant genes from halophytes are potential key players of salt tolerance in glycophytes. *Environmental and Experimental Botany*, *124*, 39–63.

Hoang, T. M. L., Tran, T. N., Nguyen, T. K. T., Williams, B., Wurm, P., Bellairs, S., & Mundree, S. (2016). Improvement of salinity stress tolerance in rice: Challenges and opportunities. *Agronomy (Basel)*, *6*(4), 54.

Hodawadekar, S., & Marmorstein, R. (2007). Chemistry of acetyl transfer by histone modifying enzymes: Structure, mechanism and implications for effector design. *Oncogene*, *26*(37), 5528–5540.

Hu, Y., Lan, W., & Miller, D. (2017). Next-generation sequencing for MicroRNA expression profile. *Bioinformatics in microRNA research*, 169-177.

Ijaz, S., Iqbal, J., Abbasi, B. A., Tufail, A., Ullah, Z., Sharifi-Rad, J., & Uddin, S. (2023). Biofortification: Lessons from the Past and Strategies for Future Food Security. In *Legumes Biofortification* (pp. 521–545). Springer International Publishing.

Ijaz, S., Iqbal, J., Abbasi, B. A., Yaseen, T., Rehman, S., Kazi, M., & Mahmood, T. (2024). Role of OMICS-Based Technologies in Plant Sciences. OMICs-based Techniques for Global Food Security, 45-66.

Inbar-Feigenberg, M., Choufani, S., Butcher, D. T., Roifman, M., & Weksberg, R. (2013). Basic concepts of epigenetics. *Fertility and Sterility*, *99*(3), 607–615.

Isayenkov, S. V., & Maathuis, F. J. (2019). Plant salinity stress: Many unanswered questions remain. *Frontiers in Plant Science*, *10*, 80.

Jablonka, E., & Lamb, M. J. (2002). The changing concept of epigenetics. *Annals of the New York Academy of Sciences*, *981*(1), 82–96.

Jadidi, O., Etminan, A., Azizi-Nezhad, R., Ebrahimi, A., & Pour-Aboughadareh, A. (2022). Physiological and molecular responses of barley genotypes to salinity stress. *Genes*, *13*(11), 2040.

Jodder, J. (2021). Regulation of pri-MIRNA processing: Mechanistic insights into the miRNA homeostasis in plant. *Plant Cell Reports*, *40*(5), 783–798.

Karan, R., DeLeon, T., Biradar, H., & Subudhi, P. K. (2012). Salt stress induced variation in DNA methylation pattern and its influence on gene expression in contrasting rice genotypes. *PLoS One*, *7*(6), e40203.

Ketehouli, T., Idrice Carther, K. F., Noman, M., Wang, F.-W., Li, X.-W., & Li, H.-Y. (2019). Adaptation of plants to salt stress: Characterization of Na+ and K+ transporters and role of CBL gene family in regulating salt stress response. *Agronomy (Basel)*, *9*(11), 687.

Khraiwesh, B., Zhu, J.-K., & Zhu, J. (2012). Role of miRNAs and siRNAs in biotic and abiotic stress responses of plants. *Biochimica et Biophysica Acta (BBA)-. Gene Regulatory Mechanisms*, *1819*(2), 137–148.

Konate, M., Wilkinson, M. J., Mayne, B. T., Pederson, S. M., Scott, E. S., Berger, B., & Rodriguez Lopez, C. M. (2018). Salt stress induces non-CG methylation in coding regions of barley seedlings (Hordeum vulgare). *Epigenomes*, *2*(2), 12.

Kumar, V., Khare, T., Shriram, V., & Wani, S. H. (2018). Plant small RNAs: The essential epigenetic regulators of gene expression for salt-stress responses and tolerance. *Plant Cell Reports*, *37*, 61–75.

Lee, J.-H., Bollschweiler, D., Schäfer, T., & Huber, R. (2021). Structural basis for the regulation of nucleosome recognition and HDAC activity by histone deacetylase assemblies. *Science Advances*, *7*(2), eabd4413.

Ma, J., Bai, X., Luo, W., Feng, Y., Shao, X., Bai, Q., Sun, S., Long, Q., & Wan, D. (2019). Genome-wide identification of long noncoding RNAs and their responses to salt stress in two closely related poplars. *Frontiers in Genetics*, *10*, 777.

Mancia, A. (2022). Genome-wide DNA Methylation protocol for epigenetics studies. In *Marine Genomics: Methods and Protocols* (pp. 19-41). Springer.

Mansour, M. M. F., Emam, M. M., Salama, K. H. A., & Morsy, A. A. (2021). Sorghum under saline conditions: Responses, tolerance mechanisms, and management strategies. *Planta*, *254*, 1–38.

Maryum, Z., Luqman, T., Nadeem, S., Khan, S. M. U. D., Wang, B., Ditta, A., & Khan, M. K. R. (2022). An overview of salinity stress, mechanism of salinity tolerance and strategies for its management in cotton. *Frontiers in Plant Science, 13*, 907937.

Mazalmazraei, T., Nejadsadeghi, L., Mehdi Khanlou, K., & Ahmadi, D. N. (2023). Comparative analysis of differentially expressed miRNAs in leaves of three sugarcanes (Saacharum officinarum L.) cultivars during salinity stress. *Molecular Biology Reports*, *50*(1), 485–492.

Miryeganeh, M. (2021). Plants' epigenetic mechanisms and abiotic stress. *Genes*, *12*(8), 1106.

Mohamed, N. B. (2022). *Modification of DNA methylation marks in plants*. University of Leeds.

Müller, M. M., & Muir, T. W. (2015). Histones: At the crossroads of peptide and protein chemistry. *Chemical Reviews*, *115*(6), 2296–2349.

Naveed, M., Sajid, H., Mustafa, A., Niamat, B., Ahmad, Z., Yaseen, M., Kamran, M., Rafique, M., Ahmar, S., & Chen, J.-T. (2020). Alleviation of salinity-induced oxidative stress, improvement in growth, physiology and mineral nutrition of canola (Brassica napus L.) through calcium-fortified composted animal manure. *Sustainability*, *12*(3), 846.

Nicolas-Francès, V., Grandperret, V., Liegard, B., Jeandroz, S., Vasselon, D., Aimé, S., Klinguer, A., Lamotte, O., Julio, E., & de Borne, F. D. (2018). Evolutionary diversification of type-2 HDAC structure, function and regulation in Nicotiana tabacum. *Plant Science*, *269*, 66–74.

Paliwal, S., Tripathi, M. K., Tiwari, S., Tripathi, N., Payasi, D. K., Tiwari, P. N., Singh, K., Yadav, R. K., Asati, R., & Chauhan, S. (2023). Molecular advances to combat different biotic and abiotic stresses in linseed (Linum usitatissimum L.): A comprehensive review. *Genes*, *14*(7), 1461.

Paul, A., Dasgupta, P., Roy, D., & Chaudhuri, S. (2017). Comparative analysis of Histone modifications and DNA methylation at OsBZ8 locus under salinity stress in IR64 and Nonabokra rice varieties. *Plant Molecular Biology*, *95*, 63–88.

Pérez-Pizá, M. C., Ibañez, V. N., Varela, A., Cejas, E., Ferreyra, M., Chamorro-Garcés, J. C., Zilli, C., Vallecorsa, P., Fina, B., & Prevosto, L. (2022). Non-thermal plasmas affect plant growth and DNA methylation patterns in glycine max. *Journal of Plant Growth Regulation*, *41*(7), 2732–2742.

Ramezanifar, H., Yazdanpanah, N., Golkar Hamzee Yazd, H., Tavousi, M., & Mahmoodabadi, M. (2022). Spinach growth regulation due to interactive salinity, water, and nitrogen stresses. *Journal of Plant Growth Regulation*, *41*(4), 1654–1671.

Rashid, M. M., Vaishnav, A., Verma, R. K., Sharma, P., Suprasanna, P., & Gaur, R. (2022). Epigenetic regulation of salinity stress responses in cereals. *Molecular Biology Reports*, •••, 1–12.

Riemann, M., Dhakarey, R., Hazman, M., Miro, B., Kohli, A., & Nick, P. (2015). Exploring jasmonates in the hormonal network of drought and salinity responses. *Frontiers in Plant Science*, *6*, 1077.

Ro ca, M., Mihalache, G., & Stoleru, V. (2023). Tomato responses to salinity stress: From morphological traits to genetic changes. *Frontiers in Plant Science*, *14*, 1118383.

Sabagh, A., Çiğ, F., Seydoşoğlu, S., Battaglia, M. L., Javed, T., Iqbal, M. A., & Awad, M. (2021). Salinity stress in maize: Effects of stress and recent developments of tolerance for improvement. *Cereal Grains*, *1*, 213.

Sabbour, M. M., & Hussein, M. (2024). Evaluation responses of potato to salinity and nano nitrogen fertilizer and pests infestations. *Egyptian Journal of Agronomy*, *46*(2), 343–353.

Saleem, A., Zulfiqar, A., Alamoudi, S. A., Ashkan, M. F., & Alhomaidi, E. (2022). Plant growth-promoting bacteria improve Triticum aestivum L. growth and photosynthetic activity in sulfonamide-contaminated soil. *Turkish Journal of Agriculture and Forestry*, *46*(6), 822–837.

Saleem, A., Zulfiqar, A., Zulfiqar, U., & Hussain, S. (2024). Iron Uptake Strategies in Rice: Exploring YSL Genes and Nutritional Enhancement. *Plant Molecular Biology Reporter*, •••, 1–19.

Samarskaya, V. O., Spechenkova, N., Ilina, I., Suprunova, T. P., Kalinina, N. O., Love, A. J., & Taliansky, M. E. (2023). A Non-Canonical Pathway Induced by Externally Applied Virus-Specific dsRNA in Potato Plants. *International Journal of Molecular Sciences*, *24*(21), 15769.

Sanjuan-Badillo, A., & Martínez-Castilla, P., LGarcía-Sandoval, , RBallester, , PFerrándiz, , CSanchez, , M. d. l. PGarcía-Ponce, , BGaray-Arroyo, , A., & RÁlvarez-Buylla, , E. (2024). HDACs MADS-domain protein interaction: A case study of HDA15 and XAL1 in Arabidopsis thaliana. *Plant Signaling & Behavior*, *19*(1), 2353536.

Sanobar, N., Lin, P.-C., Pan, Z.-J., Fang, R.-Y., Tjita, V., Chen, F.-F., Wang, H.-C., Tsai, H.-L., Wu, S.-H., & Shen, T.-L. (2021). Investigating the viral suppressor HC-pro inhibiting small RNA methylation through functional comparison of HEN1 in angiosperm and bryophyte. *Viruses, 13*(9), 1837.

Saradadevi, G. P., Das, D., Mangrauthia, S. K., Mohapatra, S., Chikkaputtaiah, C., Roorkiwal, M., Solanki, M., Sundaram, R. M., Chirravuri, N. N., & Sakhare, A. S. (2021). Genetic, epigenetic, genomic and microbial approaches to enhance salt tolerance of plants: A comprehensive review. *Biology (Basel), 10*(12), 1255.

Schneider, A., Chatterjee, S., Bousiges, O., Selvi, B. R., Swaminathan, A., Cassel, R., Blanc, F., Kundu, T. K., & Boutillier, A.-L. (2013). Acetyltransferases (HATs) as targets for neurological therapeutics. *Neurotherapeutics; the Journal of the American Society for Experimental NeuroTherapeutics, 10*(4), 568–588.

Sharma, M., Sidhu, A. K., Samota, M. K., Gupta, M., Koli, P., & Choudhary, M. (2023). Post-Translational Modifications in Histones and Their Role in Abiotic Stress Tolerance in Plants. *Proteomes, 11*(4), 38.

Shin, Y. K., Bhandari, S. R., Jo, J. S., Song, J. W., Cho, M. C., Yang, E. Y., & Lee, J. G. (2020). Response to salt stress in lettuce: Changes in chlorophyll fluorescence parameters, phytochemical contents, and antioxidant activities. *Agronomy (Basel), 10*(11), 1627.

Simoes, W. L., Coelho, D. S., Mesquita, A. C., Calgaro, M., & Silva, J. S. d. (2020). Physiological and biochemical responses of sugarcane varieties to salt stress. *Revista Caatinga, 32*, 1069–1076.

Singh, D., Singh, C. K., Tomar, R. S. S., Sharma, S., Karwa, S., Pal, M., Singh, V., Sanwal, S. K., & Sharma, P. (2020). Genetics and molecular mapping for salinity stress tolerance at seedling stage in lentil (Lens culinaris Medik). *Crop Science, 60*(3), 1254–1266.

Singroha, G., Kumar, S., Gupta, O. P., Singh, G. P., & Sharma, P. (2022). Uncovering the epigenetic marks involved in mediating salt stress tolerance in plants. *Frontiers in Genetics, 13*, 811732.

Soltabayeva, A., Ongaltay, A., Omondi, J. O., & Srivastava, S. (2021). Morphological, physiological and molecular markers for salt-stressed plants. *Plants, 10*(2), 243.

Sun, L., Xue, C., Guo, C., Jia, C., Yuan, H., Pan, X., & Tai, P. (2022). Maintenance of grafting reducing cadmium accumulation in soybean (Glycine max) is mediated by DNA methylation. *The Science of the Total Environment, 847*, 157488.

Sun, X., Xu, L., Wang, Y., Yu, R., Zhu, X., Luo, X., Gong, Y., Wang, R., Limera, C., & Zhang, K. (2015). Identification of novel and salt-responsive miRNAs to explore miRNA-mediated regulatory network of salt stress response in radish (Raphanus sativus L.). *BMC Genomics, 16*, 1–16.

Tam, S., De Borja, R., Tsao, M.-S., & McPherson, J. D. (2014). Robust global microRNA expression profiling using next-generation sequencing technologies. *Laboratory Investigation, 94*(3), 350–358.

Thakur, R., & Prasad, P. (2024). Epigenetic regulation of abiotic stress responses in plants. *Biochimica et Biophysica Acta (BBA)-General Subjects*, 130661.

Ueda, M., Matsui, A., Tanaka, M., Nakamura, T., Abe, T., Sako, K., Sasaki, T., Kim, J.-M., Ito, A., & Nishino, N. (2017). The distinct roles of class I and II RPD3-like histone deacetylases in salinity stress response. *Plant Physiology, 175*(4), 1760–1773.

Ueda, M., & Seki, M. (2020). Histone modifications form epigenetic regulatory networks to regulate abiotic stress response. *Plant Physiology, 182*(1), 15–26.

Ullah, F., Xu, Q., Zhao, Y., & Zhou, D. X. (2021). Histone deacetylase HDA710 controls salt tolerance by regulating ABA signaling in rice. *Journal of Integrative Plant Biology, 63*(3), 451–467.

Ullah, M. A., Abdullah-Zawawi, M.-R., Zainal-Abidin, R.-A., Sukiran, N. L., Uddin, M. I., & Zainal, Z. (2022). A review of integrative omic approaches for understanding rice salt response mechanisms. *Plants, 11*(11), 1430.

Ullah, Z., Iqbal, J., Abbasi, B. A., Akhtar, W., Kanwal, S., Ali, I., & Mahmood, T. (2023b). Assessment of Gus Expression Induced by Anti-Sense Os PPO Gene Promoter and Antioxidant Enzymatic Assays in Response to Drought and Heavy Metal Stress in Transgenic Arabidopsis thaliana. *Sustainability, 15*(17), 12783.

Ullah, Z., Iqbal, J., Abbasi, B. A., Ijaz, S., Sharifi-Rad, J., Yaseen, T., & Mahmood, T. (2023a). How Genome Editing Can Be Helpful in the Biofortification of Legumes. In *Legumes Biofortification* (pp. 207–232). Springer International Publishing.

Wang, P., Yang, W., Zhao, S., & Nashun, B. (2021). Regulation of chromatin structure and function: Insights into the histone chaperone FACT. *Cell Cycle (Georgetown, Tex.), 20*(5-6), 465–479.

Weyen, J. (2021). Applications of doubled haploids in plant breeding and applied research. *Doubled haploid technology: volume 1: general topics, alliaceae, cereals*, 23-39.

Xie, Z., Allen, E., Fahlgren, N., Calamar, A., Givan, S. A., & Carrington, J. C. (2005). Expression of Arabidopsis MIRNA genes. *Plant Physiology, 138*(4), 2145–2154.

Xu, R., Wang, Y., Zheng, H., Lu, W., Wu, C., Huang, J., Yan, K., Yang, G., & Zheng, C. (2015). Salt-induced transcription factor MYB74 is regulated by the RNA-directed DNA methylation pathway in Arabidopsis. *Journal of Experimental Botany*, *66*(19), 5997–6008.

Yadav, A., Mathan, J., Dubey, A. K., & Singh, A. (2024). The emerging role of non-coding RNAs (ncRNAs) in plant growth, development, and stress response signaling. *Non-Coding RNA*, *10*(1), 13.

Yang, L., Xu, M., Koo, Y., He, J., & Poethig, R. S. (2013). Sugar promotes vegetative phase change in Arabidopsis thaliana by repressing the expression of MIR156A and MIR156C. *elife, 2*, e00260.

Yousef, F., Shafique, F., Ali, Q., & Malik, A. (2020). Effects of salt stress on the growth traits of chickpea (Cicer arietinum L.) and pea (Pisum sativum L.) seedlings. *Biological and Clinical Sciences Research Journal, 2020*, e029.

Yung, W. S., Li, M. W., Sze, C. C., Wang, Q., & Lam, H. M. (2021). Histone modifications and chromatin remodelling in plants in response to salt stress. *Physiologia Plantarum*, *173*(4), 1495–1513.

Zafar, Z. U., Manzoor, H., Rasul, S., Noreen, S., Ali, Q., Iqbal, M., Javed, M., Gul, H. S., Ahmad, Z., & Shahzad, F. (2017). *Strategies to improve crop salt and drought tolerance: Success and limitations. Quality and quantum improvement in field crops*. Agribios.

Zhang, Q., & Dai, W. (2019). Plant response to salinity stress. In *Stress physiology of woody plants* (pp. 155–173). CRC Press.

Zhang, S., Liu, Y., & Yu, B. (2015). New insights into pri-miRNA processing and accumulation in plants. *Wiley Interdisciplinary Reviews. RNA*, *6*(5), 533–545.

Zheng, C., Ye, M., Sang, M., & Wu, R. (2019). A regulatory network for miR156-SPL module in Arabidopsis thaliana. *International Journal of Molecular Sciences*, *20*(24), 6166.

Zulfiqar, A., Fatima, R., Ahmed, S., Saleem, A., Sardar, R., Ahmad, M. N., & Yasin, N. A. (2023). Mechanistic insights into the interaction of fluoride resistant bacteria with wheat roots toward enhancing plant productivity by alleviating fluoride stress. *Fluoride*, *56*(3).

Zulfiqar, A., Mushtaq, I., Ahmed, S., Ali, M., Saleem, A., Sadar, R., & Ahmad, M. N. (2022). Fluoride resistant bacteria alleviate fluoride stress in Triticum aestivum l. Through modulating gas exchange characteristics and enhanced plant growth. *Fluoride*, *55*(3), 280–294.

Zulfiqar, A., Naseer, S., Saleem, A., Sabar, M., Ahmed, S., Sardar, R., Shahzadi, F., & Raza, Q. (2023). Genetic diversity studies for grain iron and zinc content analysis for Elite rice (Oryza sativa L.) genotype by using SSR markers. *Journal of Food Composition and Analysis*, *115*, 104816.

Chapter 20
Epigenetic Memory and Transgenerational Stress Adaptation in Plants

Siraj Uddin
https://orcid.org/0000-0001-7797-3853
Postgraduate Science College, Quetta, Pakistan

Samina Bibi
Quaid-i-Azam University, Islamabad, Pakistan

Khalil Ur Rehman
Postgraduate Science College, Quetta, Pakistan

Siraj Ahmed Kakar
Postgraduate Science College, Quetta, Pakistan

Ghulam Murtaza
Yunnan University, Kunming, China

Rashid Iqbal
The Islamia University of Bahawalpur, Pakistan

ABSTRACT

Recent studies highlight a distinctive appearance of "intelligence" within the kingdom plantae. Regardless of the lack of a nervous system, plants have developed a mature molecular machinery that assist to remember, to compose decisions, and take responses according to the stress stimulus. Being sessile organisms, plants can utilize these multifaceted systems to boost up their resource strategies and develop their strength for survival among hard environmental conditions. Along with these capabilities, these sessile plants also gain potential to transmit their acquired experience to their next generation via epigenetic modifications, thereby extending typical genetic inheritance. This chapter explores the sensory system by which plants observe, recognize, remember, and respond to their surroundings. As a result, these acquired adaptations can be transferred to their next generations to equip them more efficiently against any climatic challenges.

DOI: 10.4018/979-8-3693-8648-4.ch020

1. INTRODUCTION

Climate change standpoints as one of the most serious challenges of the modern era, intensely influencing agriculture and natural ecosystems. Rising temperatures, variable weather patterns, changed precipitation levels, and increased frequency of extreme events like heat waves, droughts, and floods, create significant stress for crops, threatening food security worldwide. These environmental pressures not only impact yields and crop quality but also intensify plant vulnerability to pests and diseases. As a response, researchers are focusing on innovative ways to develop climate-resilient crops, with sustainable agricultural practices and advanced breeding techniques playing essential roles (Ahmad et al., 2022a; Cammarano et al., 2022; Delarue, Benhamed, & Fragkostefanakis, 2025).

Plants, as stationary organisms, cannot escape their environment and thus have evolved sophisticated coping mechanisms for adapting to stress. Among these, epigenetic mechanisms have gained recognition for their role in helping plants to respond and remember the stressful conditions, allowing for an adaptive advantage that can sometimes be passed down to the next generation (Figure 1). Epigenetic modifications, which include DNA methylation, histone modification, and RNA-mediated regulation, do not alter the genetic code itself but rather influence gene expression in ways that can be stable and consistent over time (Gallusci et al., 2017; Guarino et al., 2022). The concept of "epigenetics" originated with William Harvey in the 17th century, who used "epigenesis" to describe how organisms develop gradually from simple beginnings (Vijver et al., 2002; Ahmad et al., 2022b). This idea evolved, and by 1942, Conrad Waddington redefined "epigenetics" as the complex interplay between an organism's genetic code (genotype) and its physical traits (phenotype) (Waddington, 1942).

Figure 1. Plant response to environmental stresses and transgenerational adaptation via epigenetic memory

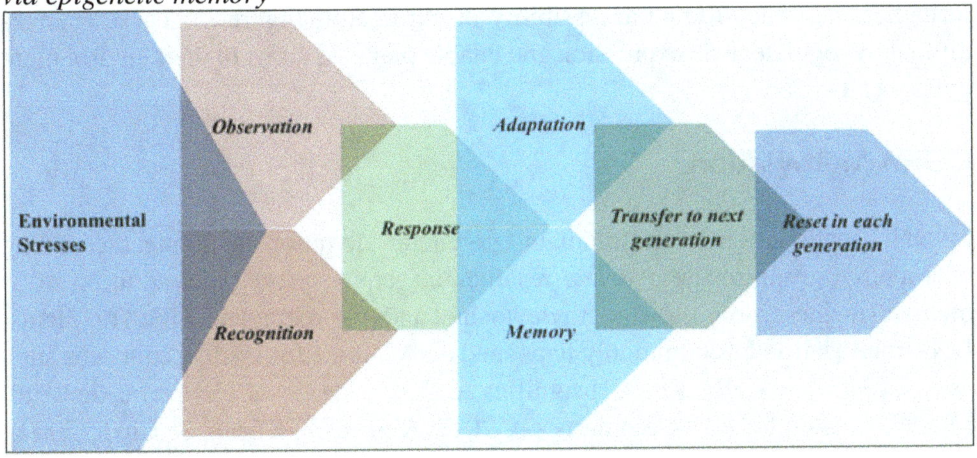

Epigenetic memory refers to the ability of plants to record environmental signs through modifications that can guide future responses to similar conditions. This stress memory relies on specific epigenetic mechanisms that regulate genes like Dicer-like 4 (DCL4) and Retrotransposon-like 1 (RTL1), allowing plants to modify their genetic response to better withstand persistent or combined stresses (Faltýnek, Lacková, & Owsianková, 2020). Transgenerational stress adaptation is a critical area of study, as it explores how stress responses established in one generation can be inherited, providing offspring with a form of "pre-adaptation" to anticipated environmental challenges (Ibañez, Quadrana, & Development, 2023; Rutowicz et al., 2015; Ahmad et al., 2024a)

This chapter delves into the epigenetic memory mechanisms in plants, highlighting the role of DNA methylation, histone modifications, and RNA-mediated processes in stress adaptation. It provides a comprehensive review of how these epigenetic changes contribute to resilience in plants, supporting transgenerational stress adaptation and offering new perspectives for enhancing crop resilience in an era of climate uncertainty (Asensi-Fabado, Amtmann, & Perrella, 2017; Manavella et al., 2023; Ahmad et al., 2024b).

2. MECHANISM OF EPIGENETICS

Epigenetics refers to heritable changes in chromatin function that do not involve alterations in the underlying DNA sequence. These modifications play pivotal roles throughout the plant life cycle, mediating responses to both internal and external stimuli (Duan et al., 2018). Epigenetic mechanisms include DNA methylation, histone modifications, and the involvement of noncoding RNAs (ncRNAs), all of which influence the structure and accessibility of chromatin (Figure 2). These changes ultimately regulate gene expression and enable plants to adapt to their environment dynamically (Table1).

I. DNA Methylation

DNA methylation is one of main epigenetic alteration involving the adding of a methyl group to the cytosine residues in gene sequence like addition of 5-methylcytosine (5-mC), N6-methylcytosine and N4-methylcytosine. In plants, DNA methylation has commonly happened across three following sequences such as CG, CHG, and CHH, where H signifies A, C, or T(da Costa, Cerqueira, de Brito, Mielke, & Gaiotto, 2024; Lennartsson, 2017; Mal, Koh, Cheah, & Chan, 2009). The foundation of DNA methylation is mainly facilitated by pathway called RNA-directed DNA methylation (RdDM). All these modifications are mainly maintained by various methyltransferases according to environmental stimulus (Cao & Jacobsen, 2002; Chang Liu et al., 2018; Zemach et al., 2013). For example, the active DNA demethylation in *Arabidopsis thaliana* is assisted through a base excision repair pathway via enzymes such as DEMETER (DME), ROS1, DML2, and DML3(R. Liu & Lang, 2020; Huiming Zhang, Lang, & Zhu, 2018). These enzymes are working to target some specific regions through various complexes (Increased DNA Methylation (IDM) complex and SWR1 chromatin remodelling complex) (Chang et al., 2020; Nie et al., 2019). DNA methylation has a substantial role in stress replies, controlling stress-responsive genes vigorously via methylation shifts (Chang et al., 2020; R. Liu & Lang, 2020).

II. Histone Modifications

In plants, another epigenetic mechanism against various stresses is Histone modifications. This type of modification regulates the chromatin structure via methylation, phosphorylation, acetylation and ubiquitination. For example, modification in histone 3 (H3) or histone 4 (H4), causes the alteration in the transcription processes (da Costa et al., 2024; Chang Liu et al., 2018). The transcription processes are mainly activated or inhibited by different types of enzyme and protein factors. For example, H3K27me3 is one of the repressive marks that displays high deposition on numerous resilient genes, catalysed by the polycomb repressive complex 2 (PRC2). These modification are mainly regulated by methyltransferases such as CURLY LEAF (CLF), MEDEA (MEA) and SWINGER (SWN) (He et al., 2018; Mozgova & Hennig, 2015). On the other hand, H3K4me3 is an active mark catalysed by the

COMPASS complex, including enzymes like ARABIDOPSIS TRITHORAX-LIKE PROTEIN 1 (ATX1) and ATX2 (Chang et al., 2020; Saleh et al., 2008). During gene expression regulations, various types of protein are in action as a writer, eraser, and reader to recognise and interpret the histone modification (Cao et al., 2010).

III. Noncoding RNAs in Epigenetics

Along with DNA methylation and histone modification, plants also use noncoding RNAs (ncRNAs) epigenetic mechanism. Noncoding RNAs are of two types, short noncoding RNAs (siRNAs and miRNAs) and long noncoding RNAs (lncRNAs). These ncRNAs regulate gene expression by modifying the chromatin status and controlling the transcription processes to affect the epigenetic landscape (Wei, Huang, Yang, & Kang, 2017; Wierzbicki, Blevins, & Swiezewski, 2021). The ncRNAs play a remarkable role in epigenetic regulation, mainly in the RdDM pathway. During this process, RNAs accelerate the DNA methylation for suppressing specific genes. For example, a transcription factor like MYB74, is mainly limited by RdDM. While during salt stress environment, MYB74 becomes active, highlighting the dynamic interaction between ncRNAs and environmental stress responses (Xu et al., 2015).

All the above three types of epigenetics mechanisms (DNA methylation, histone modification and ncRNAs) enable the plants to adjust to ecological stimuli via dynamic changes in gene expression. Marks like H3K27me3 and H3K4me3 are the prime examples of epigenetic mechanism. These marks regulate the stress-responsive pathways by balancing the activation and repression processes based on environmental signals (Saleh et al., 2008; Fromm & Avramova, 2014, Ali et al., 2022).

In plants, epigenetic changes can be triggered by abiotic and biotic factors like drought, light, temperature salinity and pathogens (Asensi-Fabado et al., 2017; Delarue et al., 2025; Xu et al., 2015). These changes or modifications contribute to phenotypic plasticity and sometimes transgenerational inheritance by remembering the past environmental stresses and transferring them on to the next generation. Some most important epigenetic regulations are flowering time during cold temperatures (vernalization), defence against biotic stresses (pathogens), tolerance against abiotic stresses (drought) and seasonal adaptation in seed (seed dormancy). These adaptive systems highlight the significance of epigenetics in plant flexibility and improvement (Asensi-Fabado et al., 2017; Chang et al., 2020; Gallusci et al., 2023; Hemenway & Gehring, 2023; Ibañez et al., 2023; Chang Liu et al., 2018).

IV. Contribution of Other Molecular Factors

Along with epigenetics regulations, some other factors including transcriptional factors, plant hormones, and metabolites contribute to developing the priming state and activating the plant defense system on right time and right place. For example transcription factors (TFs), including basic helix-loop-helix (bHLH), WRKY DNA-binding proteins, myeloblastosis (MYB), and heat shock factors (HSFs), are active in shaping the responses of plants to negative environmental stimuli. The processes of DNA methylation and chromatin remodelling significantly affect the expression levels of these TFs, which successively dominate the stress response mechanisms. For example, the phenomenon of DNA methylation, during occurrences of water shortage significantly affects the stomatal development, by improving drought resilience over consecutive generations (Dowen et al., 2012). Additionally, some phytohormones, such as abscisic acid (ABA), jasmonic acid (JA), and salicylic acid (SA), apply extensive impact on stress memory by changing chromatin structure and controlling gene expression, which helpfully contributes to improved stress resilience (Y. Han, Reyes, Malik, & He, 2020; X. Zhang, Wang, Zhuang, Gao, & Huang, 2019). Furthermore, metabolic modifications demonstrated by the accumulation of proline, support plants in coping with stress through histone methylation, thereby playing an active role in stress memory, especially during heat or cold stress (Ijaz et al., 2023; Plants et al., 2024). Collectively, these complex processes create a responsive network that authorises plants to successfully respond to environmental dangers as shown in figure 4.

Table 1. Epigenetics mechanism in plants

Epigenetic processes	Mechanisms	Plants	References
DNA Methylation (DM)	Adds a methyl group to cytosine residues. Prevents transcription factor binding and attracts chromatin-modifying proteins. Dynamic in response to environmental stresses.	Arabidopsis thaliana	(Jogam et al., 2022; Saraswat, Yadav, Sirohi, & Singh, 2017; Hao Zhang, Zhang, & Xiao, 2023)[17],
	Alters methylation patterns in response to abiotic stress such as drought and salinity. Modulates gene expression, leading to stress tolerance.	Rice	(Ferreira, Azevedo, Maroco, Oliveira, & Santos, 2015; Liang et al., 2014)
	Persistent changes in DM occur after stress removal. Inconsistent inheritance across stress events and plant species.	Maize	(Eichten & Springer, 2015; S. Li et al., 2017)

continued on following page

Table 1. Continued

Epigenetic processes	Mechanisms	Plants	References
	Pathogen infection alters DM, activating or repressing defense-related genes.	Arabidopsis	(Hawes et al., 2018; López Sánchez, Stassen, Furci, Smith, & Ton, 2016)
Histone Modifications (HM)	Dynamic regulation of chromatin and gene expression.	Arabidopsis thaliana	(Kumar, Thakur, Prasad, & Sciences, 2021; M. J. G. Miryeganeh, 2021; Nunez-Vazquez, Desvoyes, & Gutierrez, 2022; L. Wang & Qiao, 2020)
	Specific histone modifications interact with transcription factors to regulate stress-responsive genes.	Arabidopsis	(Ramirez-Prado et al., 2019; Wu et al., 2019; Zhou & Zeng, 2017)
	Histone modification patterns are stress-specific, affecting chromatin structure and gene accessibility.	Rice	(Haidao Zhang et al., 2020; A. Zhu, Greaves, Dennis, & Peacock, 2017)
Small RNA-Mediated Gene Silencing	Small RNAs (miRNAs, siRNAs) regulate gene expression under stress conditions.	Stress-responsive gene in various plant species.	(J.-J. Park, Dempewolf, Zhang, & Wang, 2017)

Figure 2. Types of epigenetic mechanisms in plants

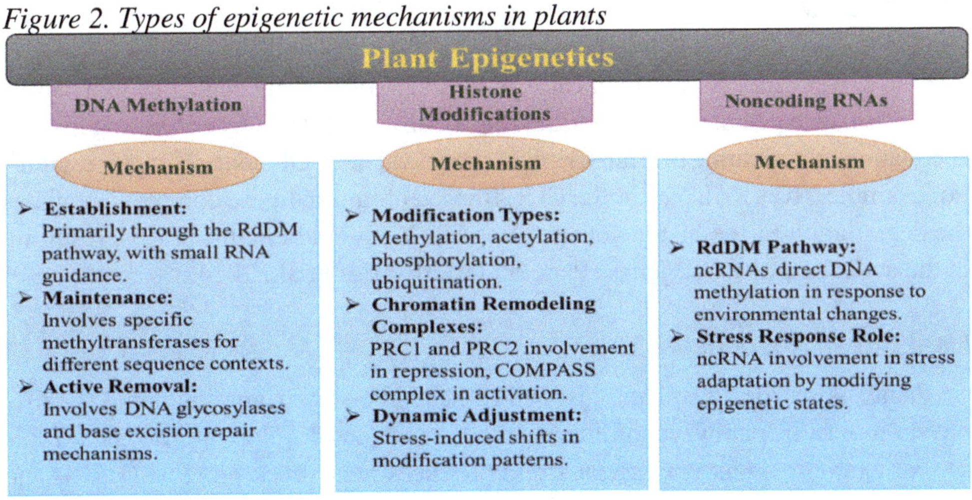

3. EPIGENETICS AND PLANT RESPONSE TO ENVIRONMENTAL STRESS

a. Abiotic Stresses and Epigenetics Mechanisms in Plants

I. Extreme Temperature

Among abiotic factors, temperature is one of the main factors playing a crucial role in plant growth and development. With the passage of time, global warming and climate changes are increasing heat and cold stresses, affecting crop productivity. Plants have ability to adopt against these abiotic stresses using epigenetic mechanism as a pivotal tool (J.-K. J. C. Zhu, 2016).

Cold Stress

Cold stress exercises a significant impact on the metabolic pathways and gene expression profiles of plants. At low non-freezing temperature (cold acclimation), plants show high resilience to succeeding freezing conditions. During cold stress, the concentrations of C-repeat binding factor family proteins (CBFs) become boosted, which activate the cold-responsive (COR) effector genes. Through this CBF-COR pathway, the plant responds to cold stress (J.-K. J. C. Zhu, 2016). Additionally, another protein named PICKLE (PKL), also contributs during cold stress by controlling genes including CBF3, RD29A, COR15A, and COR15B. The PKL protein can play a key role during two remarkable epigenetic processes such as RdDM pathways and H3K27me3. More interestingly, these epigenetic shifts may act as "memory" to facilitate plants recognise and remember the cold stresses and enhance their freezing tolerance over time (Kwon, Lee, Choi, & Chung, 2009; Yang et al., 2019). Histone acetylation also performs a substantial role against cold stress by enhancing histone acetylation by activating the transcription process using COR gene promoter regions. For example, HOS15, a part of the CUL4-based ubiquitin ligase complex, mediates the degradation of the histone deacetylase HD2C in cold circumstances, resulting in the activation of COR genes (Park et al., 2018; Ijaz et al., 2024a).

Heat Stress Responses

Plants as sessile organisms respond to heat stress by triggering some specific physio-molecular pathways and altering their growth and developing strategies (Table 2). For example, plant under heat stress produces heat shock proteins (HSPs) and plant growth regulators to cope this intense situation. During heat stress, heat shock transcription factors A1 (HsfA1s) are released which serve as a "master regulator" by activating several transcriptional networks.

These transcriptional factors control the expression of heat-responsive genes such as HSPs, which help to stabilize and fold the responsible protein under heat stress. Additionally, some other processes are also regulated during heat stress like acetylation of H3K9 (H3K9Ac) and trimethylation of H3K4 (H3K4me3) at HSP gene loci. For example, HSFA2 facilitates the accumulation of these histone marks at HSP18, HSP22.0, and APX2, connective to heat tolerance (Lämke, Brzezinka, Altmann, & Bäurle, 2016). Furthermore, RdDM pathway also plays a critical role throughout heat stress responses by reactivating or silencing the specific loci. For example, the SDC gene, which is switched off under normal conditions by RdDM, is stimulated in heat stress, signifying a mechanism to overcome epigenetic repression during abiotic stress (heat) (Sanchez & Paszkowski, 2014; Park et al., 2018; Ijaz et al., 2024b).

I. Abscisic Acid (ABA) Response

Plants also use their biochemical pathways to respond to any environmental stresses. For example, Abscisic acid (ABA) plays an essential role in regulating abiotic stresses specifically during drought and salinity conditions. This phytohormone regulates critical biological processes, such as seed germination, root growth, and stomatal closure (J.-K. J. C. Zhu, 2016). The fundamental regulatory pathways of ABA signalling are involved in binding ABA to PYR1/PYL/RCAR receptors, resulting in the inhibition of PP2Cs, and the activation of SnRK2s (J.-K. J. C. Zhu, 2016). The activation of these kinases control many biological progresses such as RNA processing, transcription, and epigenetic modification (P. Wang et al., 2013). In some studies, the epigenetic mechanisms are linked to ABA responses, like the interaction of AtNDX with PRC1 components to regulate ABA-responsive genes (Y. Zhu et al., 2020). Furthermore, histone variations also influence the expression of ABA receptor genes and related proteins (Gonzalez-Guzman et al., 2012; Mehdi et al., 2016, Haider et al., 2021a). For example, ABA-mediated responses such as germination inhibition and stomatal closure in stress-induced plants are mainly due to histone acetylation of the PYL4, PYL5, and PYL6 genes (Gonzalez-Guzman et al., 2012). Similarly, the chromatin modifying complex and histone modifications play significant roles in ABA signalling. For example, phosphorylation of BRM, affects ABA reaction (S.-K. Han et al., 2012; Peirats-Llobet et al., 2016; Saez, Rodrigues, Santiago, Rubio, & Rodriguez, 2008). Inclusively, the dynamic rule of ABA-responsive genes involves complex interactions between ABA signalling and epigenetic modifications, signifying the status of chromatin modifications in stress adaptation.

II. Nutrient Stress

Along with the above-mentioned stresses, plants also respond to nutrient stress (deficiency or toxicity) by adjusting different physiological, biochemical, and molecular processes. For example, Nitrogen is one of the most vital nutrients for plant physiology and plants adopt efficient approaches to balance nitrogen and establish growth during various stresses (R. Pandey, Vengavasi, & Hawkesford, 2021). Plant takes almost 17 elements from soil for the production of proper growth and development. For the uptake of these nutrients from soil, plants develop complex mechanisms to adapt to varying nutrient quantities in the soil (Davidian & Kopriva, 2010; Gojon, Nacry, & Davidian, 2009). Numerous studies have been conducted to understand the epigenetic regulation of nutrient stress responses (Doddavarapu, Lata, & Shah, 2024; A. Li, Hu, & Chu, 2021; REDDY, 2006; Séré, Martin, & behavior, 2020; Sirohi, Pandey, Deveshwar, & Giri, 2016; Zhang et al., 2023; Haider et al., 2021b). Normally, histone methylation, histone variants, and DNA methylation are found to be responsible for epigenetic regulation of varying level of nutrients (Table 2). For example, high concentration of nitrogen (N) represses the root nitrogen transporter gene (NRT2.1), by activation of the HIGH NITROGEN INSENSITIVE 9 (HNI9), which sets down H3K27me3 gene (Widiez et al., 2011). Similarly, in Arabidopsis, iron homeostasis are negatively regulated by PRMT5-mediated histone H4R3 symmetric dimethylation (H4R3sme2) (Fan et al., 2014), while a histone acetyltransferase (GCN5) regulates iron homeostasis by modifying the acetylation of H3K6 and H3K14 at the FRD3 gene promoter (Xing et al., 2015). Correspondingly, deficiency of phosphate has been associated to alterations in histone modifications such as H3K4me3, histone acetylation, and deposition of histone variant H2A.Z (Chen, Wu, & Schmidt, 2015). All the above examples about modifications accomplish the expression of genes essential for root hair extension and phosphate homeostasis under phosphate deficiency (Kuo et al., 2014).

III. UV Stress

Ultraviolet (UV) stress has damaging effects on plant growth, metabolism, and biomass deposit by imposing damage to biomolecules (lipids, proteins, and nucleic acids) and cellular constituents (photosynthetic apparatus), which are mainly primary targets of UV damage. The UV induced stress in plants causes reactive oxygen species (ROS) production, which in turn triggers antioxidant defence mechanisms. For example, during high UV-B radiations, a photoreceptor (UVR8) allows plants to modulate stress response in corporation with COP1 and downstream elements such as HY5. Through this collaboration, plants synthesise various phytocompounds like photoprotective flavonoids under UV-B exposure. Interestingly, low dosage of UV-B radiation promotes UV-B acclimatization and tolerance against UV-B stress. While high level of UV radiation leads to ROS production, disruption of photosynthetic processes and DNA damage. Plants have adopted effective mechanisms to develop UV-B signals and defend themselves from UV-B radiations. Likely, UV RESISTANCE LOCUS8 (UVR8) is a plant-specific UV-B photoreceptor, which interacts with CONSTITUTIVELY PHOTOMORPHOGENIC1 (COP1) to start UV-B-specific light signalling and control UV-B responsive gene expression (Haider et al., 2022a; Shoaib et al., 2024). Furthermore, N. Pandey and Pandey-Rai (2015) revealed that in *Artemisia annua,* UV-B therapy initiates DNA demethylation in the DBR2 gene promoter region, which heightens the expression of artemisinin biogenesis genes. An inclusive understanding of these UV stress response machineries and signalling pathways is indispensable for the development of UV-B-resistant plants, aimed at expanding agricultural sustainability in area with high UV-B radiation exposure.

Table 2. Epigenetic mechanism in plants against abiotic stresses

Stresses	Epigenetic Mechanisms	Response to stresses	References
Cold	Chromatin remodeling	PICKLE (PKL) mediates H3K27me3 deposition, repressing COR gene expression (*Arabidopsis* mutants hypersensitive to cold stress).	Kwon et al., 2009; Yang et al., 2019
Heat	Histone acetylation and methylation	H3K9 acetylation and H3K4 trimethylation upregulate HSP genes, enhancing heat tolerance.	Lamke et al., 2016
ABA	Histone acetylation and methylation	Histone acetylation at *PYL4, PYL5,* and *PYL6* genes regulates stomatal closure and germination.	Gonzalez-Guzman et al., 2012
Nutrient	Histone methylation	HNI9 deposits H3K27me3 on *NRT2.1* under high nitrogen, repressing gene expression.	Widiez et al., 2011
UV-B	DNA demethylation and histone acetylation	DNA hypomethylation in *DBR2* gene promotes artemisinin biosynthesis; chromatin remodeling induces flavonoid biosynthesis.	Lang-Mladek et al., 2010; Pandey & Pandey-Rai, 2015; Pandey et al., 2019

b. Biotic Stresses and Epigenetics Mechanisms in Plants

The rate of environmental changes beats the capacity of genetic mutation and recombination to create ecological genetic diversity. Epigenetic regulation offers a substitution mechanism for generating phenotypic variation in plants, in response to environmental changes. This regulation displays a vital role in plant responses to biotic stress by manipulating gene expression through epigenetic mechanisms, including small RNA-mediated post-transcriptional silencing and chromatin remodelling, are essential for inducing resistance against pests and pathogens. Alterations of histone and methylation of DNA, influence the plant defence priming like systemic acquired resistance (SAR). Furthermore, current evidence recommends that epigenetic modifications can verify inherited long-term defence priming responses across generations. This section of the chapter displays the roles of epigenetics in biotic stress resistance and reflects potential applications for advancing crop protection through epigenetic schemes.

II. Viruses

Epigenetic management plays an imperative role in plant tolerance against plant-viral infections, specifically using some useful mechanisms such as post-transcriptional gene silencing (PTGS) and transcriptional gene silencing (TGS). Through PTGS mechanism, plant degrades the viral RNA into small interfering RNAs (siRNAs) by Dicer-like proteins (DCL2 and DCL4), while TGS shows long-term defence against DNA viruses via RNA-directed DNA methylation (RdDM) pathway. These mechanisms protect plant against viral attacks and also regulate plant genes during viral diseases **(Figure 3)**. Distinct to animals, plants do not have adaptive immunity and instead depend on RNA-based defence system. As RNA viruses have mostly single-stranded RNA (ssRNA) genomes, they are converted into double-stranded RNA (dsRNA) molecules by RNA-dependent RNA polymerases before being processed into siRNAs. Recent discoveries disclose that m6A-specific methylation in the alfalfa mosaic virus (AMV) genome influences infection in *Arabidopsis*. The demethylase ALKBH9B removes m6A modifications, impacting AMV infectivity without affecting the cucumber mosaic virus (CMV). Destroying ALKBH9B fostered m6A abundance in AMV and limited the systemic invasion, establishing its role in RNA silencing and plant immunity (Ashapkin, Kutueva, Aleksandrushkina, & Vanyushin, 2020; Martínez-Pérez et al., 2017; Ramirez-Prado, Abulfaraj, Rayapuram, Benhamed, & Hirt, 2018).

III. Microbes

DNA methylation in plants is fundamental to plant immune responses against microbial pathogens. In plant, the primary layer of defence is pathogen-associated molecular pattern (PAMP)- activated immunity (PTI), which recognizes microbial patterns via pattern recognition receptors (PRRs) and stimulates basal defences. When pathogens try to respond to this with effector molecules, simultaneously, plants respond through effector-triggered immunity (ETI), implying substantial epigenetic regulation. In *Arabidopsis*, mutants defective in CG or non-CG methylation showed improved resistance to *Pseudomonas syringae*, showing the role of DNA methylation in pathogen responses (Deleris, Halter, & Navarro, 2016; Dowen et al., 2012; Ramirez-Prado et al., 2018). Similarly, flagellin detected by the PRR FLS2 turn on immune responses by altering DNA methylation. Treatment with the bacterial peptide flg22 produces changes in methylation at retroelements and defence genes using the RdDM pathway. Furthermore, DNA methylation also controls plant immune receptors, such as RMG1, which is supervised by the RdDM pathway and demethylated by ROS1 during immune activation (Yu et al., 2013).

Epigenetic priming develops plant immunity to retain a memory, permitting faster and stronger responses to returning stress (Table 3). Priming can be short-term (individual) or transgenerational (across generations), influenced by epigenetic marks. For example, hypomethylation enhances resistance facilitated by salicylic acid (SA) against biotrophic pathogens like *Hyaloperonospora arabidopsidis*. While during hypermethylation, the modification reduces SA-mediated defences, increasing resistance to necrotrophic fungi such as *Plectosphaerella cucumerina*. The epigenetic regulatory mechanisms, particularly those involving DNA methylation, intricately adjust the salicylic acid and jasmonic acid signalling pathways, thereby optimizing plant defence strategies in accordance with the specific type of pathogen or stress encountered (López Sánchez et al., 2016; Roberts, López Sánchez, Regulation, & Plants, 2019).

Epigenetic quantitative trait loci (epiQTLs) assist a significant role in developing disease resistance mechanisms within plant species. In *Arabidopsis*, the presence of epiQTLs are linked with resistance against clubroot diseases, caused by *Plasmodiophora brassicae*. EpiQTLs function to reduce the genome-wide activation of defence-related genes with no detrimental impacts on overall growth or resilience to stressors. Additionally, in potato, the application of β-aminobutyric acid (BABA) has shown to enhance resistance against *Phytophthora infestans* by assisting histone modifications at those genes that regulate defence responses. This priming development estiblished a bivalent chromatin configuration, which recognizes the activation of pathogenesis-related (PR) genes upon successive confronts with pathogens (Ashapkin et al., 2020; Furci et al., 2019; Meller, Kuźnicki, Arasimowicz-Jelonek, Deckert, & Floryszak-Wieczorek, 2018).

IV. Pests

The soybean cyst nematode (*Heterodera glycines*) produces a multinucleated feeding site, the syncytium, in soybean roots, triggering significant DNA methylation shifts in the host. A whole-genome bisulfite sequencing (WGBS) analysis of infected and non-infected soybean roots uncovered significant DNA methylation changes during SCN infection. For example, DNA methylation changes at CG context, with 718 hypermethylated (hyperDMRs) and 1,408 hypomethylated (hypoDMRs) regions were recognized (Verstraeten, 2021). This finding revealed that

SCN-induced methylation significantly impacted gene expression, emphasizing DNA methylation as a key regulator during SCN infection (Ashapkin et al., 2020).

Furthermore, it is reported that, SCN-resistant soybean and SCN-susceptible soybean lines vary in DNA methylation across infection despite high genetic similarity. Resistant plants (R) present fewer methylation variations, mainly activating oxidative stress responses, while susceptible plants display extensive methylation alterations associated to hormone signalling pathways (SA, JA, ABA). Susceptible plants (S) also exhibit altered gene expression due to methylation changes overlapping with protein-coding genes and transposable elements. These outcomes suggest that resistance concerns precise epigenetic regulation, whereas susceptibility is linked to broader methylation disruptions (Ashapkin et al., 2020; Hannan Parker, Wilkinson, & Ton, 2022; Rambani et al., 2020).

V. Parasitic Plants

Parasitic plants, like *Cuscuta* species, use haustoria to extract nutrients and signalling molecules from host plants (Table 3). These haustoria enable the movement of proteins, DNA, RNA, and small RNAs (sRNAs) between host and parasite. Transcriptome analyses of *Cuscuta pentagona* and its hosts (*Arabidopsis thaliana* and tomato) revealed that a significant portion of host transcripts moves into the parasite. Haustorial selectivity appears to be host-regulated, potentially indicating active defence mechanisms in tomato, such as exudation of defensive compounds at the infection site (Ashapkin et al., 2020; Gaut, Miller, & Seymour, 2019; Guo, Qin, Ma, & Qin, 2022; Kaiser, Vogg, Fürst, & Albert, 2015).

Cuscuta campestris takes benefits from small RNAs (sRNAs) to suppress host plant defences in parasitism. Through deep sequencing analysis, 76 *Cuscuta* sRNAs were identified including 43 miRNAs which were upregulated at the host-parasite line. These miRNAs target key host defence genes like SEOR1 (involved in blocking phloem transport) and AFB3 (related to hormonal defense signaling). Controlling these genes prepare host plants more vulnerable to parasite attacks. This trans-specific movement of sRNAs highlights the ability of *Cuscuta* to epigenetically influence host gene expression, showing the key role of epigenetics in the persistent host-parasite arms race (Alakonya et al., 2012; Shahid et al., 2018).

Figure 3. Plant defence mechanism against biotic stresses

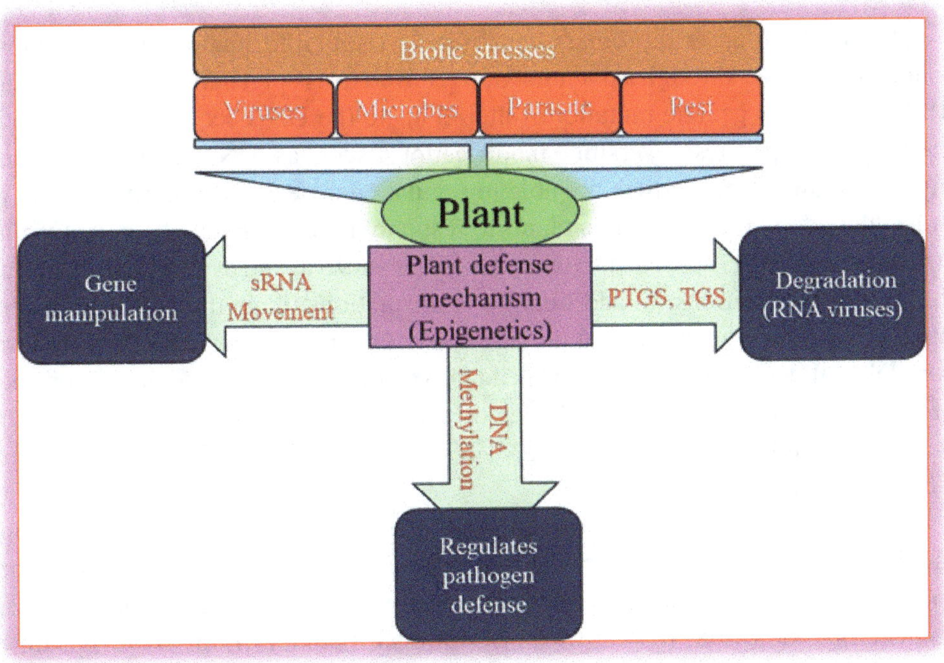

Table 3. Epigenetic mechanism in plants against biotic stresses

Stress Type	Epigenetic Mechanisms	Plant response to stresses	References
Viruses	PTGS, TGS	RNA viruses trigger PTGS through siRNAs (DCL2, DCL4); TGS uses RdDM for DNA virus defense	Ashapkin et al., 2020; Ramirez-Prado et al., 2018
Microbes	DNA Methylation (CG, CHG, CHH)	DNA methylation regulates pathogen defense; mutants in methylation pathways alter resistance.	Dowen et al., 2012; Hannan Parker et al., 2022; Ashapkin et al., 2020
PAMPs	DNA Demethylation	Flg22-induced DNA demethylation activates defense genes in response to bacterial pathogens.	Yu et al., 2013; Ashapkin et al., 2020
Soybean Cyst Nematode	DNA methylation	SCN infection reprograms DNA methylation, affecting gene expression in resistant/susceptible plants.	Rambani et al., 2020; Li et al., 2016
Cuscuta Parasitism	sRNA Movement (miRNA, siRNA)	Mobile RNAs between *Cuscuta* and hosts influence gene expression to benefit the parasite	Shahid et al., 2018; Johnson & Axtell, 2019

4. EPIGENETIC MEMORY IN PLANTS

Plants exhibit an extraordinary capability to retain information regarding environmental stresses through a phenomenon referred to as epigenetic memory (Figure 4). In contrast to genetic mutations (alterations to the DNA sequence), epigenetic memory incorporates reversible modifications such as DNA methylation and histone modifications, which are able to regulate gene expression. These types of molecular adaptations authorise plants to respond more accurately to any returning stressors. Using this approach will improve their survival and evolutionary viability during harsh conditions. In addition, these adaptations retain the potential to continue across the generations and preparing descendants to withstand the incoming parallel challenges in the future. This adaptive attribute highlights the importance of epigenetic memory in the resilience of plants (Gallusci et al., 2023; Kinoshita, Seki, & Physiology, 2014; Lagiotis, Madesis, & Stavridou, 2023; Nguyen, Vu, & Cheong, 2022).

A main mechanism triggering epigenetic memory is represented by histone modifications. In plant species like *Arabidopsis*, histone marks (H3K4me3) are sustained on stress-responsive genes succeeding to rescue from drought stress, and act as a molecular marker of transcriptional memory. Similarly, exposure to salinity stress results in a decline of H3K27me3 deposition, as a result the plants are unable to display a strong response during secondary stress occurrences. These histone modifications represent the complex molecular networks through which plants understand how to interact with environmental stressors (N. Liu, Fromm, & Avramova, 2014; Sani, 2013).

In addition to histone modifications, DNA methylation also plays a key role in epigenetic memory, specifically concerning its transgenerational characteristic. Stress-induced alterations in DNA methylation can persist across generations, presenting adaptive improvements. For example, drought stress in rice plant activates inheritable patterns of methylation that improve water-use efficiency and sustain stable seed yields (Issac, 2018; Sun et al., 2021). Likewise, exposure to heavy metals in rice makes persistent modifications to methylation and provides tolerance for the upcoming three generations (Cong et al., 2019; Ou et al., 2012). These remarks highlight the persistent influence of epigenetic memory, which enhances the adaptive capacity of plant at individual as well as population level as shown in figure 4.

Epigenetic mechanisms are also associated to plant immunity, with defence priming acting as a salient description. Plants exposed to primary stress frequently show boosted resistance to subsequent attacks (Kinoshita et al., 2014). For example, tobacco plants effectively prevent the incoming threats by storing a reservoir of toxic nicotine following insects attacks (Baldwin & Schmelz, 1996; Wilkinson, 2020). This priming is supported by molecular alterations like, the upregulation of

WRKY transcription factors, facilitated through histone acetylation and H3K4me3 modifications (Jaskiewicz, Conrath, & Peterhänsel, 2011; Roychowdhury et al., 2024). Similarly, genome-wide explorations have uncovered that stress-induced differentially methylated regions (DMRs) activate formerly silenced pathogenesis-related genes, in that way enhance the resistance against pathogens. In Arabidopsis, mutants deficient in methyltransferases show improved immunity, indicating that DNA methylation plays an active role in the regulation of plant defense mechanisms (Dowen et al., 2012; Hannan Parker et al., 2022; X. J. P. o. t. N. A. o. S. Zhang, 2012).

The interplay between biotic and abiotic stress responses further explains the complexity of epigenetic memory (Cong et al., 2019). Epigenetic modifications, such as hypomethylation within specific genomic regions, encourage plants to resist pathogens while simultaneously retaining their capacity to persist with abiotic stresses. This cross-talk highlights the versatility of epigenetic regulation, facilitating plants to manage multiple stressors concurrently (Ashapkin et al., 2020; Singh et al., 2023). For example, the bacterial flagellin-derived peptide flg22 activates both pattern-triggered immunity (PTI) and the repression of the RdDM machinery, assisting the expression of silenced sequences that enhance plant defences (Boller & Felix, 2009; Huang & Jin, 2022). Such dual functionality demonstrates the resilience of epigenetic systems in the management of plant stress mechanisms.

A noteworthy example of epigenetic memory is vernalization, in which plants "retain" the effects of extended cold introduction to delay flowering until conditions are ideal for growth. This mechanism takes the involvement of Polycomb group (PcG) proteins, which perform modifications on histones to prevent the expression of the FLOWERING LOCUS C (FLC) gene (Manzoor et al., 2023). The epigenetic adaptations induced during the interval of cold exposure sustain even after the environmental stress has ended, hence ensuring appropriate timing for flowering (M. Miryeganeh & Armitage, 2024). Conspicuously, the vernalization memory can also be transferred maternally, thus further proving the transgenerational aspect of epigenetic memory (Gallusci et al., 2023; Kinoshita et al., 2014; Nguyen et al., 2022). A comprehensive insight of these main mechanisms provides the path for the advancement of stress-resilient agricultural crops, which is an essential advancement toward sustainable farming practices in the context of climate change.

5. TRANSGENERATIONAL EPIGENETIC REGULATION

Plant exposed to stress cane respond by epigenetic modification and this modification can persist and influence the subsequent generations as a stress-resilience with phenotypic plasticity. (Figure 4). This transgenerational epigenetic regulation concerns heritable changes in chromatin design or DNA methylation landscapes introduced by environmental stressors, allowing an evolutionary improvement in dynamic ecosystems (Holeski et al., 2010; Verhoeven & Van Gurp, 2012). Such intergenerational transmission of epigenetic marks is necessary for acclimatization and long-term survival of plants in ecological exposures (Gallusci et al., 2023; Lagiotis et al., 2023; Nguyen et al., 2022). These modifications depend on specialized mechanisms, specifically during reproduction, to transmit stress-induced modifications (Gallusci et al., 2023). During vegetative propagation or clonal reproduction, stress memory can be transferred somatically through mitosis, where chromatin marks such as histone modifications and DNA methylation profiles are maintained in meristematic cells. For example, the vernalization-induced repressive histone mark (H3K27me3) on the *FLC* gene is conserved through replication-linked mechanisms (J. Liu & He, 2020; Luo & He, 2020). Studies in some plant species (*Arabidopsis*, *Populus*, and *Taraxacum*) demonstrate that stress-induced DNA methylation changes can be stably inherited via clonal propagation, although the stability and functional significances of these marks are under investigation (Lagiotis et al., 2023).

Along with clonal reproduction, epigenetic inheritance can occur via sexual reproduction, by transmitting the stress-induced DNA methylation changes to next generations. For example, osmotic stress-induced methylation modifications have been mentioned to transfer via female gametes (Kleine et al., 2021; Nguyen et al., 2022). Additionally, some data indicates a parental preference in the resetting of these marks, with an affinity to erase methylation displays originating from male gametes (Gallusci et al., 2023; Nguyen et al., 2022).

Transgenerational epigenetic regulation can be explained with some useful examples in plant species. For example, when herbivory is simulated artificially in Mimulus plant, it causes heritable modifications in gene expression. In this way plants cane regulating trichome density, by enhancing the production of trichomes in the progeny (Holeski, Chase-Alone, & Kelly, 2010). Similarly, in dandelions (*Taraxacum officinale*), parental introduction to environmental stressors stimulates modifications in genome-wide DNA methylation patterns, which influence the progeny traits such as, root-to-shoot biomass ratio, leaf morphology, phosphorus content, and complete stress resilience. Extensively, the chemical inhibition of DNA methyltransferase in *T. officinale* blocks these transgenerational effects, showcasing the involving of DNA methylation as the foundational mechanism (Verhoeven & van Gurp, 2012). In the case of rice, during tissue culture regeneration, stress causes wide changes in DNA methylation, primarily characterized by the loss of methylation. These types of changes can be retained in regenerated samples and their following progeny (Stroud et al., 2013).

However, convincing evidence for transgenerational epigenetic impacts is derived from analyses concerning defence priming in *Arabidopsis thaliana*. After bacterial infection to Arabidopsis plants, their progeny revealed intensified resistance to subsequent contaminations by oomycetes. Remarkably, this resistance persisted in second-generation descendants of unprimed parental plants. This happening convincingly indicates epigenetic regulation as the inherent mechanism in plants (Luna, Ton, & behavior, 2012).

Conclusively, transgenerational epigenetic regulation allows plants to pass stress-induced changes to their offspring. Although more investigation is required to verify how this stress induced epigenetic changes precisely affect traits, while existing studies emphasize the role of DNA methylation and chromatin changing aspects in developing plant resistance across generations against environmental challenges. Recognizing these processes can uncover the mechanism of plants adaptation over generations and provide the way to develop the stress-tolerant crops using epigenetic engineering (Cong et al., 2019; Fitz-James & Cavalli, 2022; Gallusci et al., 2023; Hauser, Aufsatz, Jonak, & Luschnig, 2011; Kinoshita et al., 2014; Nguyen et al., 2022; Haider et al., 2022b Roychowdhury et al., 2024).

Table 4. Epigenetic changes and their role in plant stress memory and transgenerational adaptation

Stress Type	Plant name	Epigenetic Mechanisms	Next Generational adaptations	References
Drought	Soybean	DNA methylation	Continuous drought stress leads to altered DNA methylation, enhancing drought tolerance across generations by modulating stress-responsive pathways.	(Kim, Chae, Oh, Nguyen, & Cheong, 2020)
Salt	Rice	Small RNA-mediated modifications	Salt stress primes progeny for increased salt tolerance by altering gene expression patterns, facilitating a faster stress response in later generations.	(Zeng et al., 2020)
Heat	Tomato	Small RNA-mediated modifications	Mild heat stress primes plants for improved thermotolerance in future generations through small RNA modifications and gene activation.	(X. Wang et al., 2020)
Water deficit	Cowpea	DNA methylation	Water deficit and subsequent rewatering lead to DNA methylation changes, improving water use efficiency and stress tolerance in the next generation.	(Tankari et al., 2021)
Drought	Foxtail millet	Gene activation/ repression	Drought stress leads to the activation of protective genes in progeny, enhancing resistance to future drought events through epigenetic changes.	(Lukić et al., 2023)
Salinity	Soybean	Histone modification	Salinity treatment results in histone marks that modulate ion homeostasis and stress responses, strengthening salinity tolerance in offspring.	(Rossatto et al., 2023)
Salinity	Rice	Gene activation/ repression	Salinity stress at the seedling stage enhances stress resilience in subsequent generations by inducing specific gene activation and improved photosynthesis.	(Habibi et al., 2023)

Figure 4. Plants exposed various environmental stresses and encounter these stresses by various molecular, physiological and epigenetics strategies. The phenomenon of stress memory matures through the initial stimuli exposure, persists throughout the regaining phase, and is reactivated during succeeding stress events, by this means enabling faster and more precise defines replies along with improved tolerance.

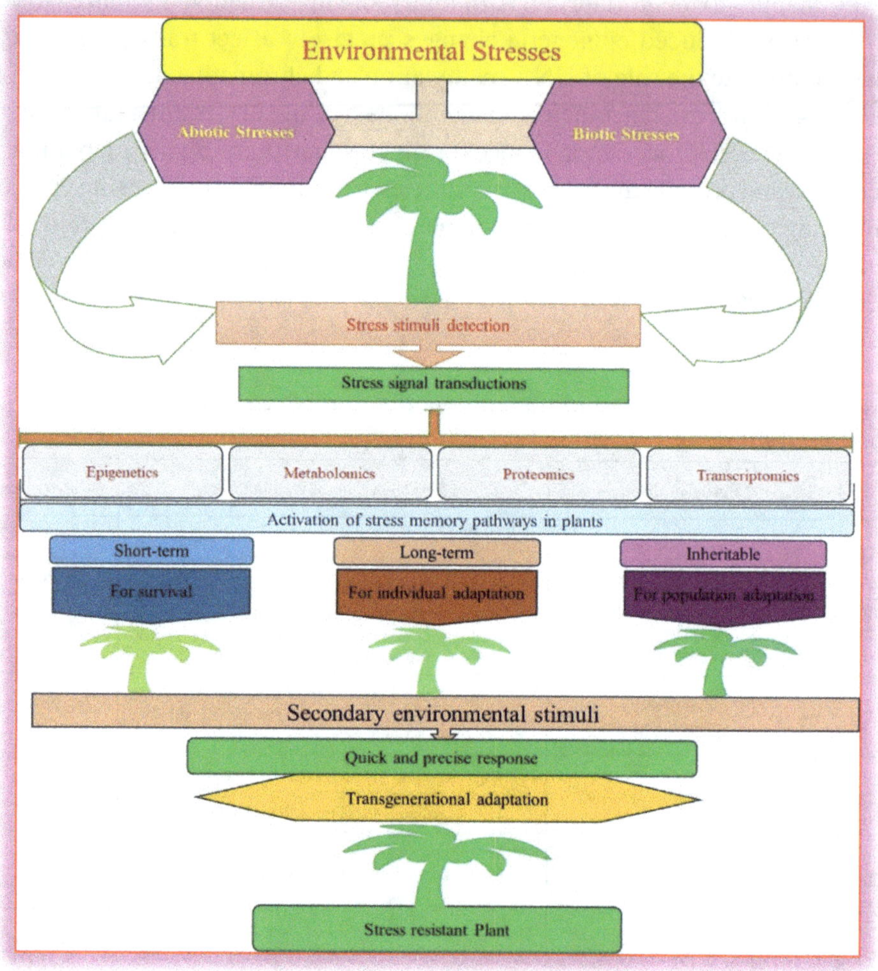

6. SUMMARY AND FUTURE PROSPECTS

Epigenetic memory in plants forms a fascinating mechanism that considerably expands their capacity to react and adapt against various environmental stresses.

As sassile organisms, plants are able to use molecular signals (DNA methylation, histone modifications, and small RNAs), to launch a memory of stress through primary exposure, reserve that memory during recovery, and afterwards reactivate it in response to secondary or future stress measures. This competency not only supports immediate survival but also extends adaptive benefits across generational profiles.

Current studies underscore the fundamental role of epigenetic modifications in the regulation of stress-responsive genes, metabolic alterations, and hormonal pathways, thereby explaining their impact on plant resilience against both biotic and abiotic stressors. The capacity to utilize these mechanisms within agricultural settings is substantial, specifically regarding the development of stress-resilient and climate-adaptive crops.

Nevertheless, the complex interaction between genetic and epigenetic factors, the sequential dynamics linked to memory retention, and the trade-offs that occur between stress resistance and susceptibility to alternative stressors remain domains that require more reflective investigation. To navigate these challenges, future research should highlight the following factors and mechanisms:

I. **Mechanistic Insights:** Explain the molecular pathways and factors that control the establishment, maintenance, and reactivation of epigenetic memory.
II. **Stress Interaction:** Investigate the mechanisms by which plants share multiple stress signals and prioritize their reactions, remarkably in the context of synchronized biotic and abiotic stresses.
III. **Population-Level Studies:** Inspect the consequences of epigenetic memory within plant populations and ecosystems to obtain a deeper understanding of community-wide resistance.
IV. **Advanced Tools and Techniques:** Utilize single-cell epigenomics, transcriptomics, and genome-wide association inquiries to uncover the mechanisms causing stress memory with unparalleled resolution.
V. **Breeding Applications:** Translate epigenetic insights into actionable strategies for breeding crops with improved stress tolerance, operating practices such as epigenome editing and priming strategies.

Epigenetic memory describes a promising possibility for grabbing the critical challenges posed by climate change and environmental variability. By integrating cutting-edge molecular approaches with ecological frameworks, investigators are poised to prepare innovative answers that enhance crop productivity and sustainability.

REFERENCES

Ahmad, Y., Haakim, Z., Iqbal, J., Abbasi, B. A., Mahmood, T., & Kazi, M. (2024b). Technological Innovations for Abiotic Stress Resistance in Horticultural Crops. OMICs-based Techniques for Global Food Security, 233-244.

Ahmad, Y., Haider, S., Iqbal, J., Abbasi, B. A., Yaseen, T., & Mahmood, T. (2022). The mechanisms of genome editing technologies in crop plants. In *Principles and Practices of OMICS and Genome Editing for Crop Improvement* (pp. 295–313). Springer International Publishing.

Ahmad, Y., Haider, S., Iqbal, J., Naseer, S., Attia, K. A., Mohammed, A. A., & Mahmood, T. (2024a). In-silico analysis and transformation of OsMYB48 transcription factor driven by CaMV35S promoter in model plant–Nicotiana tabacum L. conferring abiotic stress tolerance. *GM Crops and Food: Biotechnology in Agriculture and the Food Chain, 15*(1), 130–149. PMID: 38551174

Alakonya, A., Kumar, R., Koenig, D., Kimura, S., Townsley, B., Runo, S., & Sinha, N. (2012). Interspecific RNA interference of SHOOT MERISTEMLESS-like disrupts Cuscuta pentagona plant parasitism. *The Plant Cell, 24*(7), 3153–3166. PMID: 22822208

Ali, I., Salah, K. B. H., Sher, H., Ali, H., Ullah, Z., Ali, A., & Mahmood, T. (2022). Drought stress enhances the efficiency of floral dip method of Agrobacterium-mediated transformation in Arabidopsis thaliana. *Brazilian Journal of Biology, 84*, e259326. PMID: 35703626

Asensi-Fabado, M. A., Amtmann, A., & Perrella, G. (2017). Plant responses to abiotic stress: The chromatin context of transcriptional regulation. *Biochimica et Biophysica Acta (BBA)-. Gene Regulatory Mechanisms, 1860*(1), 106–122. PMID: 27487458

Ashapkin, V. V., Kutueva, L. I., Aleksandrushkina, N. I., & Vanyushin, B. F. (2020). Epigenetic mechanisms of plant adaptation to biotic and abiotic stresses. *International Journal of Molecular Sciences, 21*(20), 7457. PMID: 33050358

Baldwin, I. T., & Schmelz, E. A. (1996). Immunological" memory" in the induced accumulation of nicotine in wild tobacco. *Ecology, 77*(1), 236–246.

Boller, T., & Felix, G. (2009). A renaissance of elicitors: Perception of microbe-associated molecular patterns and danger signals by pattern-recognition receptors. *Annual Review of Plant Biology, 60*(1), 379–406. PMID: 19400727

Cammarano, D., Jamshidi, S., Hoogenboom, G., Ruane, A. C., Niyogi, D., & Ronga, D. (2022). Processing tomato production is expected to decrease by 2050 due to the projected increase in temperature. *Nature Food*, *3*(6), 437–444. PMID: 37118037

Cao, X., & Jacobsen, S. E. (2002). Role of the Arabidopsis DRM methyltransferases in de novo DNA methylation and gene silencing. *Current Biology*, *12*(13), 1138–1144. PMID: 12121623

Chang, Y. N., Zhu, C., Jiang, J., Zhang, H., Zhu, J. K., & Duan, C. G. (2020). Epigenetic regulation in plant abiotic stress responses. *Journal of Integrative Plant Biology*, *62*(5), 563–580. PMID: 31872527

Chen, C.-Y., Wu, K., & Schmidt, W. (2015). The histone deacetylase HDA19 controls root cell elongation and modulates a subset of phosphate starvation responses in Arabidopsis. Academic Press.

Cong, W., Miao, Y., Xu, L., Zhang, Y., Yuan, C., Wang, J., & Ou, X. (2019). Transgenerational memory of gene expression changes induced by heavy metal stress in rice (Oryza sativa L.). *BMC Plant Biology*, *19*, 1–14. PMID: 31248374

da Costa, G. S., Cerqueira, A. F., de Brito, C. R., Mielke, M. S., & Gaiotto, F. A. (2024). Epigenetics Regulation in Responses to Abiotic Factors in Plant Species: A Systematic Review. *Plants*, *13*(15), 2082. PMID: 39124200

Davidian, J. C., & Kopriva, S. (2010). Regulation of sulfate uptake and assimilation—The same or not the same? *Molecular Plant*, *3*(2), 314–325. PMID: 20139159

Delarue, M., Benhamed, M., & Fragkostefanakis, S. (2025). The role of epigenetics in tomato stress adaptation. *New crops, 2*, 100044.

Deleris, A., Halter, T., & Navarro, L. D. N. A. (2016). DNA methylation and demethylation in plant immunity. *Annual Review of Phytopathology*, *54*(1), 579–603. PMID: 27491436

Doddavarapu, B., Lata, C., & Shah, J. M. (2024). Epigenetic regulation influenced by soil microbiota and nutrients: paving road to epigenome editing in plants. *Biochimica et Biophysica Acta (BBA)-General Subjects*, 130580.

Dowen, R. H., Pelizzola, M., Schmitz, R. J., Lister, R., Dowen, J. M., Nery, J. R., & Ecker, J. R. (2012). Widespread dynamic DNA methylation in response to biotic stress. *Proceedings of the National Academy of Sciences of the United States of America*, *109*(32), E2183–E2191. PMID: 22733782

Eichten, S. R., & Springer, N. M. (2015). Minimal evidence for consistent changes in maize DNA methylation patterns following environmental stress. Academic Press.

Faltýnek, D., Lacková, Ľ., & Owsianková, H. (2020). Once again about the hapax grammar: Epigenetic linguistics. *Linguistic Frontiers*, *3*(1), 23–27.

Ferreira, L. J., Azevedo, V., Maroco, J., Oliveira, M. M., & Santos, A. P. (2015). Salt tolerant and sensitive rice varieties display differential methylome flexibility under salt stress. *PLoS One*, *10*(5), e0124060. PMID: 25932633

Fitz-James, M. H., & Cavalli, G. (2022). Molecular mechanisms of transgenerational epigenetic inheritance. *Nature Reviews. Genetics*, *23*(6), 325–341. PMID: 34983971

Furci, L., Jain, R., Stassen, J., Berkowitz, O., Whelan, J., Roquis, D., & Ton, J. (2019). Identification and characterisation of hypomethylated DNA loci controlling quantitative resistance in Arabidopsis. *eLife*, *8*, e40655. PMID: 30608232

Gallusci, P., Agius, D. R., Moschou, P. N., Dobránszki, J., Kaiserli, E., & Martinelli, F. (2023). Deep inside the epigenetic memories of stressed plants. *Trends in Plant Science*, *28*(2), 142–153. PMID: 36404175

Gallusci, P., Dai, Z., Génard, M., Gauffretau, A., Leblanc-Fournier, N., Richard-Molard, C., & Brunel-Muguet, S. (2017). Epigenetics for plant improvement: Current knowledge and modeling avenues. *Trends in Plant Science*, *22*(7), 610–623. PMID: 28587758

Gaut, B. S., Miller, A. J., & Seymour, D. K. (2019). Living with two genomes: Grafting and its implications for plant genome-to-genome interactions, phenotypic variation, and evolution. *Annual Review of Genetics*, *53*(1), 195–215. PMID: 31424971

Gojon, A., Nacry, P., & Davidian, J. C. (2009). Root uptake regulation: A central process for NPS homeostasis in plants. *Current Opinion in Plant Biology*, *12*(3), 328–338. PMID: 19501015

Gonzalez-Guzman, M., Pizzio, G. A., Antoni, R., Vera-Sirera, F., Merilo, E., Bassel, G. W., & Rodriguez, P. L. (2012). Arabidopsis PYR/PYL/RCAR receptors play a major role in quantitative regulation of stomatal aperture and transcriptional response to abscisic acid. *The Plant Cell*, *24*(6), 2483–2496. PMID: 22739828

Goswami, N., Tripathi, D., Rai, K. K., Singh, S., Pandey-Rai, S., Pandey, N., & Rai, S. K. (2019). Epigenetic control of UV-B-induced flavonoid accumulation in Artemisia annua L. *Planta. International Journal of Plant Biology*, *249*(2).

. Guarino, F., Cicatelli, A., Castiglione, S., Agius, D. R., Orhun, G. E., Fragkostefanakis, S., . . . Lieberman-Lazarovich, M. (2022). An epigenetic alphabet of crop adaptation to climate change. Academic Press.

Guo, C., Qin, L., Ma, Y., & Qin, J. (2022). Integrated metabolomic and transcriptomic analyses of the parasitic plant Cuscuta japonica Choisy on host and non-host plants. *BMC Plant Biology*, *22*(1), 393. PMID: 35934696

Habibi, N., Aryan, S., Amin, M. W., Sanada, A., Terada, N., & Koshio, K. (2023). Potential benefits of seed priming under salt stress conditions on physiological, and biochemical attributes of micro-tom tomato plants. *Plants*, *12*(11), 2187. PMID: 37299165

Haider, S., Iqbal, J., Naseer, S., Shaukat, M., Abbasi, B. A., Yaseen, T., & Mahmood, T. (2022). Unfolding molecular switches in plant heat stress resistance: A comprehensive review. *Plant Cell Reports*, *41*(3), 775–798. PMID: 34401950

Haider, S., Iqbal, J., Naseer, S., Yaseen, T., Shaukat, M., Bibi, H., & Mahmood, T. (2021). Molecular mechanisms of plant tolerance to heat stress: Current landscape and future perspectives. *Plant Cell Reports*, 1–25. PMID: 33890138

Haider, S., Iqbal, J., Shaukat, M., Naseer, S., & Mahmood, T. (2021). The epigenetic chromatin-based regulation of somatic heat stress memory in plants. *Plant Gene*, *27*, 100318.

Haider, S., Raza, A., Iqbal, J., Shaukat, M., & Mahmood, T. (2022). Analyzing the regulatory role of heat shock transcription factors in plant heat stress tolerance: A brief appraisal. *Molecular Biology Reports*, *49*(6), 5771–5785. PMID: 35182323

Han, S. K., Sang, Y., & Rodrigues, A., Wu, M. F., Rodriguez, P. L., & Wagner, D. (2012). The SWI2/SNF2 chromatin remodeling ATPase BRAHMA represses abscisic acid responses in the absence of the stress stimulus in Arabidopsis. *The Plant Cell*, *24*(12), 4892–4906. PMID: 23209114

Han, Y., Reyes, A. A., Malik, S., & He, Y. (2020). Cryo-EM structure of SWI/SNF complex bound to a nucleosome. *Nature*, *579*(7799), 452–455. PMID: 32188938

Hannan Parker, A., Wilkinson, S. W., & Ton, J. (2022). Epigenetics: A catalyst of plant immunity against pathogens. *The New Phytologist*, *233*(1), 66–83. PMID: 34455592

Hauser, M. T., Aufsatz, W., Jonak, C., & Luschnig, C. (2011). Transgenerational epigenetic inheritance in plants. *Biochimica et Biophysica Acta (BBA)-. Gene Regulatory Mechanisms*, *1809*(8), 459–468.

Hawes, N. A., Fidler, A. E., Tremblay, L. A., Pochon, X., Dunphy, B. J., & Smith, K. F. (2018). Understanding the role of DNA methylation in successful biological invasions: A review. *Biological Invasions*, *20*, 2285–2300.

Hemenway, E. A., & Gehring, M. (2023). Epigenetic regulation during plant development and the capacity for epigenetic memory. *Annual Review of Plant Biology, 74*(1), 87–109. PMID: 36854474

Holeski, L. M., Chase-Alone, R., & Kelly, J. K. (2010). The genetics of phenotypic plasticity in plant defense: Trichome production in Mimulus guttatus. *American Naturalist, 175*(4), 391–400. PMID: 20180699

Huang, C. Y., & Jin, H. (2022). Coordinated epigenetic regulation in plants: A potent managerial tool to conquer biotic stress. *Frontiers in Plant Science, 12*, 795274. PMID: 35046981

Ibañez, V. N., & Quadrana, L. (2023). Shaping inheritance: How distinct reproductive strategies influence DNA methylation memory in plants. *Current Opinion in Genetics & Development, 78*, 102018. PMID: 36525825

Ijaz, S., Iqbal, J., Abbasi, B. A., Tufail, A., Ullah, Z., Sharifi-Rad, J., & Uddin, S. (2023). Biofortification: Lessons from the Past and Strategies for Future Food Security. In *Legumes Biofortification* (pp. 521–545). Springer International Publishing.

Ijaz, S., Iqbal, J., Abbasi, B. A., Yaseen, T., Rehman, S., Kazi, M., & Mahmood, T. (2024). Role of OMICS-Based Technologies in Plant Sciences. OMICs-based Techniques for Global Food Security, 45-66.

Jaskiewicz, M., Conrath, U., & Peterhänsel, C. (2011). Chromatin modification acts as a memory for systemic acquired resistance in the plant stress response. *EMBO Reports, 12*(1), 50–55. PMID: 21132017

Jogam, P., Sandhya, D., Alok, A., Peddaboina, V., Allini, V. R., & Zhang, B. (2022). A review on CRISPR/Cas-based epigenetic regulation in plants. *International Journal of Biological Macromolecules, 219*, 1261–1271. PMID: 36057300

Kaiser, B., Vogg, G., Fürst, U. B., & Albert, M. (2015). Parasitic plants of the genus Cuscuta and their interaction with susceptible and resistant host plants. *Frontiers in Plant Science, 6*, 45. PMID: 25699071

Kim, Y. K., Chae, S., Oh, N. I., Nguyen, N. H., & Cheong, J. J. (2020). Recurrent drought conditions enhance the induction of drought stress memory genes in Glycine max L. *Frontiers in Genetics, 11*, 576086. PMID: 33193691

Kinoshita, T., & Seki, M. (2014). Epigenetic memory for stress response and adaptation in plants. *Plant & Cell Physiology, 55*(11), 1859–1863. PMID: 25298421

Kleine, T., Nägele, T., Neuhaus, H. E., Schmitz-Linneweber, C., Fernie, A. R., Geigenberger, P., & Leister, D. (2021). Acclimation in plants–the Green Hub consortium. *The Plant Journal*, *106*(1), 23–40. PMID: 33368770

Kumar, V., Thakur, J. K., & Prasad, M. (2021). Histone acetylation dynamics regulating plant development and stress responses. *Cellular and Molecular Life Sciences : CMLS*, *78*, 4467–4486. PMID: 33638653

Kuo, H. F., Chang, T. Y., Chiang, S. F., Wang, W. D., Charng, Y. Y., & Chiou, T. J. (2014). Arabidopsis inositol pentakisphosphate 2-kinase, A t IPK 1, is required for growth and modulates phosphate homeostasis at the transcriptional level. *The Plant Journal*, *80*(3), 503–515. PMID: 25155524

Kwon, C. S., Lee, D., Choi, G., & Chung, W. I. (2009). Histone occupancy-dependent and-independent removal of H3K27 trimethylation at cold-responsive genes in Arabidopsis. *The Plant Journal*, *60*(1), 112–121. PMID: 19500304

Lagiotis, G., Madesis, P., & Stavridou, E. (2023). Echoes of a Stressful Past: Abiotic Stress Memory in Crop Plants towards Enhanced Adaptation. *Agriculture*, *13*(11), 2090.

Lämke, J., Brzezinka, K., Altmann, S., & Bäurle, I. (2016). A hit-and-run heat shock factor governs sustained histone methylation and transcriptional stress memory. *The EMBO Journal*, *35*(2), 162–175. PMID: 26657708

Lang-Mladek, C., Popova, O., Kiok, K., Berlinger, M., Rakic, B., Aufsatz, W., & Luschnig, C. (2010). Transgenerational inheritance and resetting of stress-induced loss of epigenetic gene silencing in Arabidopsis. *Molecular Plant*, *3*(3), 594–602. PMID: 20410255

Lennartsson, A. (2017). Dynamics and function of DNA methylation during development. In *Chromatin regulation and dynamics* (pp. 65–94). Academic Press.

Li, A., Hu, B., & Chu, C. (2021). Epigenetic regulation of nitrogen and phosphorus responses in plants. *Journal of Plant Physiology*, *258*, 153363. PMID: 33508741

Li, S., Xia, Q., Wang, F., Yu, X., Ma, J., Kou, H., & Liu, B. (2017). Laser irradiation-induced DNA methylation changes are heritable and accompanied with transpositional activation of mPing in rice. *Frontiers in Plant Science*, *8*, 363. PMID: 28377781

Li, Z., Fu, X., Wang, Y., Liu, R., & He, Y. (2018). Polycomb-mediated gene silencing by the BAH–EMF1 complex in plants. *Nature Genetics*, *50*(9), 1254–1261. PMID: 30082786

Liang, D., Zhang, Z., Wu, H., Huang, C., Shuai, P., Ye, C. Y., & Xia, X. (2014, December). Single-base-resolution methylomes of Populus trichocarpa reveal the association between DNA methylation and drought stress. *BMC Genetics*, *15*, 1–11. PMID: 25080211

Liu, C., Lu, F., Cui, X., & Cao, X. (2010). Histone methylation in higher plants. *Annual Review of Plant Biology*, *61*(1), 395–420. PMID: 20192747

Liu, C., Xin, Y., Xu, L., Cai, Z., Xue, Y., Liu, Y., & Qi, Y. (2018). Arabidopsis ARGONAUTE 1 binds chromatin to promote gene transcription in response to hormones and stresses. *Developmental Cell*, *44*(3), 348–361. PMID: 29290588

Liu, J., & He, Z. (2020). Small DNA methylation, big player in plant abiotic stress responses and memory. *Frontiers in Plant Science*, *11*, 595603. PMID: 33362826

Liu, N., Fromm, M., & Avramova, Z. (2014). H3K27me3 and H3K4me3 chromatin environment at super-induced dehydration stress memory genes of Arabidopsis thaliana. *Molecular Plant*, *7*(3), 502–513. PMID: 24482435

Liu, R., & Lang, Z. (2020). The mechanism and function of active DNA demethylation in plants. *Journal of Integrative Plant Biology*, *62*(1), 148–159. PMID: 31628716

López Sánchez, A., Stassen, J. H., Furci, L., Smith, L. M., & Ton, J. (2016). The role of DNA (de) methylation in immune responsiveness of Arabidopsis. *The Plant Journal*, *88*(3), 361–374. PMID: 27341062

Lukić, N., Schurr, F. M., Trifković, T., Kukavica, B., & Walter, J. (2023). Transgenerational stress memory in plants is mediated by upregulation of the antioxidative system. *Environmental and Experimental Botany*, *205*, 105129.

Luna, E., & Ton, J. (2012). The epigenetic machinery controlling transgenerational systemic acquired resistance. *Plant Signaling & Behavior*, *7*(6), 615–618. PMID: 22580690

Luo, X., & He, Y. (2020). Experiencing winter for spring flowering: A molecular epigenetic perspective on vernalization. *Journal of Integrative Plant Biology*, *62*(1), 104–117. PMID: 31829495

Mal, M., Koh, P. K., Cheah, P. Y., & Chan, E. C. Y. (2009). Development and validation of a gas chromatography/mass spectrometry method for the metabolic profiling of human colon tissue. *Rapid Communications in Mass Spectrometry: An International Journal Devoted to the Rapid Dissemination of Up-to-the-Minute Research in Mass Spectrometry*, *23*(4), 487–494. PMID: 19140133

Manavella, P. A., Godoy Herz, M. A., Kornblihtt, A. R., Sorenson, R., Sieburth, L. E., Nakaminami, K., & Pikaard, C. S. (2023). Beyond transcription: Compelling open questions in plant RNA biology. *The Plant Cell, 35*(6), 1626–1653. PMID: 36477566

Manzoor, A., Ashraf, T., Shah, H., Maqbool, R., Kaul, R., & Dar, A. (2023). MicroRNAs and other Non-coding RNAs in Plant Epigenetics. In *Plant MicroRNAs and Stress Response* (pp. 329–359). CRC Press.

Martínez-Pérez, M., Aparicio, F., López-Gresa, M. P., Bellés, J. M., Sánchez-Navarro, J. A., & Pallás, V. (2017). Arabidopsis m6A demethylase activity modulates viral infection of a plant virus and the m6A abundance in its genomic RNAs. *Proceedings of the National Academy of Sciences of the United States of America, 114*(40), 10755–10760. PMID: 28923956

Mehdi, S., Derkacheva, M., Ramström, M., Kralemann, L., Bergquist, J., & Hennig, L. (2016). The WD40 domain protein MSI1 functions in a histone deacetylase complex to fine-tune abscisic acid signaling. *The Plant Cell, 28*(1), 42–54. PMID: 26704384

Meller, B., Kuźnicki, D., Arasimowicz-Jelonek, M., Deckert, J., & Floryszak-Wieczorek, J. (2018). BABA-primed histone modifications in potato for intergenerational resistance to Phytophthora infestans. *Frontiers in Plant Science, 9*, 1228. PMID: 30233606

Miryeganeh, M. (2021). Plants' epigenetic mechanisms and abiotic stress. *Genes, 12*(8), 1106. PMID: 34440280

Miryeganeh, M., & Armitage, D. W. (2024). Epigenetic responses of trees to environmental stress in the context of climate change. *Biological Reviews of the Cambridge Philosophical Society*. PMID: 39192567

Mozgova, I., & Hennig, L. (2015). The polycomb group protein regulatory network. *Annual Review of Plant Biology, 66*(1), 269–296. PMID: 25621513

Nguyen, N. H., Vu, N. T., & Cheong, J. J. (2022). Transcriptional stress memory and transgenerational inheritance of drought tolerance in plants. *International Journal of Molecular Sciences, 23*(21), 12918. PMID: 36361708

Nie, W. F., Lei, M., Zhang, M., Tang, K., Huang, H., Zhang, C., & Zhu, J. K. (2019). Histone acetylation recruits the SWR1 complex to regulate active DNA demethylation in Arabidopsis. *Proceedings of the National Academy of Sciences of the United States of America, 116*(33), 16641–16650. PMID: 31363048

Nunez-Vazquez, R., Desvoyes, B., & Gutierrez, C. (2022). Histone variants and modifications during abiotic stress response. *Frontiers in Plant Science*, *13*, 984702. PMID: 36589114

Pandey, N., & Pandey-Rai, S. (2015). Deciphering UV-B-induced variation in DNA methylation pattern and its influence on regulation of DBR2 expression in Artemisia annua L. *Planta*, *242*, 869–879. PMID: 25998525

Pandey, R., Vengavasi, K., & Hawkesford, M. J. (2021). Plant adaptation to nutrient stress. *Plant Physiology Reports*, *26*(4), 583–586.

Park, J., Lim, C. J., Shen, M., Park, H. J., Cha, J. Y., Iniesto, E., & Yun, D. J. (2018). Epigenetic switch from repressive to permissive chromatin in response to cold stress. *Proceedings of the National Academy of Sciences of the United States of America*, *115*(23), E5400–E5409. PMID: 29784800

Park, J. J., Dempewolf, E., Zhang, W., & Wang, Z. Y. (2017). RNA-guided transcriptional activation via CRISPR/dCas9 mimics overexpression phenotypes in Arabidopsis. *PLoS One*, *12*(6), e0179410. PMID: 28622347

Peirats-Llobet, M., Han, S. K., Gonzalez-Guzman, M., Jeong, C. W., Rodriguez, L., Belda-Palazon, B., & Rodriguez, P. L. (2016). A direct link between abscisic acid sensing and the chromatin-remodeling ATPase BRAHMA via core ABA signaling pathway components. *Molecular Plant*, *9*(1), 136–147. PMID: 26499068

Rambani, A., Pantalone, V., Yang, S., Rice, J. H., Song, Q., Mazarei, M., & Hewezi, T. (2020). Identification of introduced and stably inherited DNA methylation variants in soybean associated with soybean cyst nematode parasitism. *The New Phytologist*, *227*(1), 168–184. PMID: 32112408

Ramirez-Prado, J. S., Abulfaraj, A. A., Rayapuram, N., Benhamed, M., & Hirt, H. (2018). Plant immunity: From signaling to epigenetic control of defense. *Trends in Plant Science*, *23*(9), 833–844. PMID: 29970339

Ramirez-Prado, J. S., Latrasse, D., Rodriguez-Granados, N. Y., Huang, Y., Manza-Mianza, D., Brik-Chaouche, R., & Benhamed, M. (2019). The Polycomb protein LHP 1 regulates Arabidopsis thaliana stress responses through the repression of the MYC 2-dependent branch of immunity. *The Plant Journal*, *100*(6), 1118–1131. PMID: 31437321

Rao, K. M., Raghavendra, A. S., & Reddy, K. J. (Eds.). (2006). *Physiology and molecular biology of stress tolerance in plants*. Springer Science & Business Media.

Roberts, M. R., & López Sánchez, A. (2019). Plant epigenetic mechanisms in response to biotic stress. *Epigenetics in Plants of Agronomic Importance: Fundamentals and Applications: Transcriptional Regulation and Chromatin Remodelling in Plants*, 65-113.

Rossatto, T., Souza, G. M., do Amaral, M. N., Auler, P. A., Pérez-Alonso, M. M., Pollmann, S., & Braga, E. J. B. (2023). Cross-stress memory: Salt priming at vegetative growth stages improves tolerance to drought stress during grain-filling in rice plants. *Environmental and Experimental Botany*, *206*, 105187.

Roychowdhury, R., Mishra, S., Anand, G., Dalal, D., Gupta, R., Kumar, A., & Gupta, R. (2024). Decoding the molecular mechanism underlying salicylic acid (SA)-mediated plant immunity: An integrated overview from its biosynthesis to the mode of action. *Physiologia Plantarum*, *176*(3), e14399. PMID: 38894599

Rutowicz, K., Puzio, M., Halibart-Puzio, J., Lirski, M., Kotliński, M., Kroteń, M. A., & Jerzmanowski, A. (2015). A specialized histone H1 variant is required for adaptive responses to complex abiotic stress and related DNA methylation in Arabidopsis. *Plant Physiology*, *169*(3), 2080–2101. PMID: 26351307

Saez, A., Rodrigues, A., Santiago, J., Rubio, S., & Rodriguez, P. L. (2008). HAB1–SWI3B interaction reveals a link between abscisic acid signaling and putative SWI/SNF chromatin-remodeling complexes in Arabidopsis. *The Plant Cell*, *20*(11), 2972–2988. PMID: 19033529

Saleh, A., Alvarez-Venegas, R., Yilmaz, M., Le, O., Hou, G., Sadder, M., & Avramova, Z. (2008). The highly similar Arabidopsis homologs of trithorax ATX1 and ATX2 encode proteins with divergent biochemical functions. *The Plant Cell*, *20*(3), 568–579. PMID: 18375658

Sanchez, D. H., & Paszkowski, J. (2014). Heat-induced release of epigenetic silencing reveals the concealed role of an imprinted plant gene. *PLOS Genetics*, *10*(11), e1004806. PMID: 25411840

Sani, E. (2013). *Genome-wide transcriptional changes and chromatin modifications associated with plant stress memory* (Doctoral dissertation, University of Glasgow).

Saraswat, S., Yadav, A. K., Sirohi, P., & Singh, N. K. (2017). Role of epigenetics in crop improvement: Water and heat stress. *Journal of Plant Biology*, *60*, 231–240.

Séré, D., & Martin, A. (2020). Epigenetic regulation: Another layer in plant nutrition. *Plant Signaling & Behavior*, *15*(1), 1–6. PMID: 31674259

Shahid, S., Kim, G., Johnson, N. R., Wafula, E., Wang, F., Coruh, C., & Axtell, M. J. (2018). MicroRNAs from the parasitic plant Cuscuta campestris target host messenger RNAs. *Nature*, *553*(7686), 82–85. PMID: 29300014

Shoaib, N., Pan, K., Mughal, N., Raza, A., Liu, L., Zhang, J., & Pan, Z. (2024). Potential of UV-B radiation in drought stress resilience: A multidimensional approach to plant adaptation and future implications. *Plant, Cell & Environment*, *47*(2), 387–407. PMID: 38058262

Siddique, A. B., Parveen, S., Rahman, M. Z., & Rahman, J. (2024). Revisiting plant stress memory: Mechanisms and contribution to stress adaptation. *Physiology and Molecular Biology of Plants : an International Journal of Functional Plant Biology*, *30*(2), 349–367. PMID: 38623161

Singh, V. K., Ahmed, S., Saini, D. K., Gahlaut, V., Chauhan, S., Khandare, K., & Kumar, J. (2024). Manipulating epigenetic diversity in crop plants: Techniques, challenges and opportunities. *Biochimica et Biophysica Acta (BBA)-. General Subjects*, *1868*(2), 130544. PMID: 38104668

Sirohi, G., Pandey, B. K., Deveshwar, P., & Giri, J. (2016). Emerging trends in epigenetic regulation of nutrient deficiency response in plants. *Molecular Biotechnology*, *58*, 159–171. PMID: 26829932

Stroud, H., Ding, B., Simon, S. A., Feng, S., Bellizzi, M., Pellegrini, M., ... & Jacobsen, S. E. (2013). Plants regenerated from tissue culture contain stable epigenome changes in rice. *elife, 2*, e00354.

Sun, C., Ali, K., Yan, K., Fiaz, S., Dormatey, R., Bi, Z., & Bai, J. (2021). Exploration of epigenetics for improvement of drought and other stress resistance in crops: A review. *Plants*, *10*(6), 1226. PMID: 34208642

Tankari, M., Wang, C., Ma, H., Li, X., Li, L., Soothar, R. K., & Wang, Y. (2021). Drought priming improved water status, photosynthesis and water productivity of cowpea during post-anthesis drought stress. *Agricultural Water Management*, *245*, 106565.

Van Speybroeck, L., De Waele, D., & Van De Vijver, G. (2002). Theories in Early Embryology: Close Connections between Epigenesis, Preformationism, and Self-Organization. *Annals of the New York Academy of Sciences*, *981*(1), 7–49. PMID: 12547672

Verstraeten, B. (2021). *Insights into the interactions between rice and root-knot nematodes from an epigenetic perspective* (Doctoral dissertation, Ghent University).

Waddington, C. H. (2012). The epigenotype. *International Journal of Epidemiology, 41*(1), 10–13. PMID: 22186258

Wang, L., & Qiao, H. (2020). Chromatin regulation in plant hormone and plant stress responses. *Current Opinion in Plant Biology, 57*, 164–170. PMID: 33142261

Wang, P., Xue, L., Batelli, G., Lee, S., Hou, Y. J., Van Oosten, M. J., & Zhu, J. K. (2013). Quantitative phosphoproteomics identifies SnRK2 protein kinase substrates and reveals the effectors of abscisic acid action. *Proceedings of the National Academy of Sciences of the United States of America, 110*(27), 11205–11210. PMID: 23776212

Wang, X., Li, Z., Liu, B., Zhou, H., Elmongy, M. S., & Xia, Y. (2020). Combined proteome and transcriptome analysis of heat-primed azalea reveals new insights into plant heat acclimation memory. *Frontiers in Plant Science, 11*, 1278. PMID: 32973837

Wei, J. W., Huang, K., Yang, C., & Kang, C. S. (2017). Non-coding RNAs as regulators in epigenetics. *Oncology Reports, 37*(1), 3–9. PMID: 27841002

Widiez, T., El Kafafi, E. S., Girin, T., Berr, A., Ruffel, S., Krouk, G., & Lepetit, M. (2011). High nitrogen insensitive 9 (HNI9)-mediated systemic repression of root NO3− uptake is associated with changes in histone methylation. *Proceedings of the National Academy of Sciences of the United States of America, 108*(32), 13329–13334. PMID: 21788519

Wierzbicki, A. T., Blevins, T., & Swiezewski, S. (2021). Long noncoding RNAs in plants. *Annual Review of Plant Biology, 72*(1), 245–271. PMID: 33752440

Wilkinson, S. W. (2020). *Investigating the mechanisms of long-lasting within generation jasmonate induced resistance* (Doctoral dissertation, University of Sheffield).

Wu, R., Wang, T., Richardson, A. C., Allan, A. C., Macknight, R. C., & Varkonyi-Gasic, E. (2019). Histone modification and activation by SOC1-like and drought stress-related transcription factors may regulate AcSVP2 expression during kiwifruit winter dormancy. *Plant Science, 281*, 242–250. PMID: 30824057

Xu, R., Wang, Y., Zheng, H., Lu, W., Wu, C., Huang, J., & Zheng, C. (2015). Salt-induced transcription factor MYB74 is regulated by the RNA-directed DNA methylation pathway in Arabidopsis. *Journal of Experimental Botany, 66*(19), 5997–6008. PMID: 26139822

Yang, R., Hong, Y., Ren, Z., Tang, K., Zhang, H., Zhu, J. K., & Zhao, C. (2019). A role for PICKLE in the regulation of cold and salt stress tolerance in Arabidopsis. *Frontiers in Plant Science, 10*, 900. PMID: 31354770

Yu, A., Lepère, G., Jay, F., Wang, J., Bapaume, L., Wang, Y., & Navarro, L. (2013). Dynamics and biological relevance of DNA demethylation in Arabidopsis antibacterial defense. *Proceedings of the National Academy of Sciences of the United States of America, 110*(6), 2389–2394. PMID: 23335630

Zemach, A., Kim, M. Y., Hsieh, P. H., Coleman-Derr, D., Eshed-Williams, L., Thao, K., & Zilberman, D. (2013). The Arabidopsis nucleosome remodeler DDM1 allows DNA methyltransferases to access H1-containing heterochromatin. *Cell, 153*(1), 193–205. PMID: 23540698

Zeng, X., Gao, Z., Jiang, C., Yang, Y., Liu, R., & He, Y. (2020). Histone deacetylase 9 functions with polycomb silencing to repress flowering locus C expression. *Plant Physiology, 182*(1), 555–565. PMID: 31641076

Zhang, H., Guo, F., Qi, P., Huang, Y., Xie, Y., Xu, L., & Bian, H. (2020). OsHDA710-mediated histone deacetylation regulates callus formation of rice mature embryo. *Plant & Cell Physiology, 61*(9), 1646–1660. PMID: 32592489

Zhang, H., Lang, Z., & Zhu, J. K. (2018). Dynamics and function of DNA methylation in plants. *Nature Reviews. Molecular Cell Biology, 19*(8), 489–506. PMID: 29784956

Zhang, H., Zhang, X., & Xiao, J. (2023). Epigenetic regulation of nitrogen signaling and adaptation in plants. *Plants, 12*(14), 2725. PMID: 37514337

Zhang, X. (2012). Dynamic differential methylation facilitates pathogen stress response in Arabidopsis. *Proceedings of the National Academy of Sciences of the United States of America, 109*(32), 12842–12843. PMID: 22837399

Zhang, X., Wang, X., Zhuang, L., Gao, Y., & Huang, B. (2019). Abscisic acid mediation of drought priming-enhanced heat tolerance in tall fescue (Festuca arundinacea) and Arabidopsis. *Physiologia Plantarum, 167*(4), 488–501. PMID: 30977137

Zhou, B., & Zeng, L. (2017). Conventional and unconventional ubiquitination in plant immunity. *Molecular Plant Pathology, 18*(9), 1313–1330. PMID: 27925369

Zhu, A., Greaves, I. K., Dennis, E. S., & Peacock, W. J. (2017). Genome-wide analyses of four major histone modifications in Arabidopsis hybrids at the germinating seed stage. *BMC Genomics, 18*, 1–9. PMID: 28173754

Zhu, J. K. (2016). Abiotic stress signaling and responses in plants. *Cell, 167*(2), 313–324. PMID: 27716505

Zhu, Y., Hu, X., Duan, Y., Li, S., Wang, Y., Rehman, A. U., & Gong, Z. (2020). The Arabidopsis nodulin homeobox factor AtNDX interacts with AtRING1A/B and negatively regulates abscisic acid signaling. *The Plant Cell*, *32*(3), 703–721. PMID: 31919300

About the Contributors

Muhammad Waseem is Associate Professor at College of Tropical Agriculture and Forestry, Hainan University, Haikou, Hainan, China. He is also severing as Assistant Professor of Botany at the Department of Botany, University of Narowal, Pakistan. He got a doctoral degree (Ph.D., specialization in Plant Molecular Biology) from the School of Life Sciences, Chongqing University, Chongqing, PR China. He also have four years of postdoctoral researcher. His interests include transcriptional regulation of flowering time regulation, fruit ripening and development, and abiotic stresses such as salinity and drought. In addition, he has hands-on experience in next-generation sequencing data analysis, transcriptomics, and data visualization. Currently, Dr. Waseem is actively in investigating the flowering time regulation in Brassica crops. He has been awarded with a National Science Foundation Project, RFIS worth 200000 RMB. In brief, he has published more than 40 research, 5 review, 9 book chapters in various high-impact factor journals including New Phytologist, International Journal of Macromolecules, Journal of Biotechnology, Planta, Scientific reports, Pakistan Journal of Botany, GM Crops & Food, Environmental and Experimental Botany, International Journal of Molecular Biology, Genes, Plant Physiology, and Frontiers in Plant Science. Moreover, currently, he is editing a few special issues as a leading guest editor in Frontiers in Plant Science, International Journal of Molecular Biology, GENES, and Functional Plant Biology.

Pingwu Liu, specializes in cruciferous vegetable germplasm innovation, cultivation techniques, and biological control. With a Ph.D. in Agronomy from Huazhong Agricultural University, he has an extensive academic career. Since 2020, he has been serving at Hainan University's College of Horticulture, where he has led multiple significant research projects. His research includes developing early-maturing ornamental Brassica, gene excavation for vegetable traits, and creating disease-resistant sugarcane varieties. Professor Liu's contributions to both graduate and undergraduate education, particularly in vegetable genetics, breeding, and production quality control, have been widely recognized. He has secured numerous research grants from prestigious institutions such as the National Natural Science Foundation and the Guangxi Natural Science Foundation.

***

Nasir Abbas is an accomplished agricultural researcher specializing in Plant Breeding and Genetics, dedicated to advancing crop improvement for enhanced food security and sustainability. His foundational expertise stems from a Master of Science (M.Sc.) degree in Plant Breeding and Genetics, earned from the prestigious University of Agriculture Faisalabad (UAF), Pakistan – a leading institution renowned for its contributions to agricultural sciences in the region. Building upon his strong academic grounding at UAF, where he gained rigorous training in classical and quantitative genetics, experimental design, statistical analysis, and field-based breeding techniques.

Banzeer Ahsan Abbasi received her graduate degree in Botany from Quaid-i-Azam University, Islamabad, Pakistan. Dr. Banzeer is currently working as an Assistant Professor at Department of Botany, Rawalpindi Women University, Rawalpindi, Pakistan. She has published more than 60 research articles in various international journals and 10 chapters in reputed publishers (ELSIEVER, RSC and SPRINGER). Currently, she is supervising 2 Ph.D, 6 MPhil and 6 BS students. She has expertise in the synthesis of novel nanomaterials from medicinal plants and their multiple biological and envirnemental applications. Her research is also focusing on phytochemistry, natural products of medicinal plants and Phylogenetics. She has a very decent research background with research published in well-reputed journals with H-index 28, i-10 index 44 and overall citations 3317.

Sana Basharat is working as a postdoctoral researcher at the college of Tropical Agriculture and Forestry, Hainan University China. She obtained here doctoral degree from Department of Botany, University of Agriculture Faisalabad, Pakistan. She had published more than 12 research articles in reknown journals.

Amin Fathi is a results-oriented researcher with a PhD in Agronomy and extensive interest in exploring plant responses to environmental stress, climate change, and plant-microbe interactions. Proven expertise in tillage, nanotechnology, biochar applications, and strategic management approaches to improve plant growth and resilience. Demonstrates leadership skills as advisor assistant and editor, coupled with a solid foundation in laboratory techniques, statistical analysis, and manuscript publishing.

Arash Fazeli as an associated professor work in Ilam university, Iran. he focused on crop and medicinal plant in different area especially to find resistance cultivar to biotic and abiotic stress conditions.

Amber Jabeen has obtained her MSc and MPhil degree in Botany (Plant Sciences) from Quaid-i- Azam University, Islamabad, Pakistan with a distinction in MSc. Amber is currently enrolled in PhD Plant Sciences at department of Plant Sciences, Quaid I Azam University, Islamabad. Her research emphasizes on the green synthesis of nanoparticles mediated by algae and their chemical and structural characterization. Her work spans across nanomedicine, and therapeutic applications of nanoparticles, contributing to advancements in drug delivery systems and disease treatments. She has actively contributed to research in molecular biology. Throughout her career, she has co-authored peer reviewed publications, including studies on the biomedical applications of nanoparticles, bioremediations, and stress responses of plants.

Lubna Anjum Minhas obtained her B.S and M.Phil degrees in Botany (Plant Sciences) from Quaid-i-Azam University, Islamabad, Pakistan. Her M.Phil. research centered on cyanobacteria-mediated metal oxide nanoparticles, exploring their biomedical and environmental applications. She is the first author of four research articles and a review article, and she has contributed as a co-author to many other research articles. Additionally, Lubna serves as a reviewer for several prestigious international journals.

Muhammad Muzamil Sattar attained his BS Honors degree in Botany from Quaid-i-Azam University, Islamabad, and subsequently completed his M.Phil. in Plant Sciences at the same institution. He worked on the isolation, culturing and purification of algal species during his research work in Bs honors and he has finalized his MPhil thesis on the synthesis and characterization of Chara-based zinc oxide nanoparticles and their PVP-doped derivatives to improve biological efficacy. He has authored a book chapter and served as a co-author for a research article. His research work, as the primary author, is currently under modification.

Gul Nawaz, son of Muhammad Nawaz, was born on July 12, 2001, in Jampur City, Punjab Province, Pakistan. He earned a degree in Botany from the University of Education, Lahore, Dera Ghazi Khan Campus and is now an MS Research Scholar at the University of the Punjab, Lahore. He has completed 80+ e-learning courses from different platforms i.e. Coursera and the UN. Nawaz is an enthusaiatic and committed novice researcher in the the field of botany. He has publisheda research paper and several are in pipeline. e He is working on several articles and book chapters related to ecology, biodiversity, physiology and ethnobotany. A key member of "The Dream Team," he contributed to quiz wins in 2022 and runner-up success in 2023. He is a member of the Society for Ecological Restoration of America and the student ambassador for Oxford University's Nature Positive Universities initiative. Explore nature, critical thinking, and innovating approaches are his favourite hobbies.

Garima Khaspuria An accomplished academician and researcher with a diverse background in Biotechnology and Data Science, currently serving as an Associate Professor in Genetics at Vivekananda Global University, Jaipur, with over seven years of teaching and research experience. Holding a Ph.D. in Biomedical Genetics from Dr. A.P.J. Abdul Kalam Technical University, significant contributions have been made to genetic research, particularly in inflammatory gene polymorphisms and chronic periodontitis, with publications in esteemed journals reflecting dedication to scientific advancement. In pursuit of continuous learning, a recent Post-Graduate Program in Data Science and Business Analytics was completed from Great Learning and the University of Texas at Austin. A capstone project involved developing a predictive model for healthcare insurance cost estimation using machine learning algorithms, showcasing proficiency in Python, Tableau, and SQL. There is a keen interest in applying data analytics skills to solve real-world challenges in the agricultural and healthcare sectors. Leadership extends to organizing symposiums and mentoring students. As a member of the ELU Committee and the Institute Innovation Center at VGU, active promotion of experiential learning and innovation is evident. Interdisciplinary expertise and a passion for data-driven problem-solving make this individual a valuable contributor to both academic and industry landscapes.

Ammara Saleem is currently pursuing a Ph.D. in Botany at the Institute of Botany, University of the Punjab, where she also completed her MPhil degree. Her research interests lie in exploring the impacts of abiotic stress on cereal crops, focusing on a comprehensive range of morpho-physiological and biochemical analyses. Ammara has contributed significantly to her field through her published research papers, which have garnered strong citations in reputable scientific journals. Her scholarly work has made noteworthy contributions to the understanding of plant resilience under abiotic stress. Additionally, she serves as a dedicated reviewer for various scientific journals, leveraging her expertise to support the academic community and ensure the quality of published research.

Iffat Shaheen is a graduate student from the College of Agriculture, Bahauddin Zakariya University, Multan, Pakistan. She had published more than 3 research articles and book chapters in prestigious journals.

Nicolas Taylor completed his undergraduate studies and MSc at Massey University in New Zealand before moving to the University of Western Australia (UWA) in 2000 to pursue his PhD. Following his doctoral studies, he was awarded a European Molecular Biology Organisation Long Term Fellowship to conduct research at the Department of Plant Sciences, University of Oxford, UK. In 2006, he returned to UWA to join the newly established ARC Centre of Excellence in Plant Energy Biology, where he held ARC Post-Doctoral and Future Fellowships. His research encompasses a range of topics in agriculture, with a particular focus on the molecular genetic mechanisms that regulate crop growth and yield, as well as the impacts of environmental factors on crop health and productivity. He integrates high-throughput phenotyping tools, including FTIR and hyperspectral imaging coupled with machine learning, alongside metabolic analysis (multi-omic mass spectrometry) and field-based assessments to investigate the metabolic processes that influence stress tolerance and yield in crop plants. Nic places strong emphasis on high-quality research training, and students from his laboratory have secured PhD, post-doctoral, and lecturing positions at prestigious institutions such as the Max Planck Institute for Molecular Plant Physiology, the National University of Malaysia, the University of East Anglia, the University of Oxford, and the University of Cologne. He currently serves as the WA Node Director of the Australian Plant Phenomics Network.

Index

A

B

C

D

I

M

R

S

T

www.ingramcontent.com/pod-product-compliance
Lightning Source LLC
LaVergne TN
LVHW061531290926
849162LV00004B/759